KV-516-568

Developments in Earth Surface Processes, 8

CLIMATIC GEOMORPHOLOGY

MATEO GUTIÉRREZ

PROFESSOR OF GEOMORPHOLOGY
UNIVERSITY OF ZARAGOZA, SPAIN

Translated by
G. Benito, G. Desir, J.M. García-Ruiz, J. Gracia, F. Gutiérrez,
J. López-Martínez, C. Martí, J. Remondo, P. Silva and B. Valero

ELSEVIER

Amsterdam – Boston – Heidelberg – London – New York – Oxford
Paris – San Diego – San Francisco – Singapore – Sydney – Tokyo

ELSEVIER B.V.	ELSEVIER Inc.	ELSEVIER Ltd	ELSEVIER Ltd
Radarweg 29	525 B Street, Suite 1900	The Boulevard, Langford Lane	84 Theobalds Road
P.O. Box 211, 1000 AE, Amsterdam	San Diego, CA 92101-4495	Kidlington, Oxford OX5 1GB	London WC1X 8RR
The Netherlands	USA	UK	UK

© 2001 Original Spanish Language Edition, Ediciones Omega, S.A., Barcelona, Spain.
© 2005 English translation, Elsevier B.V.

This work is protected under copyright by Elsevier B.V., and the following terms and conditions apply to its use:

Photocopying

Single photocopies of single chapters may be made for personal use as allowed by national copyright laws. Permission of the Publisher and payment of a fee is required for all other photocopying, including multiple or systematic copying, copying for advertising or promotional purposes, resale, and all forms of document delivery. Special rates are available for educational institutions that wish to make photocopies for non-profit educational classroom use.

Permissions may be sought directly from Elsevier's Rights Department in Oxford, UK: phone (+44) 1865 843830, fax (+44) 1865 853333, e-mail: permissions@elsevier.com. Requests may also be completed on-line via the Elsevier homepage (http://www.elsevier.com/locate/permissions).

In the USA, users may clear permissions and make payments through the Copyright Clearance Center, Inc., 222 Rosewood Drive, Danvers, MA 01923, USA; phone: (+1) (978) 7508400, fax: (+1) (978) 7504744, and in the UK through the Copyright Licensing Agency Rapid Clearance Service (CLARCS), 90 Tottenham Court Road, London W1P 0LP, UK; phone: (+44) 20 7631 5555; fax: (+44) 20 7631 5500. Other countries may have a local reprographic rights agency for payments.

Derivative Works

Tables of contents may be reproduced for internal circulation, but permission of the Publisher is required for external resale or distribution of such material. Permission of the Publisher is required for all other derivative works, including compilations and translations.

Electronic Storage or Usage

Permission of the Publisher is required to store or use electronically any material contained in this work, including any chapter or part of a chapter.

Except as outlined above, no part of this work may be reproduced, stored in a retrieval system or transmitted in any form or by any means, electronic, mechanical, photocopying, recording or otherwise, without prior written permission of the Publisher.

Address permissions requests to: Elsevier's Rights Department, at the fax and e-mail addresses noted above.

Notice

No responsibility is assumed by the Publisher for any injury and/or damage to persons or property as a matter of products liability, negligence or otherwise, or from any use or operation of any methods, products, instructions or ideas contained in the material herein. Because of rapid advances in the medical sciences, in particular, independent verification of diagnoses and drug dosages should be made.

Original Title: Geomorfología Climática, published 2001 by Ediciones Omega, S.A., Barcelona, Spain. First edition English translation published 2005 by Elsevier B.V.

Library of Congress Cataloging in Publication Data
A catalog record is available from the Library of Congress.

British Library Cataloguing in Publication Data
A catalogue record is available from the British Library.

ISBN: 0-444-51794-4 (Hardback)
ISBN: 0-444-52128-3 (Paperback)
ISBN: 0928-2025 (Series)

♾ The paper used in this publication meets the requirements of ANSI/NISO Z39.48-1992 (Permanence of Paper).
Printed in Italy

Working together to grow
libraries in developing countries

www.elsevier.com | www.bookaid.org | www.sabre.org

ELSEVIER   BOOK AID International   Sabre Foundation

UNIVERSITY OF PLYMOUTH

9006859495

90 0685949 5

EARTH SURFACE PROCESSES, 8

CLIMATIC GEOMORPHOLOGY

WITHDRAWN
FROM
UNIVERSITY OF PLYMOUTH
LIBRARY SERVICES

Charles Seale-Hayne Library
University of Plymouth
(01752) 588 588
LibraryandITenquiries@plymouth.ac.uk

DEVELOPMENTS IN EARTH SURFACE PROCESSES

To my wife, children and grandchildren

Contents

Preface

Climatic Geomorphology constitutes a new look at a fairly old subject that has been largely ignored in other more recent trends in the field, particularly in much process geomorphology. But now Mateo Gutiérrez, well-known Professor of Geomorphology at the University of Zaragoza in northeast Spain, has presented us with this new volume, translated into English by ten of his colleagues. We are fortunate indeed to have his authoritative writing and artful photography of so many parts of the world to grace our understanding of landform character and evolution. Originally published in Spanish in 2001, this volume attracted the notice of many other geoscientists and was reviewed quite favourably in more than five of our main journals of geomorphology, physical geography, and environmental geology. The uniform praise led to suggestions that a translation into English might be warranted, with the results that you now see before you. The relative lack of such modern and well-illustrated texts devoted to climatic geomorphology in the English language indicates that this new book is a deserving addition to the literature.

In recent years the great interest in the climate change so obviously underway in the world, as well as the consequences of even greater climate change in the geological past, have led to considerable interest in the topic. This book provides a global overview of the different climatic zones of the Earth, and it reviews the advances in climatic geomorphology that have been achieved in the past decades. Certain landform assemblages are well known to be associated with different regional climate-process systems, a statement that underscores the obvious, despite the fact that the details of some climate-controlled processes are not yet that well understood. Thus we all recognize certain glacial, periglacial, or humid tropical landform types, even though processes such as the exact efficacy of deep glacial erosion, or long term rock-glacier formation and mechanics, or bornhardt generation may remain somewhat elusive. Furthermore, in the temperate latitudes where so many geomorphologists live and work, with the plethora of morphoclimatic process overprints, such as, for example, the Pleistocene glacial signatures impressed upon so much of the landscape of Europe and North America, it may surprise some that aspects of this now vanished glacial process still remain somewhat enigmatic.

Nevertheless, the fact that climate change has occurred so many times in the past is not in dispute, although the lability or amount of time that a given landform may take to come into equilibrium existence in a particular climate type may not be at all well understood. Thus a renewal of interest in the topic of climatic geomorphology seems entirely justified as we seek to better understand the many ways that climate forcings produce landforms.

Nonetheless, one point that Professor Gutiérrez makes in this book that must always be kept in mind is expressed by the principle of convergence or equifinality, in which different processes produce the same landform so that oversimplified climatic explanations do not work. For example, granites and some other massive crystalline and sedimentary rocks weather and erode to produce tors in a variety of climatic regimes from tropical humid to periglacial. Long-term chemical weathering of certain minerals in these rocks, with later stripping of the weathered mantles, occurs in a variety of climate regimes. Similarly, frost shattering and gelifluction processes can also produce these bedrock tors. In the same equifinality fashion, horizontally bedded sedimentary and volcanic rocks can produce similar cliffed topography in various climates wherever the free faces are maintained by undercutting of the resistant rocks, even through dramatic climate changes. Excellent examples include the buttes and mesas of the deserts and pluvial basins of the southwestern USA, or the tepui mesas over which Angel Falls flows in the Amazonian rainforest of Venezuela. Similarly, bog-burst mass movement landforms can occur wherever rainfall and edaphic conditions allow saturated conditions to generate unstable peat bogs, be it tropical humid highlands, humid temperate maritime regions, or periglacial tundras. But in spite of such landform convergence, or the palimpsest landform inheritance from past climatic regimes, understandings of climatic geomorphology still have much to offer those who utilize the appropriate concepts.

This book is structured into eight main sections that are set up first as an introduction to some of the history of geomorphology, then separate groupings of multiple chapters on glacial, periglacial, arid, aeolian, tropical, and climate-change geomorphology. The 25 different chapters in all of the sections discuss the dominant weathering, erosion, and deposition associated with the processes characteristic of each climate zone, along with the resulting erosional and depositional landforms. Strikingly, four chapters on applied aspects of glacial, periglacial, arid, and tropical landform processes and products provide an exceptionally important primer for many parts of the less developed world. This aspect will recommend the book to those in government offices or in the various development agencies world-wide who may need to better understand hazards or opportunities for assisting their citizenry to live more effectively in their particular climatic environments.

The dominant fluvial regimes of the temperate latitudes receive only limited attention in this book, which may somewhat reduce interest for those geomorphologists who specialize in such aspects in their regions. I suspect, however, that most geoscientists will still recognize the importance of the arid and tropical zones to so many people of the less developed world, and therefore will understand the reasons for the particular emphases in this book. In addition, the lack of a quantitative approach in the book, or the overall dearth of formulas may not meet the preferences of some, but still the many graphs and diagrams taken from the primary references do cover adequate quantitative materials.

Although the author of this book is well versed in the whole of climatic geomorphology, he has devoted about one third of the book to arid-zone topics, perhaps reflecting some of his main interests in interior, semi-arid Spain. Out of a total of 25 chapters, after the introductory chapter, 10 are on aeolian, arid zone, and applied arid topics. Next comes the glacial zone, with five chapters, and the periglacial and tropical zone sections with three chapters each. The last section on geomorphology and climate change also includes three chapters, which may add a measure of considerable interest to those younger geomorphologists who are going to live through what promise to be the most interesting climate changes to come in the 21st century. This book provides some mental preparation for recognizing and perhaps dealing with the most likely changes that can be expected.

The great majority of references in this book are in fact from the English-language literature, and thus these concepts originally had to be converted into Spanish for the initial volume. But because this book was then translated from the Spanish back into English by some ten different translators, some of our phraseology may be a little less artful than intended by the original authors of the primary references. If this is detected by anyone I must plead guilty myself to the difficulty presented in editing some of the literal translations back into the rather more idiomatic presentations that exactly reflect the original intentions, as well as being also quite accurate scientifically. We of the editorial staff did our best to convey the key ideas and information without error, but some few problems may have crept in inadvertently in this fashion anyway, inasmuch as we did not elect to rewrite the text from the original primary references. By and large the original text was also left fairly unmodified in the editorial process, with the exception of a few added observations and more up-to-date references that were felt to enhance the text.

The disconnect between the dominant studies of process geomorphology of the last half of the 20th century, or what has been referred to as *timeless geomorphology*, and an adequate exposition of *time-bound* landform evolution has become obvious to many geomorphologists at the beginning of the 21st century. This book, *Climatic Geomorphology*, is put forward as one approach to better understanding landform evolution from a climatic control of process such that the time-bound nature of the control is seen to exert both subtle and obvious process-directed changes to the landscape. An alternative and quantitatively robust approach to achieving this same result can be observed at many of the more advanced geoscientific meetings of the world today wherein the equations of climate-controlled process mechanics are run iteratively with computers to synthesize landscape evolution. Decades of new work in this vein are likely to be required to make adequate and realistic progress with this new technology, but this robust methodology is already producing dramatic results. It is likely that this book can serve as one of the background sources of concepts and references that will enable better understanding of these process mechanics under certain climate forcings, and the inevitable fluctuations that must result from climate change. In this fashion we may eventually come to better understand the rich variety of climatically controlled landscapes that dominate planet Earth.

The publication of this book, *Climatic Geomorphology*, represents the first of the now rejuvenated series on *Developments in Earth Surface Processes* published by Elsevier. Geomorphology in the 21st century seems to be undergoing a renaissance of theory and methodologies as we recover from the paucity of robust theoretical underpinnings following the partial collapse and/or dramatic revisions of the Davisian–Penckian–Kingian paradigms in the middle of the prior century. After research excursions deep into the reductionistic thickets of process geomorphology, we now seem poised to move on smartly with the integration of short-term, timeless, landform process studies into long-term, time-bound, landform evolution. This book is offered as a rejuvenated first in a series of what is hoped will be an ever-increasing list of titles of new developments in geomorphology.

John Shroder

Preface to the English language edition

During the past few decades, knowledge in climatic geomorphology has been substantially increased thanks to numerous detailed investigations, the application of a large number of techniques, and the acquisition of abundant absolute dates. The challenge of predicting the effects of the prophesied future global warming on morphogenetic processes and landforms has encouraged geomorphologists to study the Late Pleistocene and Holocene climatic changes from the geomorphological and geological record. The advances achieved in the field of climatic geomorphology during the past years are reflected in the publication of several specific monographs about the different morphoclimatic zones. The aim of this book is to provide an up-to-date general view of this branch of geomorphology. It includes a chapter on applied geomorphology for each morphoclimatic zone providing an approximation of the main environmental problems.

I hope that the book will be useful to geomorphology students and to those interested in the role played by climatic conditions on the generation of landforms and landscapes. I would be fully satisfied if it helps some researchers working in disciplines close to climatic geomorphology. I would like to express my warmest gratitude to the scientists involved in the review of the book: G. Benito, A. Cendrero, J.L. Goy, J. Gracia, F. Gutiérrez, J. López-Martínez, A. Gómez-Ortiz, A. Martín-Serrano, J.L. Peña, A. Pérez-González, J. Rodríguez-Vidal, J.M. García-Ruiz, C. Sancho and C. Zazo. Their wise suggestions and the unpublished data supplied have contributed to the improvement of the content of the book. Most of them have provided excellent photographs from different parts of the Planet. I would also like to thank Dr A. Gómez-Sal, Dr B. Leránoz, C. Maldonado, E. Pueyo and M. Urigüen for additional photographs providing.

I would especially like to thank G. Benito, G. Desir, J.M. García-Ruiz, J. Gracia, F. Gutiérrez, J. López-Martínez, C. Martí, J. Remondo, P. Silva, and B. Valero for their help with translation of the text into English so that Editor J. Shroder could proceed with completing the new edition.

We are very grateful to the other editors, societies, institutions and researchers for the authorizations conceded to reproduce several figures in this publication. We have not received responses from some editors and researchers to our request for a reproduction permit for which we apologize. We would be thankful for any information that would facilitate this task. Some of the figures came from journals that are now out of print or difficult to obtain.

My sincere gratitude goes to Ana Carmen Aguirre, who has typed all of the manuscripts and has rigorously prepared the figures with great enthusiasm.

Finally, thanks to Nieves for her continuous help, tolerance and infinite patience.

Mateo Gutiérrez

FIRST PART
Introduction and concepts

Chapter 1

Climatic geomorphology

1. Introduction to geomorphology

The term "geomorphology" comes from the ancient Greek words *geo* (Earth), *morphos* (shape) and *logos* (thesis). This is, therefore, the science of the form of the land surface. Several authors limit the scope of geomorphology to the study of subaerial landscapes, whereas others also include submarine topography. Even the study of extraterrestrial landforms has recently been included in the so-called planetary or extraterrestrial geomorphology (Baker, 1993).

The topography of the Earth's surface is the result of the balance between *endogenic* and *exogenic forces*. Endogenic forces trigger vertical movements that generate large mountain belts and depressions, whereas exogenic forces work progressively to denude the resulting relief. This permanent interplay of forces has been performed at different scales on the Earth's surface during the whole of geological history. The *external processes* occur in a classical sequence of *weathering*, *erosion*, *transport* and *sedimentation*. The result is the generation of different erosive and depositional landscapes, with different specific features depending on the dominant process working in the different morphogenetic environments. The energy that powers these processes comes from different sources. *Solar radiation* received by the Earth heats the land surface and the atmosphere, as it constitutes the main source of energy for meteorological processes. Such processes control rock weathering, soil formation, relief development and also the biological activity of plants and animals. In addition, *gravitational energy* supports the processes of sediment transport and mass wasting. Finally, the *internal energy* of the planet is the ultimate source of energy for virtually all tectonic processes and the associated crustal movements that generate relief (Büdel, 1968, 1977).

Traditionally, the study of geomorphology has focused on the analysis of the sequence and nature of geomorphic events involved in the present configuration of the Earth's surface through geological time. This approach is denoted by some authors as *historical geomorphology* (Chorley, 1978). Recently, the study of the working processes at smaller scales, together with the analysis of the correlative landform variability, has resulted in the so-called *quantitative* (Chorley, 1978) or *processes geomorphology* (Hart, 1986; Thorn, 1988a). As pointed out by Chorley et al. (1984) such historical studies are based on retrodiction, whereas processes analysis is mainly focused on forecasting. Both trends are the dominant ones in most current geomorphological studies. In some countries, processes analysis has been stimulated in opposition to traditional studies of relief development over long time spans. From the study of micro-landform development, however, we can obtain valuable information for a better understanding of the long sequence of events that has resulted in present-day landforms. We could compare this situation to that experienced by the discipline of geotectonics after micro-structural analysis was developed, which assisted far deeper knowledge of the structure of mountain belts (Gutiérrez, 1990).

These new trends in geomorphology have been instigated, as in other branches of the Earth sciences, by the evolution of ideas, knowledge and scientific methodology over the last centuries. The birth of geomorphology as a science took place when the explanatory focus replaced its initial descriptive motive; only a century ago descriptive geomorphology was known as "orography" or "hydrography" (Tricart, 1965). The Americans McGee and Powell were probably the first authors to use the term "geomorphology" during the decade of the 1880s (Hart, 1986).

2. History of climatic geomorphology

2.1. Geomorphology before the second half of the 19th century

The origins of geomorphology, as with many other scientific disciplines, are unclear and the first observations and interpretations came from the philosophers of ancient Greece. An extensive and detailed analysis of that primary epoch can be consulted in the excellent treatise of Chorley et al. (1964). Although the term "geomorphology" was not coined then, the first ideas and fundamentals were developed in concert with other intrinsic observations of the natural world.

As Holmes (1965) pointed out, in spite of the important advances in mathematics and the astronomy made by Greek philosophers, they considered natural phenomena to be powerful manifestations of their ancient gods or myths. In fact, many of the natural phenomena were personifications of the diverse aspects of nature (Poseidon, Zeus, Pluto, etc.). These beliefs had a considerable bearing on the development of geomorphologic theories until the 18th century (the teleological period of Chorley (1978)).

Nevertheless, in spite of the common conviction regarding the divinity of natural phenomena, some of the philosophers set aside this framework. Thales of Miletus (624–565 BC) considered natural processes as usual facts that could be understood by observation and logical approach. Likewise, Anaxagoras (500–428 BC) pointed out that the Earth revolutions were so slow, when compared with human life, that they go unnoticed by humans (Cailleux, 1961). This same philosopher indicated that water was elevated by the Sun and came back to the Earth's surface as rain. Plato (429–347 BC) imagined that all the rivers came from, and returned to, a huge cavern of water inside the Earth (Adams, 1938). This theory was possibly based upon field observations of the karstic landscape of Greece. Aristotle (348–322 BC) also developed excellent ideas on geomorphology, such as the hydrological cycle and the theory of relief development due to river incision.

As indicated by Cailleux (1961) and Ellenberger (1988), almost all of the deductions by the Romans about the natural world were based upon Greek theories. This fact can be inferred from the *Natural History of Pliny* (23–79 BC), which constituted a summary of Greek theories, superstitions and conjectures mixed together (Cailleux, 1961). Consequently, scientific advances were scarce during the Roman period.

This generalized hiatus in science lasted until the 16th century, and was also manifest in the development of geomorphology. During this millennium and a half, the ancient Greek works were transmitted to Spain from the Middle East by means of the Arabs. The Arab manuscripts were translated into Hebrew and Latin in the ancient Translators School of Toledo (Cailleux, 1961). Therefore, during this time the diffusion of knowledge was mainly carried out in the monasteries, but also in some cases by means of the individual patronage of a few kings. In this fashion, Avicena (Ibn Sina, 980–1037), an expert on Aristotle's work, developed hypotheses on the desert landscapes of Arabia. He thought that mountains could be generated by elevation of the ground or by the work of water and wind (Thornbury, 1954).

This period is also characterized by catastrophist or cataclysmic ideas partly based upon Christian thinking, with the pervasive use of Noah's Flood for the explanation of many natural processes or observations. This dogmatism produced absurd proclamations, such as the one manifested in 1654 AD by Archbishop Ussher of Ireland, which stated that the Sky and the Earth, centre and circumference, were created together by the Holy Trinity on October 26 of 4004 BC, at 9 o'clock in the morning (King, 1976a,b).

Catastrophist ideas dominated until the 19th century. But meanwhile the brilliant Leonardo da Vinci (1452–1519) exposed in his works significant considerations on the origin of the mountains (pioneering ideas about isostasy). He also indicated that rock salt was generated from the evaporation of seawater, as well as considering the importance of fluvial erosion and the generally slow work of geomorphic processes. Unfortunately his ideas were not published until the end of the 18th century, and his scientific influence during his epoch was practically nil (Ellenberger, 1988). Catastrophism consequently was the prevailing doctrine during this period.

During the 17th, 18th and part of the 19th centuries, the most significant advances in geomorphology came mainly from the studies on applied hydrology, carried out by several engineers. These were mainly French engineers, such as Perrault, De Chezy, Surrell, Guetthard, Desmarest and others who discarded in their works many academic concepts that were mostly inoperative.

Catastrophist theories had their opposite counterpart in the uniformitarian ideas developed at the end of the 18th century, with the postulation of the principle of actualism. Although previously noted by others in obscure works, this principle was first developed by Hutton (in 1726–1797) in his work *Theory of the Earth* (in Hutton, 1788) and later popularized by Lyell (in 1797–1875) in the first edition of *Principles of Geology* (Lyell, 1830), which was extensively reedited and reprinted. These Scottish geologists established the basis of modern geology, as well as geomorphologic methods. Uniformitarian theory can be synthesized in the simple sentence: "*the present is the key to the past.*" This means that different landforms have been created by processes similar to those of the present day, but operating over long periods of time. As we can see, it is totally contrary to the ideas of catastrophism. For example, Lyell described mountains that were eroded by weathering and fluvial transport, and later the transported materials were deposited in sedimentary basins which would ultimately give rise to new mountains.

During this period Playfair (in 1802) published his work *Illustrations of the Huttonian Theory of the Earth*, establishing that in areas of homogeneous lithology and structure that are subject to fluvial erosion, the valleys are proportional to the size of the tributaries contained in them. It is today is known as "Playfair's Law."

But the catastrophists, as with the cleric geologists Buckland and Sedgwick, argued that the erratics of glaciers and the alpine hanging valleys could only be explained by a universal flood (Beckinsale and Chorley, 1968). The gradual acceptance of the ideas of Lyell resulted in the end of the diluvialists. Also, the publication in 1840 of the book by the well-known Agassiz (1807–1873) on his studies of glaciers, by using the ideas of Scheuchzer, Martel, Saussure, and especially Charpentier, further limited the ideas of catastrophism. In this book the idea was launched of the occurrence of an ice age, and its impact on the erosion and sedimentation of valleys. The development of these ideas by Ramsay (in 1862), President of the Geologic Society of London, gave uniformitarism its decisive scientific support (Price, 1973).

2.2. The dawning of climatic geomorphology

The second half of the 19th century and the beginnings of the 20th were marked by pioneering scientific exploration of unknown or not very well-known regions of the Earth's globe. The aims of these expeditions were diverse, but mainly focused on the exploration of new territories for the evaluation of their mining and agricultural resources, but also their colonization and Christianization. These objectives came from those of the previous centuries that were specific to the different continents. Although there are earlier scientific documents, these scientific studies were mainly focused upon mining and geology, but detailed observations about landforms were scarce compared with the excellent works on botany and zoology.

These diverse expeditions resulted in the publication of different descriptive works on the Earth's relief, and barely contributed to the scientific framework of the early scientific research. Some German naturalists, however, started to consider Earth surface processes as illustrated in the works of Von Richtofen (in 1886) and Albrecht Penck (in 1894). During this period Martonne (in 1913) introduced the term "*climatic geomorphology*."

The Swiss geologist Agassiz had proposed his glacial theory in his pioneering work "*Etudes sur les Glaciers*" (Agassiz, 1840). Although many ideas in this work came first from others whom he beat to publication (Charpentier, Venetz, Perraudin), it focused upon the analysis of the set of materials deposited by glaciers, and represented the first well-known step in **glacial geomorphology**. Later, the English geologist Geikie, in his work *The Great Ice Age* (in Geikie, 1873), discussed the causes of glaciations, ice dynamics and the origin of glacial and post-glacial deposits. Afterwards, the early 20th century was a fruitful time for the recognition of existing glacial landforms and deposits. The work of Russell (in 1893) on the Malaspina Glacier, the paper on Spitsbergen glaciers by Garwood (in 1899) and Kendall's PhD (in Kendall, 1902) on glacier lakes illustrate this prolific period. During the second half of this century studies were progressively focused on the origin of erosive and depositional landforms of ancient glaciated zones (Price, 1973). The book of the American geologist R.F. Flint (in 1947) on *Glacial Geology and the Pleistocene*, and the two volumes of the treatise on the *Quaternary Age* by Charlesworth (1957), synthesized and reviewed the state of the art on relict glacier landforms and deposits.

The analysis of surface processes related to frost action goes back to the early 19th century, but the origin of **periglacial geomorphology** dates from the dawning of the 20th century. The term "periglacial" was introduced by the Polish scientist Lozinski (1909) to describe the processes and landforms resulting from freeze–thaw cycles occurring around the margins of the ancient Pleistocene ice sheets. These periglacial processes, however, are not only constrained to the periglacial climatic belt, but they can actually extend into other climatic zones as well. In consequence, at present this term holds a broader application than in its initial use (Thorn, 1992). At the same time the concept of solifluction was introduced by Andersson (1906), who also considered the climatic characteristics and morphologic features of these cold zones. Numerous investigations were also carried out in the expanses of Siberia by Russian scientists, but these went largely unnoticed by many European and American scientists due to language problems. During the beginning of the 20th century, the colonization of Alaska produced the first well-recognized advances in periglacial research (Cairnes, Capps, Eakin). In Europe, due to the inaccessibility of the main northern territories, periglacial geomorphology developed 30 years later (French and Karte, 1988). The first advances were mainly related to paleogeographic and environmental reconstructions of the late Pleistocene ice sheets in central and western Europe (i.e. Büdel, Cailleux, Dylik, Edelman, Poser, Tricart, Troll).

The last part of the 19th century was the most active period in the early development of **desert geomorphology**, mainly in the arid zones of the western United States of America (King, 1976a,b). During different geological expeditions, Powell (1834–1918), Dutton (1841–1912) and especially Gilbert (1843–1918) noticed a new kind of arid landscape in which the work of water was manifest. Powell (in 1875), in his main work (*Exploration of the western Colorado River, 1875*), introduced the key concept of base-level, as a forethought to the broader ideas of peneplanation. This same author established the genetic classification of fluvial networks as consequent, antecedent and superimposed drainages. Dutton largely contributed to the knowledge of isostasy, but also gave us detailed descriptions of alluvial fans, and established the earliest model of parallel retreat in the evolution of slopes in desert zones. Gilbert was certainly the main American geomorphologist of this epoch. His excellent reports on the *Geology of the Henry Mountains* (Gilbert, 1877) and on *Debris transport by overland flows* (in Gilbert, 1914) constitute relevant advances in the mechanics of fluvial processes, sediment transport, lateral river erosion and pediment formation, slope development and so forth. Gilbert can be considered to be a pioneer in surface process research. His report on the *History of the Bonneville Lake* (Gilbert, 1890) is one of the classics of geomorphology. In this work Gilbert underlined the use of ancient lake-levels and the resulting lake terraces for understanding the origin of the present-day Great Salt Lake of the State of Utah, and as a method that could be applied to other littoral zones. Also, he pondered the isostatic rebound produced after the evaporation of most of the large Pleistocene pluvial lakes.

Scientific contributions from other arid zones of the world were comparatively less important (Graf, 1988). Numerous expeditions were carried out by French scientists across the Sahara desert. They resulted in merely descriptive works, but some of them clearly introduced the climatic change problem (i.e. Chudeau, Flammond, Gautier, Urvoy). The Kalahari desert was extensively described by Passarge (in 1904). The Namib desert was analysed by Kaiser (in 1921) and Little studied part of the Arab desert in 1925. The Iranian and Thar (India) deserts were explored in the decade of the 1870s (Blanford, Oldham). Research on the Australian desert was focused on the age of the large erosional surfaces and their relation to the resulting products of weathering (in Juston, 1934). After a first descriptive period, desert research in South America began with the studies on the salt lakes of the Atacama by Frenguelli (in 1928). Also notable was the research on La Puna carried out by Walther Penck (in 1920). At the beginning of the 20th century the prevailing idea was that the wind was the main geomorphologic agent in the development of planation surfaces in deserts (i.e. Hedin, Passarge, Walther). Bryan (1922), however, highlighted the idea that wind action only introduces small modification to these planation surfaces. Research on the work of wind carried out by Bagnold in the Egyptian desert was compiled in his classic dissertation entitled *The physics of aeolian sands and desert dunes*. This work constitutes a basis for the understanding of aeolian processes and the fundamentals of aeolian geomorphology.

The work of the North American scientist Dana (in 1849) on the landforms developed in the volcanic islands of the Central Pacific was the key for the initiation of **tropical geomorphology** (Chorley et al., 1964). Previously, Thomson (in 1822) had pointed out the occurrence of numerous isolated hills upstanding from the plains located in the region of the Mozambique – Tanzania border (Douglas and Spencer, 1985a,b,c). Similar observations were made by Bornhardt (in 1900) in eastern Africa, who introduced the term *inselberg* to denote this kind of isolated hill. On the other hand, Darwin (in 1890) and Branner (in 1896) highlighted the occurrence of thick weathering horizons, but previously Buchanan (in 1807) had made the first scientific descriptions of laterite weathering horizons developed in India. During the first half of the 20th century, concurring with the exploration of new territories all around the world's climatic zones, different works on tropical geomorphology were published (i.e. Cushing, Falconer, Grund, Hayes, Hubert, Passarge, Sapper) (Douglas, 1978). In 1926 Thorbecke presided over the first meeting on climatic geomorphology (Düsseldorf, Germany) in which the bases of morphoclimatic classifications were established. The more relevant works of this period may be those undertaken by the German scientist Freise during the decade of the 1930s in the State of Rio de Janeiro (Brazil), and that of Sapper (in 1935) on the *Geomorphology of the Humid Tropics*, which constituted the most relevant review of the time.

2.3. The impact of landscape evolution models on climatic geomorphology

The three existing models of landscape evolution, developed by Davis, Penck and King, were proposed during different periods of the first half of the 20th century. The models of Davis and King were particularly constrained to a specific climate, whereas the Penckian model emphasized the role of tectonic and climatic factors together. Knowledge of all the scientific works of these authors can lead to a better understanding of the content, progressive development and criticisms represented by their models.

The life and work of **William Morris Davis** (1850–1934), emblematic geomorphologist and Professor of Geology at Harvard University, have been the subject of copious analytic studies, some of them made with considerable extension and depth (Chorley et al., 1973; King and Schumm, 1980). His more essential works, carried out until 1906, were compiled in a Special Volume (Davis, 1954), but the scientific production of this prolific scientist continued up to 1938, the date in which his last paper was published, long after the first one published in 1880.

Davis graduated from Harvard University. After a 3-year long stay in Córdoba, Argentina and a journey around the world, he began his lectures at Harvard University in 1875. His research work was copious, always written in a clear and brilliant language, and accompanied by wonderful illustrations that improved his papers (Higgins, 1975; King, 1976a,b) and constituted an excellent tool for the teaching of geomorphology. His best quality was the capacity for assembling most of the previous, and unconnected, knowledge of geomorphology. Some authors of the 20th century considered him to be the father of geomorphology (Tricart, 1965).

The major contribution of Davis was the geographical cycle, also known variously as the erosion cycle, the normal cycle, the humid cycle and so forth (Higgins, 1975), and it was considered to be one of the first paradigms in geomorphology (Hart, 1986). It was the first theory on landscape development with a general agreement from the scientific community (Thorn, 1988a,b). He published his basic model in the paper *The Rivers and Valleys of Pennsylvania*, but the definitive version was published with the title *The Geographical Cycle* 10 years later.

According to the ideas of Davis, a sector of a planated land surface may undergo rapid uplift leading to a subsequent period of valley incision under conditions of tectonic stability. The initial landscape underwent sequential stages of different relief features denoted as youth, maturity and old age. The final result was the elaboration of a *peneplain* with some isolated residual relief (Fig. 1.1) that could correspond to the initial planated land surface of a new cycle with renewed uplift. In the explanation of the cycle, Davis analysed the development of hillslopes and fluvial valleys, as well as the probable interruptions of the ideal cycle, in which erosion was reactivated in relation to a new or different base level.

Criticism of this model mainly came from the fact that the model only represented landscape development for humid temperate regions, considered to be the "normal climate" by Davis. He considered glacial and arid landscapes as "accidents" to the geographical cycle. Consequently he developed "arid" (Davis, 1905) and "glacial" (Davis, 1906) versions of the cycle, the latter divided into pre-glacial, optimum and post-glacial stages (Fig. 3.21). In spite of numerous criticisms, some of them addressed in his papers, the concept of the Davisian cycle of erosion had a general acceptance. Higgins (1975) pointed out 12 reasons for this general acceptance, with simple arguments such as simplicity, applicability, rationality and so forth.

On the other hand, the work of Davis also included different fields of geomorphology, such as glacial, coastal, arid and volcanic landscapes. His relationships with European scientists were important, especially those maintained with his German colleagues (Alfred and Walter Penck), where he carried out long and successive visits.

The concept of landscape cyclicity gained a wide acceptance among the different authors of that period, and was applied to a wide variety of other geomorphologic environments, as was the development of specific cycles for karst (Fig. 1.2), littoral, glacial, savannah and periglacial landscape cycles of erosion (Higgins, 1975; Hart, 1986). The ideas of Davis were well received among many German and Anglo-American scientists, but received little acceptance in the former USSR or countries of eastern and northern Europe, where his work went virtually unnoticed. He had enthusiastic supporters among French investigators, including Lapparent, Baulig and Martonne. The latter published the *Traité de Géographie Physique* in three volumes, in 1925–1927. This work exerted a great deal of influence on many French specialists. As mentioned previously, Martonne was also the first author who had introduced the term "*climatic geomorphology*" in 1913, (Stoddart, 1969a; Beckinsale and Chorley, 1991).

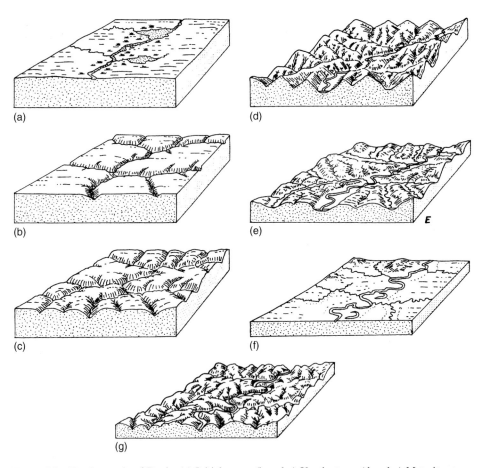

Figure 1.1. Erosion cycle of Davis. (a) Initial stage. (b and c) Youth stage. (d and e) Maturity stage. (f) Old age stage with development of peneplains and monadnocks. (g) Uplift and development of renewed erosion (drawn by Raisz in Strahler (1965)).

But Davis also had a large number of detractors. Even today, some geomorphologists consider Davis to be responsible for the more than 50 years of backwardness of geomorphology with respect other sciences (Hart, 1986). Tricart (1965) criticized the lack of methodology and the deductive approach contrary to the scientific inductive method. Other mainstream disapproval was voiced in relation to ideas of processes working in distinct climatic zones, generating different landscapes. In Germany the birthplace of climatic geomorphology (von Richtofen, Machatschek, Passarge and Troll, as representative authors), the Davisian model was largely critiqued on account of its marked overlooking of the climatic approach. In this way, Passarge (1931) pointed out: "The forces involved in landscape development are of climatic nature... For instance, vegetation depends only on the climate; but, in turn, the vegetal cover partly influences, but in a decisive way, the landscape demolition." Later, during the decade of the 1950s, most of the French investigators abandoned Davisian concepts, supporting the new principles of climatic geomorphology (Cholley, Dresch, Birot, Tricart and Cailleux).

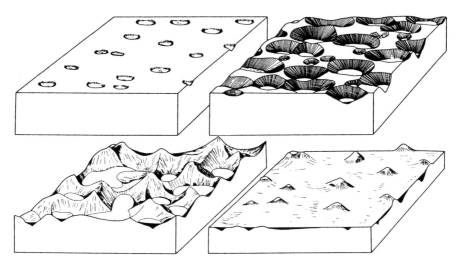

(*upper left*) 'Young' karst with extensive flat surfaces between dolines, much of the original surface remaining.
(*upper right*) 'Adolescent' karst, with larger dolines and many coalescing.
(*lower left*) 'Mature' karst (*Cockpitlandschaft*): the original surface has disappeared; large dolines and uvalas lie between the *Cockpithügeln*.
(*lower right*) 'Old' karst. Only isolated hills remain upon the residual plain.

Figure 1.2. Karstic cycle (Grund, 1914).

As we can see, the response to the Davisian model was marked from the earliest moments among some geomorphologists, but in other schools, after an initial, and fairly generalized acceptance, Davisian ideas were abandoned. The scientific opponents of Davis, however, never really offered alternative models to address the defects of the normal erosion cycle.

Only **Walther Penck** (1888–1923) elaborated another alternative model for landscape development in his work *Morphological Analysis of Landforms*. This important geomorphologist, son of the famous researcher on glaciers Albrecht Penck, after his PhD dissertation in 1910 at Heidelberg, worked for 2 years in the Andean Cordillera for the Geologic Survey of Argentina. His expertise on the geology of Central Europe, together with that acquired in the arid active tectonic zone of the Andean region, greatly influenced the development of his theory. In addition, his primarily geological education was complemented by the geomorphologic knowledge of his father. His book *Morphological Analysis of Landforms* was written in his convalescence during a fatal illness that ended with his death at the age of only 35 years (Bremer, 1983). His father eventually published the book in 1924. Their thesis was an approach to geomorphology from the point of view of a geologist who wants to find in geology an element for the interpretation of diastrophic history (Thornbury, 1954). In this way, we can consider Penck's thinking as one of the precursors of the geomorphologic method in neotectonic research.

Penck based his model on the study of the sedimentary sequences flanking the uplifted blocks of the Andes and the Alps (Chorley et al., 1973) and he deduced that the relief was generated during an initial period of slowly increasing uplift rates followed by a period of declining uplift that eventually terminated in tectonic stability. In contradistinction,

Davis had started his theory with a planated land surface undergoing abrupt uplift followed by a long period of tectonic stability. The model of Penck began with the gradual uplift of a primary surface (*Primärrumpf*) promoting the progressive development of a set of benched erosional surfaces around the margins of the primary one. This process was thought to give place to a staircase piedmont (*Piedmonttreppen*) (Fig. 1.3) in which each one of the developed benches worked as a local base-level. The relief generation was thus controlled by the balance between crustal uplift (endogenic forces) and river incision rates (exogenic forces). When crustal uplift waned, river erosion dominated and landscape progressively evolved towards a terminal planated land surface (*Endrumpf*). Parallel with the development of this model, Penck also proposed his theory of slope evolution linked to vertical crustal movements.

Penck's ideas were never popular in Germany (Bremer, 1983). His theory was really only known 8 years later by means of a critical and inaccurate interpretation of Davis (Chorley, et al., 1984). This misinterpretation was probably due to the rather obscure terminology and writing style used by Penck. The English translation of his work, in Penck (1953) resulted in an important revival of the Penckian model.

A third model for landscape development was developed by the London-born, South African geologist, ***Lester Charles King***. He was instructed by Cotton, a Davisian geomorphologist and author of three important books (1942–1944) of the mid 20th century, and then he tried to apply the Davisian theory to South Africa. King's ideas were influenced, like the Davisian and Penckian ones, by the particular geomorphology of his home landscapes, in this case in South Africa characterized by the occurrence of extensive erosion surfaces separated by large scarps.

King's ideas were first introduced in the paper *Canons of landscape evolution* (King, 1953) and later in his work *The Morphology of the Earth* (King, 1962) where they were extensively explained. Resembling the model of Davis, King's own theory is based on cyclicity. Landscape evolution was supposed to begin with rapid uplift, followed by a long period of tectonic quiescence, during which extensive pediments developed, giving rise eventually to a ***pediplain***. Some resistant isolated relief of diverse morphology, such as buttes in landscapes of flat-lying sedimentary rocks, or inselbergs and bornhardts in regions of crystalline rock can dot these extensive planated surfaces. King considered the arid and/or savannah regions with contrasted humid–dry seasons as the "normal morphoclimatic systems." Renewed uplift was to give rise to the beginning of a new cycle

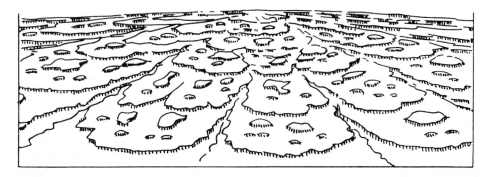

Figure 1.3. Davis's interpretation of the Penck's piedmonttreppen on a dome (Davis, 1932).

of erosion, in which previously developed pediplains continue evolving, but the associated scarp retreat rates diminish at distal locations in response to the development of new pediments (Twidale, 1992). Successive erosion cycles gave place to the staircase development of repeated pediplains at continental scale in a similar assemblage to that illustrated by the *Piedmonttreppen* of Penck.

King's model was based on the process of parallel retreat of slopes and, as aforementioned, elaborates upon the ideas of Davis and Penck (Higgins, 1975; Thorn, 1988). His ideas did not generate the relevant discussion of the predecessor models, mainly because his theory was published at a time during which emergent methods of process analysis and quantification were convulsing geomorphology with other new thinking (Higgins, 1975).

Another much discussed contribution of King's was his introduction in 1962 of the concept of *cymatogeny*. This would be added to the concepts of orogeny and epirogeny. This process, apparently based upon the *"Hebung"* of Cloos (1939), involved the upwarping and flexure of continental margins. Nonetheless, in opposition to the theory of plate tectonics because of his favoured position on the idea of an expanding Earth, King (1983) considered cymatogeny to be the dominant crustal movement.

The models of Davis, Penck and King today are the three main historical approaches to large scale landscape development. The Davis and King proposals are based upon cyclicity and have a clear climatic background, whereas the Penck approach represents a continuous contest between the rates of fluvial incision and crustal uplift. All these models are illustrated in Figure 1.4.

2.4. Process geomorphology

As another branch of science, geomorphology passed from its initial descriptive character into a more quantitative approach, for which it was necessary to use the methods of mathematics and physics. Although in the decades of the 1940s and 1950s some authors analysed the dynamics of glaciers, beaches, dunes, fluvial bars and other landforms, it was primarily during the second half of the 20th century when the revolution occurred in the process – geomorphology paradigm. During that time geomorphologists were more concerned with the workings of geomorphic processes than with the shape of the landforms (Barry, 1997). Some authors considered processes analysis to be the key topic of modern geomorphology (Hart, 1986). It has been estimated that in the year 1980, 75 per cent of the geomorphologic research in the United Kingdom was focused on process analyses at detailed scales (Gardner, 1983).

For many authors the revolution started with the paper by Strahler (1952), which can be considered as the first essay on *dynamic geomorphology* (Higgins, 1975). Others think that the works of Leopold and collaborators, partly synthesized in the volume *Fluvial Processes in Geomorphology* (Leopold et al., 1964) are the genuine pioneers and instigators of the implementation of quantitative process analysis.

Process analysis is included in the framework of process–response systems, in which the process is the working geomorphic agent and the response is the resultant landform. Process analysis may include numerous and different approaches, according to the variety of the different existing geomorphic environments (Goudie, 1981a,b). Some of the techniques

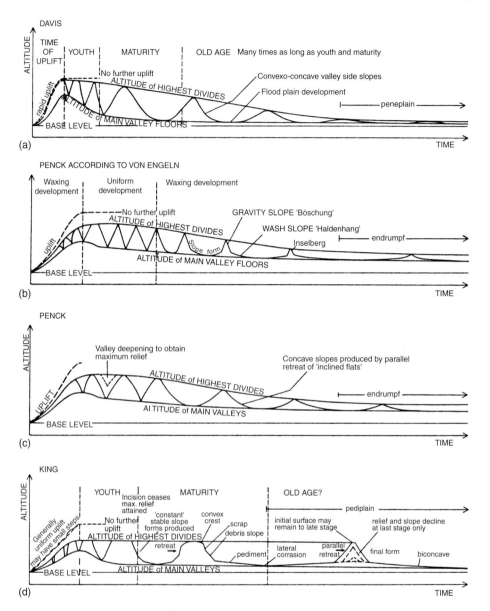

Figure 1.4. Cyclic models of landscape evolution showing the relationships between elevation and time for a fixed base-level. Also, schematic slope profile development is shown (Thornes and Brunsden, 1977).

used are very new, and in many cases not only require handmade but also complex equipment. As pointed out by Hart (1986), in many cases "necessity is the mother of invention." Process research is often difficult not only because of the complexity of the instrumentation to be used, but also because of the variety of geomorphic agents working together, as well as to our incomplete understanding of the working mechanisms of some process (Baker and Twidale, 1991). The study of processes also promoted a drastic reduction in the spatial and temporal scales being analysed. For instance, research can be focused on the analysis of micro-drainage basins and the individual slope segments. During the last few decades efforts in this analytic field have been important. It constitutes part of the basic subject matter on geomorphology, and the expected advances in it will give a better understanding to landform evolution. Today, a variety of erosion rates have already been reported for different lithologies and climates. These sets of data were compiled by Saunders and Young (1983) and Goudie (1995). For example, on abrupt glaciated slopes erosion rates are $\sim 1-4$ mm/yr, slopes in temperate zones erode about $0.01-0.1$ mm/yr, in semiarid regions, $\sim 0.1-1$ mm/yr, in badlands, $\sim 1-10$ mm/yr, and in the abrupt slopes of tropical forests, $\sim 0.1-1$ mm/yr.

Field and laboratory research on processes requires use of some of the tools of statistics for the adequate management of the obtained data. This fact gave rise to *quantitative geomorphology*, with pioneering work by the hydrologist R. Horton on the morphometry of drainage basins (Horton, 1945). The further development of quantitative analysis of landforms has been controlled by the increasing capacity of computers in data management.

As a consequence of the complexity of geomorphological processes, the use of *models* has become common as another important aspect of modern geomorphology. Scaled models partly reproduce natural processes, such as the use of wind tunnels, fluvial erosion tanks and other laboratory tests. Other models are the analogue ones, such as the ice deformation simulations by kaolin analogues, or the mathematical models in which Ahnert (1987a) and Kirby (in 1994) are among the more outstanding present-day geomorphologists. These two kinds of models may achieve all the complexity of process interaction (Baker and Twidale, 1991). At the beginning of the 21st century it now appears possible that mathematical process models iterated successively in supercomputers to simulate deep geologic time may at last make possible the integration of *timeless* process models with *timebound* landform evolution; something that has not been possible heretofore.

2.5. Structure of climatic geomorphology

On 22–23rd September 1926, in Düsseldorf, Germany, researchers held a meeting on landform development in different climatic zones, as well as on morphoclimatic classification (Thorbecke, 1927). The results of that conference constituted the first straightforward and comprehensive approach to climatic geomorphology. For some of the participants, however, their approach to problem formulation for particular climatic regions lacked a true morphoclimatic method (Beckinsale and Chorley, 1991).

In about the middle of the 20th century, Büdel (1944, 1948) published works that he afterwards expanded on and analysed in depth in his book on *Climatic Geomorphology* (Büdel, 1977), as well as in some subsequent papers (i.e. Büdel, 1980). In these works, morphoclimatic zones were defined as an assemblage of landscapes resulting from the activity of the processes working in the different types of climates. In his paper of 1948, Büdel differentiated eight different morphoclimatic zones using morphologic and climatic criteria. These were expanded to 10 in his treatise of 1977, in which mountainous environments were excluded. This author also proposed the term "climatogenic geomorphology" (Büdel, 1963), defined as the science of the reconstruction of past climates from the analysis of ancient working processes and landforms. As pointed out by Bremer (1996), however, ideas of climatic geomorphology lead directly to climatogenic geomorphology, because almost all the world regions are constituted by relict landforms from past climates.

Tricart and Callieux (1965) developed the concept of zonation in geomorphology and elaborated a map of morphoclimatic regions, giving to the vegetation a potentially dominant role in its designations. They differentiated 13 morphoclimatic regions around the world, including an azonal mountain zone. Birot (1949a) analysed the climate influence on slope development in bedrock of different lithologies. During the decade of the 1950s the climatic differentiations of karst (Lehmann et al., 1954) and granitic (Wilhelmy, 1958) landscapes were established. Peltier (1950) related a variety of exogenic processes to mean annual temperatures and total annual precipitation and proposed the differentiation of nine morphoclimatic regions; which representation was later modified by Wilson (1968, 1969). This classification was criticized for its inherent limitation from the bi-variable approach (Stoddart, 1969a). The book of Birot (1960) on *Le Cycle d'Erosion sous les Différents Climats* constituted a relevant contribution to the geomorphology of arid and humid tropical zones.

2.6. The development of climatic geomorphology

Until the second half of the 20th century, geomorphology was mainly focused on the age determination and reconstruction of the different sequential stages over which the erosive and tectonic processes built up the present-day landscapes. This was denoted by Chorley (1978) as the historic stage of geomorphology, in which its development was based upon the establishment of erosion models, the analysis of denudation chronology, and studies on structural landscapes. Later the increasing interest in process research and quantitative analysis over different climatic zones resulted in the further development of climatic geomorphology. This branch of geomorphology has made much progress during the last decades, as indicated by the profusion of the published thematic volumes on different morphoclimatic zones.

During the second half of the 20th century *glacial geomorphology* received relevant advances in the mechanics of ice-flow (i.e. Hambrey, Lliboutry, Nye, Paterson, Weertman). In the same way, research on glacial processes and landforms (i.e. Boulton, Dreimanis, Drewry, Iverson, Linton, Menzies, Sugden), but also on fluvio-glacial environments and glaciotectonics (i.e. Price, Menzies, Van der Wateren) is abundant. Aside from the papers published in specialized or general scientific journals, numerous thematic volumes have been published on tills, moraines, drumlins, glaciotectonics, fluvio-glacial and glacio-lacustrine environments. General volumes on glacial geomorphology were also produced, such as those of Embleton and King (1975), Sugden and John (1976), Menzies (1995a,b,c – 1996) and Benn and Evans (1998).

During this same period, research on *periglacial geomorphology* continued, but focused on field and laboratory analysis of processes such as frost action (rock and soil cracking, weathering, heaving, thrusting and classification), mass movement, fluvial and wind action. There was a broad group of authors dealing with these different topics (i.e. Clark, Corte, French, Lachenbruch, Lautridou, Mackay, Pissart, Washburn). Landform recognition from this variety of working processes was diverse, but well known today. It may develop in specific micro and/or meso environments that indirectly generate great gaps in the understanding of periglacial environments (Barsch and Caine, 1984). There are many relevant geomorphologists in the study of periglacial landforms and it is difficult to highlight the outstanding contributors (i.e. Barsch, Büdel, Czudek, Demek, French, Mackay, Pissart, Rapp, Tricart, Washburn). Some treatises on periglacial geomorphology have been also published, stressing the work of Washburn (1979), which was considered by Thorn (1992) to be the "bible" of this discipline. The thematic volume edited by Clark (1988) is a collection of papers covering the whole periglacial geomorphology. Finally, the treatise of French (1996) updated the present knowledge on periglacial research.

Extreme conditions in most of the present deserts of the world made it necessary to carry out numerous expeditions and group research there during the last 50 years. In this way, landscapes and landforms of these inhospitable regions have been progressively studied. Satellite imagery, however, become a much appreciated tool for research on *arid zone geomorphology* since the decade of the 1970s. It made possible large-scale desert mapping and analysis, especially of the great sand deserts that were otherwise intractable. The increasing interest in the knowledge of arid environments is manifest by the recent creation of several "desert research centres" around the world. Research on desert weathering and resultant micro-landforms has been undertaken by Cooke, Evans, Goudie, Mabbutt and Twidale. Dorn and Oberlander have mainly done rock ("desert") varnish studies, and Machette, Marion, Reeves, Summerfield and Watson developed analyses of calcrete and silcrete crusts. There is also abundant and recent literature on experimental analysis of complex process response on slope and drainage basin activities (i.e. Abrahams, Bryan, De Ploey, Parsons, Schumm, Thornes). Fluvial geomorphology on deserts has, as its main authors, Leopold, Wolman, Miller, Schumm and Shick. Badland research has been focused on slope evolution and the quantification of slope retreat (i.e. Grossman, Howard, King, Mabbutt, Oberlander, Selby, Twidale). The always challenging topic of desert glacis and pediments has been undertaken by many different authors (i.e. Birot, Dresch, King, Mabbutt, Mensching, Tricart, Twidale). The same can be said of research on alluvial fans on which Blair, Bull, Denny, Harvey, Hooke and Mcpherson, among others, can be considered as the key authors. There is also abundant research on playa-lakes and sebkhas (i.e. Eugster, Glennie, Hardie, Krinsley, Thomas). Aeolian erosion and sedimentation have been the subject of numerous research papers in different scientific journals by Kocurek, Lancaster, Livingstone, Mainguet, McCauley, Péwé, Pye, Thomas, Tsoar, Warren among others, but also thematic volumes on aeolian geomorphology have been recently published (Pye, 1987; Pye and Tsoar, 1990; Lancaster, 1995; Livingstone and Warren, 1996). On the other hand, the recurring publication of thematic volumes and special issues on arid zone geomorphology is relevant, such as those of Tricart (1969), Mabbutt (1977), Cooke et al. (1993), Abrahams and Parsons (1994) and Thomas (1997a). These last two works are excellent special volumes that comprise the contribution of many specialists on drylands.

Tropical geomorphology has not been a systematic research topic until very recent times (Gupta, 1993). Some authors (Stoddart, 1969a,b; Twidale and Lageat, 1994) do not even consider it a true morphoclimatic zone. Studies on weathering processes and laterite development have been carried out by Berry, Birot, Goudie, Loughnan, McFarlane, Ollier, Ruxton and Thomas, as the most relevant authors. Research on water erosion in tropical forests has increased during the three last decades, and research was mainly focused on experimental basins. Studies on fluvial transport were undertaken by Douglas, Spencer and Walling. Mass movements are common in deep-weathered slopes of tropical zones as can be inferred from the works of Brand, Lal, So and Tricart. Büdel (1957) carried out the analysis of erosion surfaces and inselbergs in the tropics. This author developed his theory of double planation differentiating an upper denudation surface and a lower weathering front, in which the constant removal of the weathered horizon results in the exhumation of the so-called *etchsurfaces*. The development of these ideas was mainly achieved by Bremer, Ollier, Thomas, and Twidale. Carbonate outcrops are subject to intense dissolution in tropical climates generating typical cone karst and tower karst landscapes (i.e. Balazs, Lehmann, Sweeting). Books on tropical geomorphology are scarce in relation to those dealing with other morphoclimatic zones. The works of Birot (1973) and Tricart (1974b) offer important contributions, but it is the thematic volume edited by Thomas (1994) that is the most relevant treatise on tropical processes and landforms, subsequent to his earlier contribution in 1974. Another important paper is that of Faniran and Jeje (1983) which offers a state of the art on this subject in the early 1980s. Finally, the thematic volumes of Twidale (1982b), Gerrad (1988) and Vidal and Twidale (1998) also consider problems on granite geomorphology closely related to tropical geomorphology.

2.7. Applications and future trends in climatic geomorphology

Despite the fact that climatic geomorphology has always played a relevant role in applied studies, it was the decade of the 1960s when a relevant applied nature was begun as a consequence of the increasing development of process research. One of the more outstanding fields is the research on natural hazards derived from the dynamics of surface processes, such as salt weathering, fluvial and aeolian erosion, flooding, mass movement and karstic subsidence, among others. Another field with relevant development during the last few decades is *environmental geomorphology*. In response to the constant population increase and exploitation of natural resources, the physical environment is undergoing related changes. Humans are now an important geomorphic agent, and some authors even contemplate the term *anthropic* or *anthropogenic geomorphology*. It is necessary to live according to nature, and it is also essential to gain a better understanding of the processes operating in the global ecosystem. It is in this subject matter where climatic geomorphology should play an outstanding role.

Today, geomorphology has become an encyclopaedic body of information and research (Dury, 1978), as indicated in the *Encyclopedia of Geomorphology* edited by Fairbridge (1968b). This same author opined that at the end of the 20th century, scientific production in geomorphology would be close to 200,000 papers per year, which clearly indicates the amazing, but expectable, development of this science.

It may be expected that research on climatic geomorphology will be focused in the following years on: (1) constant progress in geomorphic process research; (2) increasing interdisciplinary problem approaches; (3) more accurate advances in remote sensing data management; (4) increasing research on paleogeomorphology in relation to global climate prediction and (5) applied studies in environmental geomorphology (Gutiérrez, 1990).

As previously mentioned, the panorama of geomorphology has moved from the initial generic studies of denudational chronology to the current emphasis on surface processes research. This fact has triggered a metamorphosis and breaking up of the discipline (Chorley, 1978; Thorn, 1988a). As indicated by the first author, geomorphology today has been submerged in a *processes syndrome*. The subjacent problem is related to translating the detailed-scale studies of processes research to broad-scale landscape evolution (Barsch, 1990). This is the motive as to why geomorphology should carry on research at a wide variety of temporal and spatial scales, in an attempt to better understand the different workings of landscape development at different scales (Chorley, 1978). In addition there is hope that modern computer simulations will also be able to merge surface process research with landscape evolution studies.

3. Climatic geomorphology: processes and morphoclimatic zonation

Climatic geomorphology can be defined as the discipline that identifies climatic factors such as the intensity, frequency and duration of precipitation, frost intensity, direction and power of wind, and it explains the development of landscapes under different climatic conditions (Ahnert, 1996). It is noteworthy that in this recent definition the magnitude and frequency of surface processes are taken into account. Initially, in the pioneering work of Peltier (1950) only two climatic parameters (mean annual temperature and total annual precipitation) were considered in establishing their relation with five key geomorphic processes: chemical weathering, frost action, pluvial erosion, mass movement and wind action (Fig. 1.5). For Peltier, "this set of graphics simply represents the schematic illustration of a concept." As a whole, he distinguished two different morphogenetic elements, which are weathering processes and the transport agents of the resultant materials. From this perspective this author postulated nine different morphogenetic regions (Fig. 1.6), which may be distinguished by a characteristic assemblage of geomorphic processes. Distinctions made by Peltier were mainly qualitative and very subjective (Derbyshire, 1973b). A similar approach was that developed by Leopold et al. (1964). Wilson (1968, 1969) also developed a comparable approach, but he changed the frost action graphic for a mechanical weathering one, modified the fields for relative intensities of geomorphic processes defined by Peltier, and distinguished six climatic regimes denominated as climate-process systems. Wilson underlined the monthly variation of temperature and precipitation (seasonality), as well as its influence on the activity of geomorphic processes (Fig. 1.7).

Tanner (1961) used potential evapotranspiration instead of temperature as a climatic variable, because this factor gives us a more realistic idea of moisture availability and, in addition, is linked to the proportion of vegetal cover. Likewise, Tanner also assumed the four main morphogenetic regions to be glacial, temperate, arid and selva, but also added tundra, savannah, and semiarid regimes.

An interesting approach is that proposed by Common (1966) in his work about landslides and morphoclimatic regions. He considered that mass movement is a part of the denudational process and may be used to recognize different landscapes. Instead of using a large number of involved variables, Common elaborated three global maps based upon particular characteristics of precipitation and temperature of geomorphic interest (i.e. precipitation regimes, intervals of total precipitation and temperature).

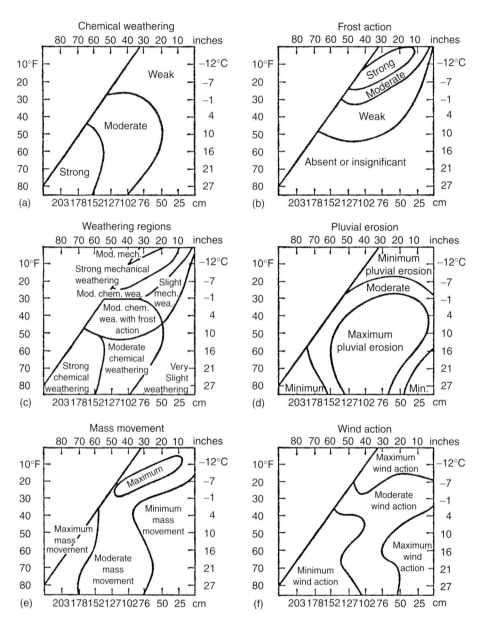

Figure 1.5. "Peltier diagrams" illustrating the relative importance of some geomorphic processes as a function of mean annual rainfall and temperature (Peltier, 1950).

Worldwide scale approaches to climate and vegetation controls on weathering intensity and products have been also undertaken (Strakhov, 1967). This author recognized a clear climatic zonation of weathering, and identified lixiviation or solution maxima in the tropical forest and taiga–podsol zones, where precipitation is also higher. Weathering is more intense and deeper in the tropics in response to the acceleration of the chemical

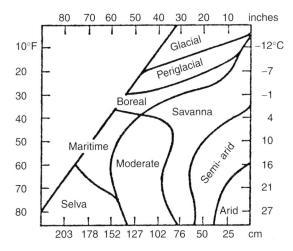

Figure 1.6. Morphogenetic regions: mean annual rainfall and temperature (Peltier, 1950).

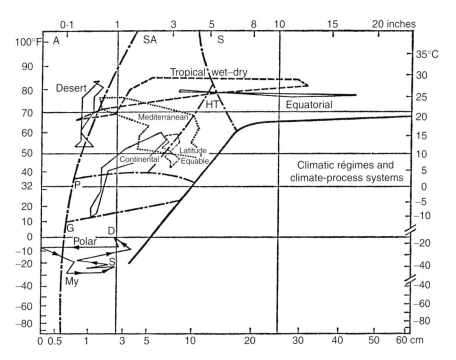

Figure 1.7. Climatic regimes and climate-processes. Graphs show monthly rainfall and temperature data reported by Strahler (1965). Scales are modified to emphasize arid regions. System boundaries are given by dash dot-lines. Systems indicated by letter abbreviation: A – arid; SA – semiarid; S – selva; HT – humid–temperate; P – periglacial and G – glacial (Wilson, 1968).

reactions by the higher temperatures. On the contrary, in the tundra and desert zones weathering is small and shallow due to scarce precipitations. The north–south global cross-section presented by Strakhov (1967) led to differentiation of a set of weathering profiles linked to the various climatic zones (Fig. 1.8). This same author also elaborated a global map showing the geographical distribution of weathering profiles.

Another attempt at morphoclimatic zonation is based on the regional differences in fluvial erosion, but this parameter is greatly influenced by extra-environmental factors. Several regression lines relating the mean annual precipitation (Langbein and Schumm, 1958; Fournier, 1960; Wilson, 1973) or mean annual runoff (Douglas, 1967; Dendy and Bolton, 1976) with sediment production in metric ton/km^2/yr have been proposed. Most of the authors assume that maximum fluvial erosion occurs around 300 mm of precipitation, whereas others highlight the occurrence of another denudational peak up to 1500 mm. Corbel (1964) synthesized the set of erosion data for the different climatic zones, and considered that erosion is smaller in the tropics. Fournier (1960) obtained very different conclusions from the analysis of 78 drainage basins between 2460 and 1,060,000 km^2. For this he correlated the climatic parameter p^2/P, where p is the maximum monthly precipitation, and P the total annual precipitation, with the suspended load transported by the rivers. Opposite to Corbel, Fournier (1960) found that the increment of erosion is directly linked to the increase of precipitation. This author elaborated the map of worldwide erosion distribution presented in Figure 1.9 and illustrated the occurrence of highest values in the humid-seasonal tropics that decrease towards equatorial and arid regions. The main conclusions of Fournier (1960) are retained in the above-mentioned work of Strakhov (1967).

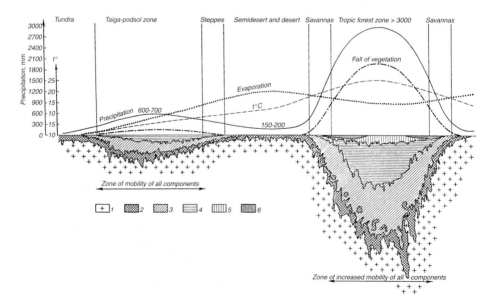

Figure 1.8. Weathering zones in relation to latitude. 1 – Fresh rock; 2 – Rock debris with small chemical weathering; 3 – Zone of dominant hydrolysis; 4 – Kaolinite zone; 5 – Zone of iron and alumina oxides; 6 – Soil armour or ferricrete (Strakhov, 1967).

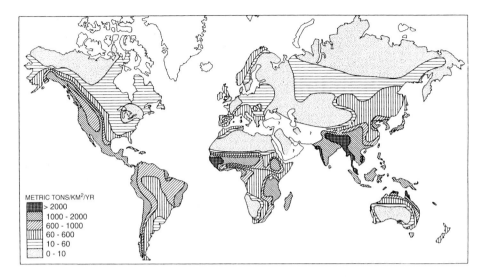

Figure 1.9. World distribution of fluvial erosion (Fournier, 1960).

Chorley et al. (1984) pointed out that a simple morphoclimatic classification may be developed considering parameters such as mean annual temperature, mean annual precipitation and seasonality, expressed by the temperature of the warmest month or by the number of months with precipitation up to 50 mm (Fig. 1.10). The diagram of this figure is composed of eight different regions that may be divided in two main groups:

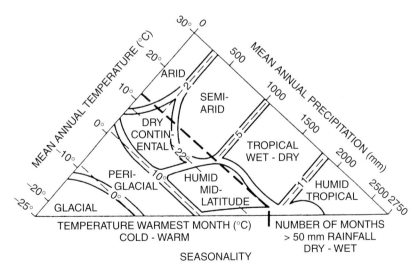

Figure 1.10. Main present morphoclimatic zones classified according to mean annual temperature (°C), mean annual precipitation (mm), mean number of wet months (>50 mm) and mean temperature of warmest month (°C) (Chorley et al., 1984).

(a) First-order morphogenetic regions, considered as non-seasonal by Beckinsale and Chorley (1991), comprising the *glacial, arid* and *humid tropical* regions. These are characterized by the occurrence of non-seasonal processes with low erosion rates, except those related to sporadic events such as surging glaciers, desert storms and mass movements. The central zones of these non-seasonal regions persist latitudinally during climatic changes.

(b) Second-order morphogenetic regions, considered as seasonal ones by the same authors, comprising tropical wet–dry, semiarid, dry continental, humid mid-latitude and periglacial regions. These have seasonal processes, occasionally of high intensity accompanying important changes. In these regions it is possible to differentiate warmer climates (tropical wet–dry and semiarid) where geomorphic processes differ significantly in terms of length of the wet season; and cooler climates (dry continental, humid mid-latitude and periglacial) whose geomorphic processes differ mainly in respect of summer temperatures and also partly with the precipitation amounts. These authors detail a map modifying that elaborated by Tricart and Cailleux (1965) of the current distribution of these eight morphogenetic regional types.

Data usually managed in geomorphology are those directly provided by climatology, such as those used by Peltier (1950), Tanner (1961), Leopold et al. (1964), Wilson (1968, 1969), Chorley et al. (1984), and others. These data can be reasonably adequate to develop broad overviews about geomorphic processes. Whatever the case, the utilized mean numeric values are inadequate because most of the quoted surface processes have a discontinuous nature with relevant variations in frequency and magnitude (Ahnert, 1987b, 1996; De Ploey et al., 1991). Therefore, it seems to be necessary to use other kinds of numerical parameters such as rainfall intensity, wind velocity, frequency and duration of freeze and so forth. Once these numerical data have been selected, a frequency–magnitude analysis can be performed to characterize the morphoclimatic elements. Ahnert (1987b) proposed a magnitude–frequency index for precipitation from which can be recognized the distribution of meteorological events of geomorphologic meaning. De Ploey et al. (1991), using daily precipitation values, obtained a sort of quantification of the accumulative potential erosion, which may serve to evaluate slope erosion by laminar overland flows. This kind of approach is not only useful for the characterization of morphoclimates but also can be applied to other branches of the science (Ahnert, 1987b). This method cannot identify non-episodic climatic features such as seasonality; however, it is a first step towards a more comprehensive morphoclimatology.

There is a general agreement about the concept and focus of climatic geomorphology, but also some criticisms and objections have been made by some authors. Stoddart (1969a) considered that is not realistic to take into account only a particular set of factors, because clear misrepresentations may arise. Climatic factors are, of course, important, but not the dominant ones; landscape builds from complex interactions among climate, bedrock lithology and vegetation, and it is not possible to separate climatic geomorphology from other geomorphologic approaches. Baker and Twidale (1991) indicated that only the glacial and arid morphogenetic regions can be easily identified, whereas the other climatic landscapes can hardly be recognized and defined. Twidale and Lageat (1994) pointed out that the glacial, periglacial and arid morphoclimatic regions represent 50 per cent of the Earth surface, and the other half is occupied by humid zones (selva, savannah, maritime and boreal regions of Peltier). In these latter zones most of the landscapes may be considered of doubtful designation (Stoddart, 1969b). Another fact analysed by Twidale and Lageat (1994) is related to the climatically triggered processes and to the shared mechanisms operating in the different climatic regions. In this sense, landforms linked to fluvial erosion and sedimentation can be observed all over the climatic regions (gulling, gorges, meanders, alluvial fans, braided channels, etc.). Tafonis, gnammas and weathering pits can be recognized in the wide variety of climatic zones, and patterned soils (grounds) in periglacial and arid zones represent a problem of landform convergence or "equifinality" throughout different climatic environments (Birot, 1955). In these cases, the activity of different processes, and with different intensity, eventually give rise to similar morphologies. Consequently, these authors admitted that the impact of climate has been largely overvalued, but did not dispute that it is a key influence. Also they assessed critically the establishment of a humid tropical region on the basis of the abundance of mass movement and etched surfaces, because these morphologic features also develop in other climatic zones.

4. The zonal concept in climatic geomorphology

From a planetary perspective, the variations occurring at the lithosphere–atmosphere contact may be considered as zonal, and they form wide bands from the poles to the equator derived from the different amounts of both precipitation and insolation (Derbyshire, 1976a,b). The zonal concept has been previously used, in other branches of the physical geography, such as climatology, pedology and biogeography. The sub-division of the continental zones into climatically featured edaphic regions was first proposed by Dokuchayev (1883), who distinguished zonal, azonal and intrazonal soils. The first ones are well developed soils that differ according to the ecological-climatic area in which they were formed. The azonal soils are poorly developed, with many features close to those present in the host rock. Finally, the intrazonal soils display good development, but they are not well drained or are affected by salinity, and so forth.

The work of Troll (1944) on periglacial micro-landforms is considered to be an introduction of the zonation concept in climatic geomorphology. The term *zonal* applies to those processes or events in which geographical distribution occurs in latitudinal bands parallel to the Equator, such as the equatorial forests, coral reefs and inland ice caps. Each one of these environments constitutes a morphoclimatic area (Tricart and Cailleux, 1965). The term *azonal* refers to those processes with a world-wide geographic distribution such as fluvial, aeolian and coastal environments. Finally, the term *extra-zonal* is used to denote some processes that, even though characteristic of a certain environment, can also locally occur in other areas, as in the case of the periglacial and glacial processes and landforms occurring in low latitude mountain systems.

The mountain areas display a characteristic climatic zonation or ***altitudinal gradient***. In large mountains located at low latitudes we can find, in a transect from the bottom to the summit, the following climatic sequence: temperate, periglacial and, lastly, glacial environments. In each altitudinal step the landscape is controlled by a specific morpho-climatic zone (Büdel, 1948). This climatic altitudinal gradient is due to the variations of temperature and precipitation with elevation. Temperature diminishes gradually with altitude, the air becomes more rarified and solar radiation increases. In these conditions, the rocks are subjected to more intense thermal effects (Tricart and Cailleux, 1965). A height increment of 1000 m in the Alps is equivalent to a decrease of temperature experienced during a 1000 km journey toward the North Pole. This thermal drop is normally accompanied by altitudinal and latitudinal variations in vegetation (Büdel, 1977, 1982). Also with increasing altitude, a characteristic succession in the morphology and size of patterned soils commonly occurs, as reported for the Swiss Alps and Karakoram by Furrer (1972). Similar cases have been described in other mountainous areas, constituting a reliable proof of the altitudinal gradient. Precipitation varies with height, displaying an initial increasing trend, which after a maximum, diminishes upwards. All the big mountains are characterized by a more rainy and cloudy intermediate step. Another important effect is that displayed in the north-facing areas at mid-latitudes where snow is preserved during long periods, thereby impressing a periglacial landscape within the altitudinal variation. In the intertropical zone, the windward areas are more humid than those located to the leeward, which are generally dry (Tricart and Cailleux, 1965).

On the other hand, it can be pointed out that at scales of $10^6 - 10^7$ km^2 the variety of landscapes is better explained by climatic differences; at scales of $10^4 - 10^5$ km^2, landscape variability is dominated by structural and bedrock lithology variations; and at scales of 10^2 km^2 the landforms are the result of the activity of different erosive processes (Chorley et al., 1984).

5. The main morphoclimatic zones

The differentiation of the different morphoclimatic areas must take into account the dominant working processes in each of these areas, but it is also essential to consider the possible occurrence of inherited Quaternary and Tertiary landforms. This interest in climatic change was already manifest for Penck and Brückner (in 1901–1909) in their dissertation on the ages of the alpine glaciations. Similarly, Passarge (1904), in his work on the Kalahari desert, found clear evidence of environments of alternating arid and savannah conditions. Büdel (1948) insisted on the importance of paleo-landforms, and in 1963 proposed the term *climatogenic geomorphology* to define the discipline of the study of relic landforms developed under different climatic conditions, and for the deduction of the succession of climatic environments over time. Obviously climatic change is particularly relevant in mid-latitude areas. Also, Birot (1960) pointed out that the influence of climatic change during Tertiary and Quaternary times extended from the Arctic to the Equator. The distinction between modern and ancient inherited landforms is essential to understanding landscape evolution in climatic terms, but in fact, numerous misinterpretations of landscape development may actually arise.

The division of the Earth surface into different morphoclimatic areas, for different reasons, has always been a complex task. There are different parts of the world for which our knowledge of the main morphoclimatic controls is very limited (central Asia, South America, etc.) and therefore it is difficult to trace precise boundaries between them (Tricart and Cailleux, 1965). The increasing use of satellite imagery, however, has partially solved this problem. On the other hand, it is necessary to keep in mind that a morphoclimatic map does not exactly come from the total or partial overlapping of existing climatic, vegetation and soil charts, but evidently there are clear relationships between them. A map of morphoclimatic areas should be an original map in which the morphoclimate constitutes a peculiar category of the natural phenomena (Tricart and Cailleux, 1965; Büdel, 1977, 1982). It is also necessary to take into account that, except for the case of coastal environments, the morphoclimatic boundaries are hardly sharp. There are, however, physiographical areas, like the Black Forest, the Po plain and the Appalachian Mountains that can be clearly differentiated (Büdel, 1977, 1982). In contrast, some transitional areas (i.e. those between the savannas and semiarid areas on the African continent) can reach huge dimensions. Only those areas corresponding to the cores of the different morphoclimatic zones have no problems for their identification.

The main reviews on climatic geomorphology are those written by Büdel (1948, 1963, 1977) and Tricart and Cailleux (1965), which elaborated different maps of the postulated morphoclimatic areas. Büdel (1948), using climatic and morphologic approaches, divided the Earth into seven *climatomorphological zones*, illustrating this work with a map of the distribution of these morphoclimatic zones in the "Old World" (Europe, Asia and Africa). Tricart and Cailleux (1965) criticized this global zonation due to: (1) its lack of homogeneity; (2) the segregation of three different permafrost regions; (3) the consideration of the Mediterranean area as a single zone and (4) the use of a blurred terminology. Büdel (1963) considered that the aclimatic factors (i.e. petrovariance or change of bedrock resistance, epeirovariance or epeirogenesis, distance from base level, topography), and human influences would play a relevant role in landscape evolution if the Earth climate was uniform. These influences occur because the existing differences between climatic zones are superior to these five aclimatic factors. For Büdel, on each point on the Earth, climate determined the elemental combination of working morphogenetic processes. From the effects produced in each climatic region, in which these active complex morphogenetic processes operate, the Earth's surface can be divided into five morphoclimatic zones (Fig. 1.11) not including the altitudinal climatic gradient effects. This differentiation uses the morphologic approach as its main criteria, which in turn is correlated to the climate. These morphoclimatic zones are denoted by the modern climate, but for mid-latitude zones the influence of the past climates may be relevant, because these zones undergo a slower geomorphic evolution.

The classification proposed by Tricart and Cailleux (1965) is based on two main criteria: (1) The large climatic and biogeographic natural areas that provide the main zonal boundaries; and (2) some subdivisions, assisted by the aforementioned set of criteria, combined with palaeoclimatic differences. They differentiated the following areas, but were only concerned with low-elevation zones where the altitudinal climatic gradient is not relevant:

(1) *Cold Zone.*
    (a) Glacier domain.
    (b) Periglacial domain.
(2) *Mid-latitude Forest Zone.*
    (a) Maritime domain with Quaternary glacial and periglacial landform heritage.
    (b) Continental domain with the great influence of Quaternary and present-day ice caps, with the possible occurrence of permafrost at depth.
    (c) Mediterranean domain with a smaller influence of relict Quaternary landforms.

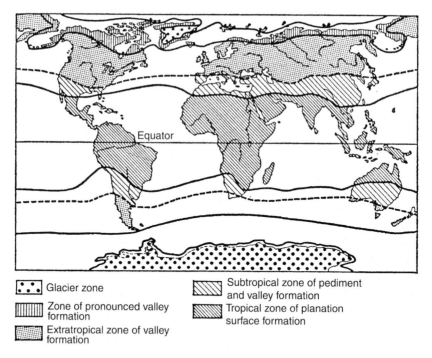

Glacier zone

Zone of pronounced valley formation

Extratropical zone of valley formation

Subtropical zone of pediment and valley formation

Tropical zone of planation surface formation

Figure 1.11.   Present-day climatomorphological zones of the Earth (Büdel, 1963).

(3) *Arid and subarid zones.* They are divided:
   (a) In relation to the moisture deficit in xerophitic bush-land steppes and deserts.
   (b) In relation to the temperature of cold and warm areas.
(4) *Humid intertropical zone.*
   (a) Savannah domain, with less dense vegetal cover, characterized by an intense chemical weathering, but discontinuous over time, and for the occurrence of significant overland flow events.
   (b) Forest domain, with important vegetal cover and intense chemical weathering.

From these primary climatic zones Tricart and Cailleux (1965) subdivided the Earth's surface into 12 different morphoclimatic regions (Fig. 1.12). But in this division the morphoclimatic influences are not clearly distinguished from the climatogenic ones (Beckinsale and Chorley, 1991). Perhaps it is also excessive to subdivide the periglacial domain into five different morphoclimatic regions characterized by the occurrence of modern or relict landforms (2–7). According to Stoddart (1969a) perhaps the map of these authors gave greater importance to the landform controlling factors than to the landscape itself. Derbyshire (1973b) indicated that vegetation is considered as a dominant geomorphic factor and that the terminology utilized is more pedologic and biogeographic than climatic or geomorphologic.

The map of morphoclimatic zones presented by Büdel (1977) (Fig. 1.13) indicated that the morphoclimatic boundaries enclose areas in which the set of active geomorphic processes depend on a particular and uniform climate. In climatic zonation it is essential

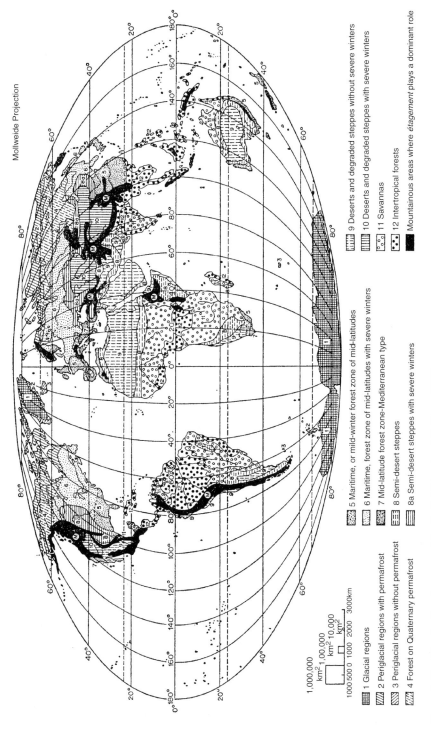

Figure 1.12. Morphoclimatic zones of the Earth (Tricart and Cailleux, 1965).

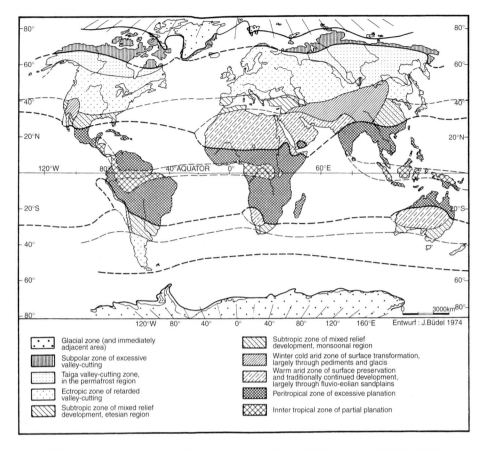

Figure 1.13. Morphoclimatic zones of the present, excluding the high mountains (Büdel, 1977).

to consider the central cores of the suspected zones, because their limits are usually not well defined. On the other hand, only the low and middle elevation zones are represented, whereas the high mountains have been misplaced. The establishment of different morphoclimatic regions is very complex because it is necessary to distinguish a large and complex set of processes implicated in its formation. It is also sometimes very difficult to differentiate the modern landforms from the relict ones (from the upper Cretaceous to the Holocene). If we understand the present active landscapes in each morphoclimatic zone, we can recognize the relict landforms in other zones. The present active processes operating in mid-latitude areas are exceptionally weak in their workings, and in these circumstances, 95 per cent of the relief of these regions is constituted by relict landforms (Büdel, 1977).

The most significant features of the main morphoclimatic areas have been collected from the work of Büdel and from the great review on this author's ideas, carried out by Kiewietdejonge (1984).

(1) *Glacier zone*, to which Büdel did not dedicate a special interest. In this zone the mechanisms operating in landscape development are more easily studied in post-glacial regions than in the presently active glacier regions.

(2, 3) *Periglacial zone of pronounced valley-cutting*. This corresponds to the subpolar and also solifluction regions of Büdel (1948), due to the importance of mass movements on periglacial slopes. Its northern area is limited by glaciers and the southern one by the solifluction boundary of the tree-cover or permafrost termination. The forest limits usually coincide with the solifluction one, whereas the permafrost boundary can expand over lower latitudes. The upper horizon of the ground is structured, thus generating patterned ground. One of the main problems in these zones is rapid fluvial incision. Slopes may develop by parallel retreat, generating tripartite hillslopes.

(4) *Ectropic zone of retarded valley-cutting*. The most diverse and possibly the most complex region, with the occurrence of a large number of different landforms. Some 95 per cent of the landscape in the European sector of this area is constituted by inherited landforms. The stepped planation surfaces (*Piedmonttrepen* of W. Penck) are very common. Glacis surfaces and not very deep, wide valleys may be characteristic. In those areas subjected to ancient glacial erosion, extensive landscapes of hills and lakes are developed. Finally, an important sheet-cover of loess derived from materials deposited during previous cold periods commonly coats the ground. This, together with the following region, are the most important ones for the study of climatogenic geomorphology.

(5, 6) *Subtropical zone of mixed relief development*. This comprises the Mediterranean and monsoonal regions which hold different morphological and climatic characteristics. Büdel did not develop descriptions of the monsoonal regions, and instead just adjusted his observations for the Mediterranean areas. This zone is a transitional area for which its evolution has been affected by tropical, periglacial, glacial, temperate humid and arid climates. In addition in this climatic type it is necessary to consider human influences, which have an extraordinary impact on the occurrence of fluvial erosion. The morphologic result is the assemblage of relict and modern landforms, commonly arranged in an altitudinal zonation. A considerable development of Tertiary planation surfaces is recognized in associated calcareous and in karstic landforms. In the fluvial valleys and intramontane basins the development of glacis is common in relation to the occurrence of fluvial terrace systems.

(7, 8) *Arid zone of surface preservation and transformation through pediments and glacis*. This area includes that of winter-cold deserts and tropical arid zones. In this latter zone a relevant intensity of wetting and drying and salt weathering processes occurs. Within this zone erosion plains with inselbergs, extensive *hamadas* and stone deserts are the usual features. In areas of limited relief the alluvial materials of glacis and terraces usually develop calcrete crusts. These alluvial materials constitute the main source area for the adjacent sand deserts. The winter-cold arid zones constitute a transformation area rather than a region for later development of planation surfaces. Frost cracking is the dominant weathering process. The most representative landforms are the glacis systems.

(9) *Peritropical zone (humid–dry) of excessive planation*. In this region thick horizons of chemical weathering commonly develop, which can reach several hundred meters thick, with the eventual development of laterites. The formation of extensive plains can be explained by the double planation mechanism of Büdel (1957) for tectonically stable areas, in which *etchplains* and *inselbergs* are characteristic.

(10) *Equatorial zone of partial planation*. This is not a very well studied zone due to the relative difficulties of research because of its dense tree cover. Chemical weathering is more intense than in the other areas. Planation surfaces and inselbergs that originated in prior humid–dry climates of the Tertiary occur in those regions that have limited development of preceding climate zonation. In the central areas of these zones planation surfaces undergo little reworking because the formative mechanisms have greatly decreased in space and time.

These classifications of analysed morphoclimatic areas reveal an important variation within the different main climatic zones: Büdel (1948) 7; Peltier (1950) 9; Büdel (1963) 5; Tricart and Cailleux (1965) 13; Wilson (1968) 6; Büdel (1977) 10 and Chorley et al. (1984) 8. These modifications follow the different approaches used by the different authors to carry out the subdivision of the Earth's surface in different morphoclimatic areas.

The monographs published on different morphoclimatic zones have been focused in the following four categories: glacial, periglacial, arid and tropical humid regions. The temperate regions have not been the object of a joint treatment due to the large number of presently existent relict landforms. Evidently, the development of the published reviews on these four climatic regions takes into account more detailed differentiations within the master climatic areas.

SECOND PART

Geomorphology of glacial zones

Chapter 2

Glaciers

1. Introduction

Since Precambrian times, several ice ages have occurred in Earth history and have been printed in the geologic registry by erosive and sedimentary features (Eyles, 1993) (Fig. 2.1). Within this general perspective, the Quaternary encompasses a glacial period that began to develop fundamentally in upper Tertiary. The remaining ice accumulations at the present time are left occupying only ~ 10 per cent of the terrestrial surface above sea level, whereas in the last glacial maximum about 18,000 years ago, they covered almost a third of the continental areas (Flint, 1971; Sugden and John, 1976). The activity of the glacier masses is, therefore, reflected by different erosive and depositional landform modelling as an important activity of the planet.

Investigations on the activity of past and present glaciers occur in different fashions. Works on glacial geology are carried out by Quaternary researchers, who fundamentally worry about glacial stratigraphy and chronology; whereas geomorphologists analyse glacier processes and model the results of ice mass performance, as well as its evolution. Glaciologists are interested in the study of the nature, physical behaviour, and work done by glaciers (Lliboutry, 1965). In most cases, glaciologic investigations and those carried out by glacial geomorphologists show a clear disconnection, although there seems to be some overlap (Sugden and John, 1976). This union of methods is necessary because glaciers constitute a means of rapid activity, with modifications in days or months, and are therefore suitable for studying present-day processes and the resulting forms.

In the past few decades a new field of work related to glacial geomorphology has been opened within planetary geology. Thus on Mars, snow and ice sheets covering 30 per cent of its surface are recognized as occurring during winter and being reduced to residual ice caps in summer, occupying a single 1 per cent. Also, the interpretation of some Martian channels as originating by valley glacier activity has been expressed. Some investigators establish for this planet a climatic change similar to that of Earth's during upper Cenozoic (Cutts et al., 1979; Baker, 1981a; Carr, 1981). These investigations of planetary geomorphology can provide important data for a greater understanding of our glacial ages.

2. Past and present extension of glaciers

The present-day glacial ice occupies an approximate surface of 15.8 million km^2 (Table 2.1), or about 10 per cent of the total emergent terrestrial surface above sea level.

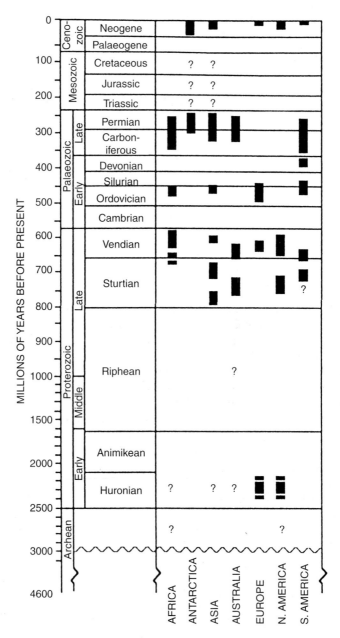

Figure 2.1. Ice ages in Earth history (Eyles, 1993).

Only 3 per cent of glaciers are outside both the great Antarctic and Greenland ice sheets that constitute the largest fresh water reserve of the hydrosphere (Lvovitch, 1967). The large polar caps cover an area of 13.6 million km², with an average thickness of 2.2 km and a maximum close to 5 km. The water equivalent volume of ice of this ice sheet is

Table 2.1. Present area occupied by ice (World Glacier Monitoring Service, 1989) and maximum expansion during the Pleistocene (after data elaborated by Flint, 1971).

Region	Present area (km^2)	Maximum expansion during Pleistocene (km^2)
Antarctica	13,593,310	13,800,000
Greenland	1,726,400	2,295,300
Mexico, United States, Canada and Alaska	276,100	16,217,091
Fennoscandian	51,046	7,169,145
Africa	10	1900
Asia (Russia included)	185,211	3,951,000
Alps and Pyrenees	2921	39,000
South America	25,908	870,000
Australasian	860	30,000
Total	15,861,766	44,373,436

considered to be ~ 30.1 million km^3. If it melts completely it would produce a rise of sea level of 59 m. The ice sheet of Greenland occupies a surface of 1.7 million km^2, with an average thickness of 1.6 km, a maximum of 3.4 km, and a volume of 2.4 million km^3 (Donn et al., 1962; Flint, 1971; Sugden and John, 1976).

The remaining ice accumulations do not surpass a half million km^2 and are scattered all over the planet, although they are fundamentally concentrated in the northern hemisphere and in mountains of middle and low latitudes (Alps, Karakoram, Himalaya). The extension of glaciers in Africa is practically negligible in contrast to the ice accumulations in the Andes, which stands out by its extension. The area in the world occupied by glaciers has recently been calculated by the World Glacier Monitoring Service in 1989 (see Hambrey, 1994). The total surface corresponds to a rather overly precise estimation of $\sim 15,861,766$ km^2.

In their maximal extension during the Pleistocene, ice sheets covered an area of more than 44 million km^2 (Flint, 1971), representing some 30 per cent of the emerged lands. Its volume on the terrestrial surface is considered to be about 100 million km^3 (Donn et al., 1962). In Table 2.1 the great difference between the area occupied by present glaciers, corresponding to an interglacial age, and the surface occupied during the last Quaternary glacial maximum can be observed.

The Antarctica and Greenland ice sheets have undergone a relatively small fluctuation, whereas, for example, the Fennoscandian and Laurentian complexes have experienced enormous advances and retreats. The last advance reached dimensions close to that of present-day Antarctica and extended below latitudes of 40°N (Fig. 2.2).

3. Mass balances of glaciers: accumulation and ablation

The mass balance in a glacier deals with the gain and loss of snow and ice and is normally measured by its equivalent in water (amount of resulting melt water). The term

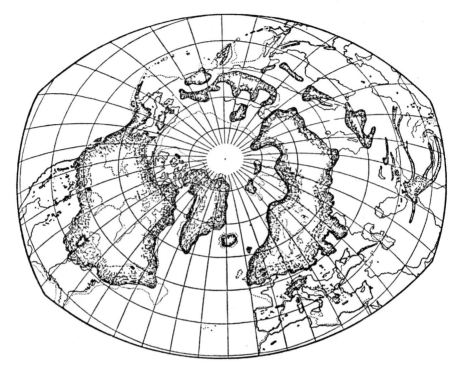

Figure 2.2. Maximal extension of Pleistocene ice sheets in the Northern Hemisphere (after Antevs and Flint; in Holmes (1944)).

accumulation refers to all those processes by which material is added to the glacier. Precipitation is the most important one, although the feeding by avalanches to valleys can sometimes be significant. Wind action plays a secondary role with its effects on snow-mass redistribution.

Ablation alludes to the set of processes by which part of the glacier mass is lost. The main one, melting, is influenced by solar radiation, rain, air temperature, cloud cover, detritus amount, and other factors (Sharp, 1988). Fine rocky particles, with a greater capacity for solar radiation absorption, penetrate when the snow around the edges melts generating small cylindrical conduits. On the contrary, thick detrital layers isolate the underlying ice from melting, which produces gravel knolls containing ice. If the detritus mantle has an irregular superficial disposition, a topography of knolls and depressions can be produced by differential melting. In the case of the presence of great blocks, the resulting morphology will be of *pedestals* or *glacial tables* (Fig. 2.3) in which the block is based on underlying ice. These tables frequently appear tilted until they finally fall, due to a greater melting of the ice oriented to the sun. Sometimes differential melting produces numerous and deep depressions, denoted as karstic glacial topography by its similarity to the landscape of limestone areas (Clayton, 1964). On the other hand, other important ablation processes in glaciers that lose their mass to lakes or seas is the generation of great

Figure 2.3. Glacial table on Godwin Austen Glacier (Karakoram, Pakistan). K2 (8611 m), the second highest summit on Earth, is visible in the background. Photo: J. López-Martínez.

ice blocks or icebergs floating in the water. This process is influenced by the degree of fracturing of the ice mass.

Any glacier is formed by two parts: the superior or upper one, where feeding or alimentation is greater than loss by melting, thus constituting the accumulation zone, and the inferior or lower one, with an opposite balance, denoted as the ablation zone. In between the two zones is the *equilibrium line* where the balance is zero (Müller, 1962).

The difference between accumulation and ablation in a year for the whole glacier constitutes the *net balance* (Andrews, 1975; Paterson, 1994) (Fig. 2.4). If the balance is positive the glacier has experienced a gain of snow and ice (annual net accumulation) and the opposite if it is negative (annual net ablation). A balance equal to zero indicates an equilibrium between accumulation and ablation. The winter balance is usually positive and the summer balance negative. The difference among them is the net balance. Balances are calculated by means of field measures or by aerial photography and hydrological methods (Benn and Evans, 1998). Few mass balances for glaciers exist and a most complete one is the Storglaciären in Sweden. In a 28 years period, 5 years resulted in positive balance, one with balance zero and in the rest ablation dominated (Ostrem et al., 1973; in Sugden and John (1976)) (Fig. 2.5).

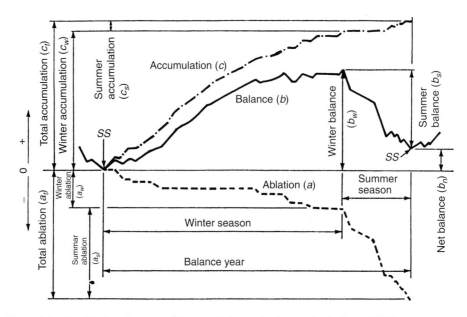

Figure 2.4. Graphical explanation of the mass balances in glaciers (in Andrews, 1975).

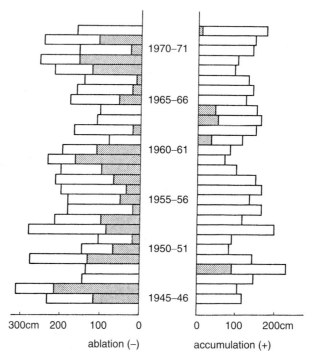

Figure 2.5. Mass balance for the period 1945–1973 in Storglaciären (Sweden) in which the annual accumulation, ablation and net balance are expressed (Ostrem et al., 1973).

4. Snow–ice transformation

Glaciers are developed in situations in which snow accumulation exceeds melting. Suitable climatic conditions are required, as well as topographic form providing ideal places for their storage (Price, 1973).

Snow fallen in the accumulation zone undergoes deep changes until it finally transforms into glacier ice. The grains constituting this ice are made up of crystals that vary from millimetre to decimetre sizes and have been formed by numerous snowflakes.

The newly fallen snow has a very low density of sometimes 0.05, a great porosity (95 per cent) and sometimes can be considered as a wind-deposited sediment (Fig. 2.6). This fresh snow changes in a short time, through pressure induced melting and refreezing, to granular or old snow with a density of 0.3 formed by loose spherical grains of snow with great permeability. Transformation through minor melting and refreezing continues with an increase of density by the modification of grains, loss of porous spaces and tightness of the packing. When density is 0.4, the firn state is reached with 50 per cent porosity and a grain size of one to several millimetres. As compaction increases, pores between grains get sealed and *firn* is transformed into glacier ice when density equals 0.80–0.85. This mass is impervious, the existing air is contained as bubbles and any increase of density is due to its compression. This snow–ice transformation is accompanied by displacements between crystals, changes in size, form, and internal deformation (Paterson, 1994).

The speed of the snow–ice transformation varies from one place to another, depending on the sensitivity of the process to the temperature and the speed of snow accumulation.

Figure 2.6. Wind-borne snow, Heritage Range, Antarctica, about 80°S. Photo: J. López-Martínez.

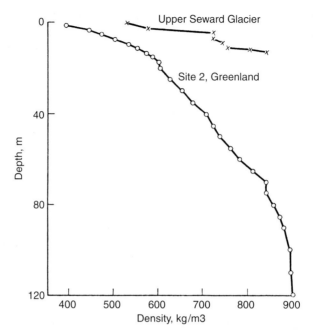

Figure 2.7. Variation of firn density with depth in the temperate Upper Seward Glacier (Yukon) and in the Greenland ice sheet (Paterson, 1994).

The presence of melting water in summer accelerates the transformation. The curves of Figure 2.7 (Paterson, 1994) show the variation of firn density as a function of depth for a dry snow glacier in Greenland and for a wet snow one in the Yukon. In the first one, firn turns to ice at 66 m depth, requiring a time of more than 100 years, whereas in the second one it happens at 13 m in a 3–5 years interval.

5. Classification of glaciers

Numerous classifications for ice accumulations have been proposed, but the most used, because they allow easy differentiation, are those based on the internal regime of the temperature of glaciers and on their geometry.

5.1. Thermal classification

Pure water is transformed into ice at 0°C under a pressure of one atmosphere, but a lower temperature is required when pressures are important. Thus, for example, at the base of a glacier of 1500 m of thickness the melting point is − 1°C. It is possible thus to differentiate *cold ice* in which temperature is below the melting point and *warm ice* in which it is so close to fusion that it can contain water (Ahlmann, 1935).

Cold ice is found when surface temperatures of the glacier are very low in winter and with little or no melting in summer, as in Antarctica, Greenland and the elevated zones of great mountains. In Antarctica air temperatures of $-89.2°C$ have been recorded; warmer temperatures of $-30°C$ are reached as well, but temperatures increase with depth as a result of the geothermal heat and of the internal deformation of the ice mass in the glacier flow (Sugden, 1982).

Warm ice is formed when sufficient heat to produce melting is available. During summer, melting of surface ice is common. This water percolates into the ice mass and can freeze again, releasing its latent heat and by each gram of refrozen water elevating 160 g of ice by 1°C. This can be one of the most important heat sources. In these circumstances the whole warm-ice glacier can be formed by ice at its pressure–melting point. At the ice base geothermal heat is usually sufficient to elevate the temperature of the ice to its melting point. This warm ice can also be at the bottom of glaciers whose superficial layers are of cold ice (Budd et al., 1970).

These ice types allow differentiation of *polar*, or *cold-based*, and *temperate* or *warm-based glaciers*. These classifications are very simplistic, because the thermal conditions of glaciers can vary in space and time (Sugden, 1977; Paterson, 1994), so that *polythermal ice* is becoming increasingly recognized. Antarctica is dominantly a polar or cold-based glacier, but some parts have warm ice at their base. In temperate glaciers the transformation of firn into ice is faster than in polar glaciers. These differences in ice temperature are of extraordinary importance in glacial geomorphology and mainly related to the presence of a thin water film in temperate glaciers lubricating the bed of the glacier and facilitating its movement. At the same time this subglacial water facilitates quite important erosive and sedimentation work.

5.2. Morphologic classification

Morphologic classifications are based upon geometry, position and size of the ice masses. Some ice accumulations are restricted by the topography, whereas others do not present any type of confinement. *Continental ice sheets* are represented by those of Antarctica (Fig. 2.8) and Greenland. They show a domal form with a convex surface produced by the ice flow. In the centre, ice accumulation thickness is larger and towards the margins the surface slopes smoothly, increasing its gradient progressively and with diminishing thickness. These ice sheets preserve or fossilize the underlying topography, although in some cases the relief of the bedrock substrate exceeds that of the ice mass, and the rock outcrops protrude through the ice as *nunataks* (Fig. 2.9).

Ice shelves (Fig. 2.10) are great masses that extend out into the sea $>500,000 \text{ km}^2$, like the Ross and Filchner-Ronne ice shelves in Antarctica that occupy great bays and have extensions corresponding to 7 per cent of the ice sheet. The ice thickness can reach up to 200 m and the cliffs up to 30 m with the loosening of icebergs (Fig. 2.11) and melting of its base. These floating ice masses contain particles of different sizes that fall to the bottom of the sea when the ice melts (*dropstones*) (Fig. 2.12). On the ocean surface remains a floating mass of split and rather thinner ice, denoted as *pack ice* (Fig. 2.13). If the size of ice domes is smaller than 50,000 km² (Sugden and John, 1976) they are denoted as *mountain, plateau* or plain *ice caps*, according to their topographic position. The first is located on a plateau zone (Fig. 2.14), being the ice cap of

Figure 2.8.   View of the Antarctic ice sheet, near the Ellsworth Mountains. Photo: J. López-Martínez.

Figure 2.9.   Nunatak in the Sentinel Range, Ellsworth Mountains, Antarctica. Photo: J. López-Martínez.

Figure 2.10. Ice platform and partially frozen sea on the western coast of Antarctica. Photo: J. López-Martínez.

Figure 2.11. Iceberg in Bransfield Strait, Western Antarctica. Photo: J. López-Martínez.

Figure 2.12. Erratic metamorphic rock block, coming from the Antarctic Peninsula, located in the background, transported by an iceberg and released by melting. The tide was high at that moment. Photo: A. Martín-Serrano.

Figure 2.13. Fractured pack ice in the proximity of the Antarctic Peninsula. Some seals are visible on the ice blocks. Photo: J. López-Martínez.

Figure 2.14. Plateau ice cap of Myrdalsjökull (Iceland). Photo: F. Gutiérrez.

Vatnajökull, in Iceland, the most characteristic example. The plain ice caps are located in areas of little relief, such as the Barnes ice cap on Baffin Island in the Arctic.

Cirque glaciers occur in the valley heads of mountainous areas (Fig. 2.15). They are small ice masses located in amphitheatres, generally of steep walls, hectometric dimensions and thicknesses of tens of metres. During a glaciation period they are the first to be developed and last ones to disappear.

Valley glaciers are laterally confined by rocky walls. If their feeding comes from cirques we are speaking of *valley glaciers* of the *alpine type* (Embleton and King, 1975a) (Fig. 2.16). They flow radially from the great massifs and can be fed by secondary valley glaciers. A clear hierarchy is given by the relation between valley width and number of tributaries. The glacier mass is long, narrow, and generally finishes in a single ice tongue or terminus. The longitudinal slopes are variable. Some glaciers exceed 100 km in length. At the present time they are common in high mountains. In the peripheral zone of the icecaps the ice masses can escape from the margins giving origin to *valley glaciers* or *outlet glaciers* (Figs 2.17 and 2.18), which are like appendices arising from the edge of ice caps. When the front of valley glaciers is constituted by an extensive plain the ice spreads out considerably over it, increasing in width and producing a *piedmont glacier*. The most classical example is the Malaspina Glacier in Alaska extending as a wide lobe in the coastal plains of the Pacific. It constitutes the ablation zone of the lower Seward Glacier, being 600 m thick and occupying a great depression that reaches 250 m below sea level (Benn and Evans, 1998). Its compressive flow produces an intense deformation of the debris bands and medial moraines (Sharp, 1958) (Fig. 2.19).

Figure 2.15. Cirque glacier of Monte Perdido. Ordesa National Park. Internal sierras of the Pyrenees. Photo: F. Gutiérrez.

Figure 2.16. Valley glacier of the alpine type. Spegazzini Glacier, Patagonia, Argentina. Photo: C. Sancho.

Figure 2.17. Outlet glacier. Svinafellsjökull (Iceland). Photo: F. Gutiérrez.

Finally, other glaciers controlled by the topography can be recognized, such as *ice fields*, accumulations with a flat surface, which are sometimes difficult to distinguish from ice caps. Other small glaciers occupy small steep topographic or lateral depressions of mountains as *slope* or *niche glaciers*.

Figure 2.18. Outlet glacier and pack ice in Whisky Bay, James Ross Island, to the East of Antarctic Peninsula. Photo: A. Martín-Serrano.

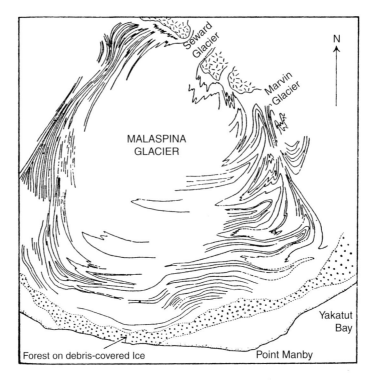

Figure 2.19. Map showing similar folds developed in the ice and moraines of the Malaspina Glacier in Alaska (Sharp, 1958).

6. Movement of glaciers

Ice is a crystalline solid that flows easily under the pull of gravity. This movement implies a continuous transference of material from the accumulation to the ablation zone. Two types of processes can be distinguished in this flow: *internal deformation* and *basal slipping* (Sharp, 1988) (Fig. 2.20). The first is a creeping flow resulting from the application of force over a long period. Intergranular sliding, re-crystallization and sliding along crystal network planes in the ice crystals are the mechanisms of this flow (Weertman, 1983).

The basal slipping indicates the displacement of the ice mass on its bed (Weertman, 1957). This movement on the bottom also implies a process of *basal plastic flow* in which the ice flows over and around greater obstacles. These advances can take place whatever the temperature of the ice is (Fig. 2.21). The other mechanism is denoted *sliding by refreezing*. The bottom of the glacier presents a rough surface with small thresholds and depressions of centimetre to metric scale. Melting by increase of pressure in the basal ice can take place if the glacier is of the temperate type, near its pressure– melting point, and a refreezing or *regelation*, when the water pressure is reduced. If the protuberances are very small, latent heat transference can take place and facilitate new melting. Refreezing layers of clear ice, laminated and of a few centimetres thickness, can occur at the base of the glacier (Kamb and LaChapelle, 1964).

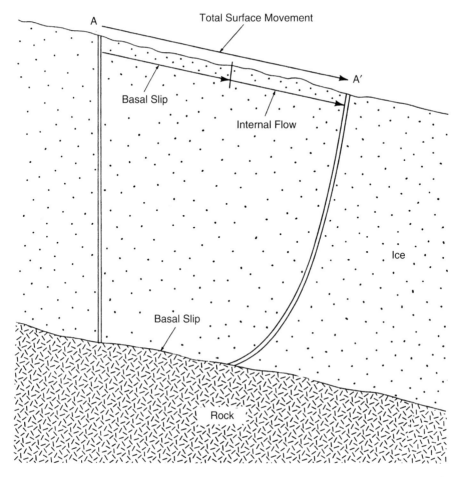

Figure 2.20. Longitudinal section of a glacier that indicates how a vertical sounding gets deformed over time. Total movement AA′ measured on the surface is the sum of internal flow and basal slipping (Sharp, 1988).

Polar glaciers show little basal slipping because they are frozen to the glacier bed, although a relatively fast sliding next to the ice–rock interface can exist. In temperate glaciers the basal slipping is favoured by the presence of a water film at the bottom, which reduces the friction between ice and rock. If the substrate of a temperate glacier is composed of unconsolidated sediments, such as basal tills, these materials can be saturated with water to become deformable tills and, therefore, the speed of basal slipping increases considerably.

The surface speeds of a glacier vary in time and space. The speeds throughout a glacier fluctuate between 3 and 300 m/yr, being able to reach values of 1–2 km/yr in steep-walled areas. The cross-sectional speeds are maximal in the centre and undergo a fast reduction towards the margins as a result of the wall friction. The edge values fluctuate between 10 and 65 per cent of the central maximal dimension. These calculations of the superficial

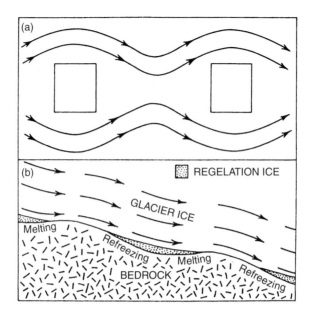

Figure 2.21. Influence of bed irregularities on the ice flow. (a) Process of basal plastic flow, sight in plan view (Weertman, 1957). (b) Sliding mechanism by refreezing, in section (Kamb and LaChapelle, 1964).

speeds of a glacier are obtained by installing stakes in the ice and locating a theodolite on fixed points of the rocky margin. In order to obtain the speed at depth, borings to the base are carried out (Fig. 2.20) and later a flexible tube is introduced into the conduit. By means of a clinometer the movement can be followed. The speed is higher in the upper part than in the lower.

Some valley glaciers experience fast and abnormal flows in which the ice mass moves as a wave at speeds of up to one hundred times the average value and with displacements up to 11 km. They are known as glaciers with spasmodic flows or *surging glaciers* (Fig. 2.22). Vibrations and intense noises accompany their sudden movements (Meir and Post, 1969; Sharp, 1988; Clarke, 1991; Dowdeswell et al., 1991; Evans, 2004a). These swellings can have a regularity of a few years to several centuries. In the glacier they cause the presence of deformed chaotic surfaces, intensely sheared margins, great changes in its thickness and strongly folded medial moraines. The intense shear at the margins of surging glaciers can produce foliation in the ice (Pfeffer, 1992). In Svalbard 504 surge-type glaciers have been described. These glaciers result from a polythermal regime and fine-grained deformable beds (Jiskoot et al., 2000). High pressures of the subglacial water motivate these fast flows and can produce catastrophic advances. The surge outbreak can also be associated with seismic activity and important modifications in precipitation.

Stress regimes (Fig. 2.23) vary throughout the length of the glacier and *compressive flow* occurs as the speed of the glacier reduces (Nye, 1952), in such places as below steep sections of the glacier. The sliding planes are curved in an ascending direction and

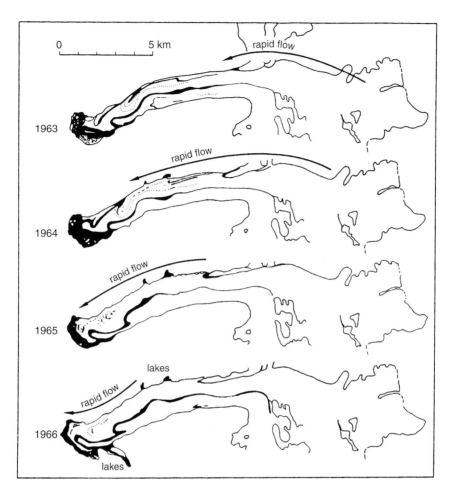

Figure 2.22. Evolution of the spasmodic surge flow of glacier Tikke (British Columbia) between 1963 and 1966, in which deformation of the moraine surface is observed (Meir and Post, 1969).

the detritus can be transported with them up to the glacier surface. In contrast, *extending flow* is located in zones, commonly with steeper gradients as at icefalls in which the ice speed increases and the shear planes are curved downwards until becoming tangent to the bed. This flow predominates through the equilibrium line.

7. Structures of glaciers

The deformation suffered by the ice masses as a result of its movement gives rise to different structure types, similar to those observed on deformed rocks. The analysis of glacial deformation is easier than that of rocks, however, because they are affected solely by the action of gravity (Menzies, 1995a) and rocks are affected by many more processes.

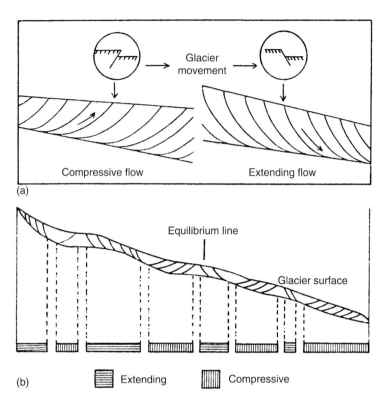

(a)

(b)   ▤ Extending   ▥ Compressive

Figure 2.23. (a) Compressive and extensive flow and sliding planes. (b) Distribution of compressive and extensive flows in a glacier (Nye, 1952).

In the accumulation areas *stratification* predominates, marked by an alternation of winter and summer ice layers containing sediment formed in the summer melting. During glacial flow, planar foliation structure originates and can be recognized by the grain of the ice with alternation of clear blue ice bands and white ice with air bubbles, with these last being the most abundant ones (Sharp, 1988). These bands vary in their dimensions from millimetres to several metres. *Foliation* has a greater development next to the walls of valleys, where it runs parallel to the rocky outcrops. It is arranged in cross-section with the direction of the ice flow far from the margins and near the glacier terminus (Rutter, 1965). Sometimes it is difficult to distinguish the stratification of the foliation where it is parallel to sedimentary layers.

Commonly glaciers appear folded and faulted (Hambrey, 1977), where it is possible to observe these structures on the glacier surface and in marginal cliffs (Fig. 2.24). *Folds* have different sizes and directions and are very evident in glaciers with surging flows (Fig. 2.22) as well as in some piedmont glaciers, such as Malaspina Glacier where great folds similar to those affecting moraines and ice bands are observed (Fig. 2.19).

In steep sloped zones of a glacier, gravity *faults* and rotational slidings can develop. At the glacier terminus and where the advance of the ice is interrupted by moraines they produce thrust planes.

Figure 2.24. Fold in the ice, shown by the alternation of clear and dark bands. Livingston Island, South Shetland Islands, Antarctica. Photo: J. López-Martínez.

Ogives or *Forbes bands* are alternating layers that extend over the surface of temperate valley glaciers (Paterson, 1994). They display a curved disposition with their convexity pointed in the flow direction. This curvature is due to the higher ice speed in its central parts (Fig. 2.25). They are spaced between 50 and 200 m apart. The bands of the ogives are constituted of clear and dark ice. Dark bands are formed by ice and sediment with an intense foliation, originated by broken ice mixed with mud and snow and later compressed. Clear bands have less foliation and contain rich white ice with air bubbles. The model that best explains the formation of band ogives at Bas Glacier d'Arolla (Switzerland) constitutes a slight variation to the "reverse faulting" hypothesis proposed by Posamentier (1978). According to this model, multiple shear zones are formed and basal ice is uplifted to the glacier surface to give rise to the dark ogive bands (Goodsell et al., 2002). Some authors consider that each pair of bands has an annual origin. In steep gradient zones of the glacier, ice advances during the winter from the great slope zones of longitudinal extension, to the lower areas of intense compression, where cracks were filled with snow originating the clear band. In summer, the cracks were opened and filled up with ice fragments and detritus formed the dark band when being recompressed (Nye, 1958; Sharp, 1988).

Crevasses are the most abundant structures of glaciers and form because of tensional stresses (Menzies, 1995a). Generally, they are straight or weakly bent and subvertical. Their length is tens to several thousands of metres and their width fluctuates from millimetres to several metres. The maximal depth is of the order of 35 m, because the ice

Figure 2.25.   In the foreground in shadow, ogives in the Mer de Glace Glacier, Mont Blanc Massif, French Alps. Photo: J. López-Martínez.

underneath shows a plastic behaviour. They constitute excellent routes for the penetration of melt water. When the system of crevasses intersects, the surface of the glacier becomes a fragile mass of dentated pinnacles known as *séracs* (Fig. 2.26).

Crevasses are grouped in systems (Fig. 2.27) (Sharp, 1960), differentiated into the *marginal* or *in chevron crevasses* formed by extensive stress generated by a higher speed of the glacier flow in the centre rather than at the margins as a result of the friction with the rocky walls. They intersect the edge of the glacier at angles of ~45°. *Transverse crevasses* (Fig. 2.28), located in zones of higher speed, are perpendicular to the flow. *Extension crevasses* occur where the valley glacier gets wider, or in bed threshold zones. These crevasses form smaller angles of ~45° with the edges. At the end of the glacier terminus, *crevasses of radial extension* are developed. Once formed, the systems of crevasses are modified by glacier flow. Marginal crevasses can undergo rotation and the transverse and extension crevasses are narrowed until they finally close, being then recognized by a blue ice vein.

Figure 2.26. Séracs in Perito Moreno Glacier at its outlet in the Argentine Lake, Patagonia, Argentina. Photo: C. Sancho.

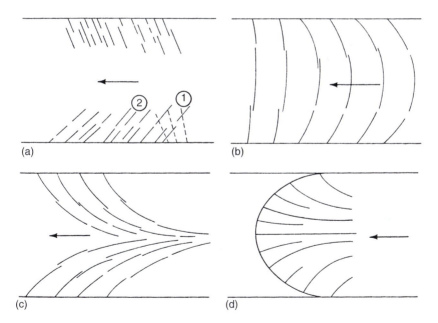

Figure 2.27. Types of crevasses in a valley glacier. (a) Marginal (1 – old rotated crevasses, 2 – cracks of new formation). (b) Transverse. (c) Splaying. (d) Terminal radial splaying. Arrow indicates the ice movement direction (Sharp, 1960).

Figure 2.28.   Transverse crevasses. Aneto Glacier, Aragonese Pyrenees. Photo: F. Gutiérrez.

Figure 2.29.   Bergschrund in the Aneto Glacier, Aragonese Pyrenees. Photo: F. Gutiérrez.

The *bergschrund* (Lliboutry et al., 1976) is the crevasse separating the glacier ice from the rock in the upper part of the glacier head. Some ice adheres to the rock wall and the rest of the glacier pulls away to form bergschrunds. They are located near very steep walls (Fig. 2.29) and sometimes extend hundreds of metres. Frequently, instead of a single crevasse, a system of bergschrund crevasses is developed.

Chapter 3
Glacial erosion

For a long time, the capacity of ice to substantially modify the preexisting relief has been well known. Knowledge about its action, in spite of the large number of publications on this issue, is not very extensive and this arises from the difficulty of studying the action of the ice on its bed, because the more effective erosive mechanisms take place under a thick ice cover. Some investigations directed to this aim by means of tunnel perforations have taken place, mainly in cirques. To this observation difficulty must be added the deficiency of knowledge on the preglacial relief, which can prevent a suitable quantitative estimation of the glacial erosion. Glaciations have also been superposed in the course of time in many areas, so that interpretations are more difficult (Price, 1973).

Glacial erosive action appears at the ice–rock (or ice-glacial deposits) interface. We know that erosive effects of ice without either movement or rocky material (clean ice) are practically negligible. For the same kind of ice in movement, if it affects fresh and little jointed rocks, the erosive power will be null or very small, although if the bottom material is loose it can substantially modify its morphology. The greater erosive effects occur under the action of thick temperate glaciers causing the extraction of glacier bed fragments and mobilizing and transporting them to a more-or-less distant place, being able in its transport to erode the rocky or loose material at the interface.

1. Erosive processes

We have already indicated that knowledge of glacial erosion is not very extensive. Normally we deduce processes by indirect methods by means of experimentation in a laboratory or by applying the most common deductive method of the Earth sciences, in which from the effects we try to classify the processes that were produced, in this case glacial erosional landforms (Goudie, 2004a).

Many factors affect the intensity of glacial erosive processes. They can be subdivided into three great groups: the ones inherent to the glacial system, those concerning the lithological and structural characteristics of the rocky substratum, and the ones relative to the geometry of the substrate (Sugden and John, 1976; Drewry, 1986).

The *importance of the glacial system* is given, on the one hand, by the basal ice temperature. When the temperature is close to the melting point the erosive effects are without a doubt much more important; in the case of glaciers with low basal temperature, the erosive action is only effective if debris is present at the interface. In addition to the influence of temperature, another factor of great significance is the basal speed of the ice mass that affects the transported amount of debris and, therefore, their erosive action and also the rock fracturing by means of differential pressure generation. A fundamental factor within the glacial system is the ice mass thickness, because it affects the friction mechanism at the ice–rock interface.

Characteristics of the rocky substrate, such as rock hardness and the presence of discontinuity planes (stratification, joints, foliation, and so forth), undoubtedly affect the action of erosive processes. Another important variable is the rocky bed permeability, because it conditions the penetration of melt waters, with the consequent variation in its erosive capacity and their possible absence in future regelation.

Finally, the *geometry characteristics of the rocky bed*, as to form, roughness and slope have to be added to the previously indicated factors with regard to variations in the intensity of the glacial erosive processes.

Not only the characteristics of the ice and the rocky substratum should be considered but also the modifications that glacial erosion undergoes in the course of time within a glacial age. Greater erosion intensity occurs in the first stages of the glacial advance (Tricart and Cailleux, 1962), in which the weathering mantle is easily exported by the ice masses in its first pulses. Once the regolith is evacuated, the ice needs a greater effort to be able to erode the non-altered substratum.

Approximate calculations of *glacial erosion rates* have been made (Andrews, 1975). On the one hand, the material transport (bottom, suspension and dissolution load) by glacial melt waters next to the glacier terminus has been determined, supposing that this material has originated in the glacial environment. Therefore, approximate calculations of glacial erosion can be made. Average rates seem to vary between 1 and 5 m every 1000 years, but this is perhaps too high and it is necessary to divide them by a factor of 2 to 10. In order to make estimations of glacial erosion rates for long time periods, theoretical calculations are also made. They have been applied to cirques, glacier valleys and fjords. The variables considered are volume, area and age of the analysed form. Volume is divided by the area and the erosion or reduction for the time undergoing the action of ice is obtained, to obtain the erosion rate. For cirques this value varies between 5 cm/1000 yr for polar cirques and 40 cm/1000 yr for the cirques of western Scotland. In any case, these calculations are to be considered as orientative only because true rates of glacial erosion are a major research topic at the present time.

1.1. Process types

The erosion mechanisms of glaciers are various and the following ones can be differentiated.

1.1.1. Abrasion

The wearing of rocks by the action of ice passing is clearly shown by its effects, represented, among others, by the microforms of glacial erosion such as striations, grooves, and other features. The abrasion process can be studied directly by means of tunnels into the contact with the ice–rock interface, by direct observation in natural subglacial cavities and also, indirectly, by laboratory experiments, analysing the behaviour of the ice sliding on different rocks and circumstances (Embleton, 1979).

Because ice at 0°C has a hardness of 1.5 in the Mohs scale, it is difficult for it to scratch or wear away the rocky surfaces over which it moves, because most minerals have higher hardness numbers than it does, from which it is deduced that the abrasive action must be due to the friction of rocky fragments held by the glacier in contact with the rocky substrate (Sugden and John, 1976).

Calculations of the *abrasion rate* have taken place at numerous sites. Values vary between 0.001 mm/yr for small glaciers and 36 mm/yr for a crystalline limestone substrate located in Argentière Glacier (French Alps) under an ice thickness of 100 m and a velocity of the glacier of 250 m/yr (Boulton, 1974). Perhaps this last number is too high, because the dissolution process of limestone, which is known to be very important in cold waters, has not been considered.

Factors affecting the *abrasion process* can be classified in two groups: those inherent to *transported fragments* and *characteristics of the rocky substrate* and the ones related to the particularities of the ice mass (Menzies, 1995b). Clean ice, no doubt due to its hardness, does not make abrasive effects on the rocky bed; on the contrary, ice at the interface shows scratches and cavities in it indicating the wearing action of the rocky bottom. The abrasion process requires for its action, the presence of rocky fragments held in the glacier base. In polar ice caps and glaciers the action of abrasion is very small, partly due to the paucity of fragments in the ice mass. In some soundings of more than two thousand meters only a few meters have been registered containing significant debris, although in soundings made in Greenland (Herron and Langway, 1979) and in Antarctica (Gow et al., 1979) abundant debris appears in the basal ice of these ice sheets. In contrast, the greater amount of material transported in temperate glaciers produces a much more important abrasive action, although in these glaciers the fragment content factor regarding the abrasion intensity should not be the only one to be considered. Another important circumstance is given by the relative hardness of the fragment with respect to that of the rocky substrate (Drewry, 1986). It is obvious that a higher hardness of fragments becomes apparent in a greater abrasive potential. To these factors it is necessary to add the morphologic characteristics of the particles. The constant fragment interaction with each other and with the rocky bed has as a consequence abrasion and reduction of angularity. This decreased effectiveness of the process can be overcome with provision of a new particle or with the breakage of previously worn fragments. The existing erosive processes at the ice–rock interface produces a fine material that, in some cases, stands between the ice mass and the rocky bed thus hindering further abrasion action; in these circumstances the presence of subglacial water favours the export of this "glacial flour" film. It is also necessary to consider the permeability of the rocky substrate, because the water presence at the interface produces an increase of flow velocity, which is also an influencing factor of abrasion, as discussed below.

The other group can be denoted as *glaciological factors* (Embleton, 1979). If the existing temperatures in the basal ice are very low, an adhesion between bed and glacial ice takes place, so that very strong stresses for sliding are necessary. If we add that under these temperature regimes the fragment proportion is, as previously indicated, very small or null, abrasive effects will practically be negligible. Basal flow speed is a factor of considerable importance because it determines the number of particles running past one point at the ice–rock contact; therefore, at higher speed a greater abrasive power is expected. Ice mass thickness has an important influence. A particle located in the ice–rock

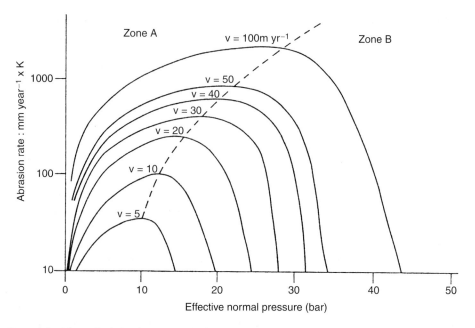

Figure 3.1. Theoretical abrasion rates and effective normal pressure for different ice speeds. In Zone A abrasion increases with the rise of pressure and in Zone B it diminishes until approaching zero, producing sedimentation of particles. *k* is a value depending on the relative hardness of fragments, the rocky substrate and the debris amount (Boulton, 1974).

contact is put under a pressure produced by the weight of the ice column and fragments that it supports. For a specific speed, abrasion intensity increases with rising pressure (Boulton, 1974), or similarly with ice thickness, until a certain threshold where the fragment-rocky substratum friction delays the advance of the fragment, so that ice flows over it instead of dragging the particle at the same speed at which the ice moves (Fig. 3.1). The chart shows the existence of two zones: one A, in which abrasion increases with rising pressure and a B zone, in which abrasion diminishes with a new increase of pressure, until it is null.

1.1.2. Fracturing

In this section we include the fractures generated by ice action as well as the ones existing before the glacier passing. It is difficult, in some cases, to specify if the origin of the joints is preglacial or is a process due strictly to the action of the ice.

We have already indicated that the existing rocky fragments in the base of the ice mass can scratch and groove the bed, but can also produce fracturing in the substrate and extract splinters when exerting pressure on the rocky material. This process becomes apparent basically by the generation of friction cracks.

Average values of basal shear obtained range round 1 bar (Embleton, 1979). Fracturing is exerted fundamentally by the action of basal fragments on the rocky bed. The phenomena of basal fracturing are more evident in the reverse-slope zones of the glacial flow. On the other hand, there is no doubt that the thermal régime of glaciers exerts a very considerable influence on the glacial fracturing process. In the case of polar glaciers great shear stresses are required, because the ice–rock contact remains as a unit due to its frozen state.

A factor of undoubted importance is the state of the material before ice advances on it. Thus the *periglacial breaking of the bottom* indicates the incidence of periglacial processes on the preparation of easily exportable material with the later advance of the ice (Boyé, 1949).

In the classic work carried out in the Yosemite Valley of California (Matthes, 1930) the importance of *sheeting* in the evacuation of material by extraction was analysed. They indicated that the optimal joint spacing for glacier mobilization ranges between 1.5 and 7.5 m. Sometimes it is difficult to determine if the origin of this topographic jointing is preglacial or if it has been generated by load loss after the disappearance of an important ice column. In some cases two sheet joint systems taking place at different times can cross (Ollier, 1984), with the crossing zones being weakness areas (Fig. 3.2). In any case, sheeting has a considerable influence on the control of the topographic form.

Existing joints, of one or another origin, are penetration routes for subglacial water; if later the freezing of this interstitial water occurs it produces the *congelifraction* process giving rise to rock breakage. Only weak temperature oscillations around the freezing point are necessary for this mechanism to be effective. *Congelifraction* seems to take place in glacier cirques where the melting water penetrates through the bergschrund. Obviously, on the scale of climatic fluctuations, this process can be important during different glacier oscillations.

Another circumstance of great interest, provided mainly by the investigators who worry about the genesis of topographic modelling from the evacuation of thick *weathering*

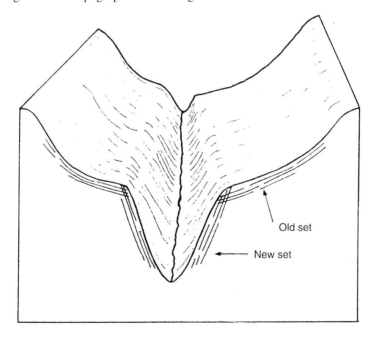

Figure 3.2.   Topography and sheeting in Vaiont glacial valley, Italy (Ollier, 1984).

profiles (Bakker, 1965; Thomas, 1994), is the presence of regolith of variable thickness generated prior to the freezings. The existence of some regions of this material that are easily eroded by the ice must be very carefully considered in the interpretation of glacial erosion modelling and in the quantification of erosion by the ice. We consider that this circumstance must be taken into account in the study of glacial sculpting in the Iberian Central System, because the presence of thick weathering deposits is common in this area, generated without a doubt in times prior to the Quaternary (Gutiérrez and Rodríguez, 1978; Molina and Blanco, 1980).

In any case, fracturing of the rocky substratum, whether inherited or of glacial origin, is a factor of extraordinary importance in glacial erosion, along with the availability of preglacial loose material.

1.1.3. Evacuation of debris

When a glacier slides on its bed, the mobilization speed of particles depends fundamentally on their size and forms and on the substrate roughness. For fragment mobilization it is necessary for the traction force to exceed the frictional resistance (Sugden and John, 1976). For the erosive action to be more effective it is also necessary to evacuate the fragments contained in the substrate, generated by the previously indicated processes. In this way the exposed rock can undergo glacial erosion mechanisms. It also should be considered that a part of the debris export is made by subglacier melt waters (Hallet, 1979; Drewry, 1986); this circumstance is significant in the case of temperate glaciers and negligible in polar glaciers. Another form of bottom material mobilization in temperate glaciers is by pressure of the ice mass on small size fragments and particles, soaked in water and located at the rock–ice contact, flowing towards areas of lower pressure.

2. Landform modelling by glacial erosion

The consequence of the action of glacial erosion processes is the generation of different forms due to the action of ice masses. These are reflected in the classification of glacial erosion forms of Table 3.1 (Sugden and John, 1976). It is difficult to carry out a classification of this type, in which a certain dose of subjectivity becomes necessary. In this classification, the three current process types are considered by distinguishing unconfined areal ice flow, linear flow in rocky channels and a third process differentiation corresponding to the interaction of glacial and periglacial activities. Another variable used for the differentiation is based upon the height or depressed position of the resulting form. Also as a classification norm the aligned or partially aligned morphology of the generated form is used. All these criteria are located on the ordinate axis, and on the abscissa a logarithmic size scale is placed.

As we will see, some of the forms are of doubtful interpretation, because strong discrepancies exist if these have been generated by erosive action of ice or by subglacial waters, or even by the combined action of specific processes of each one of these means, intimately connected. It is therefore difficult in many cases to carry out a clear distinction between glacial and fluvioglacial erosion.

Table 3.1. Classification of features of glacial erosion (Sugden and John, 1976).

PROCESS	RELIEF TYPE	RELIEF SHAPE	SCALE — Micro ← m^{-2} (1 cm) / m^{-1} (10 cm) / m^{0} (1 m) / m^{1} (10 m) / m^{2} (100 m) / m^{3} (1 km) / m^{4} (10 km) / m^{5} (100 km) / m^{6} (1,000 km) / m^{7} (10,000 km) → Macro
Areal ice flow	Eminence	Streamlined	Streamlined spur
	Eminence	Part-streamlined	Whaleback ↔ Rock drumlin; Roche moutonnée; Flyggberg
	Depression	Streamlined	Striae ↔ P-form; Groove
	Depression	Part-streamlined	Rock basin
Linear flow in rock channel	Depression	Streamlined	Trough
Interaction of glacial and periglacial	Depression		Alpine trough; Cirque
	Eminence		Residual summit or horn

Macro-scale landscapes (≈ m^{6}, 1,000 km):
- Landscape of Areal Scouring
- Landscape of Ice Sheet Linear Erosion
- Valley glacier landscape
- Nunatak landscape

Most texts differentiate two overall categories for glacial erosion modelling, dividing them into smaller and larger forms. This distinction is, however, not very realistic due to the wide dimensional interval occupied by some of them.

2.1. Striations, grooves and glacial polishing

Friction of particles at the glacial base during movement over the rocky bed produces erosive actions on it that become apparent as scratches, grooved forms and polishing of the rocky substrate surface. This action implies extraction of particles from the rocky bottom that incorporates them into the glacier.

One of the most common forms are *striations* or *straie* (Fig. 3.3), which are fine aligned grooves generally not much more than a metre long and a few millimetres wide and deep. Striations develop more easily in fine grained rocks and disappear when being exposed for long to weathering agents. In limestones this disappearance is very fast due to dissolution. So, on the slopes of the limestone pavement of Ingleborough in northern England, striations cropping out by removal of a glacial till are removed after about 10 years

Figure 3.3.   Glacial striations. Santa Elena Fort, Head of Gállego River, Aragonese Pyrenees.

(Sweeting, 1966). Not only the rocky substrate shows striations, they are also observed in fragments in glacial tills. In this way, the existence of striated clasts can help with determination of genesis, although with reservations in the interpretation of the origin of problematic deposits (Tricart and Cailleux, 1962). Care must be taken, therefore, because striations can be produced by other different processes, such as avalanches, mud flows, subglacial currents, and so forth. Striations are more common in reverse-slope areas, but they can even appear in vertical walls. Under the microscope, striations consist of numerous increasing fractures (Iverson, 1995). They indicate direction, but not the movement of glacial flow; sometimes, several directions or systems of striation intercrossing occur. The morphologic irregularities of the substrate can explain the directional variations; they can also indicate several glacial stages or phases of ice advance. In any case, and mainly through the study of the great glacial ice sheets, statistical analysis of hundreds of striations is fundamental to obtaining reliable data of the glacial flow direction for a certain time.

Another modelling type is *grooves* (Fig. 3.4). They are furrows of variable dimensions produced upon rocks greater than 1 metre long. In the Mackenzie River valley of northwestern Canada, gigantic grooves appear whose size can reach 12 km in length, 30 m depth and 100 m width (Smith, 1948). Their direction agrees with the glacial flow directions deduced from other forms, although a strictly glacial genesis is difficult to understand. Whereas striations seem to be formed by friction of silt size particles or sand on the rock, the origin of grooves is due to the action of large individual or grouped fragments. But grooves cannot only be generated this way, because many authors adduce

Figure 3.4. Grooves and glacial striations in Rongbuk Valley, Everest massif, Tibet. Photo: J. López-Martínez.

Figure 3.5.   Glacial polished surface and till. Pordoi Pass, Dolomitas, Italy.

the action of subglacial water currents or a similar preglacial morphology where these characteristics are emphasized later by the ice.

The action of constant abrasion produces *glacial polishing* (Fig. 3.5). Observed with a magnifying glass or under the microscope, a glacially polished rock has a surface furrowed by fine scratches corresponding to small striations. This polishing is more evident on fine grained rocks. It also disappears quickly when being exposed to weathering agents.

2.2. Friction cracks and P-forms

These are small forms related to glacial action appearing by fracturing of rock or by concavities where extraction of rocky substratum material might have taken place or not. The most classic works in which both differentiation and origin of the friction cracks are considered are from Gilbert (1906), Harris (1943) and Dreimanis (1953).

Friction cracks are developed better on intermediate grain-sized rocks, in which striations and polishing are less evident. Normally they appear on hard and fragile rocks such as granite, basalt, quartzite, and so forth. Four types are distinguished by Embleton and King (1975a) (Fig. 3.6):

(a) *Lunate fracture* (Fig. 3.6a), in which the horns point in the direction of the ice movement. These features are formed by two fractures, one of low angle and another subvertical; the intersection between them gives rise to a rock slab.
(b) *Crescentic gouge* (Figs 3.6b and 3.7), the horns point opposite to the ice flow. Its size ranges between 2 cm and 2 m. They usually appear in two to six systems and generally one behind another. They also have the two types of fractures mentioned above.

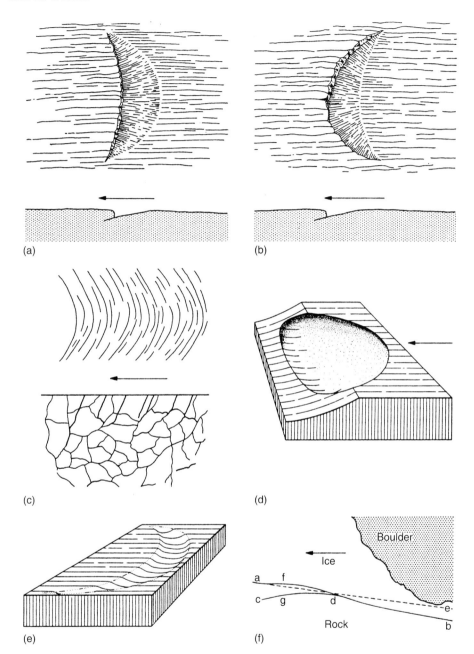

Figure 3.6. Friction cracks and sichelwannen (according to several authors, in Embleton and King (1975a). Horizontal arrow indicates the direction of the ice movement. (a) lunate fracture (plan view and section); (b) crescentic gouge (plan view and section); (c) crescentic fractures (plan view and section); (d) conchoidal fractures; (e) sichelwannen; (f) formation of a crescentic gouge; ade – original rock surface; afdb – rock surface deformed by pressure transmitted from the block contained in the ice; cgd – conchoidal fracture; fg – secondary fracture formed when the fgd rock wedge is broken.

Figure 3.7. Crescentic gouge. Threshold of Gerber cirque, Lerida Pyrenees.

(c) *Crescentic fractures* (Fig. 3.6c) are concave downstream to the glacier flow direction
and are constituted by subvertical fractures. Slab extraction, as in the two previous
examples, does not occur.

(d) *Concoidal fractures* (Fig. 3.6d) in which the fracture plane is concave upwards.

All these types are more frequently encountered in reverse-slope zones. As friction
cracks are larger than striations they are conserved more easily, because they resist
weathering and erosion effects longer. The perpendicular to half moon forms can indicate
ice flow direction. This can be determined by the inclination of the fracture of the smaller
angle diving downwards with the glacier. This rule, indicated by Gilbert (1906) and Harris
(1943), is not always fulfilled, because there are many opposite cases described in the
literature.

In general, it is agreed that friction cracks result from the pressure of a block on the
rocky substrate (Fig. 3.6f). This pressure is greater in areas of reverse-slope (Boulton,
1974), which is the reason for the greater abundance of this micromodelling in these zones.
It appears that at the beginning an elastic deformation followed by a breakage takes place,
generating a conchoidal fracture. A vertical fracture cutting through the fracture of low
inclination originates afterwards when the block advances. This type of crack has been
reproduced experimentally using a knife or a steel ball on glass.

P-forms, or plastically sculpted surfaces (Dahl, 1965), are on areas exposed to the
action of the ice. They are smaller forms never exceeding 20 m in size. The most abundant
form is the *sichelwannen* (Fig. 3.6e), a depression in the form of a half moon modelled on
hard rocks, whose horns aim downwards with the ice flow. They can be found in surfaces
with any inclination. Another type, the *cavetto forms*, are channels of steep edges (unlike

grooves) of up to half metre depth found on steep surfaces. In conjunction with these forms appear grooves, basins, folds and curved and winding channels (Kor et al., 1991). Their origin is very controversial. The most accepted hypothesis is the one explaining this micromodelling as due to subglacial water action under pressure. Other authors, however, defend the idea of abrasion processes (debris loaded basal ice), or erosive action produced by the movement under pressure of water-saturated basal till (Gjessing, 1965; Gray, 1981).

2.3. Roches moutonnées and other forms

Roches moutonnées are aligned hills, generally grouped, asymmetric, with the minor slope frequently polished and striated and the other one constituted by an irregular and fragmented, sometimes steep, surface. This disposition stays constant indicating the direction of the ice movement, moving from the minor slope side to the one of greater inclination (Fig. 3.8). These forms are developed better on crystalline rocks and are very common in areas covered by ice caps and also in thresholds of cirques. Their sizes are very variable, from less than a metre to hundreds of metres, reaching, in the case of *flyggberg*, to kilometric dimensions and heights of hundreds of metres (Sudgen et al., 1992). Within these forms greater roches moutonnées can be found, as well as others of smaller size.

The origin of these forms is not clear. In some cases (Matthes, 1930; Sugden et al., 1992) the importance of joint spacing in the generation of this modelling type is indicated (Fig. 3.9), although many cases do not conform to this rule. Others researchers, such as Carol (1947), have explained the steep side of roches moutonnées due to the action of the

Figure 3.8. Roches moutonnées. Tollstringen, Trolls massif, Norway.

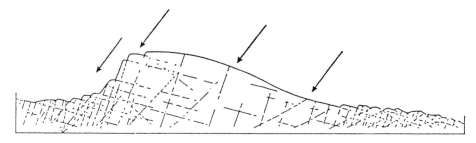

Figure 3.9. Longitudinal profile of a roches moutonnée showing its form and the influence of joints. Arrows indicate the direction of the ice pressure (Matthes, 1930).

congelifraction process, in frozen to melting temperature (Fig. 3.10). In these circumstances, when passing over a protuberance, the pressure increases and ice turns to a semiplastic state; pressure diminishes on the side of greater inclination and the melt water is frozen again, breaking the rock. These observations were made in a subglacial cavity. As indicated by Embleton and King (1975a,b), however, the question of whether or not roches moutonnées only form under temperate ice, as the subglacial congelifraction hypothesis indicates, still stands open.

Another set of forms related or associated to roches moutonnées are recognized. *Whalebacks, rock drumlins* and *streamlined spurs* are more or less extended forms of smoothed slopes. Sizes are reflected in Table 3.1. It seems that joints regulate the limit between hills in whalebacks. Rocky drumlins sometimes appear associated with elaborated fields of drumlins in glacial material. It has been questioned as to whether or not streamlined spurs are due to glacial action, but the smoothness of their slopes along with the parallelism of their direction with that of the glacial flow, inclines one to think that this modelling is the result of glacial erosion.

All previously described forms constitute positive relief, but it is well known that a landscape of glacial erosion is sprinkled with lakes of different sizes, that in most cases are *rock basins*. Their morphometry is very variable and structure plays an important role as much in their morphology as in their genesis. Their origin can be due to irregularities prior

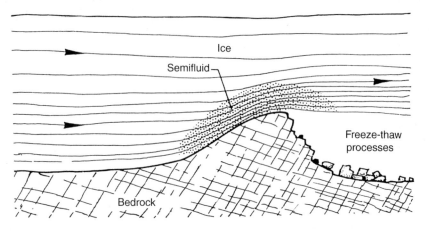

Figure 3.10. Operating processes in the genesis of roches moutonnées (Carol, 1947).

to the passage of the ice, perpetuated and emphasized by its action; at other times, glacial erosion processes are the ones that generate these closed depressions.

2.4. Cirques

Cirques are, along with glacial valleys, the most spectacular forms of glacial erosion. The texts on glacial geomorphology indicate that the word "cirque" was used for the first time in the Pyrenees by Jean de Charpentier in 1823. On the other hand, various denotations for cirque in different languages and countries do exist. We can consider the glacial cirque to be a semicircular or semielliptical depression dominated by steep slopes that is or has been occupied by ice (Fig. 3.11). Some authors consider as a characteristic of their definition the existence of a rocky basin, but not all cirques present this form (Fig. 3.11).

The presence of a rocky basin carries with it the existence of a threshold at the cirque's exit, which is sometimes rocky or composed of glacial till. The walls of the cirque usually appear fragmented and broken, unlike the basal surface, which is generally smoothed and displays smaller forms of glacial erosion.

The dimensions of cirques are very varied and fluctuate between tens of meters and kilometric widths. The biggest cirque known is the Walcott Cirque, in Antarctica, some 16 km wide with walls about 3000 m height (Price, 1973). Size depends on many variables, such as the duration of the glaciation, structural and lithological characteristics of the rocky massif on which the cirque is located, and so forth (Embleton and King, 1975a). The form is also a function, among others, of this last characteristic.

Figure 3.11. Cirque and morainic crest, developed on lower Cretaceous materials, head of Miera Valley, Cantabria.

The *morphometry* of a cirque is determined with sufficient detail by means of a set of characters that appear partially expressed in Figure 3.12 (Andrews and Dugdale, 1971). By himself Andrews (1975) indicated another value, the one of the cotangent of the angle formed by the threshold of the cirque with the final wall that indicates the intensity of glacial erosion. Another index, the occlusion degree (Evans, 1969) is defined as the number of degrees of the longer contour and for 180° indicating that the walls of the cirque are parallel. In addition to these values other parameters have been set out to describe the

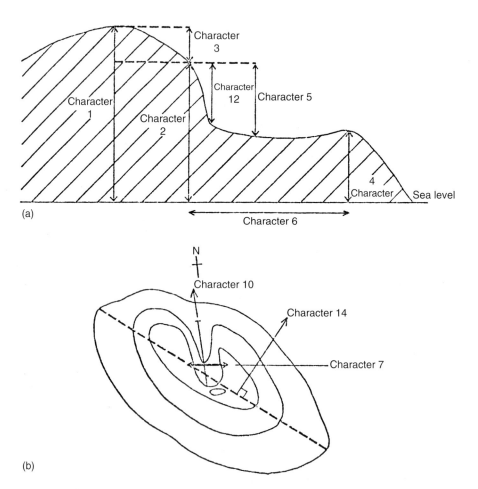

Figure 3.12. Cirque characters (Andrews and Dugdale, 1971). (a) transverse section of a cirque, (b) plan view of a cirque and its summit area. The different characters are the following ones: 1 – Maximum mountain elevation in which the cirque is located. 2 – Maximum elevation of the cirque. 3 – Difference between 1 and 2. 4 – Elevation of the cirque threshold. 5 – Maximum vertical development of the cirque. 6 – Length of the cirque's greater axis. 7 – Maximum width of the cirque perpendicular to the greater axis. 8 – Length/width relation. 9 – Length/height relation. 10 – Direction of the greater axis. 11 – Type of ice mass: (a) empty, without ice; (b) snowfield; (c) ice plate; (d) cirque glacier. 12 – Final wall height of the cirque. 13 – Relation between height of the final wall and the length of the cirque. 14 – Angle of the regional slope.

morphometric characteristics of a cirque. In general, these values lack precision because they are taken from topographic maps and, in some cases, for some parameters the subjectivity of the investigator plays a role. As indicated by Andrews (1975), however, "the morphologic analysis of cirques can provide great information about the erosion processes in the cirque, although only of a deductive type: morphology is an answer to processes...".

In order to understand better the *processes* that have given rise to its form, it is necessary to understand, among others, the structure and movement of ice in glacial cirques. The most precise observations were taken by McCall (1960) in the studies of a small cirque glacier, Vesl-Skautbreen in Norway. Two tunnels were perforated (Fig. 3.13) and among other observations, ice speeds at different points were measured and different ablation surfaces were distinguished in summer. Observations showed that flow lines of the ice dip 30° in the upper zone of the glacier and in the lower one they incline 26–28° but in the opposite direction to the slope of the cirque, their inclination diminishing downwards; this disposition indicates a synclinal structure for the ice layers. The distribution of speeds indicates faster movements in the upper zone of the glacier and slower ones in lower parts; in this area the different speed vectors reveal upward movements of the ice. All these data provide evidence that movement takes place by rotational sliding.

Erosive processes happening in a cirque are basically twofold and are due to glacial and periglacial activity. At the ice–rock contact, abrasion carries out continuous wearing as much in the walls as at the bottom and is responsible for the rocky basin existing in many cirques (Richardson and Holmund, 1996). This depression is easily explained by rotational sliding of the ice and this movement also produces the overcoming of the threshold and the evacuation of debris. The other type of process, congelifraction, is very effective on outcropping rocks over the glacier (Gardner, 1987). Here, it seems that the freeze-thaw action is more effective in temperate glaciers than in polar ones. It is deduced that the

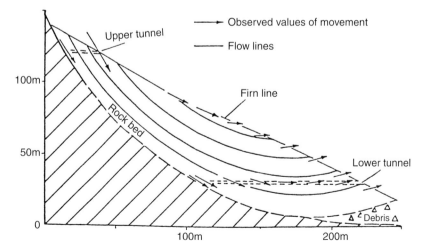

Figure 3.13. Section of Vesl-Skautbreen cirque glacier (Norway) in which the flow and speed lines are indicated. After McCall (1960), simplified.

cirque enlarges fundamentally by the action of congelifraction and deepens by the effect of abrasion. The speed of erosion in cirques obtained by different methods is of the order of 500 mm/ka (Benn and Evans, 1998). It seems that the backward movement is faster than the excavation. On the other hand, the elevations of the bottoms of the cirques have apparently not changed much during glaciation. Elevation, generally, is used to calculate approximately the level of permanent snows, because it corresponds with the 0°C isotherm in summer, in considering the lowest cirques of equal orientation (Flint, 1971).

The *origin* of cirques, according to most authors, is tied first to the snow accumulation in a preexisting depression in which congelifraction processes and particle evacuation take place by snow melting in summer (*nivation*) (Thorn and Hall, 1980; Evans, 2004c). This brings as a consequence the widening of the depression and the generation of a nivation niche. If excavation is deep enough, snow lasts from 1 year to another in the nivation niche, becoming firn and also turning this material into ice. In this way an embryonic cirque is formed. The nival accumulation stage is denoted as the "incipient phase" and when snow begins to last, the "firn phase" (Tricart and Cailleux, 1962).

Once formed, if the circumstances are appropriate, the cirque continues its development. Its *evolution* is influenced by different causes. The duration and number of glaciations exert, without a doubt, a fundamental role in their development. Also, the glacier has to evolve differently if it is located in temperate and subpolar latitudes, or in polar areas. Another great incidence variable is the structure and lithology of the rocky massif on which the cirque is based (Embleton, 1979). Sometimes cirques are located in relief produced by lithological differences, which in a monoclinal series can give origin to *stepped cirques*. The degree of fracturing and the spacing between joint planes play a fundamental role in mechanical disintegration by gelifraction. Exfoliation can be of importance in the form of the cirque. Another basic factor in the development of cirques is their orientation. In middle latitudes of the northern hemisphere, most cirques presenting good development face towards the north and east, because in that direction they are protected from solar rays and the snow lasts more easily. One of the analysed factors is the influence of climate in the variation of the cirque (Derbyshire and Evans, 1976). These authors indicate that it is not easy to establish generalizations on this aspect, in spite of the great number of regional studies on form and distribution of cirques.

The development of cirques also includes a backward movement through headward erosion of its walls, which at the limits with other cirques (Fig. 3.14) can cause knife-edged ridges known as *arêtes*. The disappearance by growth of the cirque of these arêtes can give rise to *coalescent cirques* (Fig. 3.15). The formation of pyramidal peaks or *horns* (Figs 3.16 and 3.17), usually presenting three or four faces, is a result of the backward movement of the multiple cirque walls.

2.5. Glacial valleys

Deep glacial valleys are one of the most characteristic forms of erosive activity of glaciers. In contrast, in fluvial valleys, rivers are only in contact with a small part of the valley, whereas the ice of a glacial valley occupies a greater part of the cross-sectional profile (Price, 1973). These canalised ice masses excavate deep troughs to hundreds or even thousands of metres of depth (Figs 3.18 and 3.19). This intense erosion is mainly carried

Figure 3.14. Aerial view of horns, cirques, arêtes and glacial valleys. Eastern Pyrenees.

out in mountain areas located to windward, that receive great nival precipitation and also feed thick glaciers running through preglacial valleys producing a considerable, sometimes spectacular, modification of the relief.

According to Linton (1963) four types of glacial valleys are distinguished. The *alpine type* has feeding that takes place from a cirque or set of cirques in high zones. The *iceland type* has an ice cap escaping preglacial valleys that forms the glacier. The *composed type* occurs where ice does not find sufficient previous valleys to unload all the ice, or similarly,

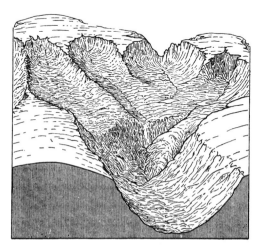

Figure 3.15. Coalescent cirques, arêtes and a glacial valley (in Holmes (1965), Fig. 4.81 modified by Davis (1906)).

Figure 3.16.   Cervino or Matterhorn, in the Alps, at the border between Italy and Switzerland, presents a typical horn morphology. Photo: J. López-Martínez.

Figure 3.17.   Horn in central Andes, El Portillo, Chile.

Figure 3.18. Glacial valley in central Andes, in which thick accumulations of frost-shattered clasts on the slope are observed feeding the bottom of the valley. Inca Bridge, Argentina.

Figure 3.19. U-shaped cross section in the glacial valley of Llanganuco, Cordillera Blanca, Peru. Photo J. López-Martínez.

from a superposition of a preglacial valley with others developed from the partial or complete destruction of preglacial watersheds. Finally, in the *intrusive or inverse type*, the ice flows up into the reverse-slope of a preglacial valley. They occur mainly in areas of little relief.

Glacial valleys are characterized by the peculiarities of their cross-sectional and longitudinal profiles. The *cross-sectional profile* results from the erosive action of the ice on old fluvial valleys (in most cases) producing a widening and deepening. Generally they present a U form (Figs 3.19 and 3.20), with steep walls and flatter bottoms than fluvial valleys, due to a later levelling produced by deposits of alluvium. However, other U-shaped valleys exist, as with the cradle valleys in periglacial environments (Tricart, 1967), but these present their slopes totally covered by debris, unlike glacial valleys whose walls are, generally, formed by outcropping rock. With relative frequency, cross sections of a V-shape are often the result of the erosive activity of subglacial channels. In spite of these reservations, the typical profile is parabolic or close to it, possibly because this form exerts the minimum resistance to friction (Flint, 1971). Sometimes, the walls of the glacial valleys show *plains* or *shoulders*, whose interpretation has produced many controversies (Derruau, 1965). It seems that the most suitable explanation is the confinement of a reduced endglacial glacier in a wide valley, shaped by a glacier corresponding to the last glacial maximum.

The investigation of cross-sectional profiles is very precarious due to its steep walls. Generally maps or photogrammetric surveys are used but in some cases these procedures do not permit us to know if the wall is naked rock or, on the contrary, is covered with

Figure 3.20. Zezere glacial valley, with cross-sectional U-shaped profile. Serra da Estrella, Portugal, constituting the western end of the Central System of the Iberian Peninsula.

deposits of different origins. These data are fundamental for a correct interpretation of the profile evolution.

The analysis of a *longitudinal profile* reveals a great number of irregularities produced by *basins* and *thresholds* differentiating it from a fluvial valley profile. The basins, once ice is removed, become lakes and, in other cases, are silted by sediments. Frequently thresholds present certain signs of glacial erosion. If the excavation of a main glacial valley is greater than that of lateral valleys, once ice has disappeared, a set of *hanging valleys* and *faced spurs*, triangular or trapezoidal, are exposed in between them (Figs 3.21 and 3.22). This postglacial topography gives rise to cascades, like the existing ones in the valleys of Yosemite in California or Lauterbrunnen in Switzerland.

The *origin of thresholds* is explained for different forms. Lithologic variations or different spacing of fractures (Matthes, 1930) (Fig. 3.23), can by themselves provide a suitable cause for the genesis of the profile steps. Also a change in the gradient of the preglacial valley is adduced, conserved and emphasized by the ice. Another cause indicated for the origin of the basins is the union of glacial valleys. Previously, the importance of deep differential weathering was indicated; this regolith can easily be excavated by glaciers in their first stages, leaving exposed the basal surface of weathering with its depressions and highs (Bakker, 1965). In any case, the basins indicate the capacity of the ice to flow against slope.

Processes and causes implied in the *genesis of a glacial valley* are, to a large extent, the same ones giving origin to other forms. Abrasion will be more effective in temperate glaciers. Mobilization of jointed blocks will take place by pushing by other blocks transported by the glacier. Another factor to consider is the *periglacial fracturing of the substrate* (Fig. 3.24) of Boyé (1949) that Cailleux (1952) developed for glacial valleys. In one first periglacial stage, the substrate materials in the bottom of the valley are fractured by congelifraction processes. When the glacial tongue advances, it plays the role of a

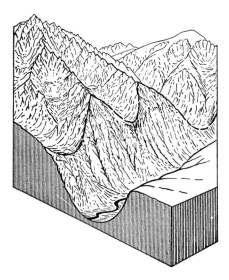

Figure 3.21. Block diagram showing hanging valleys, faceted spurs, U-shaped valleys, cirques, arêtes and horns (Davis, 1906).

Figure 3.22. Hanging valley. Western coast of Norway. Photo: J. López-Martínez.

thermal mantle and allows the disappearance of the permafrost so that clasts can easily be evacuated by the glacier. The generation of a concave sheeting subparallel to the cross-sectional section helps the attainment of the parabolic profile and of its perpetuation. The disposition of the extensive and compressive glacial flow of Nye (1952) partially explains the existence of thresholds and basins (Fig. 2.23). In compressive zones an increase of erosion takes place and once irregularities have appeared they tend to be conserved or even accentuated.

Sometimes, the thickness or height level of glaciers increases because of blocking by another glacier, or due to superior feeding rather than evacuation. In these cases, a glacier can find a preglacial breach in the rim of the watershed to unload into an adjacent valley. Such breaching can also produce glacial erosion and is able to produce a U-form and even to form rocky basins (Penck, 1905). These passes are denoted as glacial *diffluence*. In the

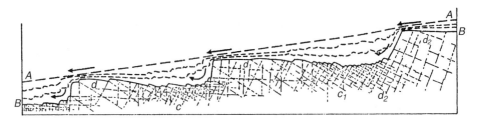

Figure 3.23. Longitudinal profile illustrating the development of a glacier staircase, because of the different joint spacing. AA- represents the preglacial profile and BB- the glacier staircase. The areas d, d_1 and d_2 are zones of difficult excavation by the glacier, due to the joint spacing; the opposite occurs in c and c_1. Arrows indicate the direction of the ice movement (Matthes, 1930).

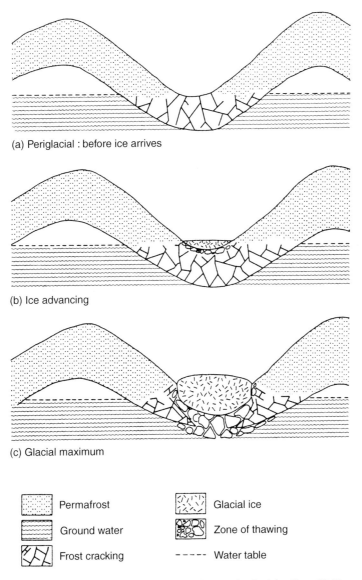

(a) Periglacial : before ice arrives

(b) Ice advancing

(c) Glacial maximum

Permafrost		Glacial ice	
Ground water		Zone of thawing	
Frost cracking		- - - - - Water table	

Figure 3.24. Mechanism of periglacial breaking of the bottom in glacial valleys (Cailleux, 1952).

case where the height level of the ice rises so that all breachings are used by diffluent ice, then the term *transfluence* can be used.

2.6. Glacial landform modelling in regions of little contrasted relief

In previous sections we have analysed the erosive action of ice in areas of strongly contrasted relief, generating cirques and glacial valleys as the most significant forms. But it is necessary to also consider that a great part of the terrestrial surface above sea level, occupied by ice in the algid glacial stages, was formed by great extensions of low energetic relief or that were even flattened. These circumstances existed as much in elevated plateaus as in low-altitude plains (Davies, 1969; Embleton and King, 1975a). Most of these areas were covered by ice caps and the topographic modelling resulting from the erosive action displays peculiarities distinguishing it from the glacial erosion forms in alpine areas.

Many investigators indicate that in these circumstances ice can play a protective role on the preglacial relief (Ambrose, 1964). Evidently, in many areas, signs of erosion and glacial sedimentation are very scarce or they do not exist (Sugden, 1974). These regions are probably related to zones in which ice in contact with the bed is below the freezing point and no movement between the ice–rock interface exists. This protectionistic role of glaciers is reflected in different forms. In some places, valleys do not bear any relation to the glacial flow direction and, in other areas, preglacial flattened surfaces are conserved with only slight modifications.

Although manifest evidence of the protective role of ice caps exists, evidence of glacial erosion is more numerous in these areas of low relief. In these regions the erosive glacial processes generate a typical *hill and lake* topography (knock and loch) (Linton, 1963) (Fig. 3.25). It is a development of confused relief where hills are developed with dimensions of tens of meters with intervening, generally fairly shallow, depressions occupied by lakes or peat bogs. Lakes are aligned and lengthened along structural control. The accompanying drainage network is very hazardous. Hills can present signs of striation and glacial polishing and roches moutonneés are common. Other times, hills are of smoothed and extended forms, resembling drumlins being the reason why its denominated topography is *rock drumlins*. The difference with the classic drumlins modelling is that these are formed by glacier till and fluvioglacial deposits. The topography of hills and lakes is typical of the

Figure 3.25. Hill and lakes relief. South Harris, Hebrides Islands, United Kingdom.

northwest of Scotland and of the Canadian and Baltic Sea shield areas (Sugden, 1978; Gordon, 1981; Rea and Evans, 1996). Glacial sedimentation is characterized by small discontinuous and thin till sheets. Another form developed in these circumstances, although not exclusive of them, is the *crag and tail* form consisting of an extended "tail of till" located down the icestream from the hill crag. The observations of Jahns (1943) concerning this type of relief indicate that erosion of a hill is more important when shielded from the ice flow. This author arrived at this conclusion by the erosion experienced by sheeting structures previous to the passing of the glacier.

2.7. Fjords

Fjord is a Norwegian term used for a long and deep sea arm characterized by more or less straight segments, steep walls and very great depths (Fairbridge, 1968a).The longest fjords are Norvestfjord/Scores by Sund in Greenland (230 km), Sognefjord in Norway (220 km) and Greely Fjord/Nansen Sound in the Canadian arctic (400 km) (Evans, 2004b). They are located along high latitude coasts (Figs. 3.26 and 3.27), and are characterized by a threshold at the outlet, totally or partially submerged, and by a deep fjord basin inside (Syvitski et al., 1987). The threshold is generally rocky and basins can reach enormous depths. The greatest depth measured is in a basin 1288 m deep in Messier Channel (Chile), but it is possible that some of those existing in Antarctica reach 2000 m (Andrews, 1975). As indicated by several authors, some fjord areas (Eastern Canadian Arctic, Greenland, Norway and New Zeeland) present a clear structural control, a cause adduced by some

Figure 3.26. Geiranger Fjord, Norway.

Figure 3.27.   Seydisfjördur Fjord, Iceland. Photo: F. Gutiérrez.

investigators as fundamental to the genesis of fjords, but considered at the present time as an adjunct factor (Powell, 2003).

Most fjords turn inland into glacial valleys. Also the walls of fjords present faced spurs and glacial hanging valleys that spilled ice into the main valley, today turned into a fjord,

Figure 3.28.   Strandflats. Alesund, Norway.

by postglacial elevation of sea level. It seems that glacial erosion motivated the excavation of basins and the existence of the threshold can be explained by the thinning and spilling of the glacier when arriving at the continental platform.

Norwegian fjords are most fundamentally characterized along their southern coast (Hardanger, Sogne and Geiranger Fjords, among others) by scarped or cliffed walls, hanging valleys and cascades, whereas more to the north of the region, between Alesund and Trondheim, slopes are much more smoothed, similar to the *sea lochs* or *firths* of Scotland. In this area are recognized rocky island zones of low altitude (*skjaergard*) or partially submerged erosion platforms (*strandflats*) (Fig. 3.28). These flattened zones are interpreted as originating by marine, subaerial and subglacial erosion and even as etch plains (Benn and Evans, 1998).

Scottish *firths* are long coastal coves associated with the glaciation in coasts of low relief. They are distinguished from fjords by their irregular form, deficiency of basins or transverse U-shaped cross sections. In any case it is difficult to distinguish, in coasts of low relief, coves generated by glacial erosion (Augustinus, 1992).

Chapter 4

Glacial transport and sedimentation

1. Glacial supply and environments

Glaciers are able to transport large quantities of rocky material for considerable distances, although not all of the mobilized particles have been produced by erosive glacial activity. There are many other sediment sources, such as clasts formed by congelifraction processes in the slopes around the glaciers. Other types of sources are snow and rock avalanches, landslides and nonglacial streams that flow into the glacier. In some cases, volcanic activity can supply pyroclastics to the ice masses, as in some glaciers that cover volcanoes in the Andes. Finally, winds can supply small particles to glaciers (Boulton, 1978; Small, 1987; Kirkbride, 1995; Benn and Evans, 1998).

In West Greenland, discharge of debris takes place predominantly on the ice-sheet margin, through the basal ice layer. The relative abundance of debris is an important control of glacial sedimentary processes and features. The sediment production varies substantially between ice-sheets and valley glaciers (Knight et al., 2002).

Glacial debris can be carried in three different environments. The material mobilized on the ice surface constitutes *supraglacial debris*, which generally is angular and scarcely modified by glacial activity. It is more abundant on valley glaciers, because their rocky walls contribute directly to the ice surface, whereas on continental ice the only rocky sources are the walls of the nunataks. In the accumulation area debris can be covered by snow and, in some mountain areas, it can cover the whole ice surface, as occurs on many of the low-albedo glaciers of the Karakoram and Himalayas (Figure 4.1) that protect the ice against melting. In most glacial valleys, the proportion of debris increases toward the front of the glacier terminus. The transport of supraglacial debris occurs as on a conveyor belt, and very big blocks can be transported for large distances.

Englacial debris occur disseminated within the ice mass, and then their proportion rarely exceeds 15 %, or in planar layers up to 5 m thick separated by clean ice, where it can represent up to 80 % of the total mass (Embleton, 1979). These layers correspond to filled cracks or crevasses, though in most cases they are subglacial debris displaced upwards along flow lines. This transport occurs in the areas of compressive flow of the valley glacier. In other cases, compression is caused by terminal moraines blocking the advance of the ice (Goldthwait, 1951; Boulton, 1970a).

Subglacial debris come from the glacial valley bottom or from the material that penetrates the cracks, crevasses, and tunnels. These bottom debris undergo abrasion and fracturing, and consequently the clast roundness is greater, and the percentage of fine material higher. Transport is by a traction mechanism (Sharp, 1988), with the particles sliding and rolling as they are pushed by the glacier. If the subglacial bed is deformable, then the particles will be pressed into the bottom or will become lodged in it. The clast

Figure 4.1. The dark, debris-covered Baltoro glacier at about 4300 m in the Karakoram Himalayas (Pakistan). Photo: J. López-Martínez.

morphologies, as well as their orientation, affect the characteristics of movement, in such a way that the elongated pieces are oriented in the direction of the flow (Price, 1973).

2. Mechanisms of glacial sedimentation

The debris within the ice mass can be carried toward the glacial margin, deposited in the bottom of the valley, incorporated in the melt waters or carried out by icebergs. There are several primary processes linked to the sedimentation of the material carried by the glaciers, and other processes that modify the sediment after deposition (Whiteman, 1995).

Supraglacial melting is the main way of glacier melting, and characterizes temperate glaciers, because cold glaciers show limited melting. Thus, in the warm-based glaciers, thawing can reach values of around 12 cm/yr, whereas in the Antarctic it is restricted to a few centimetres per year (Embleton, 1979). Supraglacial melting is much higher than that occurring at the base of the glacier. As a consequence of melting, debris is liberated to form supraglacial till, which can retain former structures developed during glacial transport (Lawson, 1979). More debris can be added progressively to the till layers by a continuous accretion. The presence of this mantle of supraglacial material delays the melting of the underlying ice. The supraglacial material can remain *in situ* or be mobilized by melting waters. In the glacial tongues most of the supraglacial till is affected by flow and rotational sliding (Dreimanis, 1988). The water-saturated till moves by gravity, affected by the gradient, roughness and water percentage. Flows of several metres per minute have been recorded, and the debris mobilized can be displaced for tens or even hundreds of metres (Sugden and John, 1976).

Three heat sources can act in *subglacial melting*: geothermal heat flow, frictional heat due to sliding of the ice, and that resulting from an increase of pressure due to the obstructions of an irregular substrate (Embleton, 1979). All this calorific supply results in melting of the subglacial ice and sedimentation of debris. The subglacial sedimentation (lodgement) implies melting under pressure and the development of basal till. The irregularities in the bed are progressively infilled with till, thus smoothing out the profile of the valley bottom (Boulton, 1971) (Figure 4.2). The subglacial till shows shear structures due to stress during the ice movement (Boulton, 1970b) (Figure 4.3). The subglacial sediments, saturated in water, are deformed by the weight of the ice (Boulton, 1975, 1982). A subglacial flow occurs in such a way that the squeezed till can move toward cracks or basal cavities. This plastic flow of saturated till can have an annual periodicity in some temperate glaciers. In summer, water penetrates by cracks, thus saturating the till and enabling the flow. In winter this process stops. The wet till rearranges its fabric. In addition, the ice movement erodes the subglacial till, causing the development of furrows and drumlin-like forms, the so-called *fluted moraines* (Sugden and John, 1976).

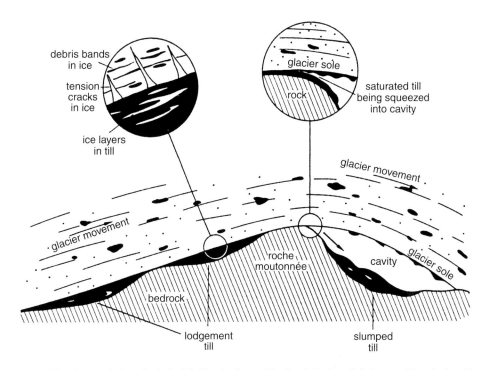

Figure 4.2. Accumulation of subglacial till at the base of Svalbard glaciers, Spitsbergen. Note the lee side subglacial cavity that is infilled by till (Boulton, 1971; modified in Sugden and John (1976), Figure 11.3).

direction of ice movement

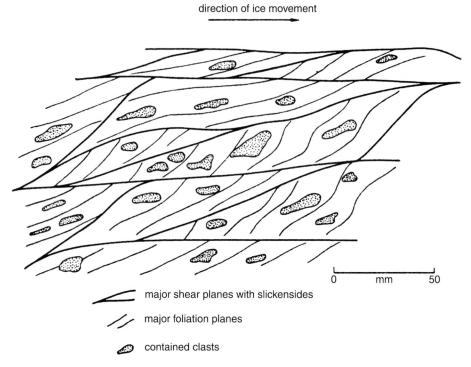

major shear planes with slickensides

major foliation planes

contained clasts

Figure 4.3.   Schematic showing the shear structures in lodgement till caused by stress under moving ice (*Source*: Boulton, 1970b, Figure. 8).

At the edges of the glacier the melting also results in sedimentation of supraglacial, englacial and subglacial debris. This *marginal discharge* is the origin of moraine ridges. The associated till usually does not show any fabric. If the glacier advances, the material deposited at the frontal margin can form *push moraines*. The till is folded and fractured causing imbricated structures. The resulting fabric has an orientation parallel to the direction of the ice movement (Croot, 1987; Van der Wateren, 1987).

3. Characteristics of till

Till is a poorly sorted deposit, composed of a variety of grain sizes, with big blocks embedded in a fine, sometimes clayey matrix. Till generally does not show any stratification and, in general, includes a large variety of rocks, the clasts of which may have facets and striae (Sugden and John, 1976). Some scientists use the term moraine as equivalent to till, but the word moraine has a morphological or landform implication, whereas till is a sedimentological term (Price, 1973).

Glaciers have a great capacity for transport, as *erratic blocks* demonstrate (Figure 4.4). These are blocks located above bedrock surfaces or on till material deposited by glacial activity. The size of these blocks is occasionally dramatic, such as the Schollen erratic

Figure 4.4. Erratic blocks, Markarfljót (Iceland). Photo: F. Gutiérrez.

block, in Germany, which is 4 km × 2 km × 120 m (Price, 1973). The Okotoks erratic block, near Calgary (Canada), has a weight of 118,000 t (Sharp, 1988). In some cases, where lithology enables recognition of the provenance, the distance travelled can be assessed. Some of these erratic blocks have been transported up to 1200 km (Figure 4.5) (Flint, 1971).

Grain size distribution provides valuable data on till development. Most tills show a bi-modal distribution, with a peak corresponding to coarse-gravel size, and other peak related to the matrix (Dreimanis and Vagners, 1971) (Fig. 4.6). For any lithology, the grain size diminishes with the distance from the source area, and a dominance of mineral grains versus rock fragments indicates a higher degree of till evolution. This feature underlines the existence of mechanisms of classification in the glaciers that are related to the physical breaking of the fragments. If they have a calcareous composition, then the till can be substantially modified by carbonate dissolution (Sugden and John, 1976).

The term *fabric* refers to the arrangement of the particles within the till and it is sometimes called a *glaciotectonite* (Elson, 1989). Many times, the arrangement of the particles is clearly visible, particularly the orientation of the greatest axes of the largest fragments (Figure 4.7). The analysis of the direction and dip of the elongate clasts is used to determine the direction of the ice flow (Flint, 1971; Benn, 1995). These studies on the till fabric must be made in several stations of the same area, because only one sample can lead to mistaken interpretations. Furthermore, these investigations can help to distinguish two or more types of till in a sequence.

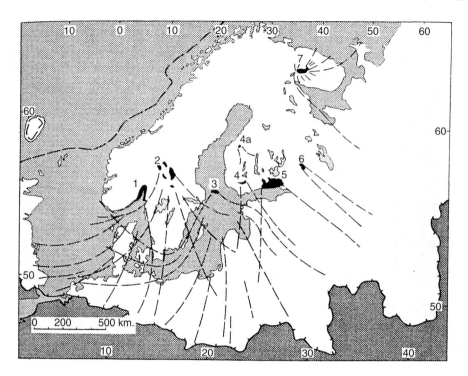

Figure 4.5. Paths of erratic blocks in the Scandinavian ice sheet. (1) Oslo District bedrock. (2) Dala porphyries. (3) Aland Islands Rapakivi granites and quartz porphyries. (4) Lappajärvi Karnaite. (5) Viipuri Rapakivi granites. (6) Lake Ladoga Rapakivi granites. (7) Umptek and Lujavr-Urt nepheline syenite. (Compiled by Flint, 1971, from several authors).

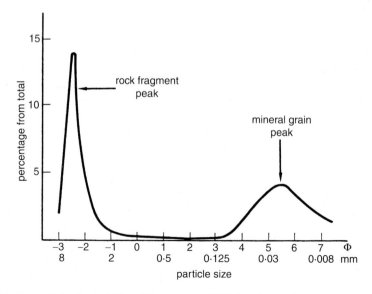

Figure 4.6. Bi-modal distributions of particle sizes in a till (Dreimanis and Vagners, 1971, in Sugden and John (1976), Figure 11.12).

Figure 4.7. Till deposit with a clear arrangement of the largest blocks. Headwaters of the Gállego River in the Central Spanish Pyrenees.

4. Landforms resulting from glacial sedimentation

Approximately 10 % of the Earth's continental surface is covered by deposits related to the activity of ice masses (Price, 1973). They can reach up to 400 m in thickness in the Spokane Valley (USA) (Flint, 1971) or constitute a thin covering that fossilizes the substrate. Moraines are constituted of till or a mixture of till and fluvioglacial deposits. These accumulations show different forms, depending on the glacial dynamics, on their position in relation to the ice, and on the processes that have operated.

According to the position that moraines occupy in relation to the glacier, they can be classified as terminal, lateral or medial moraines. The first two indicate positions of advance or retreat of the glacier, whereas the medial moraines (Figure 4.8) originate by the junction of two valley glaciers. These latter moraines can be superficial and of limited depth. Lateral moraines can rest against the glacial valley side wall or in some cases form separate ridges out from the valley wall (Figure 4.9 and Figure 4.10). In both cases they are fed essentially by debris coming from the valley walls. In some cases, the lateral moraines join with the frontal ones, resulting in moraine arcs or ridges that can dam the meltwaters, thus causing the formation of ice-dammed lakes. These moraine ridges are eroded quickly by the fluvial and fluvioglacial discharges, leaving small ridges in which, very commonly, it is possible to distinguish the stages of glacial retreat from these remnants of terminal moraine arcs (Figure. 24.10).

Figure 4.8.   Central moraine of the Upper Baltoro glacier, at about 5000 m, in the Karakoram Himalaya of Pakistan. Photo: J. López-Martínez.

Figure 4.9.   Lateral morainic ridge in the Miera glacial valley of Cantabria, Spain.

Figure 4.10.   Lateral morainic ridge. Passo de Pordoi in the Dolomites of Italy.

The identification of the different moraine types can be based on the activity of ice masses and the environments (subglacial and marginal) in which they developed (Chorley et al., 1984) (Figure 4.11). Subglacial forms originate under an active temperate glacier due to the supply of particles resulting from the melting of the glacier base. One of the

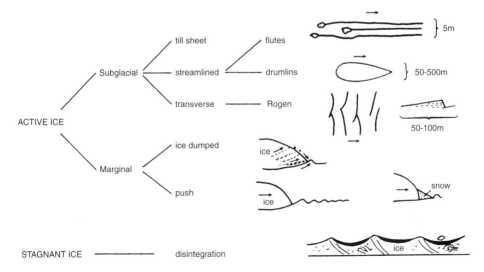

Figure 4.11.   A classification of glacial moraines (Chorley et al., 1984). (Reproduced with permission).

most extensive morphologies is that of the *till sheets*, which are also known as *ground moraines*. They are large plains covered with till, whose depth varies between half a metre and tens of metres, and they can cover a surface area of thousands of km^2 (Kemmis, 1989). They cover large areas of the European and North American plains that were occupied by ice-sheets. In detail, their topography is very irregular, with rock substrate outcrops, small hills and lakes, and lobate forms of the moraines. Some till sections show intercalations of gravels and sands, an indication of alternating glacial and fluvioglacial environments. The thickness is variable and the existence of an extensive thick accumulation is interpreted as a stagnation of the glacier, whereas a thin covering can indicate a fast ice retreat. At present these till sheets are being destroyed by fluvial erosion, and incision enables recognition of sections of the deposit. These accumulations are the result of different advances and retreats of the Pleistocene ice sheets, and consequently their sedimentological study provides very valuable data on the Quaternary geology of these regions.

Other subglacial forms show a parallel alignment to the ice flow. The *fluted moraines* are an alternation of rectilinear ridges and furrows over the till, located in the borders of retreating glaciers. Their length is of tens or hundreds of metres and their heights do not generally exceed 5 m. They originate in areas sheltered by large blocks deposited in the ice bed, which create an area of low pressures downstream (Boulton, 1976; Gordon et al., 1992). The longitudinal profile of the forms is similar to that of elongate drumlins and they are considered by some authors to be linear, very narrow drumlins (Prest, 1968). One of the unsolved problems of interpretation is the regularity of the spacing between the ridges in some moraines.

Drumlins are elongate hills with a major axis parallel to the ice movement. They have an inverted spoon or semi-ellipsoid form. Their length varies between 10 and 3000 m, their height is between 5 and 50 m, and their width is variable according to the degree of length (Menzies, 1979; Chorley et al., 1984). Their longitudinal profile shows a greater gradient upstream, coinciding with the greater height, whereas downstream it develops a pointed morphology. Drumlins rarely appear isolated, and usually are grouped in drumlin fields in an *en echelon* pattern, resulting in the so called "basket of eggs" topography (Figure 4.12). They develop basically in areas of old icesheets, where thousands of drumlins occur, as is the case of the centre of Finland, north Ireland, the state of New York, and south Ontario (Benn and Evans, 1998) (Figure 4.13). In general, they are constituted by till, although in some places sands and gravel can be intercalated, or they can even include a rocky nucleus.

There are many works related to the origin of the drumlins that can be synthesized into two groups of theories. One of them promotes the idea of ice erosion over

Figure 4.12. "Basket of eggs" topography caused by ice movement from left to right (after Holmes (1965), Figure 497).

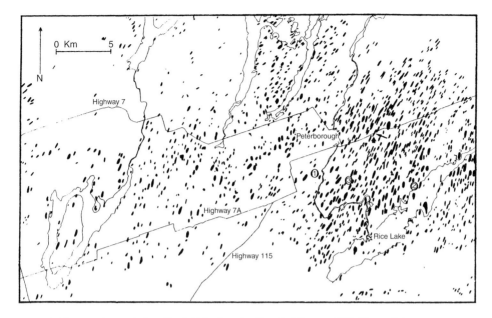

Figure 4.13. Peterborough drumlin field in Ontario, Canada (Sharpe, 1967, Figure 2a).

a pre-existing till mantle, whereas others suggest an accumulation of subglacial till in streamlined forms. In the first case two glaciations or important advances of the ice masses are necessary, in order to explain deposition and then erosion (Embleton and King, 1975).

The existence of rock or block thresholds suggests that they can develop by accretion around these thresholds, because some drumlins show concentric layers. The initial stage of their evolution would be the morphologies of a hill or crag and tail. The form of the drumlin is related to the variations of basal pressure in a moving glacier, as the orientation of clasts seems to indicate. The highest part of the drumlin is the area of low pressure (Figure 4.14), and the particles are carried to these summit areas from the high-pressure, low topographic areas (Evenson, 1971). This interpretation corroborates the fact that in the drumlin fields the mantle of till is thinner in the areas between drumlins. Boulton (1987) explained drumlins according to rheological properties of the subglacial sediments and their water content. As a consequence of glacial stresses, the coarse materials are slightly strained, and fines flow easily. Some interiors can be affected, showing a high degree of deformation, unlike those with a stable nucleus.

The concentration of drumlins in certain areas seems to indicate that there are specific conditions for their generation. Toward the margin of the glacier the ice becomes thinner, and consequently a critical state of dilatancy can be reached in the subglacial till (Figure 4.15a) (Smalley and Unwin, 1968). It occurs when the particles of the till subject to stresses are displaced and disengaged with respect to the others. This process results in an expansion or increase of volume, and a greater strength to shear stress. Not all tills are dilatant materials, because a minimum proportion of blocks are required. Such compact tills resist posterior stresses, in such a manner that the rest of the till flows around it. The C stress is critical for the generation of drumlins (Figure 4.15b). If it is greater than (a),

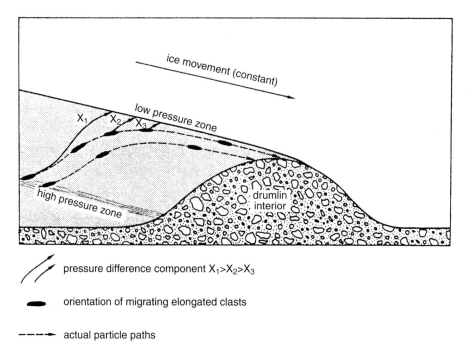

Figure 4.14. Theoretical flow model showing particle migration paths on the flanks of a growing drumlin (Evenson, 1971, in Sugden and John (1976), Figure 12.4).

then the drumlins do not develop, and the till is carried by the glacier. If the stress is lower than (b), then a continuous deformation of the till occurs, finally developing the terminal moraines. According to this hypothesis, drumlins originate upstream of the end moraines.

Nevertheless, in spite of a number of mechanisms argued as the cause of drumlins, the origin of these forms is still problematic. Probably they are polygenetic forms. In addition, no drumlins are known to be forming currently.

Other types of moraines generated in active ice and in subglacial environments are the *valley-transverse moraines* or *Rogen moraines* (Figure 4.11), the latter of which takes its name from a Swedish lake where they were first described. They are moraine ridges, transverse to the ice flow and asymmetrical, with the low-gradient slope located upstream. Their form is arcuate, slightly concave due to the ice movement. Their lengths vary between 100 and 3000 m and their heights between 5 and 20 m. The glacial till shows an intense deformation, with structures imbricated toward the downstream direction (Lundqvist, 1989). Also they can include rock splinters within this fabric (Moran, 1971). All this indicates an intense compressive flow under the ice, causing pressure toward the margin and strongly deforming the basal till, thus originating the transverse ridges (Price, 1973). Another possibility is that the pressure of the till builds the transverse moraine under a water column of a marginal lake (Andrews and Smithson, 1966).

Ice dumped moraines (Figure 4.11) result from the accumulation of material from the ice melt and carried by the glacier in their different environments (Bennet et al., 2000). Consequently, they are deposits that reveal the combination of many processes.

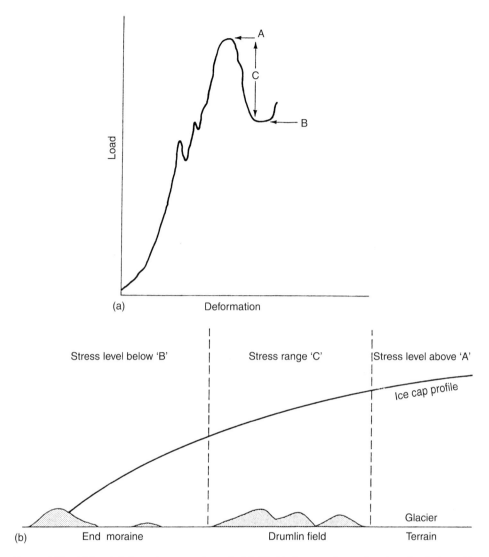

Figure 4.15. (a) Load-deformation curve for a glacial till. (b) Cross section at the edge of a glacier, with critical stress conditions for drumlin development (adapted from Smalley and Unwin (1968)).

The material is of supraglacial or englacial origin; it reaches the surface along shear planes and slides over the ice surface, or is deposited by the ice when it melts (Sugden and John, 1976). These moraines have a ridge form and are located in the glacier marginal areas. These marginal moraines include the deposits carried as medial moraines and lateral moraines (Figure 4.10) that join the end moraines forming large moraine ridges. In these moraines the rock fragments do not show a preferred orientation, although some inherited fabric can be recognized.

Those glaciers that end on land supply large volumes of material to their marginal environments. The quantity of particles deposited depends on the volume of solid material

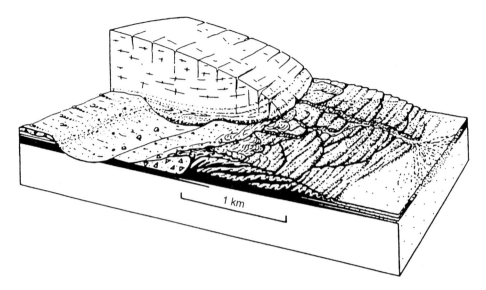

Figure 4.16. Simplified block diagram of the Holmströmbreen push moraine in Spitsbergen, with faults and overthrusting folds. It is composed of fine-grained sediments (black) and till (triangles). Vertical scale exaggerated (Van der Wateren, 1992; in Ehlers (1996), Figure 35).

carried by the glacier. The size of the accumulations also depends on the time that the glacier stays in the same place. Thus, the moraine ridges of the Franz Josef glacier of New Zealand, located in a coastal plain at the foot of the mountains, reached heights of up to 430 m during a stationary stage.

Push moraines (Figure 4.11) are the result of advances of ice at the frontal margins, forcing forward the material previously deposited by the glacier. This bulldozer effect causes the formation of small, asymmetric moraine ridges that do not usually surpass 2 m in height, with the greater gradient on the ice-facing slopes. The deposits are constituted of glacial till, as well as fluvioglacial material. As a consequence of the ice push, imbricated folds and inverse faults can occur (Sharp, 1982; Van der Wateren, 1992) (Figure 4.16). It is also well known that the glaciers tend to advance in winter when melting processes are at a minimum; these annual advances in a glacier in continuous retreat result in small ridges whose spacing is related to the melting balance (Hewitt, 1967; Worsley, 1974); these ridges are known as annual or De Geer moraines (Hoppe, 1959). Glaciers of spasmodic or surging flow can produce large push moraines when they advance, with moraine ridges of much greater size and a pronounced internal deformation.

Finally, when stagnant ice melts, *disintegration moraines* develop (Figure 4.11). In these stagnant glaciers the highest areas are covered by debris that protect the ice from melting, and the particles flow or slide toward the depressions. When the ice disappears an inversion of the relief occurs, and the oldest accumulations of the depressions result in small hills (Boulton, 1972). This landscape constitutes a *hummocky* moraine, characterized by an undulating relief composed of small hills and closed depressions.

Chapter 5
Fluvioglacial erosion and sedimentation

1. The melt waters

The waters resulting from ice melt are a significant part of the glacial system and are the main product of glacier ablation (Figure 5.1). The importance of melt water on the velocity of glacier movement and on particle flow in a subglacial environment has already been discussed. Furthermore, its important erosive and depositional role must be added, especially downstream from the ice masses, where it can cause rapid changes in the drainage network (Price, 1980) (Figure 5.2). Ice melt increases from the equilibrium line and reaches a maximum in the more distal parts of the glacier. Here the fluvioglacial subsystem overlaps the glacial one, in such a manner that the latter losses its predominance against the fluvioglacial processes in the final parts of the ice mass (Sugden and John, 1976).

Melt waters can have a surface or an internal origin, the former being much more important. The proportion of the surface ice melt diminishes with altitude, as other sources increase: summer rainstorms, contributions from tributaries and groundwaters (Benn and Evans, 1998). The basal and internal water derives from geothermal heat that, as an average, is able to melt annually an ice sheet of ~6 mm thickness. The basal sliding and deformation also produce heat and can cause the melting of a layer of 10 to 15 mm thickness (Walder, 1982). Ice melt is maximum in temperate seacoast environments, and diminishes in the high latitudes and in the interior of continents.

Ice melt results in channel developments (Figure 5.3) that occupy very diverse positions inside and out of the ice masses, and even can pass through different environments (Brodzikowski and Van Loon, 1991). According to the thermal regime, the water can flow over the surface or in within the glacier. The size of the channels can range from metric to kilometric scales. In *marginal* (Figure 5.4) and *submarginal channels*, melt waters flow along the contact between the ice mass and the rocky hillslope, and can occupy a lateral or frontal position in relation to the glacier. They are short, rarely exceeding 2 km in length. *Supraglacial channels* reach a great development over the ice sheets, with dendriform networks (Ferguson, 1973). The channels are straight or meandering, with depths ranging between a few centimetres to a few metres (Figure 5.5). Their waters can occasionally disappear into crevasses that progressively enlarge, becoming cylindrical pools called *moulins*, with a glacial karstic topography (Stenborg, 1970).

The system of *intraglacial and subglacial channels* (Figure 5.6) reflects an internal drainage, similar to that of soluble rocks, with large corridors and englacial lakes (Figure 5.3 and Figure 5.7). This series of internal pipes is originated by ice melt, whereas in karstic bedrock they develop by dissolution. The section of the pipes is circular if they develop by forced or phreatic flow, and vertical-elongate if the flow is vadose or free. The enlargement of the internal pipe network occurs mostly in summer, as a consequence of

Figure 5.1. Subglacial melting in an ice front of the Darwin Range, Tierra del Fuego. Photo: J. López-Martínez.

the temperature increase and the greater water supply to the system (Pohjola, 1994). Studies with tracers carried out in Storglaciären (Sweden) reveal that the englacial drainage is of a braided type (Hooke et al., 1988). Sometimes, these intraglacial cavities reach dramatic sizes, as in Antarctica (180 × 45 km), where they have been studied by geophysical methods (Robin et al., 1977). Finally, the *proglacial channels* (Figure 5.8) correspond to melt waters that flow out of the ice mass.

The *discharge* in these fluvioglacial environments varies greatly both in short and long term. Daily fluctuations result in a low discharge in the morning and a sudden increase at the end of the afternoon. Seasonal variations show minimum discharges in winter and maximum in summer (Menzies, 1995b). The discharge in the Argentière glacier, Mont Blanc massif, is 0.1 to 1.5 $m^3 s^{-1}$ in winter and 10 to 11 $m^3 s^{-1}$ in summer (Boulton and Vivian, 1973). Occasionally, dramatic discharges can occur in surging glaciers or due to flashy drainage of subglacial lakes or ice-dammed lakes (Benn and Evans, 1998). These discharges cause extreme floods called *jökulhlaup* in Iceland, characterized by high discharges in a few hours, followed by a fast decrease. One of the best known examples is the lake Grimsvötn, at the west of the Vatnajökull Ice Sheet, located over a volcano in

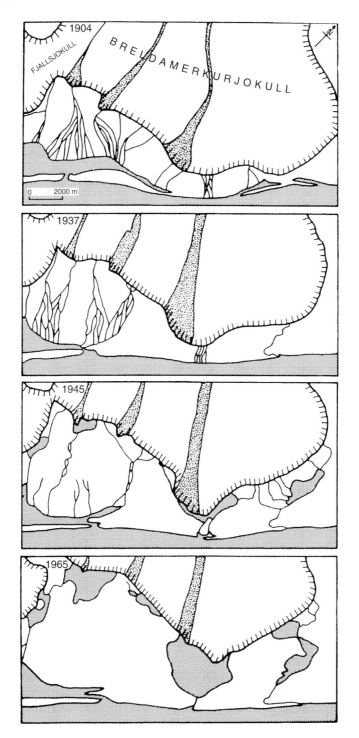

Figure 5.2. The evolution of proglacial drainage at Breidamerkurjökull in Iceland. (Price, 1980).

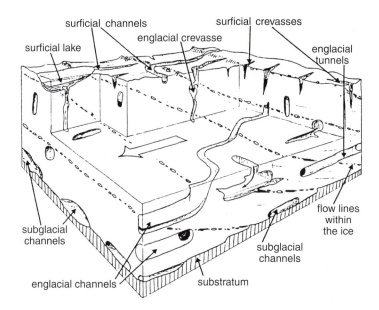

Figure 5.3.   Schematic block diagram showing the different types of channels in a temperate glacier (after Brodzikowski and Van Loon, 1991).

Figure 5.4.   Marginal channel and lateral moraine, Skaftafellsjökull in Iceland. Photo: F. Gutiérrez.

Figure 5.5. Meandering surface channel in South Shetland Islands, Antarctica. Photo: J. López-Martínez.

Figure 5.6. Subglacial channel of the Biafo glacier, located at 3200 m, Karakoram Himalayas of Pakistan. Photo: J. López-Martínez.

Figure 5.7.   Endoglacial channel and lake, exposed by the collapse of the glacier surface. Baltoro glacier, Karakoram Himalayas of Pakistan. Photo: J. López-Martínez.

Figure 5.8.   Proglacial channels, marginal lakes and sandur plain and valley in Skeidararjökull, Iceland. Photo: F. Gutiérrez.

which every 10 years or so flash floods occur with discharges of about 40,000 to 50,000 m^3 s^{-1}, flooding around 1000 km^2 of a sandy-gravel plain (sandur). Its normal discharge is about 400 m^3 s^{-1} (Thorarinsson, 1953; Björnsson, 1992).

In the geomorphic record, there are some outstanding extreme paleofloods in the Pleistocene era, such as those recently discovered in the Altai Mountains of Siberia (Baker et al., 1993), and those of Lake Missoula, in the northwestern United States (Bretz, 1923). This latter is also known as the Spokane flood. In both cases, the peak discharges reached several million cubic metres per second. The floods were triggered by advances of lobes of the ice sheet lobes that blocked the drainages, and formed large ice-dammed lakes. When the dam was destroyed by the ice dam floating upwards or being melted through, a sudden, extreme flood occurred.

Lake Missoula (Baker and Bunker, 1985) (Figure 5.9), dammed by the Purcell lobe of the Cordilleran ice-sheet during the Wisconsin glaciation, was located in northwest Montana and occupied an area of 7700 km^2, with a maximum depth of 600 m and a volume of $\sim$2500 km^3. All these data have been obtained from the now-abandoned lake shore line. The breaking of the ice produced a peak flow of 21.3 million m^3 s^{-1}, that is, about 100 times the average discharge of the Amazon River, or 20 times the discharge of all the world's rivers. The water reached velocities of 70 km hr^{-1} in the channels. The study of the flood deposits indicates the occurrence of 40 to 60 extreme floods between 15,300 and 12,700 yr BP. The flood extended westward as anastomosing channels across

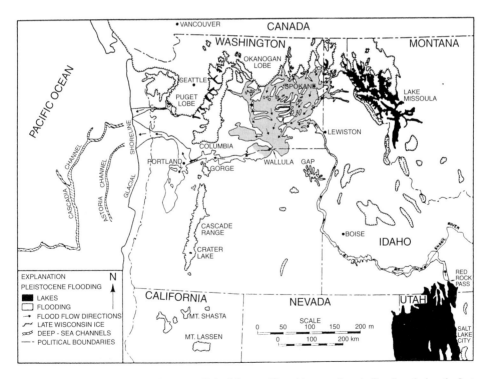

Figure 5.9. Regions of the northwestern United States affected by cataclysmic flooding during the Late Wisconsin (modified from Baker and Bunker, 1985).

the basalt plateau of the Columbia River, and in the Columbia canyon reached 300 m in depth. Over the plains it removed most of the loess cover and caused a landscape of bare, channelled rocks (Channelled Scabland) (Bretz, 1969; Baker, 1981b). Giant ripples have been recognized in gravels, with 100 m wavelengths up to 5 m height, as well as large plunge pools and deep rocky basins of up to 30 m depth. In the Columbia River canyon discharges of 10 millions m^3 have been estimated (Benito, 1997), and dramatic erosive and sedimentary landscapes were built.

2. Forms resulting from fluvioglacial erosion

Glacier melt waters are capable of carrying a large sediment load and are an important erosive agent. The velocity of the discharge is high, with common values of 8 to 15 m s^{-1}. When they are subject to high hydrostatic pressures they can erode even the rocky substrate (Drewry, 1986). These melt waters carry great quantities of suspended sediment and bedload. Thus, in Norwegian glaciers, suspended sediment can reach values of 3800 mg l^{-1}, and the bedload can be 25 % of the total sediment load during the summer peak flows. In addition a small dissolved load is mobilized, even surpassing the suspended load in winter, as in the Argentière glacier (Sugden and John, 1976).

Several microforms have been attributed to the ice melt waters. They are known as *p-forms* or plastically moulded forms. They have been described in the section on glacial erosion, because there is some discussion as to their interpretation as glacial or fluvioglacial forms, although this latter hypothesis seems to be more accepted (Allen, 1971).

The origin of p-forms can be estimated using a series of criteria: (1) If these p-forms are found on the glacier bed, in an improbable position for the trajectory of the subglaciar melt water, they can thus be of mechanical origin; (2) If the glacier bed is massive and the p-forms do not originate from pre-existing fractures, they may have a fluvial origin; (3) If the striations around the p-forms indicate ice flowed into and along the p-forms, then mechanical ice erosion has modified them (Rea et al, 2000).

The giant potholes developed in the bed of the ice melt channels are also characteristic of this environment, reaching depths of up to 20 m and widths of 15 m. In general, they are wider than deep, and show helicoidal grooves caused by gravel in a turbine-type movement (Kor et al., 1991; Dionne, 2004).

The most outstanding features of fluvioglacial erosion are the channels incised on the rocky substrate or on the loose material produced by glacial activity. The *marginal, lateral or terminal channels* can incise in the ice or on the substratum that, in general, is a glacial till. They use to be short, rarely exceeding 2 km in length. The channels subparallel to the contour levels are interpreted as marginal, though this is problematic, because they can be subglacial channels that flow close to the glacial margin (Price, 1973). The *subglacial channels* can incise up to 100 m, with 1 to 2 km width and 75 km length (Selby, 1985). They once had abrupt walls and their beds are flat and plain with some irregularities. In general, they end suddenly, and their terminus is supposed to correspond to the outlet, where an alluvial fan could be developed. They are interpreted as subglacial when they occupy anomalous positions within the drainage system. Where the ice margins have adequate gradients to evacuate the melt waters, then the channels emanate from the glacial border. These *proglacial channels* spill the waters into higher-order rivers, into lakes (Figure 5.10), or into the sea (Figure 5.11), producing considerable erosive work along

Figure 5.10. Proglacial lake. Jökullsarlon, Iceland. Photo: F. Gutiérrez.

Figure 5.11. Proglacial channels in Marambio Island, flowing into the Weddel Sea. Note in the background some icebergs. Antarctica. Photo: A, Martín-Serrano.

Figure 5.12.   Glaciotectonic structures in fluvioglacial deposits. Isaba, Western Pyrenees, Spain.

Table 5.1. A classification of fluvioglacial deposits and forms (Price, 1973).

Dominant sediment	Environment	General form	Relationship to ice	Genetic term
		Ice-contact deposits		
Sand and gravel	Fluvial	Ridge	Marginal, subglacial Englacial, supraglacial	Esker
Sand and gravel	Fluvial	Mound	Marginal, subglacial Englacial, supraglacial	Kame Kame complex
Sand and gravel	Fluvial	Spread with depressions	Marginal	Kettled sandur
		Proglacial Deposits		
Sand and gravel	Fluvial	Spread	Proglacial	Sandur
Silt and clay	Lacustrine	Spread	Proglacial/marginal	Lake plain
Sand and gravel	Lacustrine	Terraces, ridges	Proglacial/marginal	Beach
Clay, sand, and gravel	Lacustrine	Terrace	Proglacial/marginal	Delta
Sand and clay	Marine	Spread	Proglacial/marginal	Raised mud flat
Sand and gravel	Marine	Terrace, ridges	Proglacial/marginal	Raised beach
Clay, sand, and gravel	Marine	Terrace	Proglacial/marginal	Raised dalta

their courses. The *overflowing channels* that drain the lakes occur in relation to glaciolacustrine beach or delta deposits.

3. Forms resulting from fluvioglacial sedimentation

It is well known that one of the characteristics of the fluvioglacial environment is the large discharge fluctuations and, consequently, the transport capacity. These fluctuations result in sudden changes in the particle sizes and rapid variations of sedimentary structures, both laterally and vertically. In addition, they explain why many of the fluvioglacial forms have a discontinuous character. This is emphasized by the continuous erosion that affects these forms in areas of retreating glaciers (Menzies, 1995b). On the other hand, the fluvioglacial materials can be mixed with deposits of glacial till in some places, and the fluvioglacial sediments commonly are generated from reworking of glacial till (Flint, 1971). Therefore, this is an environment subject to continuous and deep changes, where the morphology is modified by erosion and sedimentation processes.

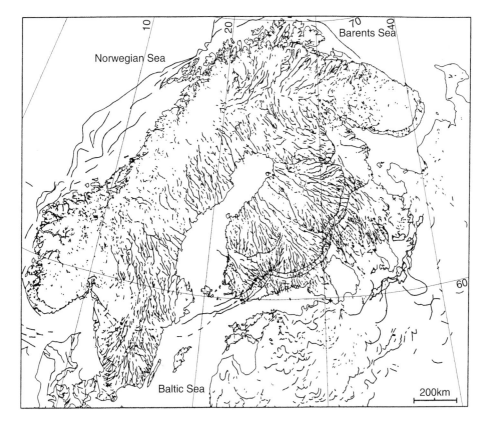

Figure 5.13. Eskers and ice-marginal features in the central part of the Scandinavian Ice Sheet. Eskers form well-organized patterns and are less common in areas of sedimentary rock, where subglacial water drains through permeable beds, (Punkari, 1997).

   Transport and sedimentation mechanisms are similar to those occurring in other fluvial, lacustrine, and marine environments. In the fluvioglacial channels, ripples, dunes, and cross stratification are common in sandy deposits. These alternate with gravels where the flow energy is greater. The sorting of the deposits is a consequence of the seasonal, annual, and even daily fluctuations of the discharge. The particles of silt and clay size are carried longer distances, and they are deposited in glaciolacustrine or glaciomarine environments (Price, 1973).

   Lacustrine sedimentation occurs in subglacial, marginal, and proglacial lakes. The deepest layers of sedimentation represent the distal sedimentation of fine particles, whereas the frontal and upward layers represent the supply of coarser proximal material.

   The deposits of fluvioglacial channels are composed of rounder clasts than the till clasts, although this differentiation is not a conclusive criterion, because some fluvioglacial deposits have undergone a short transport. On the other hand, the grain size distribution is very different from the till deposits. It is not bimodal, because the silt and clay particles have been carried away, unlike the tills.

   Fluvioglacial sediments commonly show *glaciotectonic* deformation in their original fabric. These secondary structures are related to collapse and subsidence due to the melting of ice buried by the deposits (Figure 5.12), resulting in vertical stratifications, folds and faults. Also diapiric structures occur, as well as dike injections, convolutions, and so forth, in relation to saturated fine-grained sediments affected by overloading (Van der Wateren, 1995).

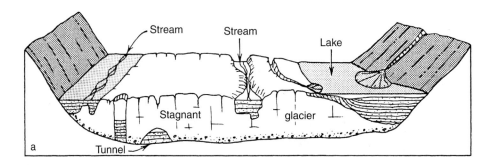

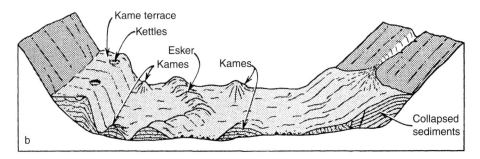

Figure 5.14. Fluvioglacial forms developed in contact with ice. (a) Ice-melt stage. (b) After glacial retreat (according to Flint, 1971).

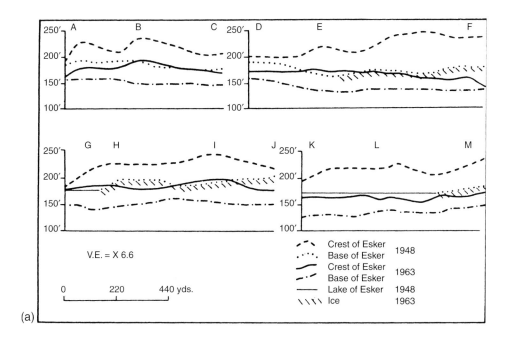

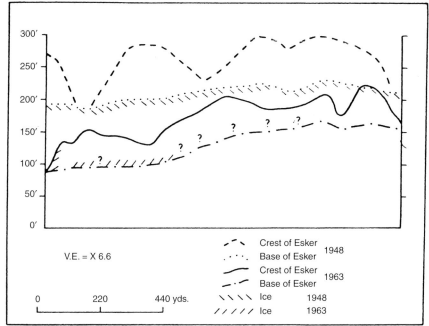

Figure 5.15. Profiles of Eskers in the Casement Glacier in Alaska (a) and in Breidamerkurjökull in Iceland (b) (Price, 1969).

Fluvioglacial deposits can be classified according to the sedimentation environment: deposits due to ice melt channels, and deposits in confined waters, such as lakes and seas. Also a classification can be established according to their position in relation to the ice mass (Table 5.1). The proglacial sedimentary deposits occur at a certain distance from the ice margin, generating smoothed morphologies, and the sedimentary structures are not deformed. The deposits located in contact with the ice show distorted structures and their morphologies are produced by the ice mass or by a later settling in relation to the melting of buried ice. They are, in addition, classified according to the prevailing sediment and the form of the accumulation.

Eskers (*osar* in Scandinavia) are ridges composed of stratified deposits with rounded gravels and sands with some blocks. Their direction is parallel to the regional flow of the ice. They are sinuous or straight, sometimes discontinuous, with sharpened or flattened

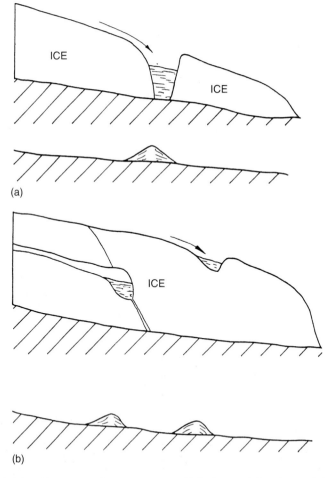

Figure 5.16. Origin of kames. (a) Crevasses open until the ice–rock interface and development of a cavity in which sediments accumulate. (b) Supraglacial and endoglacial cavities produce similar forms (modified from Price, 1973).

crests, and lengths from a few metres to 400 km (Aylsworth and Shilts, 1989; Punkari, 1997) (Figure 5.13). Their height can reach 200 m, with 3 km in width. These latter dimensions are related to the total length, in such a manner that the longest eskers are commonly the widest and highest. They can also appear as unique ridges or develop an interconnected network with confluent and diffluent eskers that join the main ridge. The slopes of the eskers can have angles of about 30°, close to the angle of repose, though when they are degraded the gradient is between 5 and 10° (Embleton and King, 1975a). The deposits dip toward the margins of the esker and can show distortions due to the underlying ice melt.

Eskers originate in subglacial, englacial, supraglacial, and ice-marginal channels (Figure 5.14). They develop above all after the peak flow at the end of summer, when the velocity of flow of the channels diminishes, and can be destroyed totally or partially with new peak flows (Sugden and John, 1976; Menzies and Shilts, 1996). The deposits of the bottom of the subglacial channels can accumulate between the ice walls, and the shape of the cross section of the tunnels plays an important role in the final esker morphology (Price, 1973). The hypothesis of subglacial channels hardly explains the existence of tunnels hundreds of kilometres long. The origin from englacial channels must be treated jointly with that of a supraglacial origin. Studies of aerial photos obtained in different years made in areas of fast ice retreat of the Casement Glacier in southeast Alaska and the Breidamerkurjökull Ice Sheet in Iceland reveal that the profiles of the esker base and crest have diminished in height (Price, 1966, 1969) (Figure 5.15). This diminution is due to the melting of buried ice during the period between the aerial photos. This indicates an englacial or supraglacial origin.

Figure 5.17. Kame terrace affected by badlands. Talon River, Jura of France. Photo: J.L. Peña.

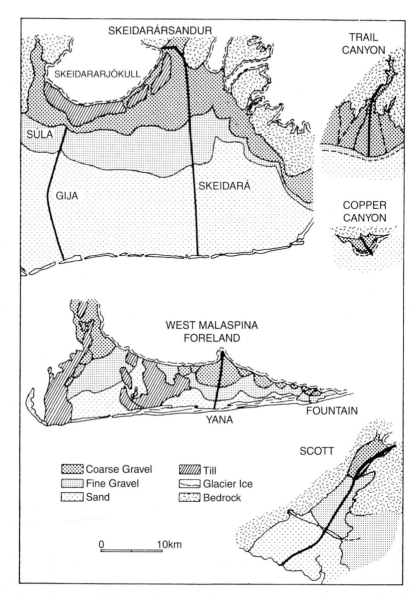

Figure 5.18. Grain-size distribution in some sandar. Skeidarásandur is in Iceland, and the Scott and Malaspina systems in Alaska (modified from Boothroyd and Nummedal, 1978).

Kames are small hills composed of deposits of stratified sands and gravels. Their size varies from a few metres to conical hills of up to 50 m in height and 400 m in length. They form by sediment accumulation in supraglacial and englacial cavities (Johnson and Menzies, 1996) (Figure 5.16). The term kame is also used to describe forms built in a particular position within or along the glacier. When a channel flows into a lake located on the ice mass, it carries sediment deposited in the margins; after the ice melt, they result in a *kame delta* (Figure 5.14). *Kame terraces* are formed by accumulation of channel deposits that run between the valley wall and the lateral ice border (Figure 5.17). The side of the deposit located close to the ice shows a steep border due to ice melt. A *kame complex* is caused by accumulation of sediment in supraglacial depressions (karstic glacial topography) and later settling by ice melt. The kames are modified by subsidence if they have underlying ice, and can generate subcircular depressions called *kettle holes*. When small hills are associated with depressions, then the term *kame and kettle topography* is used, equivalent to the kame complex (Gray, 1991).

The discharge of large volumes of water from the ice margins causes the transport and sedimentation of particles far away from the ice borders in different environments and produces proglacial deposits (Table 5.1). In these areas, fluvioglacial sedimentation prevails because of the decrease in the transport capacity of the ice melt water. The discharge is associated with a network of braided channels of great lateral mobility, depositing gravels and sands that build large depositional plains called *outwash* or *sandur* (in plural *sandar*) (Maizels, 1995) (Figure 5.8). They are similar to alluvial fans, although with a great influence of strong seasonal fluctuations. The grain sizes in the sandar diminish from proximal to distal areas (Boothroyd and Nummedal, 1978) (Figure 5.18). The longitudinal profiles are concave, whereas the cross sections are, in general, convex. Sandur morphology is continually modified by extraordinary high-flow events as well as daily variations in flow (Vandenberghe and Woo, 2002). The *sandur valleys* develop where they are confined by glacial U-shaped valleys (Figure 5.19). Those developed in the margins of ice sheets are called *sandur plains* (Figure 5.8). The proximal area in a sandur has few ice-melt channels and is characterised by an area of gravels with a number of closed depressions, resulting from the melting of buried ice; this is known as *kettled sandur* or *pitted outwash* (Price, 1969) (Figure 5.20). Far away from the ice mass, the

Figure 5.19. Myrdalsjökull Ice Sheet in Iceland and sandur valley. Photo: F. Gutiérrez.

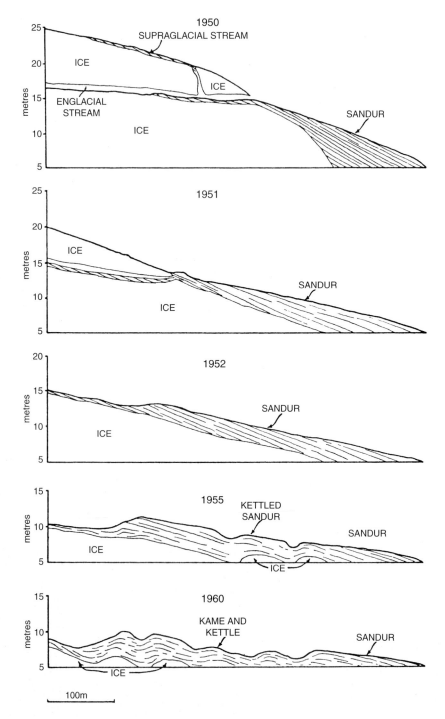

Figure 5.20. Sketch showing the forms resulting from sedimentation by supraglacial and englacial streams. The sequence has been obtained from aerial photos in Breidamerkurjökull in Iceland (adapted from Price, 1969, Figure 80).

number of braided channels increases. They are shallow (1 to 2 m in depth), change frequently in position and commonly are flooded during high discharges. Finally, the distal area has proglacial lakes. During the stages of glacial retreat, the proximal areas of the sandar are eroded and new sandar develop from the previous ones.

The presence of lakes is very common in fluvioglacial environments, particularly in marginal and proglacial positions. The discharge of materials in the lakes results in *glaciolacustrine beaches* and *deltas*. If the lake is drained afterwards, then the bottom deposits make up *plains of glaciolacustrine origin*. Some of these deposits have a rhythmic character and show seasonal variations. They are called *varves* and have been used in absolute dating with diverse results. Each varve is composed of two layers, one of coarser sand deposited in summer and the other of finer clay and silt that is deposited during the rest of the year. The value of the method of counting varves for chronology construction has been subject to criticism, because in some years more than a pair of layers can deposit, whereas other years may not have a sedimentation record due to discharge variations or deviations of the channels that flow into the lake (Smith et al., 1982).

Because tributary valleys are frequently dammed by the main valley glacier (Figure 5.21), the moraine deposits can appear interbedded with fluvial and lacustrine marginal deposits (Bordonau, 1993) (Figure 5.22). In glaciers close to the sea, the ice melt water discharged in estuaries, fiords and bays causes the accumulation of a great quantity of glacial and fluvioglacial sediment (Figure 5.23). If this sediment is uplifted as a result of glacioisostatic rebound after a deglaciation, then it can result in *raised beaches and deltas*, and *exposed glaciomarine bottom muds*, with similar morphologies to the glaciolacustrine deposits. The identification of these deposits as marine is made by their fossil content.

Figure 5.21. Lateral moraine and dammed depressions in the Miera valley of Cantabria, Spain.

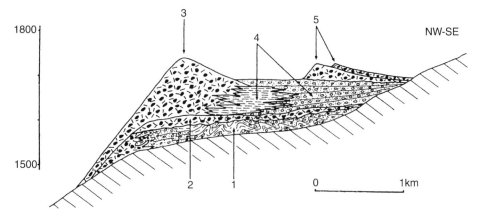

Figure 5.22. Cross section of the Cerler lateral moraine complex, Ésera valley, Central Southern Pyrenees, Spain. (3) Lateral moraine. (1 and 2) Older deposits. (4) Lacustrine deposits with interfingering deposits as a result of glacial damming of a tributary valley. A later glacier advance in the tributary valley deposited younger end moraine ridges. (5) All deposits related to the last glaciation. (Reprinted from Bordonau, 1993).

Figure 5.23. Ice melting and sediment discharge in a glacier front, South Shetland Islands, Antarctica. Photo: J. López-Martínez.

Chapter 6

Geomorphology applied to glacial regions

1. Introduction

Applied problems in glacial areas derive, on the one hand, from the activity of the actual ice masses and, on the other hand, from the characteristics of the materials deposited by the glaciers and meltwaters in past periods. Some of the functional processes that develop from ice accumulations result in different geomorphic processes that will be analysed below. From another point of view, these areas produce a number of benefits to human communities. Those derived from tourism, such as landscape observation, skiing, trekking, climbing, and other sports have undergone a considerable increase in the last few decades. Summer ice melt in mountain areas leads to the exploitation of the water falling in order to produce hydroelectric power. In addition, glacial melt water is a very important source for irrigation in large arid zones of the world, such as the Thar Desert in northwest India, irrigated by the Indira Gandhi and Great Bhakra Canals, the Peruvian–Chilean littoral desert, and the areas of Mendoza and San Juan at the foot of the Argentinean Pre-cordilleran Andes. All the discharges for these irrigation systems come from snow and ice melt in the neighbouring mountain ranges.

Other important resources derive from the exploitation of the groundwater contained in the Quaternary deposits of glacial origin. Thus, in North America, a large part of domestic and industrial water is taken from fluvioglacial aquifers (Thornbury, 1954). One of the most outstanding features of these accumulations is the variability in grain size and depth, and consequently the different layers show different transmissivity, permeability and storage capacity values (Derbyshire and Love, 1986). Till is a poor aquifer due to its clay content, whereas gravel and sand deposits are excellent groundwater reservoirs (Lloyd, 1983) (Figure 6.1). The common lateral variations in the characteristics of these sediments cause the occurrence of barriers for water flow, and they impede detailed knowledge of the aquifer. In some cases, the presence of preglacial valleys buried and fossilized by these sediments can result in areas of great water storage. For this reason, a precise knowledge of the geomorphological history of any such region is necessary.

2. Hazards caused by the activity of ice masses

2.1. Icebergs

Icebergs are a consequence of the breaking of ice masses, especially ice sheets where they arrive at the ocean (Figure 6.2). They are a risk for shipping and also for the oil drilling platforms. This latter circumstance has encouraged studies on the forecasting of iceberg

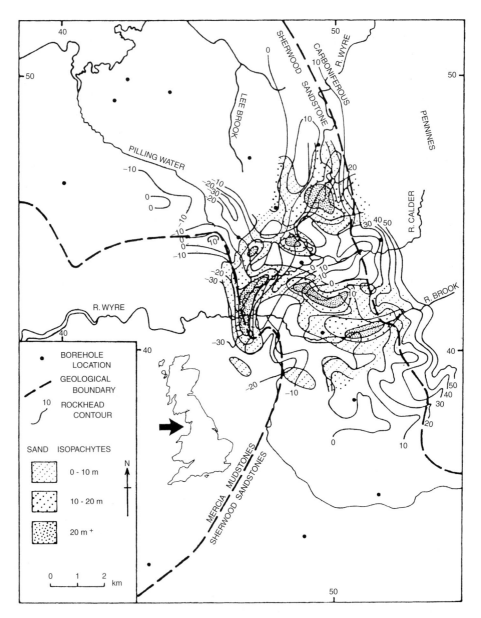

Figure 6.1. Structure contour map of the top of the Sherwood Sandstone with isopach lines for overlying glaciofluvial sands in the Fylde areas of Lancashire (Lloyd, 1983, in Eyles, 1983, Figure 15.8).

movements along the margins of the eastern Canadian coast (Marko et al., 1988). They can travel for thousands of kilometres before melting. Their movement is controlled by the surface oceanic currents, although it can be affected by deeper currents or by the wind.

Along the eastern coast of North America the icebergs come from the Greenland Ice Sheet and travel with the Labrador Current. About one hundred icebergs per month cross

Figure 6.2. Icebergs between Newfoundland and Iceland (Photo: F. Pueyo).

the latitudinal lines during the period from March to August, although they rarely go far south of 50°N (Figure 6.3). In Antarctica the icebergs travel parallel to the continent from east to west until they strike currents coming from the east that move away them from the coast (Hambrey and Alean, 1992).

2.2. Ice avalanches

The study of ice avalanches is very important, because they affect villages, roads, ski resorts, and hydropower plants. They occur in mountain areas and are much more rare than snow avalanches, although in some places they can be common and are a tourist attraction, such as the steep face of Balmhorn Glacier in Switzerland (Hambrey and Alean, 1992). Research carried out with aerial photos enables knowledge of their maximum runout distances, which is most valuable information for evaluating the risk. The maximum extent of an ice avalanche depends on the terrain roughness and the season (McClung and Schaerer, 1993). Thus, the velocity is greater in winter because the avalanche displaces over a snow cover that facilitates the displacement; the opposite occurs in summer, when the friction increases and the movement is less. Ice avalanches are forecast by new cracks and ice falls some weeks before the main event is triggered. In some cases, ice avalanches rise hundred of metres on the opposite slope.

The oldest known ice avalanche occurred in the Valais Canton of Switzerland, on 31 August 1597, where an ice mass falling from the Balmen Glacier affected the village of Eggen, which was buried with 81 inhabitants (Table 6.1). In the same canton, on 30 August 1965, an ice mass fell in the Mattmark-Saas valley during the construction

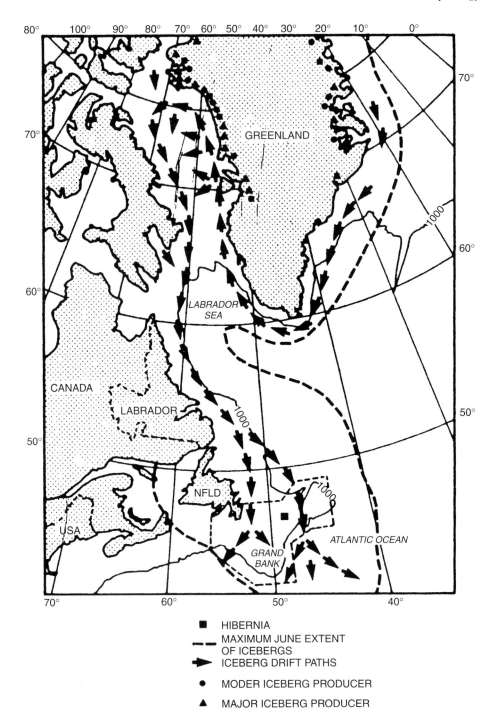

Figure 6.3.   The pattern of iceberg circulation off eastern North America (Marko et al., 1988, en El-Sabh and Murty, T.S. (eds.), 1988, p. 437, Figure 1).

Table 6.1. Some disasters related with glaciers in Switzerland (excluding those affecting mountaineers) (Hambrey and Alean, 1992).

Year	Location	Event	Fatalities
1595	Glacier de Giétro	Lake outburst	160
1597	Balmengletscher	Ice avalanche	81
1636	Weisshorn	Ice and snow avalanche	37
1782	Altels	Ice avalanche	4
1818	Glacier de Giétro	Lake outburst	40
1819	Weisshorn	Ice and snow avalanche	2
1895	Altels	Ice avalanche	6
1965	Allalingletscher	Ice avalanche	88

of a hydroelectric dam. The ice avalanche came from the Allalin Glacier tongue and moved a million cubic metres of ice, burying the construction area and causing the death of 88 workers (Hambrey and Alean, 1992; McClung and Schaerer, 1993). This disaster encouraged research on ice avalanches, especially on their triggering mechanisms. Since then, an estimation of the dam strength is made against the waves generated in the reservoir as a consequence of avalanches. The ice avalanche that occurred on 14 August 1949 on the Tour Glacier in the French Alps is well known; it caused the fall of the lower, very steep part of the glacier, displacing about 3 million cubic metres that led to the death of six hikers. The causes are interpreted as a result of the weakening of the terminus of the glacier due to alternating hot days and cold nights, resulting in important melting and re-freezing (Vivian, 1979).

If ice avalanches are relatively common in the Alps, they reach greater volumes and catastrophic effects in the Andes. In the Peruvian Cordillera Blanca some peaks contain glaciers above 6000 m, and Huascarán (6654 m) is the highest (Figure 6.4 and Figure 6.5). The Rio Santa valley located in this area was affected by two large ice and rock avalanches. The first, in 1962 caused 4000 deaths, and the second, of greater dimensions, was triggered by an earthquake of a 7.7 magnitude on the Richter scale on 31 May 1979. In this second case 50 million cubic metres of ice, rock, debris, and water were mobilized, reaching velocities of around 280 km h^{-1}, and covering 16 km of the valley bottom in 3 min. The village of Yungay was buried and between 18,000 and 20,000 people died (Figure 6.6). This avalanche seems to have been the most important in historical times because of its greater runout, velocity, and volume, and must be considered as an ice and rock avalanche (Plafker and Ericksen, 1978).

2.3. Glacial dammed lakes

These marginal lakes can originate at the junction of two valley glaciers or when a glacial tongue penetrates into an ice-free tributary valley. As a consequence of damming, the lakes are progressively filled up, particularly during the summer. If the lake is completely filled up, then the water overflows by supraglacial or ice-marginal channels. Another possibility occurs when the ice dam suddenly floats upwards and breaks, resulting in flash floods downstream. These breaks are relatively unpredictable.

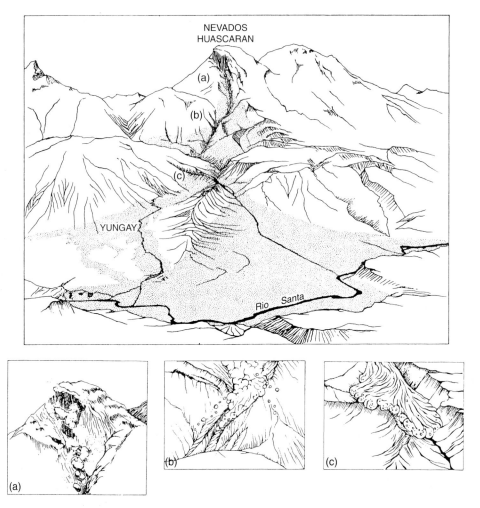

Figure 6.4. Line drawing showing the ice and debris avalanche (shaded) in Huascarán, Peruvian Andes, on May 31, 1970. (a) Diagrammatic representation of the fall. (b) Air-launching of debris near terminus of moraine ridges. (c) Splitting of the debris into the Yungay and Ranrahirca lobes (modified from Plafker and Ericksen, 1978).

The dammed lakes used to be more common in the Polar and subpolar regions and drained throughout overflowing channels, whereas in the temperate-ice dams sudden breakings are relatively common (Figure 6.7). These lakes of the temperate latitudes are smaller than those of the Polar Regions, but their proximity to inhabited areas makes the hazard higher, with very important damages (Hambrey and Alean, 1992).

Ice-dammed lakes sometimes drain periodically, thus indicating that water depth and their consequent hydrostatic pressure are the main factor controlling the drainage of the lake. Several mechanisms of ice-dam breaking have been proposed. When the hydrostatic pressure of the water in the lake is higher than the ice-dam pressure, then the basal breaking of the ice body occurs and the lake drains through a subglacial outlet. In other

Figure 6.5. The Huascarán from the Rio Santa Valley. The 1962 and 1970 catastrophic avalanches started from the rocky wall of the North Peak (left). Cordillera Blanca, Peru (Photo: J. López-Martínez).

Figure 6.6. Area where the village of Yungay was located (Cordillera Blanca, Peru) and buried by a catastrophic debris avalanche in 1970 (Photo: J. López-Martínez).

Figure 6.7. Lake dammed by the Perito Moreno Glacier of Argentina, where periodic slides and overflows occur.

cases the breaking is linked to an enlargement of cracks due to the ice flow and the high hydrostatic pressure. Finally, the channels located on the ice mass enlarge progressively, weakening the strength of the ice dam. In addition, seismic movements increase the instability of the system (Costa, 1988).

These sudden outputs of glacial waters result in extreme floods, called *jökulhlaups* in Iceland, which mobilize large quantities of sediment and build the sandar or gravel and sand plains (Maizels, 1997). Commonly this activity is catastrophic. Two types of hydrographs can be recognized in these floods (Costa, 1988a). One shows a sudden raising limb, with an acute peak flow, almost impossible to be measured adequately, followed by an abrupt decreasing limb. The time interval between the beginning of the discharge increase and the end of the decrease ranges from some minutes to hours. This hydrograph corresponds to a sudden breaking of the ice dam. The other type of hydrograph (Figure 6.8) shows a progressive increase, with an acute peak flow and a sharp decrease, corresponding to the reservoir exhaustion. The time interval between the beginning and the end of the discharge ranges from hours to days. The characteristics of this hydrograph suggest the occurrence of floods caused by the enlargement of englacial tunnels or overflowing of lakes (Church, 1988).

Most classical, or the better-studied examples of jökulhlaups occur in Iceland, although in this country the volcanic activity interacts with the glaciers. The most dramatic cases are related to the united activity of both phenomena and will be analysed later. It is interesting to note, however, that during the recent retreat of the glaciers, after the Little Ice Age (Grove, 1988), jökulhlaups have been more frequent but with a lower discharge, due to the thinning of the ice dams.

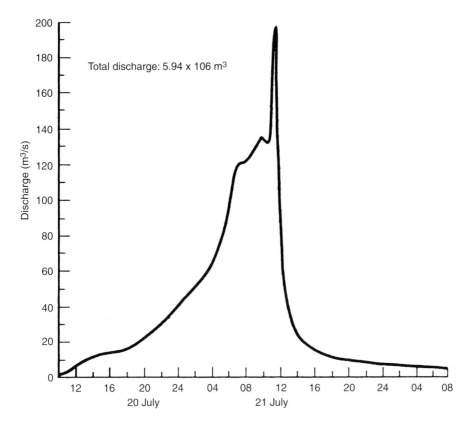

Figure 6.8. Jökulhlaup of July 1967, in Ekalugad Fjord, Baffin Island, generated by overflow of an ice dam (partially modified from Church, 1988).

A number of cases of glacier dam breakings are known in the Alps. Since the year 1600 floods have been recorded in the Oetz valley of the Austrian Tyrolean Alps (Figure 6.9). The tributary Vernagt Glacier penetrates into the river, causing the development of a lake that has broken many times (Grove, 1988; Gerrard, 1990). These breakings are also recognized in the Allalin Glacier of Switzerland and in the Gietro Glacier, with catastrophic floods throughout its history, with 140 deaths in 1595 and 50 in 1818 (Vivian, 1979).

Likewise, a wealth of documentation exists on sudden breakings within alpine glaciers. The most outstanding known catastrophe occurred on 12 July 1892, as a consequence of the breaking of a subglacial cavity in the Tête Rousse Glacier on the west-facing slope of Mont Blanc; the floods caused 175 deaths and great material losses. These breakings are also known in the Miage Glacier, with a number of floods in the last two centuries, and in the Trient Glacier, where they occur with a periodicity of 3 to 5 years in July or August (Vivian, 1979).

Other types of dammed lakes are caused by end moraines that are filled up by melt and rain waters (Figure 6.10). In these lakes the glacial till is easily eroded during the fast ice melt or rainstorms. The overflowing channel quickly erodes the deposit and consequently a rapid increase occurs in the discharge. Holes are drilled to reach the lake bottom and

Figure 6.9. Dates of disasters caused by the Fischbachenchängenfels. Oetz valley, Tyrolean Alps of
Austria (Photo: F. Gutiérrez).

Figure 6.10. Glacial-origin lake, dammed by an terminal moraine arc. Cordillera Blanca, Peru (Photo:
J. López-Martínez).

drain the lake in order to avoid the spilling and subsequent destruction of the moraine dam. A catastrophic destruction of an end-moraine lake occurred in the Santa Ana river valley of the Peruvian Andes on 15 March 1941. It resulted in a large flood that destroyed the village of Huaraz, causing the death of 6000 people (Hambrey and Alean, 1992). Numerous such examples are also known in the Himalayas as well.

2.4. Volcanism and glaciers

When the heat liberated by volcanic activity affects glaciers, a sudden ice melt can occur. The resulting water, mixed with volcanic ash, results in *lahars* or volcanic mudflows that move by gravity, affecting human settlements and activities located at the foot of the volcano. If the melt waters achieve large volumes, then quite dramatic floods can be triggered. Three examples from different parts of the world serve to illustrate the dramatic development of lahars and their significant consequences in both human lives and economic losses. The crater lake of Ruapehu volcano (2797 m), located on the North Island of New Zealand, broke out along with significant ice and snow melt on Christmas eve, 1953, causing a lahar that destroyed a railway bridge immediately before the passage of a train. The locomotive and five cars fell into the river and 151 people died (Francis, 1993).

The Mount St. Helens volcano (2950 m), located in the state of Washington, in the Cascades Range, USA, begun its eruption on 18 May 1980, though its activity had started 2 months before with earthquakes and small phreatic eruptions. The volcano was subject to close vigilance in hopes of avoiding greater problems (Crandell et al., 1975; Wesson, 1983). The big eruption caused by the landslide failure of the north face of the mountain, lowered the height of the volcano ~ 400 m and triggered large ice avalanches. The ash resulting from the explosion covered the neighbouring glaciers, which advanced in the following few years as a consequence of the protection against solar radiation. On the other hand, large lahars of up to 30 km in length developed, destroying houses and forests. Some dams restrained them, and in other cases the trees contributed to damming the waters in the bridges, resulting in floods when the bridges broke. Fifty-seven people died as a consequence of this eruption, and the economic losses were estimated to be about $1000 million dollars (Lipman and Mullineaux, 1981; Hickson and Peterson, 1990).

The Nevado del Ruiz volcano (5400 m), located in the Colombian Andes at 5° north latitude, has an ice field in its summit. On November 13th, 1985, it had a low magnitude eruption, with the emission of a few pyroclastic products. The eruption, as in previous cases, was forecast 1 year before due to seismic activity and one phreatic explosion. In spite of the low energy of the eruption, ice and snow avalanches were triggered, and the hot ash melted ice masses; the resulting waters carried a large quantity of recent and old ash, causing lahars or volcanic mudflows running toward the densely inhabited valleys (Figure 6.11). The lahars developed in two main stages. On the east slope they were channelized by the Lagunillas river, with velocities of 30 to 40 km h^{-1}, until at midnight they reached the city of Armero, located 60 km away from the volcano. The village was buried in a few minutes by the lahars and 22,000 people died. On the west-facing slope in the locality of Chinchiná, mudflows killed 2000 people (Parra and Cepeda, 1990; Voight, 1990).

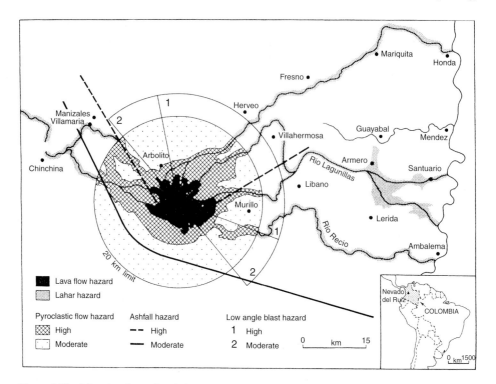

Figure 6.11. Map showing volcanic hazards on Nevado del Ruiz of Colombia, compiled in November 1985. According to Parra and Cepeda (1990) and Voight (1990) (in Chester, 1993).

An increase of temperature at the base of temperate or polar glaciers due to geothermal heat results in melting of the ice mass and, as a consequence, the development of subglacial lakes. Probably most of them are located under the Antarctic Ice Sheet (Hambrey and Alean, 1992). The largest of all known subglacial lakes is Lake Vostok, below almost 4 km of ice at the Russian Antarctica Station of the same name. The lake has an area of 15,000 km^2, an average depth of 125 m or more, and a considerable accumulation of sediment in the bottom (Kapitsa et al., 1996). The lakes under the small Iceland ice sheets have a well-documented history of volcanic activity under the glacial ice and their consequent floods. Ten percent of these ice sheets cover volcanically active rift areas.

Near the middle of the Vatnajökull Ice Sheet, the Grímsvötn subglacial lake is located in an active geothermal area where a number of eruptions have occurred in historical times (Thorarinsson, 1953). When the lake breaks very large floods (jökulhlaups) are triggered on the Skeidarársandur plains (1000 km^2 with a gradient of 1/200) at the south end of the ice sheet, thus building the greatest of Icelandic sandur. In past times an average of 12 floods every 10 years has occurred, and the mean volume of each flood has been 3 to 3.5 km^3, with their maximum discharge being 10,000 m^3 s^{-1}. Each jökulhlaup has carried about 30 million tons of sediment (Bjornsson, 1979). These floods can be triggered by an increase of geothermal heat or by an eruption, like that of 1934, wherein the total sediment transported was 150 million tons. It is very difficult to plan a permanent road network in the Skeidarársandur as a consequence of the high frequency of these floods.

Under the Myrdalsjökull Ice Sheet the Karla volcano has an eruptive activity well known since 1625. The last eruption was in 1918 and the discharge reached 200,000 m^3 s^{-1}. The southern coast advanced 200 m as a consequence of this eruption, and a progradation of 2.2 to 2.5 km has been estimated since 1660 (Bjornsson, 1979).

3. Problems derived from glaciotectonics

Deformations caused by glacial activity can produce locally severe disturbances in some geological structures, resulting in folds, fractures, and genesis of breccia in the stratigraphic sequences, or in a reduction of the mechanical resistance of rocks and sediment. Therefore, it is very important to have a detailed knowledge of these deformations, because they can have consequences to ore mineral surveys, mining works, civil engineering, and soil mapping.

In some mines of western Canada important glaciotectonic activity has been recognized (Aber et al., 1989): about 12,000,000 m^3 of coal have been mobilized by overthrusting and a large depression was generated, and later covered with sediments (Figure 6.12). Because the coal has been relocated closer to the surface, it has oxidized, thus reducing its caloric value, and in addition it has increased in water content, so for these reasons it is not exploited.

In other cases the substrate can be affected by glaciotectonic deformation, with a consequent decrease in shear strength. Under these circumstances, slides of weakened materials can affect roads and bridges, as occurs with the Upper Cretaceous argilites in the South Saskatchewan River valley of Canada (Krahn et al., 1979).

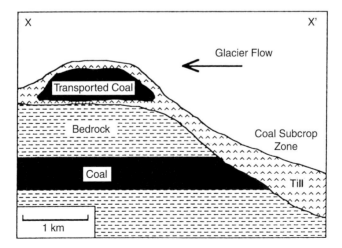

Figure 6.12. Schematic cross section showing the depression formed by glaciotectonism and the hill located downglacier that includes some of the coal transported from the depression (after Aber et al., 1989).

4. Some considerations about engineering geology in areas previously occupied by ice

The most outstanding feature of glacial-origin sequences is their high variability in grain size (Figure 6.13), inner structure, depth, geometry, and surface morphologies. Most commonly these changes occur in short distances. Obviously, this variability is responsible for a number of problems in civil engineering issues (Derbyshire and Love, 1986). For this reason, detailed geomorphological and geological maps are of extreme importance (Strachan and Dearman, 1983). Likewise, drilling campaigns are necessary to assess the depth and geometry of these glaciogenic deposits. In addition, soil mechanics tests provide understanding of the geotechnical properties and predictable behaviour of the materials.

The use of till deposits and fluvioglacial sands and gravels as material for constructing linear engineering works is very common. On the other hand, glaciomarine and glacio-lacustrine silts and clays are low resistance areas within the glacial sequences, and show serious stability problems in steep slopes (Legget, 1974).

Glaciomarine clays that have been raised in postglacial periods by glacioisostasy, can be extremely dangerous because of their tendency to fail, which is referred to as remoulding in the engineering literature. These marine sediments have a high sodium chloride content that increases cohesion of the clay particles. Where salts are dissolved

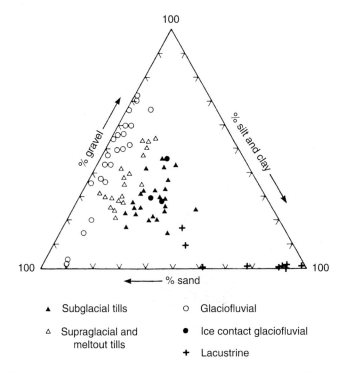

Figure 6.13. Glaciogenic sediments and grain size distribution (according to Derbyshire and Love, 1986).

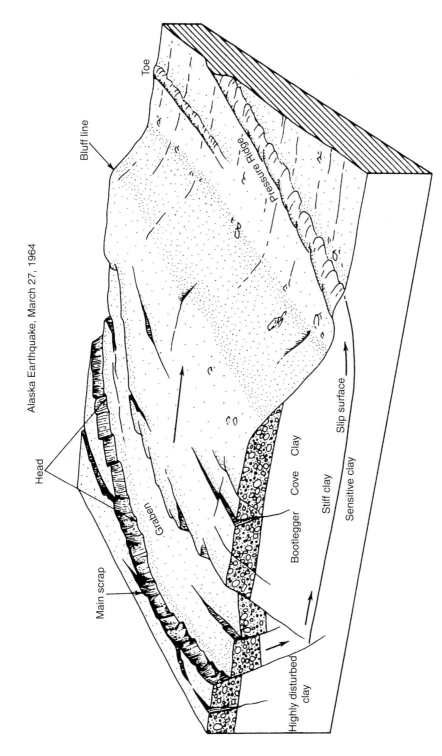

Figure 6.14. Block diagram of the slide of Anchorage, Alaska, triggered by the earthquake of 27 March 1964 (according to Hansen, 1965).

by freshwater, cohesion reduces, the clays liquefy and they can flow through long distances, even at low gradients. They are called *quick clays* due to this change in behaviour and subsequent flow, and are very common in Canada and Scandinavia (Veder, 1981; Aune, 1983).

One of the most important failures linked to the activity of quick clays is that triggered in the coastal city of Anchorage, Alaska, on 27 March 1964, which was triggered by the greatest earthquake recorded in North America. The failure seems to be due to the presence of water-saturated glaciomarine clays (Figure 6.14). The landslide went into the valleys and inlets and destroyed 75 houses of a residential area. In this type of failure

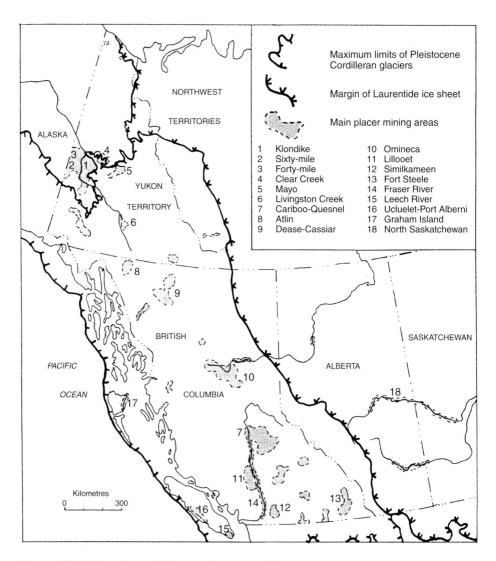

Figure 6.15. Major placer mining areas in western Canada in relation to the maximum limits of Pleistocene glaciers (compiled by Levson and Morison, 1996).

the greatest damages were related to graben areas in the upper part and in the push ridges of the frontal part (Hansen, 1965).

5. Placer deposits in glacial environments

Placers are accumulations of heavy metals that have been extracted from the rock substrate by weathering, water, wind, and glacial processes. They are important economic resources and are ores of gold, diamonds, platinum, tin, and other minerals. Placers subject to ice action are exploited in northern Asia, Canada, and Alaska, though most of these exploitations are in environments not affected by glacial activity (Boyle, 1979).

Placer deposits (Figure 6.15) are located in a wide range of sedimentary situations, such as alluvial, colluvial, glacial, fluvioglacial, coastal, and marine environments (Levson and Morison, 1996). Alluvial placers are located in active or abandoned streams, such as gullies, braided and meandering rivers, alluvial fans, and fan-deltas. Gold concentrates essentially together with loose blocks and gravels and in a conglomeratic facies (Fayzullin, 1969).

Fluvial placers are the most frequent type and are located in recent alluvial deposits and in lower terraces. In meandering rivers they locate in point bars and in the internal curve of the meanders. In braided rivers the distribution of heavy metals is discontinuous; they locate in transverse bars, curves of the channel, junctions of channels and stretches with convergent flow (Smith and Minter, 1980).

In alluvial fans, placers locate in interbedded gravels and in debris-flow sediments. These develop after glacial retreat, when the absence of plant cover facilitates slope instability. Once they fail, they are subject to erosion by glacial meltwaters.

THIRD PART
Geomorphology of periglacial zones

Chapter 7

The periglacial environment

1. Introduction

The term "periglacial" was introduced in 1909 by the Pole, Lozinski, to designate the climatic conditions and geomorphic features of the areas peripheral to the Pleistocene ice-sheets. Subsequently, the term has been extended to cover the characteristic processes and landforms of cold climates, regardless of their proximity to glaciers or glaciated areas. There are vast zones such as eastern Siberia unconnected to the glacial areas where periglacial activity predominates (Jahn, 1975). Due to the lack of precision in the definition, the term periglacial has had very wide use. Thus, the word "geocryology," commonly used by the Russian researchers, has been introduced as an approximately equivalent, alternative term (Washburn, 1979). This science is commonly defined as the study of the terrestrial materials with temperatures below 0°C, that is to say permanently or seasonally frozen grounds (Washburn, 1979; French, 1996).

Periglacial environments are characterized by freeze–thaw cycles and the presence of permafrost or permanently frozen ground. Either or both of them occur in the overall periglacial domain. In this way, some of the periglacial landforms are not associated with the presence of permafrost. The periglacial environment is distributed in polar zones and in medium and low latitude alpine areas of many mountain ranges of planet Earth (Harris, 1988). At the present time, the periglacial domain covers close to one fifth of the global land surface and an additional 20% of the area has undergone periglacial conditions during the Pleistocene cold periods (French and Karte, 1988).

Many palaeoclimatologists recognize the importance of periglacial phenomena for palaeoclimatic reconstructions and of permafrost for palaeoclimatic modelling (Isarin, 1997). Ecologists and environmental scientists have pointed out the vulnerability of present-day periglacial regions due to their increased anthropogenic use and the implications of expected climate change (Vandenberghe and Thorn, 2002).

Periglacial environments have a wide range of climatic conditions. The mean annual temperature may be close to or far below the freezing point, and the range between the maximum and minimum yearly temperatures is generally large. According to Peltier (1950), the annual precipitation varies between 130 and 1400 mm and between 50 and 1250 mm following Wilson (1969). Tricart (1967) has established a climatic classification based on the temperature, precipitation, wind conditions, and their seasonal distribution. Three types of periglacial climates are distinguished:

(a) *Dry climates with cold winters*. These conditions are found in subpolar zones of the northern hemisphere. These areas have very low temperatures in winter, very short summers, low precipitation, and strong winds. This climate propagates the existence of a perigelisol (permafrost). The morphogenetic system linked to this climate type is characterized by a strong frost action, a limited runoff influence and significant wind activity.

(b) *Cold and wet climates with strong winters.* The arctic and mountain types are differentiated within this group. The *arctic type* has large climatic irregularities due to the oceanic influence and has permafrost. The temperatures are similar to climate A with a lower annual range, precipitation is above 300 mm, and violent winds are common. Consequently the frost action is less intense or persistent than in climate A, the snow cover inhibits eolian processes, and run-off is relatively important. The *mountain type* occurs in the alpine areas of the temperate zones. Temperatures are similar to the arctic type with higher mean annuals and lower annual ranges. The annual precipitation is higher than in the arctic type. The role of slope and aspect are very important in these areas. Therefore, frost action is important, permafrost is generally absent, the wind influence is weak, and run-off-related processes are conspicuous.

(c) *Climates with small annual range in temperature.* The mean temperature is close to 0°C and the annual temperature range is around 10°C. Two types are distinguished. The climate of *high latitude islands* has marked weather instability, small temperature range, and snow precipitation above 400 mm that inhibits the effects of the wind. Freeze–thaw cycles are abundant and have limited penetration into the ground. The *low latitude mountain* climate types lack seasonal temperature variations and the daily range is very high and larger than the annual range. Precipitation is commonly high, except in the arid mountains. These climatic conditions engender the existence of permafrost, numerous freeze–thaw cycles, faint penetration of the frost action, and insignificant wind activity except in arid mountains.

In addition, the periglacial domains present two main types of vegetation, the subarctic or northern forest, and the arctic tundra. This allows distinguishing between the forest periglacial environments and the woodland-free environments (French, 1996).

2. Permafrost

Permafrost, also called pergelisol or permanently frozen ground, is defined as a soil, surficial deposit or rock mass that remains continuously below 0°C for more than 2 years, regardless of other properties like the moisture content and lithology (Muller, 1945). Glaciers and ice-sheets are excluded. Permafrost covers around 22% of the land surface in the Northern Hemisphere and extensive areas in the Andes and Antarctica (Harris, 1986). According to Washburn (1979), 14% of the land surface on the Earth is underlain by permafrost. It extends as far as 50% of Canada and 80% of Alaska (Harry, 1988). Permafrost also occurs in the bottom of shallow seas in high latitudes (*submarine permafrost*). Two types of permafrost occur on the continents; *polar permafrost* in high latitude areas and *alpine permafrost*. Alpine permafrost is developed in mountain environments and varies depending on the latitude and elevation. The altitudinal lower limit of the permafrost rises from high to low latitudes (Harris, 1988).

The continuous permafrost wedges out towards low latitudes and breaks up into patches giving way to the *discontinuous and sporadic permafrost* (Harris, 1986)

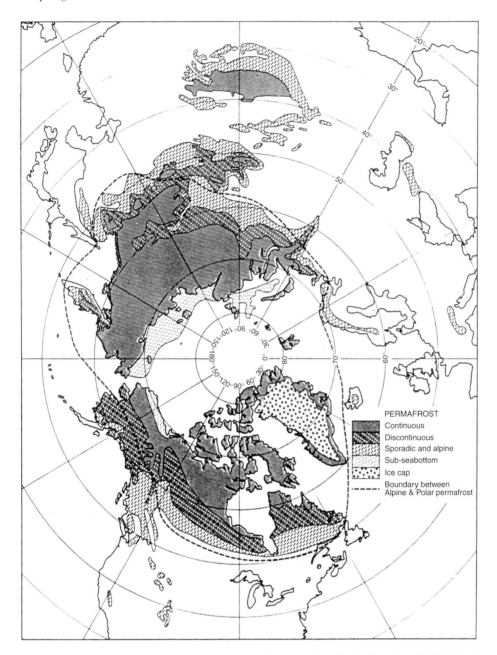

Figure 7.1. Distribution of different permafrost types in the Northern Hemisphere (compiled by Harris, 1986).

(Figure 7.1). This differentiation is based upon the percentage of permafrost beneath a certain point. The proportion of frozen ground in the continuous permafrost is higher than 80% between 30 and 80% in the discontinuous permafrost, and less than 30% in the sporadic permafrost. A N-S section of the permafrost in Canada is shown in Figure 7.2. This sketch depicts the thinning of the permafrost from high to low latitudes and the reduction of the frozen patches separated by the unfrozen ground called *talik*. The maximum documented thickness of permafrost occurs in the River Markha in Siberia where it reaches 1450 m in depth (Washburn, 1979).

The upper part of the ground overlying the permafrost subjected to seasonal freezing and thawing is called the *active layer* or *mollisol*, although some authors restrict these terms to areas where the ground is frozen solely in winter. At a continental scale the active layer thickens towards low latitudes, but it does not exceed 3 to 4 m in depth (Figure 7.2).

The frozen water that forms part of the permafrost comes from the atmosphere, run-off, or underground flow. This ice cements the mineral and organic particles and the ice may occur as individual crystals, particle coatings or pore fillings. The ice that developed in pre-existing sedimentary sequences is called *epigenetic* and the ice that formed during the sedimentation process is named *syngenetic* (Harry, 1988), such as the advance of the permafrost in deltaic and fluvial sequences.

The ice bodies in the permafrost have variable geometries (Washburn, 1979), and remains of old *glacial ice* also occur that are buried by sediments and recognizable by their deformations. The *icings* or *aufeis* are mostly tabular ice bodies of fresh water that form in winter where groundwater emerging from springs or floodwaters in river valleys freezes. This aufeis may be incorporated into the permafrost when buried. *Ice lenses* are

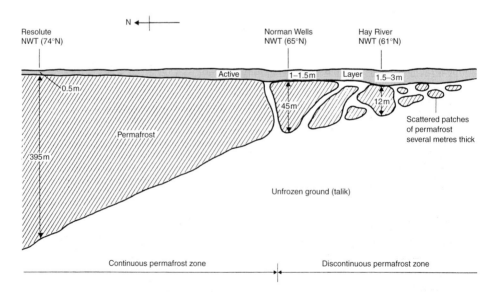

Figure 7.2. North–South section of the permafrost in Canada. It shows the southwards wedging out of the frozen ground and the differentiation between continuous and discontinuous permafrost (Brown, 1970).

horizontal lenticular layers of clear ice with variable thicknesses. They may form *in situ* by ice segregation during the freezing process. *Ice veins* and *wedges* are vertical or inclined cracks in the ground filled with ice. The wedges result from the growth of veins by the supply of meteoric water. They are V-shaped, tapering downwards, with the ice crystals having a vertical arrangement (Lachenbruch, 1962) (Figure 7.3) and their expansive growth deforms the adjacent sediment. Ice wedges are commonly 2 to 3 m wide and penetrate about 5 to 8 m when resulting from epigenetic freezing. In Siberia, syngenetic ice wedges formed in alluvium reach 5 m wide and 40 to 50 m deep (Harry, 1988). They commonly form part of polygonal systems and if there is an

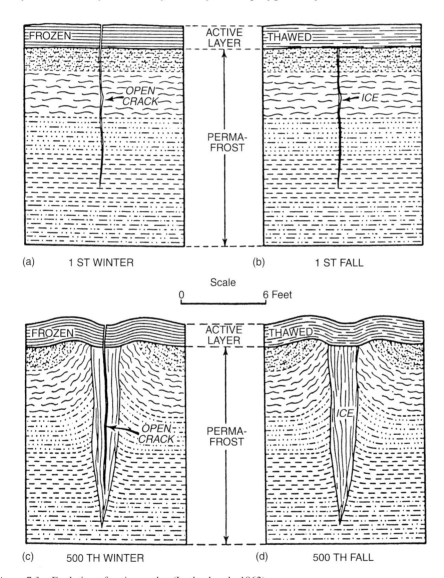

Figure 7.3. Evolution of an ice wedge (Lachenbruch, 1962).

amelioration of climate, the ice infilling may disappear and the ice-wedge cast becomes infilled by sediments. These relict or fossil wedges are indicators of the existence of permafrost in the past (Eissmann, 1981) (Figure 7.4). The ground ice may also constitute the core of mounds called *pingos*. These are fresh-water massive ice bodies with a flat and convex morphology formed under particular hydrological and hydrothermal conditions. They result from the freezing of meteoric water, water-rich sediments, or of artesian water intruded through the permafrost.

The depth reached by the permafrost is determined by the balance between the heat increase at depth related to the geothermal gradient (1°C every 30 to 60 m) and the heat lost at the surface. The basal limit of the permafrost is located at the depth where the temperature reaches 0°C (Brown, 1970) (Figure 7.5). The seasonal fluctuations in temperature that affect the upper part attenuate with depth above a level of zero thermal oscillation situated between 6 and 16 m (French, 1996).

The changes in the ground temperature cause the increase (aggradation) or reduction (degradation) of the permafrost extent (Williams and Smith, 1989). Subtle climatic variations may affect the permafrost with temperatures close to 0°C. On the other side, the climatic changes that take place during decades or centuries may cause significant modifications in the permafrost area and thickness. A great part of the permafrost is relict, formed in past Quaternary periods, and out of balance with the present climatic conditions. This is corroborated by the presence of mammoth fossils (*Mamuthus primigenius*) and other Pleistocene animals preserved in the permafrost of Siberia (Washburn, 1979). This fact demonstrates the existence of permafrost at the time when the animals died, otherwise they would have decomposed.

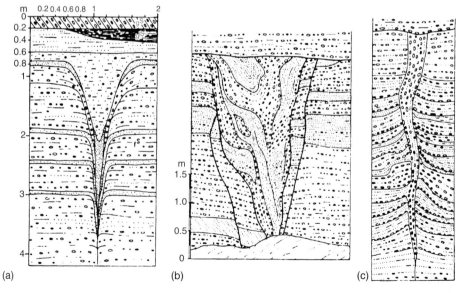

(a)   (b)   (c)

Figure 7.4. Casts of fossil ice wedges from Leizpig area, Saxony (Eissmann, 1981, in Ehlers (1996), Figure 71).

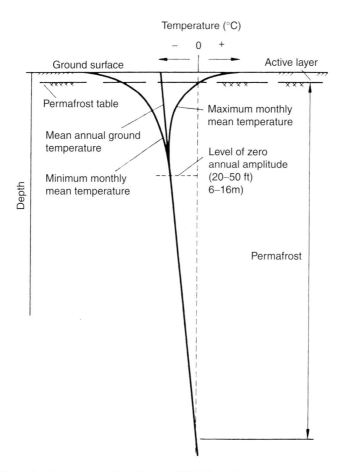

Figure 7.5. Thermal regime in permafrost (Brown, 1970, Figure 6).

3. Periglacial processes

A great part of the morphogenetic processes that operate in the periglacial domain are active in other morphoclimatic zones. However, the processes related to freeze–thaw cycles reach their maximum intensity in this environment. On the other hand, thawing in the slopes aids in the development of numerous mass movements. These processes, together with surface runoff and eolian action in cold deserts, dominate in the periglacial regions.

3.1. Frost action

This term is commonly used to designate the different processes that result from freeze–thaw action. In many situations, several processes linked to frost action act jointly, whereas in some cases they may operate associated with processes of a different origin.

The study of the freezing process may be relatively complex because it involves knowing the magnitude and duration of the temperatures below 0°C. Besides, it may affect materials with different moisture contents and variable heat conductivity (Williams and Smith, 1989). The understanding of the frost-related processes requires knowing the evolution of the freeze–thaw cycles in the ground surface throughout the year. These physical changes may correspond to seasonal variations or may have a daily regime as in the alpine periglacial environments of the low latitudes. The daily variations in these zones are due to insolation changes. In some parts of the Andes more than 300 freeze–thaw cycles have been recorded (Troll, 1944). In the Ransol meteorological station of the Pyrenees in Andorra, at an altitude of 1640 m, an annual average of 80 freeze–thaw cycles have been recognized (Raso and García Loureiro, 1998) and 117 in the University Hostel (2510 m) of the Sierra Nevada in the Betic Cordillera of Spain (Gómez Ortiz and Salvador Franch, 1997). The different processes associated with frost action and some of the most relevant effects are analysed below.

3.1.1. Congelifraction

This is the most important weathering process in the periglacial zones. It is also called *gelifraction* or *frost-shattering*. Liquid water undergoes a volume increase of approximately 9% with the freezing process. The rocks break when the stresses exerted by the interstitial water surpass their tensile strength. If the water freezes in a confined space, it may transmit a large pressure to the host rock. However, the freezing of the water in cracks does not take place in a totally confined condition because it may extrude and only part of the expansion contributes to the disintegration of the rock (Ollier, 1984). The frost expansion may cause the fracturing of the rock producing angular clasts (also called cryoclasts) or its granular disintegration. The mechanical fragmentation of the rocks is an important geomorphological agent as is demonstrated by the block fields developed on planated surfaces and the talus accumulations on the slopes of mountain areas. Gelifraction has a larger effect during the spring when the water derived from snow melting penetrates into the cracks and subsequently freezes. In the southern slopes of Sierra Nevada in the Betic Cordiller of Spain, frost shattering reaches a maximum intensity during the pre- and post-nival periods (Gómez Ortiz and Salvador Franch, 1998). The continuous production of cryoclasts produces the back-wearing of the rock slopes. According to field measurements, the retreat rates may range from 0.3 mm/yr in Spitzbergen (Jahn, 1976) to 2.5 mm/yr in the Swiss Alps (Barsch, 1977a).

The effectiveness of the congelifraction depends on the rock type and the water content. Generally, slates disintegrate to a larger extent than igneous crystalline rocks. This process is largely controlled by the structural characteristics of the rock, mainly by the density of discontinuity planes in the rock mass (Figure 7.6). The role of porosity is particularly important in sedimentary rocks (French, 1996). Numerous experiments about the cryogenic weathering of different rock types under variable conditions have been carried out. Those performed with schists indicate that fragmentation increases with the frost intensity due to a fatigue effect (Lautridou, 1988) (Figure 7.7). An increase in the disintegration of the schists with the number of freeze–thaw cycles and a decrease in the median of the cryoclasts with time are also observed.

Figure 7.6. Rounded clasts with discontinuity planes affected by gelifraction. James Ross Island, east of the Antarctic Peninsula (Photo by A. Martín-Serrano).

3.1.2. Frost heaving

The pressures produced by frost-induced expansion act in every direction. The vertical and horizontal components of the resulting displacements are called *heaving* and *translation* (Eakin, 1916). The action of the vertical movements has the greatest importance. The magnitude of the heaving has been measured in the field by introducing sticks and pins at different depths. The maximum heaving occurs in areas with high moisture content and during the spring, when the ground has a relatively higher proportion of water derived from melting. The sticks driven in the ground may undergo different degrees of heaving or may even be expelled. Values of 1 to 5 cm/yr are common although much greater values have been measured (French, 1996). The heaving increases with depth up to about 30 cm and it affects mainly to the upper part of the active layer because the lower part has a passive behaviour with respect to the frost action. Studies carried out in Sierra Nevada of Spain at 3200 m place the lower boundary of the horizon affected by frost action at 25 to 30 cm (Gómez Ortiz and Salvador Franch, 1997). Experimental works indicate that clasts surrounded by finer particles undergo heaving more easily and that the larger clasts rise more rapidly than the smaller ones (Corte, 1969, 1971). The planar particles tend to rotate and rest on edge with the major faces perpendicular to the cooling surface (Figure 7.8) (Schunke, 1974), as demonstrated in laboratory studies (Kaplar, 1965). The lifting of the clasts leads to a fining upwards of the ground particles; a very common characteristic in periglacial accumulations.

Two different explanations have been given for frost-heaving. One of them proposes that both the stones and the fine particles rise as the ground expands (Figure 7.9).

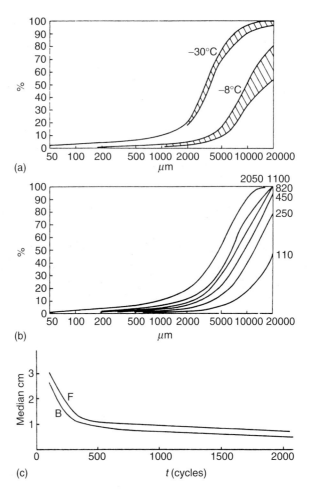

Figure 7.7. Examples of the grain-size distribution of gelifracts produced from Precambrian schists from Norman. (A) Influence of the frost intensity. (B) Changes in the grain size with time (indicated by the number of freezing cycles at $-8°C$). (C) Reduction with time of the median of the particles produced by gelifraction (frost cycles at $-8°C$) for examples B and F (modified from Lautridou, 1988).

When the larger clasts rise, they leave a space beneath their base where ice lenses form. During the thawing the finer grained material settles, whereas some ice remains beneath the clasts. Subsequently, the voids left by the thawed ice lenses are filled by finer particles (*frost-pull hypothesis*). This seems to be the most common mechanism. According to the second explanation, the water flows around the clasts and accumulates beneath their base. When the water freezes, it pushes the clasts upwards. The base of the clasts reaches the freezing point before the surrounding fine-grained material because the larger particles have a higher heat conductivity than the granular mass. During thawing, the translation component compresses the void beneath the clasts and the fine material migrates towards the void, impeding the large particles from settling back down to their initial position (*frost-push hypothesis*) (Bowley and Burghardt, 1971).

Figure 7.8. Planar clasts with vertical disposition due to frost-heaving action. Sierra de Javalambre in the Iberian Range, Teruel Province, Spain.

Other frost heaving mechanisms relate to *ice needles* or *pipkrake*. These are elongated ice crystals with sizes ranging from a few millimetres to 40 cm that form perpendicular and close to the surface. They are particularly common in alpine environments. These crystals lift clasts as they grow and subsequently collapse during thaw. This process causes a sorting of the soil particles. The *nubbins* are elongated or oval-shaped bulges surface

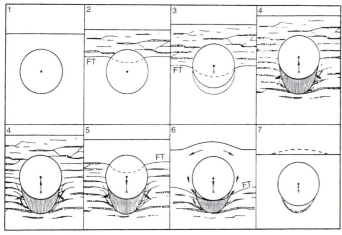

Top row = During freezing
Botttom row = During thawing

Figure 7.9. Lifting of clasts by frost action (Beskow, 1930; in Washburn, 1979, Figure 4.3) FT = Freezing front, Striped pattern = ice lenses.

Figure 7.10.   Gaps around clasts in Sierra Pelarda of the Iberian Range, Spain.

a few centimetres in size in the ground. They are thought to be created by ice needle action. The *spaces around clasts* (Figure 7.10) result from their lifting by frost action and the later settlement during the thaw (Washburn, 1979).

3.1.3. Mass displacement

This process corresponds to the deformation of unconsolidated material and displacement of particles. Commonly the movements of particles have a predominant vertical component although they may be also affected by horizontal displacements. Numerous mechanisms are adduced to explain these internal movements although *cryostatic pressure* seems to be the main cause (Washburn, 1956). Cryostatic pressure is related to the propagation of freezing-induced pressures to unfrozen parts of the ground situated between the freezing front and the top of the permafrost. The freezing of the ground surface starts in autumn and the freezing front advances downwards during the winter. Since different parts of the ground have variable moisture contents, the ground freezes irregularly generating differential volumetric expansions. The pressures transmitted to the unfrozen ground may cause its liquefaction or the generation of dome-shaped bulges in the ground surface. The high magnitude of the cryostatic pressure has been demonstrated in laboratory experiments (Corte, 1969; Pissart, 1970). It is believed that these stresses are the main genetic cause of the *cryoturbations* and *involutions* (Sharp, 1942b), although other origins are considered. These are chaotic structures characterized in section by disharmonic folds, intrusions, injections and faults developed during the freezing of the ground in zones affected by seasonal freezing (Figure 7.11). Involutions

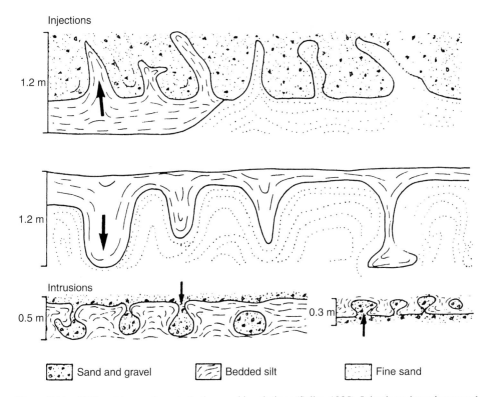

Injections

1.2 m

1.2 m

Intrusions

0.5 m

0.3 m

Sand and gravel   Bedded silt   Fine sand

Figure 7.11. Different types of cryoturbations and involutions (Selby, 1985). Injections, intrusions; sand and gravel, stratified silt, fine sand.

and other deformations also occur in other morphoclimatic regions, mainly arid, but the causes responsible for the volumetric variations have a different nature (French, 1996).

3.1.4. Frost cracking

This cracking results from thermal contraction when the temperature of the frozen ground decreases. Commonly the frost fissures and cracks tend to form polygonal systems of four, five, or six sides (Lachenbruch, 1962). Two main types of frost cracks are differentiated. Those filled with ice or *ice wedges* grow by the supply of snow, percolating water, groundwater, and water vapour to the cracks (Figure 7.3). These features are typical of humid environments and require specific temperature conditions for their growth and preservation. In Alaska the southern limit of active ice wedges corresponds to a mean air temperature between -6 and $-8°C$ (Péwé, 1966). Once the ice melts, the wedges can be filled by detrital material. The preserved wedge casts constitute valuable paleoenvironmental indicators. In dry environments with precipitation below 100 mm, the lack of moisture precludes the infilling of the wedges with ice. In these situations the wedges may be filled by eolian sand forming *sand wedges* (Péwé, 1959; Jahn, 1975) (Figure 7.12). The paleoclimatic meaning of the sand wedges is similar to the ice wedges because they have the same origin.

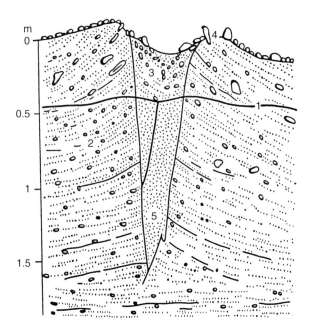

Figure 7.12.   Sand wedge. (1) Top of the permafrost below the active layer. (2) Crudely stratified sands and gravels. (3) Collapsed sand a gravel particles. (4) Clasts. (5) Structureless sand (Péwé, 1959, in Jahn, 1975, Figure 34).

In addition to frost cracking there are other mechanisms responsible for the generation of fissures in periglacial environments. The cracking due to desiccation is a common process in the generation of small size polygons. The desiccation is not only due to evaporation, but also to the loss of moisture in zones of ice formation as demonstrated in experimental studies (Washburn, 1979). Another mechanism is cracking due to dilation or the differential expansion of surficial materials. The resulting lifting produces radial cracks and the subsequent thawing-induced settlement leads to the generation of concentric fracture systems, such as occurs in the evolution of pingos (Benedict, 1970; Jahn, 1975).

3.1.5. Frost sorting

Frost sorting involves the displacement and segregation of the particles of roughly uniform size. Several of the previously explained processes like the lifting of clasts, the action of the ice needles and the mass displacement contribute to the sorting of particles. Laboratory experiments reveal three types of sorting processes (Corte, 1969, 1971). One type is related to heaving; the differential upward movement of the larger particles gives place to horizons with different grain sizes. This is called *vertical sorting*. When the freezing front advances, the finer particles move following parabolic trajectories leaving the coarser particles behind. This type is known as *lateral sorting*. Finally, the coarser particles move by gravity towards the margins of the bulges formed in the ground surface, whereas the finer particles remain in the inner part. This *mechanical sorting* is also found in the polygons where the clasts situated inside the bulged cell move toward the cracks filling them. The experimental studies indicate that the sorting increases with moisture and when freezing is slow.

3.2. Chemical weathering

The presence of extensive accumulations of frost-shattered clasts on surfaces of variable inclination is very common in periglacial environments, indicating that congelifraction is the main weathering process. Although this process primarily generates large clasts, it may also produce clay-size particles, as has been demonstrated with experiments in climatic chambers (Guillien and Lautridou, 1970).

Because water remains in a solid state except during the melting period, the activity of chemical weathering is generally limited (Hall et al., 2002). Besides, the kinetics of most of the chemical reactions is slowed at low temperatures. However, rounded blocks of diabase (basic igneous rock) and sandstones with exfoliation due to hydration and oxidation processes have been locally recognized (Czeppe, 1964). In some cases chemical weathering features are interpreted by some authors as indicators of past nonperiglacial conditions. In Mediterranean mountain environments as in the Sierra Nevada of Spain, the chemical weathering during the warm season on mica-schists is considered to be important in the production of silts (A. Gómez Ortiz, personal communication). In coastal environments, the presence of sodium-rich saline efflorescences resulting from the evaporation of seawater transported by the wind is relatively common on the rocks. Melt waters help the percolation of these saline waters into the rocks and the subsequent precipitation of the dissolved ions may cause their fragmentation and granular disintegration. This salt weathering process (haloclasty) is attributed to the genesis of honeycomb and tafoni weathering, together with the wind action that helps to evacuate the particles resulting from the disintegration of the rock (Washburn, 1969; Selby, 1972).

Nevertheless, the activity and intensity of chemical weathering in the periglacial zones is not very well known due to scarce research carried out in this field. Surprising results have been obtained in an 8-year long study conducted in an area with mica-schists and amphibolites in the north of Sweden (Rapp, 1986). Here 48% of the weathering products were evacuated in solution by the runoff. Other investigations indicate very low values for this type of transport.

The solubility of carbonate rocks increases with the carbon dioxide content of the water, which in turn reaches the maximum solubility in water at 0°C. For this reason the conditions of the periglacial environments favour the dissolution of carbonate rock. The solution rates, however, are lower than in warm areas due to the scant biotic activity, which is generally the main source of CO_2, and the fact that the water may remain in a solid state during the greater part of the year. The ice and the snow supply liquid water by melting, and the karstification of the carbonates produces karren, doline fields and cave systems, especially in alpine environments that may lead to the generation of collapse sinkholes. In areas of continuous permafrost the melting water is not able to penetrate into the ground and flows on the surface. Where the permafrost is discontinuous, small caves may develop with a rare development of stalactites and stalagmites (Sweeting, 1972).

3.3. Mass movements

These types of processes correspond to the movement of detrital material on slopes due to gravity. In periglacial environments these processes reach a great importance due to the high water content of the active layer that contributes to the reduction of the shear strength of the wet material. The permafrost behaves as an impervious material so that its top acts as a hydrostatically loaded surface that favours the displacement of the overlying active layer (French, 1996). The different types of mass movements act with variable intensity depending on the climatic, topographic and lithological conditions.

In steep glaciated mountains, sea cliffs, and fluvial scarps the water percolates through the rocks and the frost-induced expansion may favour the generation of *rock-falls*. These slope movements are particularly common during the melting period and undercutting may enhance their activity. The result is the formation of talus or scree accumulations at the foot of the rock cliffs. The scarp retreat rates calculated with different methods give values of 0.3 to 0.6 mm/yr (André, 1993), well below those estimated for temperate and semiarid areas.

Solifluction (Andersson, 1906) is one of the most widespread processes in periglacial areas. It is a slow downslope flow of water-saturated detrital material. Although this type of movement can occur under a variety of climatic conditions, it is especially effective in periglacial regions where the term *gelifluction* is used (Baulig, 1957). *Frost-creep* may act in association with this process. It involves the dilation of the soil perpendicular to the surface during freezing and vertical settlement with the thawing (Washburn, 1967) (Figure 7.13). Three different types of movements are distinguished. The potential frost-creep (PFC) is that related to the frost-induced heave of the soil (P_1 to P_2). This expansion of the soil is perpendicular to the slope, which constitutes the cooling surface (Taber, 1929). A gelifluction component (G) may also act and the retrogressive movement (R) is opposite to the former ones and is related to the cohesion among particles (Davison, 1889). During a frost–thaw cycle the trajectory of the particles would be from P_1 to P_4. These movements act jointly but the resulting landforms are designated with the term gelifluction because this is generally the dominant process.

Frost-creep increases with the number of freeze–thaw cycles, slope angle, moisture content of the soil, and decreases with depth (Washburn, 1979). Gelifluction may be active with slopes from 1°. Its activity starts with the thawing, reaches its maximum intensity in summer and decreases in autumn, due to the progressive water loss from the ground by evaporation and interstitial flow. Furthermore, gelifluction activity also increases with slope angle and water content of the soil and decreases progressively with depth. Experimental studies indicate that this process is commonly restricted to the upper 50 cm of the active layer (French, 1996). Laboratory simulation experiments indicate that gelifluction occurs only during thaw consolidation of the upper parts of the soil profile, while thawing of the deeper layer provokes little downslope displacement (Harris and Davies, 2000). Other experiments carried out by these researchers (Harris et al., 2003) revealed that gelifluction is not a time-dependent viscosity-controlled flow, but it presents elasto-plastic behaviour. Gelifluction is strongly controlled by the grain size and texture of the soil. The coarse-grained and highly porous soils favour the drainage of the ground, whereas the fine-grained and low permeability soils become

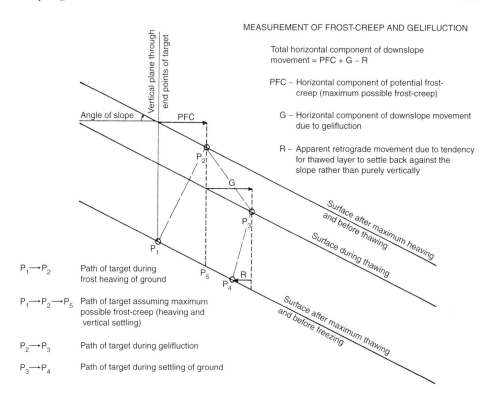

MEASUREMENT OF FROST-CREEP AND GELIFLUCTION

Total horizontal component of downslope
movement = PFC + G – R

PFC – Horizontal component of potential frost-
creep (maximum possible frost-creep)

G – Horizontal component of downslope movement
due to gelifluction

R – Apparent retrograde movement due to tendency
for thawed layer to settle back against the
slope rather than purely vertically

Vertical plane through end points of target

Angle of slope

PFC

Surface after maximum heaving
and before thawing

Surface during thawing

Surface after maximum thawing
and before freezing

$P_1 \rightarrow P_2$ Path of target during
frost heaving of ground

$P_1 \rightarrow P_2 \rightarrow P_5$ Path of target assuming maximum
possible frost-creep (heaving and
vertical settling)

$P_2 \rightarrow P_3$ Path of target during gelifluction

$P_3 \rightarrow P_4$ Path of target during settling of ground

Figure 7.13. Sketch showing the trajectory of particles affected by potential frost-creep (PFC), gelifluction (G) and retrograde component (R) (Washburn, 1967, in Washburn, 1979, Figure 6.8).

more easily water saturated. Although it may be thought that the vegetation slows down the flow of the soil, the measures obtained in vegetated areas provide relatively higher values than in vegetation-free zones. This circumstance seems to be due to the higher water content of the soils with vegetation because it retains water and inhibits run-off (Washburn, 1967). Besides, the gelifluction rates are generally higher on the shaded slopes than on the sunny slopes because of the higher water content. A higher solar radiation, however, accelerates the thaw. Numerous field experiments have been carried out to quantify the gelifluction rates under variable conditions. For example, on slopes ranging between 5 and 15°, values between 1 and 12 cm/yr have been measured (Washburn, 1979). On the other hand, the flow of the soil tends to orient the major axis of the particles parallel to the slope direction.

Large blocks in a solifluidal or gelifluction soil may move downslope and rotate, orientating their major axis parallel to the slope. These features, called *ploughing blocks* in periglacial regions, may leave a furrow along their path and accumulate upthrust soil material at the front (Tufnell, 1972) (Figure 7.14).

Slope movements may be common in areas of active gelifluction and they commonly occur in the middle and lower parts of the slopes (Figure 7.15). These are commonly shallow movements confined to the active layer where the shear strength is reduced by the

Figure 7.14. Ploughing block. Bonaigua Pass, Lérida Province, Pyreneees.

saturation or high water content of the soil. Generally the top of the permafrost determines the position of the failure surface and favours the failure of the overlying material. The shallow debris slides and mudslides commonly disintegrate and transform into *debris flows* and *mud flows* (Figure 7.16). Usually, the head scars of these complex slope movements (slide-flows) are less than 5 m long and the overall length rarely exceeds 100 m (French, 1996). Although thawing is normally the main cause of their genesis, sometimes they may be triggered by summer rains.

3.4. Nival processes

Nival activity is accomplished mainly by rapid snow movements called avalanches and by a combination of much slower processes caused by melt waters at the edge of, and beneath, melting snowfields termed "nivation."

Avalanches are rapid downslope movements of snow and ice that are common in mountainous terrains of temperate zones and unusual in polar regions. They commonly occur on slopes with gradients between 25 and 50° and can reach speeds above 100 km/h. On steeper slopes they are uncommon because the high gradient does not allow the accumulation of a thick snow cover. Several criteria have been used to classify avalanches (López-Martínez, 1988) (Figure 7.17). An avalanche may be composed of dry or wet snow, or the snow may be mixed with detritus, and it may move on the ground or in the air as powder. Avalanches may be confined by a channel (couloir) or they may move

Figure 7.15. Multiple shallow debris slides at the foot of a slope affected by active gelifluction. The colluvium overlays a Cretaceous clastic sequence. James Ross Island, east of the Antarctic Peninsula. Photo by A. Martín-Serrano.

Figure 7.16. Mud flows at the foot of a gelifluction slope. James Ross Island, east on the Antarctic Peninsula. Photo by A. Martín-Serrano.

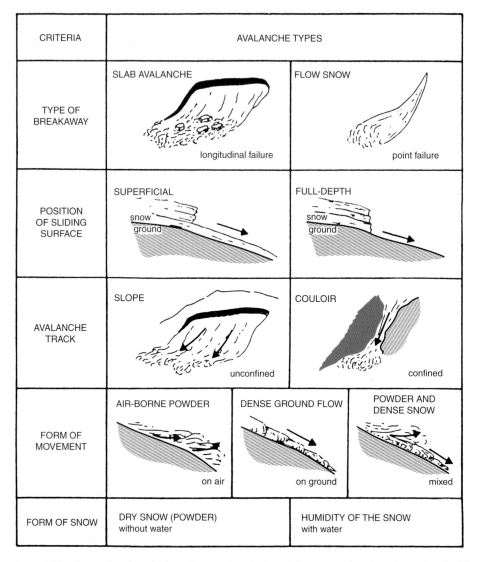

CRITERIA	AVALANCHE TYPES	
TYPE OF BREAKAWAY	SLAB AVALANCHE — longitudinal failure	FLOW SNOW — point failure
POSITION OF SLIDING SURFACE	SUPERFICIAL — snow / ground	FULL-DEPTH — snow / ground
AVALANCHE TRACK	SLOPE — unconfined	COULOIR — confined
FORM OF MOVEMENT	AIR-BORNE POWDER — on air / DENSE GROUND FLOW — on ground	POWDER AND DENSE SNOW — mixed
FORM OF SNOW	DRY SNOW (POWDER) without water	HUMIDITY OF THE SNOW with water

Figure 7.17. Avalanche classification. Compiled and adapted from several authors (reproduced with permission of López-Martínez, 1988).

unconfined (Figure 7.18). The instability of the snow cover may be favoured or triggered by several causes. Avalanches are especially common after heavy snow precipitation that overloads the previously deposited snow cover. The frozen snow or hoar-frost layers constitute favourable potential sliding surfaces. In addition, the percolation of snow melt waters facilitates the generation of avalanches (Embleton, 1979).

According to Furdada (1996), there are two main types of avalanches. *Powder avalanches* move in the air and have a limited geomorphic effect although they can knock

Figure 7.18. Unconfined avalanches of wet and dense snow with surface sliding surfaces. Pic d'Arcalis in Andorra, Pyrenees. Photo by M. Urigüen.

down trees and destroy houses. *Slab avalanches* are rigid bodies of cohesive snow that move on a sliding surface controlled by an old snow layer (surface avalanche) or by the snow–ground interface (full-depth avalanche). The full-depth slab avalanches generally incorporate detrital material and may cause important erosional effects. Avalanches are especially common during the thaw period and winter (Owens, 2004). The movements of water-saturated snow masses or slush avalanches that flow in arctic areas along fluvial valleys during the spring should also considered as avalanches. They can transport blocks up to 100 t on slopes of 5° (Hestners, 1985).

Avalanches are highly variable both in space and time. There are zones where they are very rare and locations where they occur regularly. Some geomorphic and vegetation evidence helps to identify avalanche-prone areas and enable production of avalanche hazard maps. Many of the talus slopes and talus cones are partially built by avalanches (Figure 7.19 and Figure 7.20). The avalanches destabilize and incorporate loose particles and erode the channels along which they mobilize.

The term *nivation*, introduced by Matthes (1900), designates a combination of several processes such as gelifraction, gelifluction, frost-creep and run-off related to the water derived from the melting of snow (Thorn and Hall, 2002). The degree of nivation depends in part on the presence or absence of permafrost in the underlying material. In areas with permafrost the melt waters may refreeze. In permafrost-free zones the snow cover has an insulating effect protecting the ground from freeze–thaw cycles (Embleton and King, 1975a,b). Thin snow covers favour the processes of congelifraction. Nivation is most active in subarctic and alpine environments and the *nivation cirques* or *niches* are the most outstanding geomorphic effect (Thorn, 1988b).

Figure 7.19. Avalanche channels (*couloirs*), talus slopes and cones. Main Cordillera of the Andes, Punta Vacas, Mendoza Province, Argentina.

Figure 7.20. Talus slopes at the foot of acicular basaltic peaks. James Ross Island to the East of the Antarctic Peninsula. Photo by A. Martín-Serrano.

3.5. Fluvial activity

The literature about fluvial activity in periglacial environments is very scarce. Traditionally it was considered to be not very relevant because the water remains in the solid state throughout most of the year. Its geomorphic effect was thought to be limited because rivers in periglacial areas without allogenic supply only have flowing waters during the thaw period (French, 1996) (Figure 7.21).

Rivers in periglacial areas are fed by snowmelt, glacier melt, rainfall, groundwater seepage, a limited amount of ground ice melt, and inflows from extra-periglacial areas. Most rivers are strongly influenced by snowmelt and their seasonal flow pattern has been described as the nival regime that consists of a period of high flows sustained by melt water with prominent diurnal runoff (Vandenberghe and Woo, 2002).

The melting of the snowfall during the winter season may take place very rapidly. The resulting high run-off gives rise to floods whose hydrographs commonly have a steep rising limb, an acute peak flood and a prolonged falling limb (Arnborg et al., 1967). During the summer period the greater part of the discharge waters may be derived from lingering snowfields and glaciers. On the other hand, during this season rainstorms may occur in arid periglacial zones. Even with a low mean annual precipitation, most of the run-off concentrates in a short time period. It is considered that between 30 and 90% of the annual surface runoff flows during the 2 or 3 weeks of the melting peak (Clark,

Figure 7.21. Creek with melt waters and slopes with heterometric colluvium covering Cretaceous terrigenous sediments. James Ross Island, to the east of the Antarctic Peninsula. Photo taken by A. Martín-Serrano.

1988). Finally, the runoff disappears from the beginning of the freezing period with the onset of the winter.

In periglacial regions the fluvial transport of sediments is generally low and commonly derives from deposits of glacial origin. There is generally a predominance of bedload that may reach 90%, a negligible dissolved load ($<1\%$) and a suspended load ranging from 5 to 25% (Church, 1972). The transported volumes estimated with variable methods in different catchments indicate values of surface lowering between 0.03 and 0.5 mm/yr (Arnborg et al., 1967). In accordance with the high bedload, the fluvial channels are dominantly of the braided-type, with unstable and highly variable bars that are covered during high-stage periods. In general, changes in fluvial behaviour of channel form often are associated with specific thresholds and basin characteristics (Schumm, 1979; Mol et al., 2000). The water of the rivers exerts thermal erosion and contributes to the thawing of frozen riverbanks and generation of collapses by undercutting. The continuous thermal erosion leads to the progressive widening of the channel bed giving place to U-shaped channels (French, 1996). On the other hand, large rivers may reach a talik beneath the channel favouring its entrenchment.

Figure 7.22. Loess deposits with buried soils. Neerijse, Belgium.

3.6. Eolian action

In the Arctic and Antarctic regions the extreme cold and arid conditions impede the growth of vegetation so that the wind may be an important activity. During the Pleistocene, the transport and sedimentation of eolian particles had a great importance in Europe (Kasse, 1997) and in the centre of the United States (Embleton and King, 1975a,b). While the cover sands constitute continuous deposits in central Europe, in Great Britain they are patchy. Between the Last Glacial Maximum and the beginning of the Holocene, three phases of eolian sand deposition have been identified. These phases do not coincide strictly with stadial or glacial conditions. Aridity, scarce vegetation cover and delayed responses of the eolian activity to climatic change determine the timing of eolian phases (Kasse, 2002). Besides, the wind plays a relevant role in the redistribution of the snow during the winter. In summer it melts more rapidly on the sun-facing slopes. These changes in the snow cover are very important for nivation processes and the generation of landforms such as asymmetric valleys and oriented lakes that will be assessed below. In these cold polar regions, the wind action operates predominantly in summer, causing the deflation of fine particles from the land surface generating periglacial pavements on alluvial plains and till accumulations. Overdeepening due to differential eolian erosion may give rise to deflation basins (Seppälä, 2004). The abrasion caused by the blasting of wind-transported particles on the ground produces ventifacts or faceted clasts, yardangs in rock outcrops (elongated mushroom shapes), microyardangs in loose sediments (small-scale knife-shaped and aligned landforms), and tafonis close to the base of rock slopes (French, 1996).

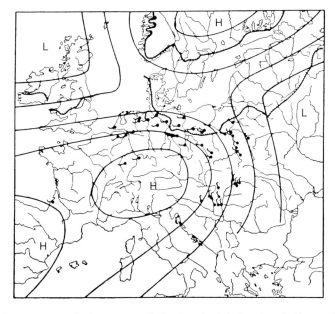

Figure 7.23. Summer atmospheric pressures (isobars) and wind directions in Europe during the Late Glacial period. H and L: High and low pressures (modified from Poser, 1950).

    The combined effect of the eolian sedimentation of snow and mineral particles generate nivo-eolian deposits (Koster and Dijkmans, 1988). As in ice masses, its differential melting produces small hillocks and depressions that may disappear from one year to another. The most prominent depositional features, however, correspond to sand dunes and especially to unstratified, wind-blown accumulations of silt particles called *loess* (Péwé, 1955; Péwé et al., 1995) (Figure 7.22). These are supplied by dust clouds and may be deposited at great distances from their source. Loesses are massive and very well-sorted deposits with 50 to 60% of the particles between 0.01 and 0.05 mm (silt) and the rest of sand and clay size. They generally have a gray colour that turns into a yellowish tone with weathering. Gelifluction structures, cryoturbations and ice-wedge casts may be common. During the Pleistocene, extensive loess accumulations were formed around the southern margins of the North American and Eurasian ice sheets. There is, however, also loess generated in nonperiglacial environments such as those occurring in tropical and midlatitude desert areas. The dominant wind and atmospheric conditions (isobars) of past periods can be inferred with paleocurrents obtained from the internal structures of old eolian deposits (Poser, 1950) (Figure 7.23).

Chapter 8
Periglacial landforms

The previously described cold-climate processes give rise to a wide variety of periglacial landforms. There are two main environments where these features are especially abundant. The high-latitude areas with markedly seasonal contrasts and the middle- and low-latitude alpine zones like the Andes, Himalaya, or the mountains of equatorial Africa characterised by strong daily fluctuations. The differences in the climatic conditions have an important influence in most of the processes responsible for the generation of periglacial landforms (King, 1976a,b).

The variations in the climatic belts during the Pleistocene have caused changes in the distribution of the periglacial environments. It is important to differentiate between relict and active periglacial landforms. The identification of relict landforms allows us to infer the existence of past periglacial conditions in certain areas (Washburn, 1979).

1. Patterned ground

Patterned ground corresponds to a group of small-scale landforms with circular, polygonal, and stripped geometries that raised the attention of the first explorers of the periglacial regions to the detriment of other more widespread landforms (French, 1996). Patterned ground is not specific to periglacial areas alone, however, and similar morphologies are developed in other environments, especially in the hot desert (Hunt and Washburn, 1966). Therefore, a situation of equifinality or morphological convergence takes place because different processes under distinct conditions give place to similar landforms. It is very important to bear in mind this circumstance when trying to deduce past environmental conditions from relict landforms and depositional features. In the subdesertic central sector of the Ebro Depression around Zaragoza city (NE Spain), several researchers (Johnson, 1960; Brosche, 1971, 1972) interpreted structures found in fluvial terrace deposits overlying Tertiary evaporites as ice wedges and cryoturbations. Recent studies, however, have demonstrated that they were generated by subsidence-related process caused by the karstification of the soluble bedrock (Zuidam, 1976a,b). Only those morphologies generated by the thermal cracking of frozen ground are specific of periglacial environments because they indicate the presence of permafrost or intense cold conditions.

Periglacial patterned ground generates in the active layer and is classified, according to the classical morphologic systematisation of Washburn (1956) into circles, polygons, nets, steps, and stripes. The presence or absence of particle sorting is considered for each of the types. The circles, polygons, and nets develop mainly on horizontal surfaces. These cells tend to lengthen with slopes between 2 and 7° becoming stripes (Sharpe, 1938; Büdel, 1960) (Figure 8.1).

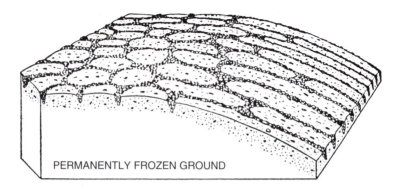

PERMANENTLY FROZEN GROUND

Figure 8.1. Block diagram showing sorted polygons on subhorizontal slopes that change into sorted stripes with the increase in the slope angle (Sharpe, 1938, in Holmes, 1965, Figure 283).

The size of the *circles* ranges from 0.5 to 3 m. Those with no sorting have a bulging and vegetation-free central part with cracks. The circles commonly have a high proportion of fine material and in cross section show structures that indicate uplift in the central areas. The sorted circles have a rim with stones surrounding the inner part composed of fine-grained material. The planar stones tend to rest on edge with the major faces oriented vertically (Furrer, 1968). Both the sorted and nonsorted circles may occur isolated (Figure 8.2) or grouped (Figure 8.3). The most important processes attributed to the

Figure 8.2. Relict-elongated sorted circle in the bottom of a funnel-shaped doline. The clasts of the margin are smaller than those situated in the central part. Nonsorted steps may be also observed on the slopes, Sierra de Javalambre, Teruel Province, Spain.

Figure 8.3. Sorted circles. Southern Shetland Islands (Antarctica). Scale 1 m. Photo: J. López Martínez.

genesis of the circles are differential frost-heaving, mass displacement, and frost sorting (Jahn, 1975).

The *polygons* with diameters below and above 1 m are differentiated because they may have a different origin (Washburn, 1979). The small nonsorted polygons may have diameters as low as 5 cm and the large ones may reach 100 m. They may develop in soils with particles of variable sizes and the vegetation tends to concentrate in the borders emphasising the polygonal pattern. Some polygons have ice wedges along the margins. These have a domed cell during the growth stage and a depression with ponds during the recession stage (Jahn, 1972). The borders show an opposite behaviour. Some polygons may host small pingos and a mesh of smaller polygons in the main cell. In the sorted polygons the margins are formed by stones that surround a cell of finer material (Figure 8.4). The maximum size of the large sorted polygons does not exceed 10 m and the minimum size of the smaller ones is around 10 cm. The marginal stones decrease in size with depth and they wedge out downwards or widen merging with a continuous layer of clasts (Jahn, 1975) (Figure 8.5). Cracking is considered to be essential in the genesis of the polygons. It seems that the large ones are generated by frost cracking whereas the smaller ones result from desiccation or dilation cracking (Washburn, 1979).

Figure 8.4. Sorted polygons. Southern Shetland Islands, Antarctica. Scale 1 m. Photo: J. López Martínez.

The *nets* are intermediate forms whose mesh is neither dominantly circular nor polygonal and can be sorted or nonsorted. The *hummocks* are included in this type (Figure 8.6). These are mounds of earth covered with vegetation that may reach 2 m in diameter and 50 cm in height. Their internal structure displays cryoturbations. They occur above the tree line (timber line) and their genesis seems to be related to mass displacements (Schunke and Zoltai, 1988).

The *steps* are a type of patterned ground that corresponds to benches that develop in slopes with a gradient between 3 and 20°. The margins of the nonsorted steps are covered with vegetation (turf-banked terraces) whereas in the sorted ones the borders are made up of imbricated clasts (stone-banked terraces) (Sharp, 1942a). Very likely the nonsorted steps derive from hummocks and the sorted steps from sorted circles and polygons (Washburn, 1979). Their origin is considered to be related mainly to differential mass displacement.

The *stripes* are strips oriented parallel to the steepest slope gradient. The nonsorted stripes consist of parallel lines of vegetation-covered ground and intervening strips of relatively bare ground. The sorted stripes are parallel lines of stones and intervening strips of fine material with the later tending to be several times wider than the coarser lines

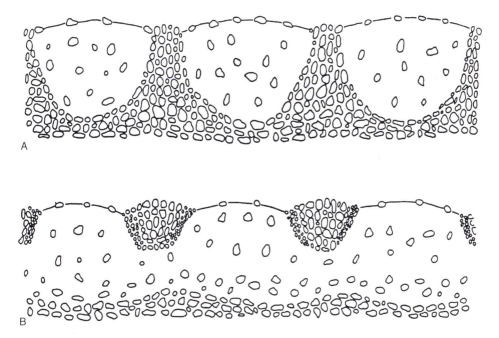

Figure 8.5. Cross section of sorted polygons. (A) Border of stones that widens downwards merging with a layer of clasts. (B) Border of stones that wedges out downwards (Jahn, 1975).

Figure 8.6. Hummocks. Headwaters of the River Ter Basin, Eastern Pyrenees.

Figure 8.7.   Relict-sorted stripes. Sierra de Javalambre, Teruel Province, Iberian Range, Spain.

(Figure 8.7). They are generally straight and may reach 120 m in length (Washburn, 1969). The clasts are usually on edge with the major axis parallel to the stripes. The size of the clasts generally decreases with depth and the sorting penetrates up to 1 m. In slopes from 1° the sorted polygons change progressively into sorted stripes (Figure 8.1) due to the sealing by mass movements of the sides perpendicular to the slope gradient (Furrer, 1968).

Although attempts have been made to unravel the origin of the patterned ground in periglacial environments, its genesis is not well understood. There is not doubt that they are polygenic and that similar forms may be due to different processes. Besides the activity of certain processes may give place to different landforms (Washburn, 1979). Despite the abundant literature devoted to the topic, the origin of periglacial patterned ground is still an unsolved challenge.

2. Ice-cored mounds

The ice-cored mounds comprise a group of small hills generated by frost action. The numerous terms proposed for their differentiation in the abundant literature have created some confusion. These mounds may be perennial like palsas and pingos, or seasonal, like the hydrolaccoliths or blisters. The latter ones reach 8 m in diameter and 3 to 50 cm in height and have been little studied because of their ephemeral development restricted to the winter season when they are commonly covered by snow. Their genesis is related to the freezing of groundwater above the permafrost (Pollard, 1988).

2.1. Palsas

These mounds occur in boggy areas with discontinuous or sporadic permafrost (Zoltai, 1971). Their surface is criss-crossed with open fissures caused by dilation or desiccation (Embleton and King, 1975b). At the present time they are occur in areas with a mean annual temperature below 0°C, as in Canada, Iceland, Sweden, and alpine regions (Kershaw and Gill, 1979). Palsas may be isolated or form in groups and are differentiated based on the morphology of dome-shaped, ribbon-shaped, and platform palsas. The former ones are between 0.5 and 7 m in height, 10 to 30 m wide and 15 to 150 m long. The larger palsas tend to be less conical than the smaller ones. The ribbon-shaped palsas have a lower height and may be parallel or perpendicular to the contour lines, reaching several hundred meters in length. The platform palsas rise 1 to 1.5 m above the surrounding bog and may cover several square kilometres. Their generation is related to differential thawing. Finally, the palsa complexes are formed by several types of palsas at different growth stages and show numerous closed depressions generated by thawing (Seppälä, 1988). Some authors use the term string bogs to designate the association of different types of palsas.

Based on their internal structure, two types of palsas are differentiated. The palsas with a peat core have an outer layer of peat that thaws and dries in summer. Around 80 to 90% of the core is composed ice in small crystals and lenses of segregated ice whose thickness increases with depth from 5 to 10 mm to 5 to 10 cm. The base is generally formed by silt-sized mineral particles. The silt core palsas also have an outer peat layer and a silt and clay core with ice lenses and veins with a reticular arrangement (Seppälä, 1988). Lithalsas are similar mounds, but without any peat cover (Harris, 1998). We can see that no definitive agreement exists with regard to terminology (Pissart, 2000).

The origin of the palsas is largely related to the lower heat conductivity of the dry peat with respect to the saturated and frozen peat, with a conductivity around twenty times higher (Washburn, 1979). In winter the peat wets and increases its heat conductivity favouring the inhomogeneous penetration of the freezing and the formation of segregated ice. The consequent differential cryostatic pressure gives place to elevations in the surface. The outer peat layer that dries in summer with low heat conductivity has an insulating effect preventing the thawing of the inner part of the palsa. The palsas may disintegrate gradually from the margins due to a rise in temperature. Finally, they may become shallow ponds and a new stage of peat accumulation may take place (Friedman et al., 1971). The remnants of lithalsas present encircling ramparts (Pissart, 2000). Seasonal palsas linked to oozes have been reported in the Paramera of Avila Province (Central Spain) at 1200 m in altitude (Molina and Pellitero, 1982). Although most of the palsas of the Scandinavian countries formed 1000 to 3000 years ago, absolute datings indicate that some of them were formed recently (Seppälä, 1988).

2.2. Pingos

"Pingo" is the Eskimo term for mound. They are also called hydrolaccoliths and in Siberia are called bulgunniaks. The pingos are domed perennial ice-cored mounds that generally bulged out in large plains. Recently, submarine pingos have been recognised on some subsea platforms. It is not known whether they have grown under the sea or have formed on land and have been invaded latterly by marine waters. The highest documented pingo in emerged land above sea level is the 48- m high Ibyuk pingo, located close to the Mackenzie Delta in Canada. Most of the pingos are less than 20 m high and their diameters range between 30 and 600 m. There is an inverse relationship between the diameter and height of the pingos. In some cases they are elongated, more than a kilometre in length and maximum height of 9 m. Because they keep a constant diameter, the gradient of the slopes increases during the growth and never exceeds 45° (Pissart, 1988). Pingos show two types of cracks. Dilation radial cracks converge in the top and result from the upward push produced by the ice core during its growth stage (Washburn, 1979). The opening of the cracks may lead to the partial thawing of the underlying ice giving place to a crater-like subsidence depression in the centre. The concentric cracks, not so conspicuous as the former ones, result from the thawing of the ice core during the negative growth stage of the pingo (Müller, 1959). In contrast to the palsas, pingos have a massive ice core that penetrates several meters beneath the land surface (Lundquist, 1969). The cover material is made up of loose sediments like gravel, sand and silt, but pingos with a rock cover of sandstone and shale are also known. This cover layer may reach up to 14 m thick. The growth rate varies from very low values to 1.5 m/yr (Mackay, 1973). On the other hand, all the known pingos are younger than 10,000 years old and some of them are a few hundred years old. Finally, pingos disintegrate progressively due to a rise in temperature giving place to thaw lakes.

Two methods of pingo formation have been suggested. The so-called *closed-system* origin explains the pingos of the Mackenzie Delta (Canada) and Central Yakutia (Siberia) developed in areas with thick continuous permafrost. They form in relation to lake depressions. During the development of the permafrost, the freezing of entrapped water within a lake creates a massive ice core by downward percolation and aggradation. The volume increase generates a cryostatic pressure that domes the lake sediments and may even cause the extrusion of water towards the surface. If the water reaches the surface it may give place to aufeis and when it contains gas it may produce explosive activity. Generally, the water does not reach the surface but it freezes generating a massive ice core (Mackay, 1979) (Figure 8.8). The cryostatic origin of the pingos in the Mackenzie Delta is supported by the fact that 98% of the 1380 mapped pingos are located at the edge or inside of contemporaneous or old lakes (Stager, 1956).

A great proportion of the *open-system* pingos are located in areas of relatively thin permafrost of Alaska and Greenland. They are generated by groundwater flowing under artesian pressure through a thin permafrost or in taliks within the permafrost. This water freezes as it forces its way upwards forming an ice core that domes the surface. The hydrostatic pressure result from height differences (Holmes et al., 1968) (Figure 8.9) so that they form in topographic lows like valley bottoms or distal sectors of low angle slopes (Müller, 1959).

3. Slope morphology and evolution

The slopes in periglacial environments show a wide range of morphologies because they develop under variable conditions of temperature, moisture, lithology and vegetation

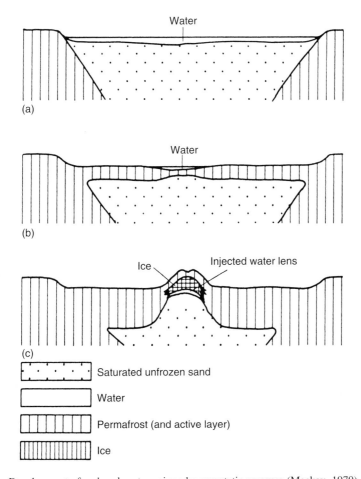

Figure 8.8. Development of a closed-system pingo by cryostatic pressure (Mackay, 1979).

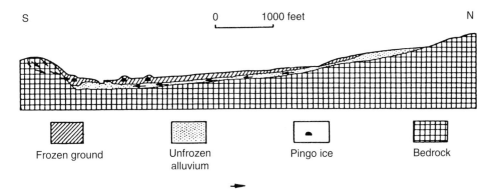

Figure 8.9. Origin of an open-system pingo by hydraulic pressure (Holmes et al., 1968).

cover. Several processes may be involved in their genesis, such as gelifraction (frost-shattering), mass movements, runoff or nivation. All of them may operate independently or in a combined way with a highly variable intensity. On the other hand, many of the resulting landforms are not specific to periglacial regions as they also generate in other morphoclimatic zones. Besides, some of the periglacial slope morphologies were formed in the past (relict) and may be out of balance with the currently existing conditions.

3.1. Gelifluction slopes

The differential movements related to the downslope displacement of colluvial deposits and rocks caused by frost-creep and gelifluction processes give place to several landforms whose differentiation is based on geometric criteria (Washburn, 1979). *Gelifluction sheets* have a large lateral extent and a festooned lower edge. They develop on slopes with gradients higher than 1 to 3°. *Gelifluction benches* are characterised by a terraced form and the major dimension is commonly parallel to the contour lines. *Gelifluction lobes* with a tongue-like shape develop, like the benches, on slopes with gradients up to 20 to 25°. The lobate geometry is due to the higher displacement rate along the central axis of the lobe. The gelifluction features with a pronounced linear geometry oriented parallel to the slope gradient are called *gelifluction streams*. These mass movements, having some similarities to debris flows, may flow on glacial ice. Their displacement is favoured by the presence of water at the base where the ice surface is at the thawing point (Figure 8.10).

Figure 8.10.   Gelifluction deposits flowing on glacial ice. Whisky Glacier, James Ross Island, east of the Antarctic Peninsula. Photo: A. Martín-Serrano.

All these landforms develop more easily on slopes with scarce vegetation and on sunny facing slopes where the thawing is favoured by the higher solar radiation. The gelifluction deposits are usually poorly sorted and may show a crude stratification. The clasts are commonly angular and show an oriented fabric with the major axis oriented parallel to the direction of movement. These characteristics may be also occur in deposits affected by gelifluction in other morphoclimatic regions. Those, however, that develop in periglacial conditions are more angular and are scarcely affected by chemical weathering.

3.2. Cryoplanation terraces and cryopediments

These are low-angle, planated surfaces developed in periglacial regions. They have received numerous descriptive and genetic names in the geomorphological literature, especially in areas such as Siberia (*goletz*) and Alaska (*altiplanation*) where such landforms are most numerous. *Cryoplanation terraces* develop in the upper and middle part of slopes on hills and mountains up to 3000 m high. They show a terraced profile and outcrops of bedrock and tors (rocky residual relief) (Czudek, 1964) (Figure 8.11). These terraces occur on slopes with gradients lower than 25° and at the foot of small scarps. The enlargement of the highest terrace may eventually give place to a planation surface in the summit. The width of the terraces ranges from 5 m to more than 1 km, and the length from 30 m to more than 10 km. The gradient varies between 1 and 14° and the height of the scarp may reach 50 m. The scarps show an abrupt link with the terraces and may have an accumulation of snow. The terraces are covered by gelifluction deposits up to 3 m thick derived from gelifraction processes (Priesnitz, 1988). These benches may show a structural control and their origin is related to nivation processes. These involve gelifluction and overland flow derived from snow melt waters that mobilise the particles produced by congelifraction (Bryan, 1946). These processes cause the retreat of the scarp with the consequent enlargement of the terraces and increase in their retention capacity of snow (Demek, 1969). The presence of patterned ground on the surface of cryoplanation bench treads and on pediments has often been raised as a potential indicator of contemporary inactivity (Thorn and Hall, 2002). Excellent examples of cryoplanation terraces have been recognised in the Sierra Nevada (Betic Cordillera, Spain) and in the Pyrenees (Gómez Ortiz, 1996) (Figure 8.12).

Cryopediments are low-angle erosional surfaces developed at the foot of the slopes in valley margins. Generally, only one level is recognised although in some cases several stepped levels may occur. They have larger dimensions than the altiplanation or cryoplanation terraces, reaching lengths up to several tens of kilometres. The slope of the cryopediments varies between 1 and 12° and shows a straight or slightly concave longitudinal profile. They are frequently covered by a thin veneer (<2 m) of detrital material and the bedrock may crop out locally. Gelifluction landforms are frequent in the proximal sectors of the cryopediments and patterned ground commonly occurs in the distal areas (Priesnitz, 1988). Their origin is related to frost action in the source area where the detritus, as on the cryoplanation terraces, is mobilised by gelifluction and overland flow. The latter process is more active in the cryopediments and in consequence the transport of particles is longer. In the distal sectors the particles are transported by shallow braided channels. The continuous production and transport of

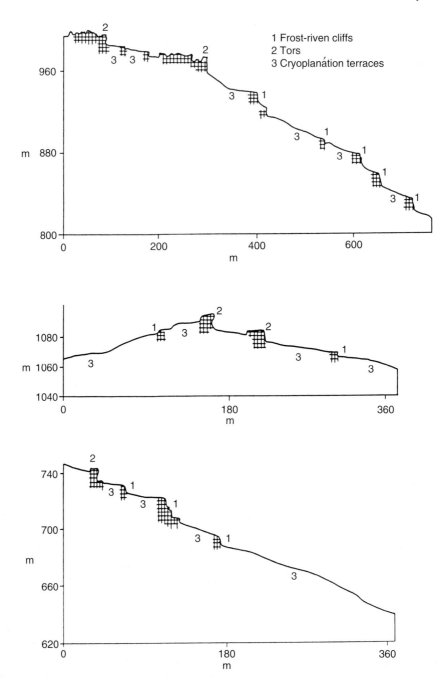

1 Frost-riven cliffs
2 Tors
3 Cryoplanátion terraces

Figure 8.11. Slope profiles with cryoplanation terraces and tors in the Hruby Jesenik Mountains, Czech Republic (partially modified from Czudek, 1964).

Figure 8.12. Cryoplanation terrace covered with angular clasts. Planell de la Carabassa (2690 m). Sierra de Colomer, Eastern Pyrenees, Spain. Photo: A. Gómez Ortiz.

particles leads to the slow retreat of slopes and the enlargement of the cryopediments (Czudek and Demek, 1970a). From another point, cryopediments also show a clear morphologic resemblance with pediments developed in hot deserts although they have a quite different origin (Demek, 1969).

3.3. Talus slopes and cones

These are more or less continuous accumulations of angular clasts (scree) in slopes. Although very common in periglacial environments and especially in alpine areas, they also occur in other morphoclimatic zones, such as deserts (Steijn et al., 2002). A *talus* is an approximately wedge-shaped heap of fragmented rock supplied by an upper cliff or rock face that generally covers the middle and lower part of the slopes (Figure 8.13 and Figure 8.14). Their thickness is highly variable and may reach 30 m (Brunner and Scheidegger, 1974). These accumulations commonly have a concave longitudinal profile and a higher gradient in the upper part. The slope of the talus corresponds to the repose angle of the coarser particles and ranges between 25 and 40° (French, 1996). The talus commonly show a clear particle-size sorting along their longitudinal profile. The size of the clasts increases towards the base of the accumulation as larger clasts have a higher momentum and travel further down slope (Washburn, 1979). Besides, the larger clasts tend to show their major axis oriented parallel to the slope gradient.

The fragmentation of the rocks in the upper rock face of the slope may be due to several processes, although gelifraction is obviously the dominant process in periglacial regions.

Figure 8.13. Talus slopes and debris cones in Punta Vacas. Main Cordillera of the Andes, Mendoza Province, Argentina. Note the progradation of the talus (scree slopes) on the valley bottom.

Figure 8.14. Talus slopes and debris cones. Fluvioglacial deposits in the foreground. Punta Vacas, Main Cordillera of the Andes, Mendoza Province, Argentina.

Figure 8.15. Talus cone in Punta Vacas, Main Cordillera of the Andes, Mendoza Province, Argentina.

The continuous supply of frost-shattered clasts leads to the retreat of the rock face which is largely controlled by the rock type and structure. Retreat rates around 1 mm/yr have been measured in Laponia and Spitzbergen (Rapp, 1957) and 1 to 3 mm/yr in Great Britain (Ballantyne and Kirkbride, 1987). The clasts in the talus move by rolling, creep and small slides. The displacement is usually restricted to the upper half meter and the velocity may be highly variable with values ranging from 1 to 500 cm/yr (Washburn, 1979).

The frost-shattering processes may act preferentially along planes of weakness generating steep channels in the rocky slopes. These channels or torrents (couloirs) are commonly preferential paths for snow and rock avalanches. At the foot of these channels the rock fragments loose confinement and generate *talus cones*, similar to the talus slopes but with conspicuous convex transverse profiles (Figure 8.15). Several adjacent cones may coalesce showing morphologies similar to those of the talus slopes.

3.4. Block-fields, block-slopes and block-streams

These accumulations of large angular particles develop on or below outcrops of hard rocks affected by systems of discontinuity planes with a dominantly decimetric and metric spacing. In periglacial environments, the weathering processes acting on these materials, mainly frost shattering caused by freeze–thaw cycles, give place to boulder-sized, decimetric and metric clasts together with finer particles. These block accumulations may have a finer interstitial matrix favouring the growth of vegetation. Where the matrix is washed by interstitial water flows, the accumulation shows an openwork and clast-supported texture. Based on their topographic position, these accumulations are called block-fields (felsenmeers), block-slopes or block-streams. The latter developed in valley bottoms and incisions excavated in the slopes (Washburn, 1979). These accumulations are usually found in high-latitude alpine zones.

Block-fields are extensive accumulations of angular blocks embedded in a fine-grained matrix that cover more than half of the surface. They develop in flat or gently sloping surfaces and reflect a strong frost-shattering activity. A limit of 5° is established to differentiate between block-fields and block-slopes (Smith, 1953; Dahl, 1966).

Block-slopes (Figure 8.16) are relatively common accumulations of boulder-size particles generated by frost-action on or below rocky slopes. The major axis of the clasts are commonly oriented parallel to the maximum slope direction and show a vertical sorting with the coarser particles at the top and the finer ones towards the base. An increase in the proportion of finer particles and in the size of the blocks towards the foot of the slope is also observed. Differential movements in the deposits produce lobate and

Figure 8.16. Block-slope composed of Silurian quartzite particles. Tremedal Paleozoic Massif, Sierra de Albarracín, Iberian Range Teruel Province, Spain.

bench-like shapes in the middle and lower parts of the slopes (Gutiérrez and Peña, 1977) (Figure 8.17).

Block-streams (Figure 8.18) are formed by angular frost-shattered blocks accumulated in valley bottoms. They result from the supply of blocks coming from the valley slopes. One of the most magnificent examples is found in the Tremedal Paleozoic Massif (Figure 8.19) in the Iberian Range (NE Spain) where most of the main and lateral valleys have block-streams (Gutiérrez and Peña, 1977). The largest one reaches 2.6 km in length and 0.25 km in width. The maximum thickness of these accumulations of quartzitic blocks is not known although exposures up to 4 m thick have been measured. The longitudinal profile of the valleys show highly variable slopes and locally show transverse steps that could be structurally controlled. The surface of the block-streams has small closed depressions up to several meters in diameter. Their origin could be due to the removal of fines by subsurface wash and the consequent differential settlement of the blocks. Beneath the upper blocks that show an openwork texture, the skeletal blocks are embedded in a heterometric matrix. The lack of fines in the upper part of the deposit is attributed to the wash-down of the matrix by percolation waters (Andersson, 1906; Smith and Smith, 1945; Smith, 1953; Potter and Moss, 1968). This circumstance prevents the growth of vegetation. The clasts in the block-streams of the Tremedal Massif do not show any clear fabrics unlike other examples that show lobate forms and marked upslope dips in the major axis of the blocks (Potter and Moss, 1968; Caine, 1972). The longitudinal displacement of the blocks has been negligible because none of the block-streams extend beyond of the valley margins. The blocks of these relict landforms are profusely covered by lichens and are locally being colonised by vegetation. At the present time the water flows thought the deposits as throughflow and emerges in the transverse steps. The processes involved in the generation of these depositional landforms include gelifraction, frost-creep, gelifluction, frost-sorting, and the wash down of interstitial fines (Washburn, 1979).

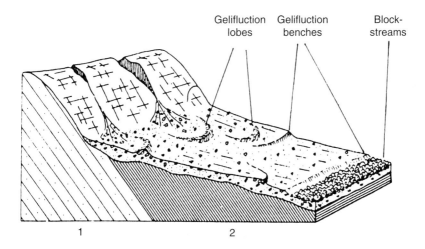

Figure 8.17. Block diagram showing depositional periglacial landforms in the slopes and valley bottoms of the Tremedal Paleozoic Massif, Iberian Range, Teruel Province, NE Spain. 1: Lower Valentian quartzites. 2: Valentian ampellitic shales (Gutiérrez and Peña, 1977).

Figure 8.18. Block-stream composed of Silurian boulder-sized quartzite particles. Tremedal Paleozoic Massif, Iberian Range, Teruel Province, NE Spain.

3.5. Rock glaciers

Rock glaciers are one of the most remarkable geomorphic features of the alpine periglacial zones and a visible expression of mountain permafrost. Active rock glaciers are defined by Barsch (1996) as lobate or tongue-shaped bodies of perennially frozen, unconsolidated material, supersaturated with interstitial ice and ice lenses that move downslope or downvalley by creep as a consequence of the deformation of the ice contained in them (Figure 8.20). There is long debate about whether classifying the rock glaciers as periglacial or glacial landform (Tricart and Cailleux, 1962), and indeed, there is evidence that some rock-glacier forms are quite polygenetic (Giardino, 1983). Some geomorphologists consider that rock glaciers may have an ice core of glacial origin (Potter, 1972). At the present time, numerous authors consider that the ice-cored rock glaciers form part of the glacial system, whereas those with interstitial ice are characteristic of the periglacial systems with permafrost. There may be also some confusion in differentiating between block-slopes, block-streams, certain landslides, and rock glaciers since a continuous transition may be observed between these landforms (Corte, 1976; Giardino, 1983).

Talus rock glaciers contain frost-shattered clasts from the talus slopes commonly fed by rock walls of glacial cirques and valleys (Serrat, 1979; Gutiérrez and Peña, 1981) (Figure 8.21). According to their morphology, the *tongue-shaped rock glaciers* have a length/width ratio higher than 1, the *lobate rock glaciers* have a length/width ratio smaller

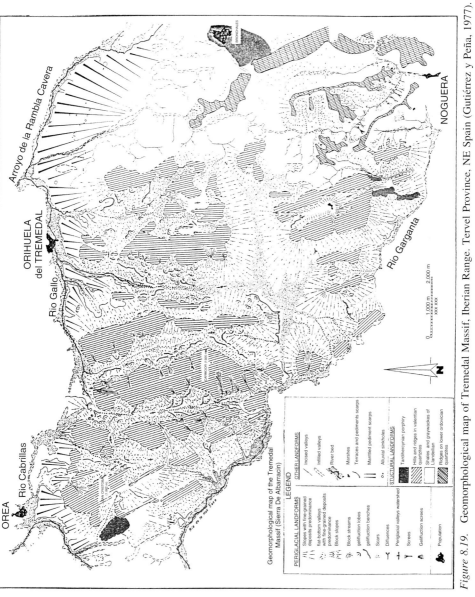

Figure 8.19. Geomorphological map of Tremedal Massif, Iberian Range. Tervel Province, NE Spain (Gutiérrez y Peña, 1977).

Figure 8.20. Rock glaciers. Clot de Claror, Aranser Valley, Eastern Pyrenees. Photo: A. Gómez-Ortiz.

than 1, and the *spatulate rock glaciers* similar to the lobate rock glaciers with a broader lower part (Wahrhaftig and Cox, 1959; Barsch, 1996) (Figure 8.22). The *complex rock glaciers* are those formed by the aggregation of several individual rock glaciers (Washburn, 1979; Barsch, 1996).

The width of the majority of rock glaciers range between 100 and 500 m and the length may exceed 1 km. Regarding the texture of the detrital material, the rock glaciers are formed by a 2 to 5 m thick upper bouldery layer underlain by a fine-grained ice supersaturated sandy and silty material (Barsch, 1996). The total thickness may reach more than 50 m. Rock glaciers show steep fronts, with inclinations between 35 and 45° (Barsch, 1988). The side slopes are also steep and the upper bouldery surface has a microrelief of ridges and furrows which are bent downvalley (Figure 8.20 and Figure 8.23). The volume transported in a rock glacier generally exceeds one million cubic meters, 40 to 50% correspond to detrital material and the 50 to 60% to interstitial ice and ice lenses. The displacement of the rock glaciers results from the plastic or creep deformation of the ice and is largely controlled by factors like the slope, ice content, temperature, or grain size of the inner clasts. The velocity of active rock glaciers, commonly higher in summer than in winter, varies from a few centimetres per year (Barsch and Hell, 1975) (Figure 8.24) to 5 m/yr. The highest values have been measured in the Obergurl rock glacier (Austrian Alps) at abrupt changes in slope. The movement of the rock glaciers is quicker than in the glaciers and slower than in the slope deposits affected by gelifluction.

Discharge of active rock glaciers is controlled by strong seasonal and diurnal variations. Snowmelt water and summer storms water are released in a very rapid way, provoking flooding events. The discharge of glaciers is very similar to that one of active

Figure 8.21. Geomorphological map of a glacial cirque in the Eastern Pyrenees, Bonaigua area. (Gutiérrez and Peña, 1981).

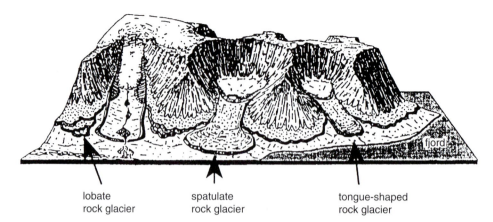

lobate   spatulate   tongue-shaped
rock glacier   rock glacier   rock glacier

Figure 8.22. Morphologic types of rock glaciers developed in mountain polar environments (Modified from Humlum, 1982, in Barsch, 1996, Figure 1.6).

rock glaciers. However, average yearly mean specific discharge in active rock glaciers is much lower than water discharge released from glaciers (Krainer and Mostler, 2002).

According to their functionality, *active rock glaciers* are those that are moving at the present time. Active rock glaciers moving at rates of 22 cm/yr have been recently reported in the Pyrenees (Spain) (Serrano and Agudo, 1998). *Inactive rock glaciers* do not show movement today but still contain frozen material (Barsch, 1988). Generally, the initial

Figure 8.23. Rock glacier with ridge and furrow topography in the headwaters of the Garona River Basin, Eastern Pyrenees.

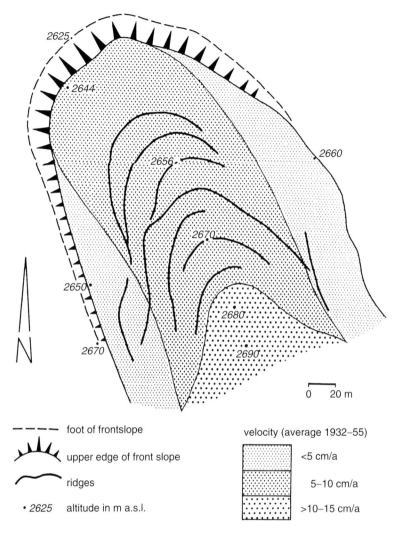

Figure 8.24. Zonation based on the mean displacement rate of the rock glacier Murtél, (Corvatsch, Swiss Alps) from measurements taken during the period 1932–1955 (23 years). The average estimated velocity for the whole rock glacier is 7.1 cm/yr (Barsch and Hell, 1975, in Clark (Ed.), 1988, Figure 4.6).

evidences of inactivity correspond to the colonisation of the front by vegetation and the development of a talus apron at the foot of the front. In the *relict or fossil rock glaciers* (Barsch, 1977b) all the ice has been melted and consequently show subsidence features and a subdued surface relief.

The rock glaciers develop preferably in mountain areas with continental and semiarid climate since wetter conditions favour the generation of glacial bodies (Höllerman, 1983). Given of the scarcity of absolute datings, Barsch (1996) considers that an age of several thousand years up to 10,000 years (postglacial) seems to be an appropriate temporal range for active rock glaciers.

3.6. Grèzes litées

The grèzes litées are slope deposits made up of alternating layers of well-sorted angular stones and finer material primarily composed of sand, silt, and clay material (Tricart, 1952) (Figure 8.25). They are also called *éboulis ordonnés* by some authors. The term *groise litées* is used when they contain a high proportion of boulder-sized particles. The coarse-grained layers have a matrix-free, clast-supported texture. Where the deposit is composed of calcareous particles, the less pervious fine-grained layer is cemented by calcium carbonate. The boundary between layers is generally abrupt and channel-shaped bases may be observed. The grèzes litées may occur in slopes with any orientation although they are more frequent in sunny-facing slopes. They are found on slopes with gradients ranging from 7 to 45°, may reach 40 m in thickness and the layers thicken towards the foot of the slope (Guillien, 1951). Grèzes litées are relatively common in mid-latitudes but they have not been documented in polar zones. Their genesis requires freeze–thaw cycles for the production of frost-shattered clasts, enough water for the sorting of the particles, and

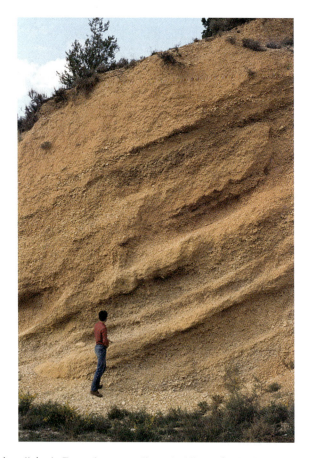

Figure 8.25. Grèzes litées in Entrambasaguas, Sierra de Albarracín, Iberian Range, Spain.

scarce or nonexistent vegetation cover (DeWolf, 1988). The stratification and sorting were considered to be related to the action of melt waters from snow accumulations, responsible for the sedimentation of the coarse material and the eluviation of the finer particles (Guillien, 1951). Francou (1988) has studied stratified slope deposits in the Peruvian Andes at 4400 to 4900 m in altitude. The slopes have an inclination between 33 and 35° and show large solifluction lobes in the surface that move a few centimetres each year. Cuttings in the front of these lobes reveal that their movement resembles the unrolling of a carpet that buries the clasts that armour the lobes and generate a stratified deposit similar to the grèzes litées (Figure 8.26). The frost-heaving lifts the clasts separating them from the fines. Besides, the clasts are displaced by frost-creep and pipkrake moving at a higher rate

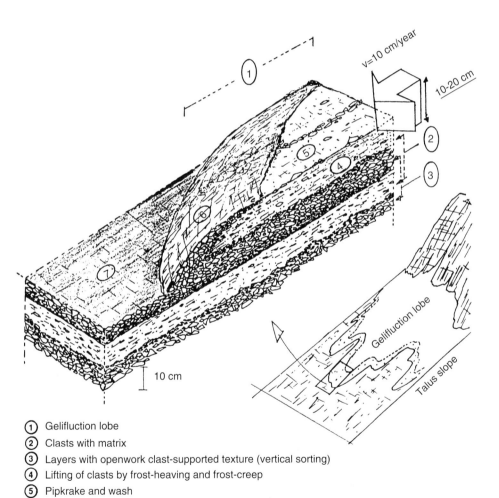

① Gelifluction lobe
② Clasts with matrix
③ Layers with openwork clast-supported texture (vertical sorting)
④ Lifting of clasts by frost-heaving and frost-creep
⑤ Pipkrake and wash
⑥ Fall and wash (longitudinal sorting)
⑦ Clasts in equilibrium

Figure 8.26. Origin of the stratified slope deposits in periglacial environments (Francou, 1988).

than the fines that form part of the solifluction lobes. Equifinality can be the cause of many difficulties in such interpretations (Steijn et al., 2002). Grèzes litées are abundant in the central Pyrenees, especially above heights of 600 to 800 m. It is thought that they have formed during more than one phase although the main period corresponds to the Late-Glacial (Peña et al., 1998).

3.7. Nivation landforms

Nivation niches or *nivation hollows* develop in slopes beneath a snow cover (Figure 8.27). The melt waters percolate through the ground beneath and adjacent to the snow patches. When this water reaches the freezing point, mainly during the night, it causes frost weathering of the substrate and the particles produced by this process may be transported downslope by overland flow, gelifluction or frost-creep. Gelifraction due to freeze–thaw cycles is the main process involved in the generation of the nivation niches. An accumulation of detritus commonly develops at the lower edge of the niche (Thorn, 1988b). In addition, the eluviation of the fine particles from the niche may lead to the development of periglacial pavements. Nivation processes are especially pronounced in low latitude mountain environments, as in some sectors of the Andean Cordillera, where more than 300 freeze–thaw cycles have been recorded (Troll, 1944). The continuous activity of nivation processes results in the widening of the niches that start as small meter-size nivation hollows and may reach 1 km by the coalescence of several niches. The enlargement processes work preferentially out on the margins where the melting of the snow and gelifraction are more active, whereas in the centre the insulating effect of the

Figure 8.27.   Nivation niches in the Hardanger Massif (Norway).

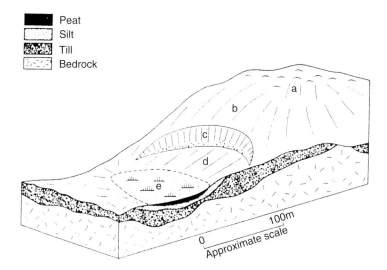

Peat
Silt
Till
Bedrock

Figure 8.28. Block diagram showing a nivation niche close to Knob Lake (Quebec): a: Bedrock outcrops in the top of the hill; b: Detritus-covered slopes (10°–15°); c: Scarped amphitheatre (30–40°) in colluvial deposits with accumulations of blocks at its foot; Low gradient slope (approx. 5°); Swampy bottom on gelifluction deposits (Henderson, 1956, in Embleton and King, 1975, Figure 5.3).

snow inhibits the freeze–thaw cycles. The niches that reach large dimensions are called nivation cirques or thermocirques (Selby, 1985). The nivation niches have an amphitheatre-like scarp with a gentle slope at its foot that facilitates the removal of detritus (Henderson, 1956) (Figure 8.28). They develop more easily in rock outcrops than in slopes covered with vegetated soils. The generation of nivation hollows in a few years beneath snow patches has been reported. The niches may be embryonic cirques when nivation is replaced by glacial processes as the accumulated snow persists enough time for its transformation into ice (Tricart and Cailleux, 1962).

Protalus ramparts (Washburn, 1979) or *nivation ridges* (Butzer, 1976) are linear or curvilinear ridges formed from the accumulation of frost-shattered debris at the foot of lingering snow patches (Figure 8.29). The rock fragments derived from rock scarps above snowfields or from the backwall of a nivation niche fall on top of the snow and roll and slide to its lower edge. Once the snow melts, the resulting ridge becomes disconnected from the previous slope profile. In some cases different parallel ridges are differentiated reflecting the parallel retreat of the slope or the reduction of the snowfield (Sharpe, 1938) (Figure 8.30).

3.8. Slope evolution

The slopes developed in periglacial environments are classified according to their shape and morphogenetic processes. The main types are concave cliff–talus slopes with talus accumulations, benched slopes that result from cryoplanation processes and the

Figure 8.29.   Protalus ramparts in Turbón calcareous Massif, Central Pyrenees, Spain. (Photo: J.L. Peña).

convex–concave slopes with a continuous cover of gelifluction material (French, 1996) (Figure 8.31).

   The cliff–talus slopes evolve by the parallel retreat of the free face and the accumulation of detrital particles in the lower segment of the slope leading to the increase in its length and the decrease of its inclination through time (Carson and

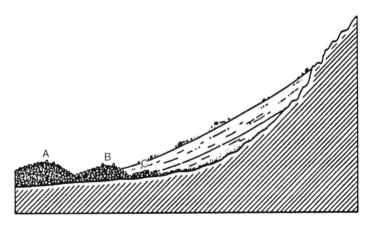

Figure 8.30.   Sketch showing a sequence of protalus ramparts generated by the accumulation of frost-shattered debris at the foot of a retreating snowfield (Sharpe, 1938, in Fairbridge (Ed.), 1968, p. 377, Figure 9).

Figure 8.31. Slopes on detrital Cretaceous sediments with a continuous cover of nonsorted deposit affected by active gelifluction. The foot of the slope has been undermined by fluvial action during the thaw season. James Ross Island, east of the Antarctic Peninsula. Photo: A. Martín-Serrano.

Kirkby, 1972). Generally, scarp retreat in periglacial environments is slower than in other morphoclimatic zones, although this process depends largely on the litho-structural characteristics of the rock face and on microclimatic conditions determined by factors such as the orientation (Souchez, 1966).

The benched-slopes result from cryoplanation processes that produce the flattening and decrease in inclination of the profile. The erosion in the upper parts of the slope and sedimentation in the lower ones, mainly in the valley bottoms, involve parallel retreat of the scarps and widening of the benched slope segments. These ideas were introduced by Peltier (1950) in his "cycle of periglacial erosion" (Figure 8.32).

The convex–concave slopes are characterised by a profile without irregularities. They are mantled by a continuous cover of frost-shattered clasts (Figure 8.33) that may be affected by gelifluction, surface wash and rilling in sectors with high gradient. The incision of these slopes by gully systems may give place to *tripartite slopes* or *triangular slope facets* (Büdel, 1982) (Figure 8.34). The gullies responsible for the dissection of the slopes develop small low-gradient debris cones in the lower part of the slope. Once the slope facet is disconnected from the upper part of the slope, it hardly receives overland flow and becomes a relict landform with a high preservation potential. The alternation of accumulation and incision periods in these slope systems gives place to sequences of tripartite slopes.

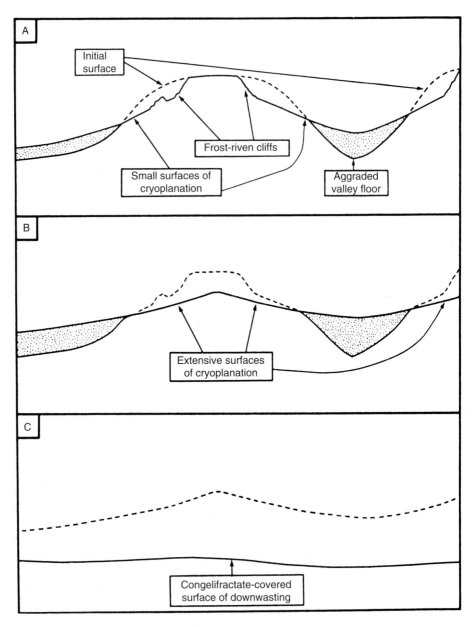

Figure 8.32. The periglacial cycle of erosion. (A) Youth, (B) Maturity, (C) Senility (following Peltier, 1950, in Small, 1970, Figure 6.5). (A) Initial surface, Scarps affected by frost shattering, Cryoplanation surfaces, Aggradation valley. (B) Large cryoplanation surfaces. (C) Surface mantles with frost-shattered clasts.

Figure 8.33. Convex and concave slopes and possibly relict slopes in Sierra de Javalambre, Teruel Province, Iberian Range, NE Spain.

4. Fluvial landforms

The high-gradient, braided channels with a high bedload and large width–depth ratio are the most common in periglacial environments. In summer, the high-discharge flows produce significant mechanical erosion in the channel banks. The run-off waters

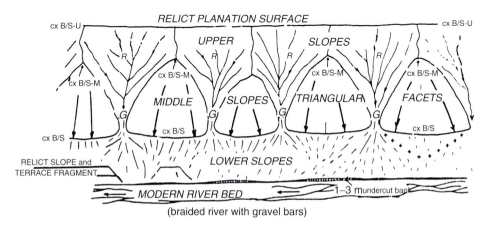

Figure 8.34. Tripartite slopes in the southeast of Spitsbergen. R: rills in the upper part of the slope, G: gullies dissecting the middle and lower part of the slope (Modified from Büdel, 1982, Figure 31).

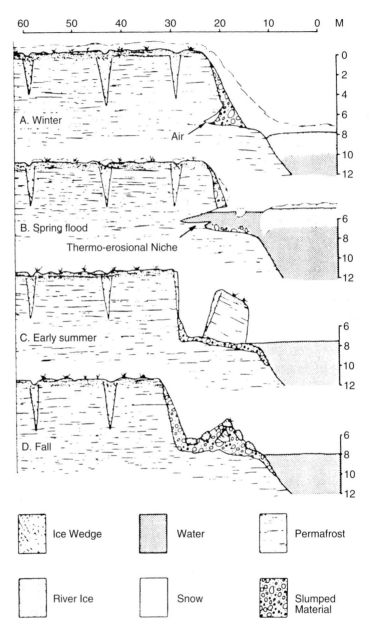

Figure 8.35. Thermal erosion on a river bank (Walker, 1973, in Fookes and Vaughan, 1986, Figure 8.7).

may thaw the permafrost and undercut the channel banks giving place to *thermo-erosion niches* that may reach 3 m in height and 1 to 3 m in width (Czudek and Demek, 1970b; Walker, 1973) (Figure 8.35). These niches eventually collapse supplying additional bedload to the river. In Central Yakutia, frozen river banks are affected by both thermal and mechanical erosion. Using some river features of the flood epochs, such as geometry, temperature and discharge, a one-dimensional model was made to predict erosion in river banks with different ice content. The results predicted by the model agree with the field observations made in the Lena River (Costard et al., 2003).

Few studies have been devoted to the morphology of the valleys developed in periglacial areas and most of the literature is focused on the problem of *asymmetric valleys*. Quite a lot of information also exists about relict Pleistocene valleys developed under past periglacial conditions from central and Eastern Europe. The asymmetric valleys are those that have one of the slopes distinctly more steep than the opposite. The asymmetry may be determined by lithological and structural factors and it is not restricted to periglacial areas. In these regions the shaded slopes are the steepest ones (north-facing slopes in the Northern Hemisphere). The earlier and longer thaw in the sun-facing slopes favour gelifluction and mass movement processes in some cases. The detritus accumulated at the foot of these slopes forces the fluvial currents to migrate towards the shaded margins of the valleys, giving place to steeper slopes by undercutting. Besides, summer thawing affects the permafrost in the shaded slopes to a lesser extent, thus contributing to preserving their steeper gradients (Gravis, 1969; Czudek, 1973). A different explanation of the asymmetric valleys is related to the preferential accumulation of snow in downwind-oriented slopes (French, 1971). During the summer the mass wasting processes caused by the meltwaters act preferentially on the downwind facing slopes, thus generating steeper gradients.

The lower-order segments of mountain drainage basins may show different morphologies. The U-shaped valleys (Tricart, 1967) with gentle longitudinal profiles develop in sectors where the surface run-off has a limited morphologic influence. The slope profiles have a convex–concave geometry because gelifluction is the dominant processes in the transport of detritus. The bottoms of the U-shaped valleys are commonly covered by non-sorted gelifluction deposits. These valleys are abundant in extensive European regions that were affected by periglacial conditions during the Pleistocene and are now mostly dry. Generally, the U-shaped valleys that are not longer than 3 km frequently grade into *flat-bottomed valleys*. In Iceland the flat-bottomed valleys are around 5 to 100 m wide, 5 to 20 m deep and have gradients between 4 and 11% (Schunke, 1975). Their bottom is primarily composed of nonsorted material derived from the slopes that has been subject to some longitudinal fluvial transport. The link between the flat bottom and the slopes commonly shows a concave geometry that indicates a limited capacity of the fluvial current to evacuate the gelifluction deposits supplied from the slopes. In some cases the link between the valley bottom and the slopes shows an angular geometry that reflects the effect of the fluvial erosion at the foot of the slopes during high stage events (Tricart, 1967). The flat-bottomed and U-shaped valleys are generally dry at the present time (Jahn, 1975). This circumstance seems to be related to the permeability and thickness of the valley fill deposits that favour the percolation and subsurface flow of the water through the alluvium. Nevertheless, it is important to take into account that dry valleys may also develop in other contexts like karst regions regardless of the climatic conditions.

5. Thermokarst or cryokarst

These terms refer to the landforms created by subsidence caused by the thawing of the upper part of the permafrost (Figure 8.36). These phenomena may give place to a very irregular topography. The word thermokarst is not widely accepted since the resulting landforms do not have much resemblance to the exokarstic features occurring in karst terranes (Pissart, 2004). The thermokarst depressions are mainly found in low-relief areas like alluvial plains, because their persistence is unlikely in zones with rugged topography (Williams and Smith, 1989; Kääb and Haeberli, 2001). The subsidence due to thawing of ground ice may be very active and consequently these periglacial landscapes change very rapidly (Harry, 1988).

The causes of the thawing are highly varied. A climate change towards milder temperatures may affect the thermal equilibrium of the permafrost in extensive areas.

Figure 8.36. Small collapse affecting slope deposits resulting from the thawing of ground ice during the summer. James Ross Island, east of the Antarctic Peninsula. Photo: A. Martín-Serrano.

The disappearance of the vegetation cover with an insulating effect due to fires, deforestation or construction may trigger the thawing of the permafrost. Other causes of a geomorphological nature are thermal erosion due to stagnant water, or mass movement processes such as slumps or mud flows that remove the active layer exposing the permafrost (French, 1987).

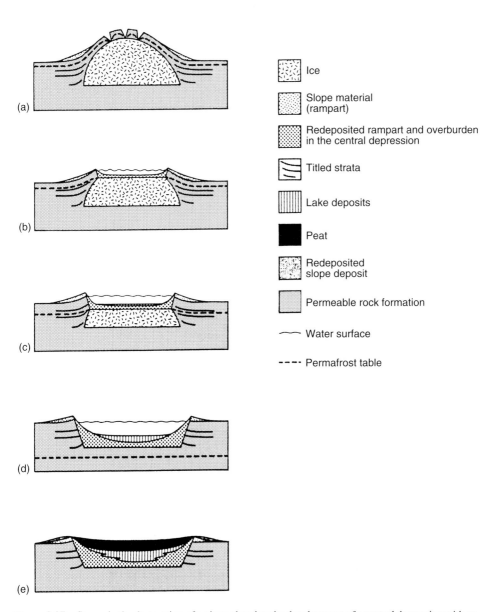

Figure 8.37. Stages in the destruction of a pingo showing the development of a central depression with an uplifted edge (Modified from De Gans, 1988).

The most common landforms are the *thaw lakes*. They are shallow lakes enclosed in basins that result from subsidence due to the thawing of frozen ground, up to 4 m deep and 2 km wide (French, 1996). The thaw lakes develop primarily in flat areas and are silted up quite rapidly by peat and fine-grained sediments rarely exceeding 3000 years old. Where they form groupings with parallel alignment, they are called *oriented lakes*.

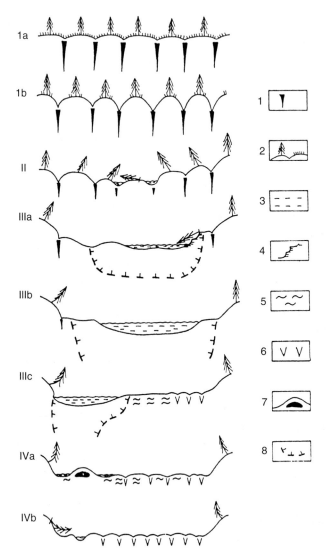

Figure 8.38. Main development stages of the alasas. (1a) Original surface with syngenetic ice wedges; (1b) initial stage of thermokarst–baydjarakhs; (II) dujoda; (IIIa) youthful alas; (IIIb) mature alas; (IIIc) senile alas; (IVa) khonu with pingo; (IVb) khonu with depression resulting from the collapse of the pingo. 1, collapsed wedges; 2: pines and grass; 3, water; 4, slope movements; 5, alas deposits; 6, epigenetic ice veins; 7, pingo; 8, upper limit of the permafrost (Czudek and Demek, 1970b, based on Soloviev publications).

The oriented lakes are unusual and the best example is located in the Arctic coastal plain of Alaska. The length of these elliptical or rectangular lakes reaches 5 km, being two or thee times larger than the width and may be up to 6 m deep (Black and Barksdale, 1949). Complex shapes due to the coalescence of several lakes are also recognised. Their uncertain origin has been related to differential thawing of the permafrost under the influence of prevailing winds that accumulate sediments in the shores perpendicular to the wind direction. The deposits insulate these shores from thawing and from wave action as they reach a very gentle profile. This situation favours the development of drift currents with high erosion capacity towards the extremes of the lakes leading to their enlargement. The mean retreat rates are very high, reaching 25 m/yr (Carson and Hussey, 1962).

Collapsed pingos may host crater-like enclosed subsidence depressions surrounded by an uplifted edge whose sediments show centrifugal dips. These ponds are filled with peat and other lacustrine–paludal deposits (De Gans, 1988) (Figure 8.37). These landforms have been identified in areas from Europe and Asia that were under periglacial conditions in the past. They may be very difficult to differentiate from other thermokarst depressions or some fluvioglacial kettles.

Where the frozen ground is exposed to the surface, it may thaw and give rise to *thaw collapses*. These are a series of linked and curved depressions opened on the downslope side that form a festooned scarp. As the uppermost part of the permafrost thaws, it becomes saturated in water and flows, thus generating a head scarp several meters high. This is one of the most rapid erosion processes of the periglacial regions as the retreat of these scars may reach 7 to 10 m/yr (French and Egginton, 1973).

The term *alas* derives from central Yakutia (Siberia) (Soloviev, 1962, 1963 in Jahn, 1975; Czudek and Demek, 1970b; Soloviev, 1973) and designates closed depressions with steep margins and a flat floor commonly covered with grass vegetation. They are rounded or elliptical in shape and in some cases host shallow lakes. Their length ranges from 0.1 to 15 km and are 3 to 40 m deep. They form from long-term localised melting of the permafrost and the coalescence of several depressions may generate large linear throughs known as alas valleys tens of kilometres long. As the thawing progresses several landforms develop (Figure 8.38). The growth of ice wedges produces linear depressions separated by domed cells called *baydjarakhs*, 3 to 4 m high and 3 to 20 m in diameter (stage I). The progressive melting of the ice produces closed depressions with conical mounds called *duyodas* (stage II). The generation of the alas (stage III) takes place when the depression has scarped edges and a flat bottom with grass vegetation and is sometimes covered with a shallow lake. If the permafrost reactivates, ice veins and pingos may be developed in the bottom of the depression (*khonu*) that may finally collapse (stage IV). In Yakutia, 40 to 50% of the original surface has been modified by alas systems that are several thousand years old and have been related to the Climatic Optimum (2500 to 9000 BP). Currently, more than 10% of the territory has active permafrost.

Chapter 9

Some aspects on applied geomorphology in periglacial regions

1. Introduction

The thermal equilibrium of the permafrost is highly sensitive in both alpine and high-latitude zones and it may be easily disturbed by human activity (Ritter, 1978). Permafrost terrains poses many problems, not only to cartographers but also to engineers and builders, miners, oil and gas producers, climatologists, archaeologists, and everyone with an interest in polar and alpine regions (Heginbottom, 2002). A significant part of the mountain environments are suffering a harmful anthropogenic pressure. Activities such as defore-station or overgrazing contribute to alter the development of some morphogenetic processes increasing the potential hazard of avalanches, soil erosion, slope movements, and floods (Gerrard, 1990) in an environment with a greater amount of potential victims. This situation is largely due to the spectacular increase experienced by the influx of tourists into the mountain zones in developed countries during the last decades. In some sectors of developing countries, the lack of space in the low areas leads to the development of marginal and unstable areas. On the other hand, in some mountain regions of these countries the human impact on the environment is becoming a worrisome problem.

In the arctic and subarctic regions, the limitations that the geomorphological environ-ment poses to economic development are mainly related to processes caused by the freezing of the ground surface (gelifraction and frost-heaving) and the thawing of the permafrost. These circumstances have contributed to restrain the development of these areas and society has been obliged to respond to these problems with new technical solutions (Walker, 1986).

The occupation of high-latitude regions by humans is relatively old (Harris, 1986; Cooke and Doornkamp, 1990). The influx of people with advanced technology with the aim to exploit natural resources is, however, very recent. These settlers mainly come from the circumpolar countries of the northern hemisphere (Russia, United States, Canada, and the Scandinavian countries). Russia is probably the nation that has acquired the largest experience in the development of permafrost areas from the construction of the Tran-siberian railway during the transit of the 19th and 20th centuries to the construction of modern cities like Norilsk. In North America the development started with gold mining in Alaska and the Yukon Territory during the end of the 19th century. It had a significant impetus after the Second World War with the construction of airfields and communication routes and reached the maximum apogee from the oil discovery in 1968 in the Prudhoe Bay in Northern Alaska (Mackay, 1973).

The permafrost areas do not cause many engineering problems unless the frozen ground is disturbed (USGS, 1983). Once a substantial disturbance is introduced into the system, thermokarst develops very rapidly and it may take many years to re-establish the initial equilibrium (Harry, 1988). These perturbations may be due to variable causes like climate changes that result in the increase of freezing or thawing. Little is known about the causes of some variations undergone by the permafrost (Williams and Smith, 1989). The response of the permafrost to changes in climate variations is generally slow due to the low heat conductivity of the ground. It is thought that the expansion of the thermokarst topography in some sectors of Russia is due to a climatic warming that has taken place during the last decades. Subtle climatic changes may produce considerable effects in areas with mean temperatures close to 0°C that cover extensive areas. These changes cause important modifications in the hydrologic systems, mainly affecting the run-off. Human-induced alterations derived from deforestation, construction of linear infrastructures (roads, railways, aerodromes, pipelines) or buildings and the drainage of lakes cause the thawing of the ground ice. Consequently the ground becomes unstable and susceptible to affects by mass movements, subsidence and frost-heaving (USGS, 1983; French, 1996).

The mitigation of the problems derived from human impact on the periglacial environments requires a good background on the processes affecting the seasonally and permanently frozen ground, evaluating their activity and understanding their relationship with the environmental variables. This knowledge largely comes from the study of the natural landforms and from experience acquired during the last few decades about the behaviour of the frozen ground in response to human activities (Harris, 1986). Another source of knowledge derives from the practice of freezing the ground to endow it with greater strength (Williams and Smith, 1989). This technique, used in the excavation of mine galleries and tunnels (the underground of several Japanese cities and the tunnel beneath the Seine River in Paris), provides valuable information about the behaviour of frozen ground. Nevertheless, some of the processes related with the freeze–thaw cycles are very complex and are not very well understood, such as the transfer of heat and moisture in the frozen ground or the rheology of the materials. An adequate understanding of the processes helps to implement proper construction practices in order to reduce the induced detrimental effects. At the present time, in 1:250,000 scale maps of sensitivity in ground classification and 1:50,000 susceptibility maps to ground modifications are produced. These documents help to select the zones where human-induced ground disturbances and the thermokarst subsidence have a lower impact (French, 1996).

2. Snow avalanches

Snow avalanches are very common in alpine mountain environments and constitute an important hazard in regions where human activities are well developed, especially tourism and recreational practices in developed countries (Gerrard, 1990) (Figure 9.1). In British Columbia, 25,000 avalanches have been recorded affecting highways (McClung, 2003). Snow avalanches may incorporate different types of detritus (rock fragments, earth, trees) increasing their destructive capacity (Keylock, 1997). They may be triggered by earthquakes (Voight and Pariseau, 1978) (Figure 9.2), the passage of skiers, thunder, vibrations caused by human activities, transit of animals, sonic booms from jet planes, and other factors.

Figure 9.1. Channeled snow avalanche accumulated in the distal sector with fan morphology, Viso Massif, Western Alps.

Snow avalanches have produced numerous disasters involving a large number of casualties and affecting human activities with consequent financial losses (Davis, 1992). At the beginning of October 218 BC, Hannibal and his army, while going to Rome, where caught unaware by a snow avalanche in the Col de la Traversette. The avalanche killed 18,000 soldiers, 2000 horses and several elephants. During the First World War between 4000 and 8000 soldiers lost their lives in the Tyrolean Alps due to avalanches. The well-known disasters produced by avalanches that buried the mining camps of Blons (Austrian Alps) (11 January 1954), trapped 300 persons, and in Camp Leduc (British Columbia) (18 February 1965), 27 miners were killed. In the Czechoslovakian Carpathian Mountains, an avalanche produced by a large amount of snow accumulated by strong winds (8 March 1956) killed 16 woodcutters in Tatra (Bolt et al., 1975). In 1910 an avalanche ran over two trains killing 96 people in Wellington (Rocky Mountains). Finally, in 1978, an avalanche swept away 60 skiers that were waiting for a ski lift in the Col de Messer (Switzerland) (Bryant, 1991).

At the present time detailed studies are carried out in some areas where avalanches interfere with human activities with the aim of evaluating and mitigating the risk. The thickness of the snow mantle is measured because the avalanche hazard increases with this parameter. Besides, the wind may redistribute the snow generating over-accumulations in downwind areas. On the other hand, the temperature has a considerable influence as it affects the physical–chemical changes (metamorphism) of the snow. Also, the stratigraphy of the snow cover (snow profiles) is studied to characterize the snow beds and to identify discontinuity planes that may constitute potential failure surfaces (McClung and Schaerer, 1993; Furdada, 1996). There are standardized procedures for

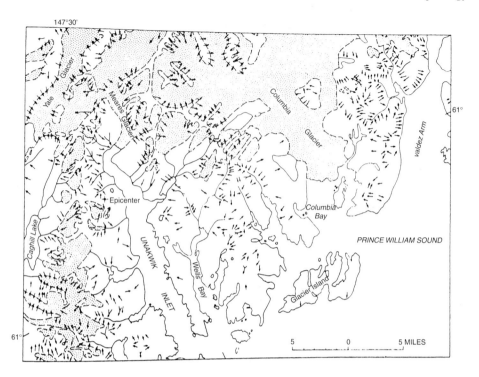

Figure 9.2. Distribution and direction of movement of the slides and avalanches triggered by a 8.5-magnitude earthquake (Richter scale) that occurred in March 27, 1964 in Alaska. The mapped area is close to the epicentre of the seismic event. From the 2036 mapped slides and avalanches, 20 involved rock, 58 snow and rock, and the rest solely snow. The shaded areas correspond to glaciers (Hackman, 1965; in Voight and Pariseau (1978), p. 10, Figure 6).

the description of several characteristic of the snow relevant for producing avalanches, such as the resistance to the penetration, shape and size of the crystals, density, water content, surface roughness, and other factors (UNESCO, 1981). The localities and trajectories of past avalanches help to identify some of the hazard-prone areas. This information is obtained from interviews, historical documents, and evidence of past avalanche activity inferred from vegetation and geomorphic features. It is also important for hazard zonation to investigate the morphology of the slopes and their orientation with respect to the insolation and prevalent winds. Avalanches show a higher frequency in slopes between 25 and 50° and have convex longitudinal profiles. In addition, studies about the return period of avalanches and their run-out distances are needed to quantitatively evaluate the avalanche hazard of certain areas (López-Martínez, 1988; McClung and Schaerer, 1993; Furdada, 1996). After McClung (2003) the factors that control the magnitude of snow avalanches appear to be: terrain steepness, starting zone characteristics, track confinement and scale, and snow supply.

The disasters caused by avalanches have decreased in recent years, thanks to hazard assessment studies and the application of prevention and correction measures. Some measurements classified as passive include the installation of wind deflectors and palisades to avoid the preferential accumulation of snow, the construction of terraces or obstacles (fences, dikes, berms) on the slopes to stabilize the snow cover, dissipate the kinetic energy of the avalanches, or deflect them (Figure 9.3 and Figure 9.4). False tunnels are also built in locations where roads and railways cross avalanche tracks. Active methods lie in triggering avalanches artificially to avoid the development of large avalanches at an uncertain time. Traditionally, the most widely used technique was the use of explosives in the starting zone (Bolt et al., 1975; McClung and Schaerer, 1993).

The incorporation of avalanche hazard mitigation strategies to land-use planning requires the elaboration of hazard and risk maps. To make specific decisions these maps need to have a large scale, generally larger than 1:50,000. They allow us to avoid the most hazardous areas, identify the high-risk zones for the application of correction measures where properties and activities are endangered by avalanches, and to design warning and evacuation plans. Some of these maps differentiate the zones of the avalanche tracks (starting, travel, and stop zones). They are generally in colour and red indicates the high-hazard zones (building is prohibited), the blue, medium hazard (building is allowed with some restrictions), and the white, hazard-free areas (Gerrard, 1990). Furdada (1996) has carried out an avalanche hazard map in a sector of the Eastern Pyrenees using a GIS. The methods used for the compilation of data have been aerial photograph interpretation, field survey, and interviews of the local people. This work is an important contribution in the management of the natural hazards and land-use planning in the Spanish mountain areas.

Figure 9.3. Palisades in the slopes to hold the snow and retain snow avalanches. Pordoi Pass, Dolomitic Alps.

Figure 9.4. Palisades in the starting zone of snow avalanches in Val d'Isère, French Alps (Photo: F. Gutiérrez).

3. Mass movements

The slopes in mountain regions are particularly unstable due to their high inclination and the presence of strongly fractured rocks and unconsolidated Pleistocene deposits such as glacial till, fluvioglacial sediments or gelifluction deposits. Mass movements are commonly triggered by large pluvial precipitations, the rapid melting of snow cover, river under-cutting, high-magnitude earthquakes, deforestation, and other factors. Large slope movements may cause the damming of river channels or fluvial valleys with consequent upstream flooding. The floods derived from the catastrophic failure of these natural dams may be highly destructive (Jones et al., 1983). The high slope movement hazard in mountain areas makes the construction and conservation of linear infrastructures difficult (Selby, 1993). The elaboration of hazard maps and their incorporation in land-use planning may generate significant cost savings and avoid the loss of human lives.

The communication routes that cross talus slopes and talus cones should preferably follow the lowest sector of the slope where the frequency and energy of mass movements is lower. Cuttings into these deposits may easily affect the stability of the accumulation of frost-shattered clasts. In some mountain areas like the Peruvian Andes, some roads require the daily removal of clasts (Gerrard, 1990).

There is a very large occurrence of large rock avalanches in mountain regions involving the displacement of several millions of cubic metres at velocities higher than 100 km/h. The rock avalanche triggered in Mount Rainier in Washington by the 1964 earthquake travelled 7 km with a height difference of 2000 m, and the Sherman Glacier rock avalanche covered 8.25 km^2 in Alaska. The physical damage caused by these rock avalanches was limited. The translational rockslide of Mayunmarca village, in the central Peruvian Andes that occurred in April 1974, however, killed 451 people. This mass movement was caused by undercutting of a slope by the River Mantaro (Voight and Pariseau, 1978). The historical and geological record demonstrates that mass movements commonly develop several times in a particular location showing a recurrent character. Inzing village and several sectors of the Voralberg in the Austrian Alps have suffered from recurrent and catastrophic mass movements during the last centuries (Gerrard, 1990). Despite the numerous protection measures, on 26 July 1966, Inzing village was affected by a channelled debris avalanche that travelled at 100 km/h killing three people and burying 4 km of road and 500 m of railway. A great part of the protection structures were destroyed.

In some circumstances active rock glaciers may constitute a geomorphological hazard. The damage caused by their activity results from the downslope flow of the frozen material, the subsidence and generation of depressions (several metres or tens of metres in size) caused by the thawing of the underlying ice and the eluviation of fine particles, and the fall of surficial blocks related to the higher speed of the upper bouldery layer (this process may reach rates of 1 block per minute) (Giardino and Vick, 1987). These sites should be avoided for the installation of structures like poles or towers for electric lines or ski lifts. In the Rocky Mountains of Colorado, USA, communication routes and pipelines that cross active rock glaciers need periodic maintenance work and in some cases the courses have had to be modified. Tunnels excavated in rock glaciers as in the case of Mount Mestas, Rocky Mountains, Colorado, are affected by deformations and collapses due to the hydrostatic pressures related to thawing of the ice (Giardino, 1983).

The development of slides affecting the active layer is a very common process in permafrost areas during the melting season. These mass movements generally have a shallow failure surface and may reach high speeds (Harris, 1986). In riverbanks, the generation of thermoerosion niches allows the formation of falls, topples, and debris flows with speeds reaching up to 10 m/s (Harris, 1981). In addition, the presence of periglacial deposits with low shear strength is relatively common in mid-latitude areas subjected to periglacial conditions during the Pleistocene period. The fulfilment of construction works in these unstable materials may pose numerous problems (Derbyshire, 1977).

4. Engineering problems in permafrost areas

The thermal equilibrium of the permafrost is easily disturbed by human activities causing its degradation or expansion. The reduction or removal of the vegetation cover favours the degradation of the permafrost. In recent times the construction of buildings, roads, railways, pipelines, etc., is one of the main causes of the alteration of the permafrost. Important disturbances may also be due to changes in the natural hydrological conditions produced by the diversion of rivers, construction of reservoirs, or desiccation of peat bogs (Haugen and Brown, 1971). The main effect derived from these activities is the degradation of the permafrost, usually an irreversible process. The frozen ground, due to its low heat conductivity, needs years or decades to attain the equilibrium for the new thermal conditions imposed by the man action (Cooke and Doorkamp, 1990).

The study of the permafrost areas before the development of any activity capable of disturbing the equilibrium of the frozen ground may help to avoid problems and generate substantial cost savings. The analysis of the area of interest with aerial photographs allows us to differentiate types of ground based on criteria like the vegetation that reflects soil characteristics such as moisture content and texture. Subsequently, field studies should be carried out to examine the distribution and nature of the permafrost and the active layer, drainage characteristics of the ground, surficial materials, temperature and humidity of the soil and atmosphere, and other features (Cooke and Doornkamp, 1990).

The methods used for the avoidance and mitigation of geotechnical problems are classified into active and passive (Muller, 1945). In ground sensitive to thaw passive methods are the most frequently used. These methods are based on keeping the materials in their initial thermal state to avoid settlements. The active methods generally involve the replacement of the material sensitive to the freezing by others more resistant to this process. Additionally, structures capable of resisting the stresses derived from alterations in the thermal regime causing heaving and settlement deformations are also designed.

4.1. Buildings

The first constructions in periglacial regions were founded on wooden piles as basements. From the use of concrete for the supporting structures, numerous buildings started to be affected by cracks and differential settlements (Harris, 1986). One of the most outstanding examples of construction problems in permafrost areas is Atlavik village in the Mackenzie River Delta (Cooke and Doornkamp, 1990; French, 1996). The village was founded in 1912, and in 1950 the permanent population reached 400 people, rising to 1500 in summer with the arrival of Eskimos and Indians. The later growth of the village with the use of inadequate construction techniques entailed numerous stability problems in the buildings. Atlavik was settled on fine-grained and poorly-drained deltaic sediments with a high proportion of interstitial ice. The new development caused local uplifts and settlements in the ground and the generation of flood-prone depression. The adverse situation led to the abandonment of the site and its relocation in an emplacement with more suitable geotechnical conditions. The construction of a new village called Inuvik, 48 km away from Atlavik, started in 1955 and at the present time is inhabited by some 1500 people. The selection of the site, that combines all the suitable conditions for the foundation of a new locality, was based on detailed studies. This example demonstrates the need of having a precise knowledge of the permafrost dynamics and its local conditions for the avoidance of damages in buildings.

The capacity of the ground to bear a load without undergoing deformation depends on the type of material and may change significantly during the freezing and thawing periods

(Swinzow, 1969). The deformation of the building structures takes place during these periods as the underlying and adjacent ground changes its strength due to natural and human-induced temperature variations. During the freezing period the buildings may rise and break. The thawing of the ground ice involves a reduction of the volume and strength of the ground leading to the settlement of the structures. Flowage of sediments may take place in low permeability clay- and silt-sized soils that retain a great part of the melting waters. Besides, as the water cannot percolate downwards through the permafrost, the fine-grained sediments of the active layer may become highly plastic causing differential settlements in the buildings. Conversely, the coarse-grained sediments are fairly stable since they easily drain the interstitial water. In case there is progressive warming through the years in addition to the seasonal thawing, the deformation increase gradually together with the thickening of the active layer (Lobacz and Quinn, 1966) (Figure 9.5). On the other hand, the stability of the substratum is much higher when ice occupies the interstitial pores than when it forms masses (Harris, 1986). Consequently, the foundations are a very important issue in the construction of buildings in permafrost areas.

The top of the permafrost may rise beneath and next to uninhabited buildings as they produce a shadow effect isolating the ground from the solar radiation (Muller, 1945) (Figure 9.6). When the buildings irradiate heat, an asymmetric thawing bulb develops beneath the building. The asymmetry of the bulb is related to differences in the solar radiation between the sunny and shaded faces of the house. These variations entail differential vertical movements that may produce uplifts or settlements in different corners

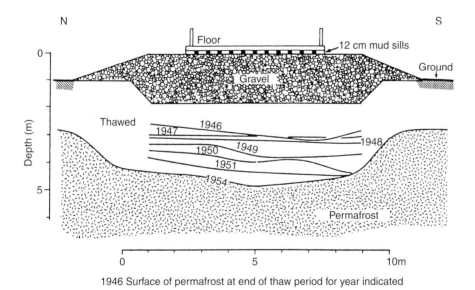

Figure 9.5. Gradual degradation of the permafrost and thickening of the active layer beneath an artificial gravel bed at the base of building 4 in Alaska Field Station, 4.5 km to the northeast of Fairbank, Alaska. The active layer was 0.9 m thick before the construction of the building in 1946 and increased to 1.75 m later than the removal of the tree and shrub cover. The lines indicate the position of the permafrost at the end of the thawing period on different dates (Lobacz and Quinn, 1966, in Harris (1986), Figure 4.1).

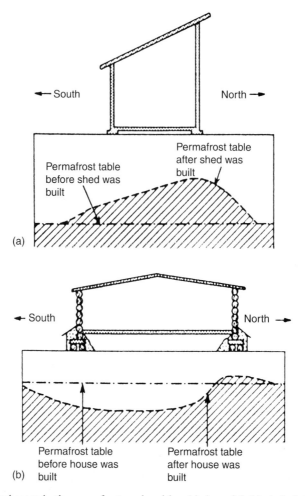

Figure 9.6. Disturbances in the permafrost produced by: (a) An uninhabited shed. (b) A house with heating (Muller, 1945, in Cooke and Doornkamp (1974), Figure 9.7).

of the buildings (Lobacz and Quinn, 1966) (Figure 9.7). These movements may cause breakages in the foundations and cracks in the walls of the buildings. The thawing-induced settlements take place in a progressive fashion and stops once the thermal equilibrium is reached. These processes may last several decades.

The construction techniques are varied and in areas with continuous permafrost are directed towards maintaining the frozen layer intact. The base of the buildings is frequently built about 60 cm above the land surface on piles of wood, steel or reinforced concrete anchored in the permafrost (Harris, 1986) (Figure 9.8). In this way the heat irradiated by the building is dissipated as the air may circulate between the ground and the base of the structure avoiding the thawing of the permafrost during the winter. During the summer the thawing is reduced since the buildings shade the ground. Artificial refrigeration may be also installed around the piles (cryoanchors). In addition, the upper

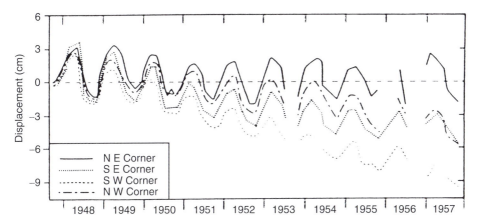

Figure 9.7. Differential uplifts and settlements recorded in the four corners of building 9 in Alaska Field Station, 4.5 km to the northeast of Fairbanks, Alaska. The movements result from seasonal freeze–thaw cycles (up to 5 cm) and the long-term gradual thawing of the permafrost (up to 10 cm in 10 years) (Lobacz and Quinn, 1966, in Harris (1986), Figure 4.3).

Figure 9.8. Several buildings in the Argentinean Comodoro Marumbio Base in James Ross Island, east of the Antarctic Peninsula. The buildings are supported by protruding piles allowing the free circulation of the air beneath the base of the structures with the aim of dissipating the irradiated heat (Photo: Martín-Serrano).

layer of the foundation is provided with ventilation pipes to assist in the cooling of the permafrost. Finally, some buildings are constructed on a gravel bed (Figure 9.5) that favours drainage and prevents swamping and thermal erosion.

4.2. Lineal infrastructures

During the end of the 19th century the colossal work of the Trans-Siberian Railway was developed for the wood transport towards the west. Subsequent to the Second World War numerous roads, railways, and airfields were built in Canada, Alaska, Siberia, and Mongolia (Figure 9.9). The roads and railways with permanent service are commonly built in areas with discontinuous or sporadic permafrost where the ground temperature is around $-2°C$ and the structures are more easily damaged (Harris, 1986).

The problems that affect the linear infrastructures in permafrost areas are varied and complex. They include differential vertical movements related to the removal of the vegetation cover that involve the alteration of the permafrost equilibrium and the consequent thermokarst. Floods may destroy bridges during the melting period. Mass movements may bury large road and railway stretches. Besides, the melt waters accumulated on the infrastructures may freeze producing icing (Ritter, 1978).

The simplest way of transport in the permafrost areas are the so-called winter roads (Harris, 1986). These are routes located on snow mantles or frozen rivers and lakes. Special vehicles are required for transit on in these temporal roads. The railways and roads

Figure 9.9. The Argentinean Comodoro Marumbio Base built on James Ross Island, east of the Antarctic Peninsula. The photo shows the presence of a muddy soil (mollisol) in the airfield that causes significant problems during the summer season (Photo: Martín-Serrano).

with continuous use require costly special construction techniques. Preferably they should follow well-drained areas with a gentle topography to avoid cuttings. Generally, the key problems are the adequate selection of the trace and the necessity of having available grounded gravels for the fill. The volume of rock or gravel needed for the fill is usually very large since this is commonly 0.6 to 1.5 m thick (Linell and Johnston, 1973). The fill is installed to support the load and vibrations induced by the transport and is penetrated by the permafrost. The thickness required for the fill may be reduced with the use of insulating materials like polystyrene layers (Figure 9.10).

In permafrost areas the large rivers are commonly crossed with ferries. However, in some cases the construction of bridges is necessary. The lateral migration of the river may modify the talik developed beneath the channel causing subsidence of the bridge piers (Figure 9.11). In case the channel shifts significantly, the permafrost may undergo local aggradation involving an uplift (French, 1996). For these reasons, one of the main protection measures is stabilization of the river channel.

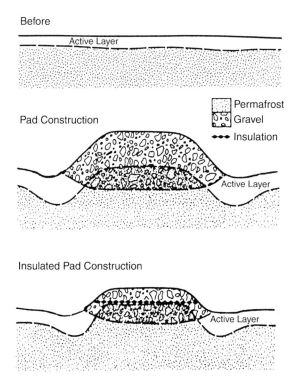

Figure 9.10. Effects on the top of the permafrost beneath a road caused by gravel fills with or without an insulator. The thickness of gravels required for fill is reduced with the use of an insulator (following Harris, 1986).

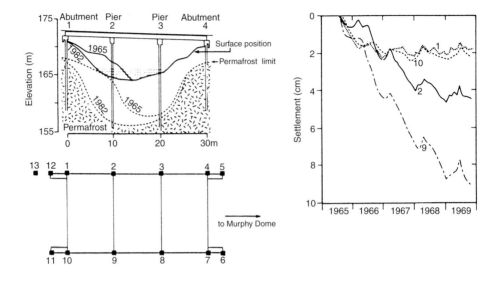

Figure 9.11. Cross section of the Goldstrean Creek Bridge and displacement through time of some piers (Crory, 1985; in Harris (1986), Figure 5.9).

4.3. Oil and gas pipelines

Pipelines are relatively abundant in periglacial regions such as Siberia, Canada and Alaska. The most well known one is the 1300-km long Trans-Alaskan Pipeline that was built after the discovery of a large oil field in 1968. It starts in Prudhoe Bay, on the arctic coast, and finishes in Valdez in the Alaska Gulf. The oil is transported by ship from this harbour to refineries located in the western coast of the United States and in Japan. The pipeline was built between 1969 and 1977 at a final cost of 7000 million dollars, eight times higher than the initial estimate. About 70 % of the pipeline was built on permafrost zones (Harris, 1986).

Numerous and complex problems were faced during the construction of the Trans-Alaskan Pipeline. One of the most relevant ones was related to the temperature of the oil that was introduced into the pipe at 58°C. A pipeline buried beneath the surface and transporting hot fluids causes the thawing of the permafrost. With no protection measures, the thawing of the permafrost expands through time around the pipe (Lachenbruch, 1970) (Figure 9.12). The speed of the thawing expansion decreases with time and never reaches an equilibrium state. Consequently, the ground progressively looses its mechanical strength and eventually is unable to support the pipe. The lack of bearing capacity of the ground leads to the generation of settlements at the surface and deformations in the pipe. In addition, differential uplifts and settlements are very common and of variable magnitude as the pipes cross zones with different types of material. Where it crosses ice wedges the detrimental effects due to thawing occur rapidly. Additional problems also rose from the destruction of the vegetation cover during the construction (Ritter, 1978).

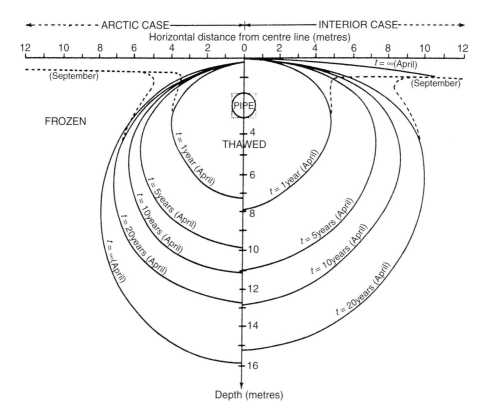

Figure 9.12. Theoretical growth of a thaw cylinder around a heated pipe installed in silty ground. The pipe is 1.21 m in diameter, is located at a depth of 2.42 m and maintains a temperature of 80°C. The curves to the left correspond to conditions similar to those of the arctic coast of Alaska, whereas the curves to the right correspond to conditions close to those at the southern limit of the permafrost (Lachenbruch, 1970, in Cooke and Doornkamp (1974), Figure 9.8).

As a consequence of these difficulties, in permafrost zones, the pipeline was built suspended on a structure with pillars that penetrated in the ground from 8 to 20 m in depth (Harris, 1986). In some stretches these pillars are provided with a refrigeration system to avoid the thawing of the permafrost. The pipe was buried in permafrost-free zones and in areas devoid of potential problems (well-drained sand and gravelly areas). Similar problems are found in the construction of gas pipelines. When possible these are usually buried to reduce the explosion hazard and to help the temperature of the pipe raise the resistance of the steel to the gas pressures (Harris, 1986).

4.4. Mining

In arctic areas, where mineral resources are abundant, some classical mining practices have been modified due to the peculiar conditions that the permafrost imposes. The main problems are related to the freezing state of the rocks and minerals and with the ice content of the minerals. In addition, the transport of the minerals is generally restricted to the thawing season (Harris, 1986).

The exploitation of placer deposits requires the thawing of the detrital deposits previous to mechanical treatment. The ground ice is thawed with water under pressure. In former times hot water was used but this method was costly and of small efficiency. These practices have a high cost and, therefore, are only applied to high-value ores like gold, platinum, cassiterite, and wolframite. In Alaska, the main mined placer deposits are the sandurs developed at the foot of the mountains by melt waters derived from the Pleistocene glaciers (French et al., 1983).

In open-cast mines the mineral-bearing rocks have a lower ice content than the placer deposits. One of the most relevant problems is related to the stability of the excavation fronts whose mechanical strength is reduced by the melting of the ground ice (Harris, 1986). The extraction of the mineral and the accompanying rock is carried out with perforation and explosive networks. Boreholes drilled with the use of water may have problems because it may freeze. For this reason, dry holes are employed on some occasions. The continuous melting in these excavations makes it necessary to drain the surficial water to facilitate the works. Obviously, the profitability of these costly workings is highly dependent on the variations of the mineral price.

The presence of permafrost affects the strength and stability of the country rock in underground mine workings. The entrances to the galleries are highly unstable due to the thawing of the permafrost and have a high rock-fall hazard (Linell and Lobacz, 1978). Most of the mines in permafrost zones are ventilated with air at 5 to 10°C that causes the thawing of the roofs and walls affecting to their stability. The melt waters generates a continuous water flow that has to be extract with pumps (Harris, 1986). In some cases the water may penetrate in a sudden manner into the mine galleries causing floods capable of causing tragic effects.

4.5. Other activities

Water is essential for the maintenance of settlements in periglacial regions. It is necessary to have good quality drinkable water. Water for industrial use is commonly derived from the treatment of waste waters (Harris, 1986). In these regions it is generally difficult to obtain perennial water at low cost. Underground water may be found beneath or above the permafrost or within the talik (French, 1996). The water obtained has to endure the whole year and for this reason it is generally stored in deep reservoirs to mitigate the effect of the thick ice layer that forms during the winter season.

A common problem faced in this region is also the disposal of waste waters and solid wastes. These may pass on diseases and pollute the environment. Several methods are used for their elimination (physical–chemical, biological, grinding, incineration, etc.) (Harris, 1986).

In some areas with continuous and discontinuous permafrost as in Yakutia (Siberia), there is extensive grain agriculture. Here, the thawing caused largely by agricultural practices produces thermokarst phenomena (Czudek and Demek, 1970b) and machinery frequently sinks in thaw holes. Usually, the alas topography makes agricultural works difficult. Vegetable crops were introduced into these areas during the last decades of the 20th century for self-sufficiency. These cultivations are carried out in buildings to withstand the strong frosts. The ground is generally composed of sand and gravel for a good drainage and dripping irrigation is used (Harris, 1986). Thawing-related problems are common. In alpine areas, the lower permafrost limit is commonly below the tree line. Consequently, wood exploitation may produce undesirable thermokarst processes.

FOURTH PART
Geomorphology of arid regions

Chapter 10
The arid zones

1. Introduction

The arid regions are areas of scarce precipitation and as a consequence the vegetation cover is reduced or nonexistent (Figure 10.1). They present a variable distribution, due to their location in high- and low- latitude areas, in the inner part of continents and littoral zones and also in zones of high altitude. Cold deserts in medium latitudes produce a particular problem, different to those in warm zones, because they are affected by the activity of periglacial processes.

The exploration of many of the arid regions is related to colonisation by Europeans of these areas, as they searched for new natural resources. The times of this exploration were highly variable and are basically circumscribed by modern and contemporary times. The first documents known are of a general character and they emphasise geographical and naturalistic aspects. At end of the 19th century the first geomorphological works of great scientific rigor were carried out, such as those made by Gilbert, Powell, and Dutton in their study of the deserts of the southwest United States. Despite these precedents, however, the studies have been generally of a descriptive character with the use of local terms, which has produced a certain confusion of ideas among geomorphologists. In the late 20th century with the advent of dynamic geomorphology, which analysed geomorphic processes by means of the application of physical and chemical techniques to explain existing forms, there has been a considerable advance in the knowledge and evolution of landforms of the arid zones.

The geologic, climatic and vegetation characteristics of warm desert environments are highly variable (Goudie, 2002). These overlay geologic formations of different composition and age, with a full landform stability or, to the contrary, with a marked neotectonic activity. The climates of deserts are quite variable and characterised generally by scarce precipitation and high temperatures with a great daily amplitude. Vegetation is sparse or nonexistent but well adapted to the high range of temperatures. They fundamentally obtain water through their roots, which can penetrate up to 20 m deep (Nicholson, 1993).

On the other hand, desert environments are not suitable conditions for humans due to the extreme climates and the scarcity or absence of water and food. Human behaviour in these areas varies from place to place, because in some places zones of population are lacking and others, on the contrary, are densely populated. It is important to note that 15% of the worldwide population lives in these regions (Heathcote, 1983). Moreover, the growth of the population is very important in many of these zones, with increments of 50% or more in the last 50 years. It was calculated that the population would reach 1200 million by the year 2000. Therefore, the use by humans of scarce hydric and vegetable resources

Figure 10.1. Region of strong aridity devoid of vegetation. Negev Desert, Eilat, Israel.

passes on to later generations numerous environmental problems of difficult and expensive resolution (Cooke et al., 1982; Goudie, 1990a).

Dry regions have been defined in different ways according to the function of the required aim. At the same time, different criteria have been used for their differentiation, among the more common of which have been based on vegetation, systems of drainage, soil degradation, and water or eolian erosion. Nevertheless, the criterion most used is the availability of water in the ecosystem, because it so affects the flora and fauna and the dynamics of the geomorphic processes. Water availability becomes highlighted for its water balance, and in the warm deserts the loss of humidity by evaporation and transpiration (ET), together with the subterranean and superficial loss of drainage (D), is greater that the precipitation (P). This degree of difference is obtained by means of aridity indexes, based on relations between precipitation and temperature. The aridity varies enormously for similar precipitation. For example, the Sahel, the Great Plains of the West United States and the driest areas of central Europe all have an annual average rainfall of 400 to 500 mm; nevertheless, the existing vegetation in these regions varies from the steppe to dense forest (Nicholson, 1993). Thornthwaite (1948) established his index of humidity (Ih) taking into account the evaporation and transpiration. Meigs (1953) carried out the division of the arid zones, based on this index, and differentiated three types of arid climates: hyper-arid (Ih < −56), arid (−56 ≤ Ih − 40), semi-arid (−40 ≤ Ih − 20), ranging the sub-humid climate between − 20 and 0. UNESCO (1979) scientists published their distribution of the arid zones of the world based on the data of 1600 meteorological stations. This work has been modified recently for the World Atlas of Desertification (UNEP, 1992; Middleton and Thomas, 1997), elaborated in the Program of Environment of the United Nations. Both are based on the degree of bioclimatic aridity that is defined for the relationship between precipitation and potential evapotranspiration. They distinguish four types of aridity: hyper-arid (P/ET < 0.05), arid (0.05 ≤ P/ET < 0.20), semi-arid (0.20 ≤ P/ET < 0.50) and, finally, dry sub-humid between (0.50 ≤ P/ET < 0.65). Grove (1977) performed a simple differentiation, keeping precipitation exclusively in mind, and distinguished semi-arid (200 to 500 mm), arid (25 to 200 mm) and hyper-arid zones, in which a seasonal distribution of precipitation doesn't exist and where there are periods over 12 months without measurable rain.

The area occupied by all the arid zones of the world varied according to the function of the climatic classification utilised. According to the World Atlas of Desertification (1992) and Middleton and Thomas (1997) 37.3% of the earth is occupied by arid zones, corresponding to 17.7% for the semi-arid regions, 12.1% for the arid areas, and 7.5% for the hyper-arid zones (Figure 10.2 and Table 10.1). There are problems of delimitation between these regions, due to the great interannual variability of the precipitation and because the transition from one to the other is gradual. Because the limits are defined as a function of vegetation, the establishment of precise limits can be difficult, inasmuch as the precipitation that directly affects the development of vegetation is variable and, on the other hand, human action modifies it considerably in some regions (Thomas, 1989).

The major extension of arid zones occurs in a wide fringe developed in the north of Africa and Asia (Figure 10.2) and comprises the deserts of the Sahara, Arabia, Pakistan, Indian, and Central Asia. In South Africa occur the coastal desert of Namib–Angola and those of the Kalahari and Karroo. The Australian continent, which lacks hyper-arid zones, is occupied by deserts over 75% of its surface. In the western United States and in Mexico there is another important nucleus of aridity. Finally, in western South America one finds the coastal desert of Peru–Chile and the Andean and Patagonian arid regions.

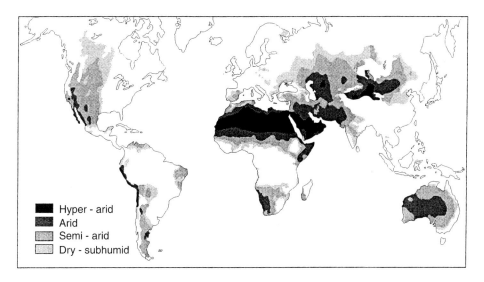

Figure 10.2. Distribution of the arid zones in the world (UNEP, 1992).

Table 10.1. Aridity zones by region in millions of hectares (UNEP, 1992).

Zone	Region						Total	%
	Africa	Asia	Australasia	Europe	North America	South America		
Cold	0.0	1082.5	0.0	27.9	616.9	37.7	1765.0	13.6
Humid	1007.6	1224.3	218.9	622.9	838.5	1188.1	5100.4	39.2
Dry subhumid	268.7	352.7	51.3	183.5	231.5	207.0	1294.7	9.9
Semi-arid	513.8	693.4	309.0	105.2	419.4	264.5	2305.3	17.7
Arid	503.5	625.7	303.0	11.0	81.5	44.5	1569.2	12.1
Hyper-arid	672.0	277.3	0.0	0.0	3.1	25.7	978.1	7.5
Total	2965.6	4256.0	882.2	950.5	2190.9	1767.5	13012.7	
% Continent	56.9	37.5	69.4	12.2	23.9	18.9		
% Total global	34.8	32.9	12.6	2.4	10.4	6.9		

2. Causes of aridity

Aridity is controlled by a conjunction of climatic, orographic and oceanographic factors. In some zones the scarce disposition of water is basically due to a concrete cause, whereas in others the aridity results from the superimposition of various factors (Dresch, 1982; Thomas, 1989,1997b; Cooke et al., 1993; Nicholson, 1993). The following are the main controls.

The *zonal factor* occurs in relation to the trajectory of anticyclonic cells that control the tropical and subtropical deserts. Occasionally these deserts can be penetrated by storm systems, although precipitation is scarce. Zones of great aridity, surrounded by small semi-arid stripes, characterise them. The most representative example is the Sahara desert.

Continentality refers to the distance from the ocean of these deserts. In temperate latitudes, the rain-producing fronts gradually lose their humidity along their way towards the interior of the continents. In these regions cold winters are common. In them the more arid zones occupy little surface and are bordered by semi-arid areas of great extension. The mid-latitude deserts of Central Asia have these characteristics.

The *orographic effect* occurs because of the presence of rain-shadow zones situated on the leeward side of mountainous chains, in regions with dominant winds, as on some west coasts or in the Trade Winds belts. When air masses descend the slopes of a chain of mountains they heat up and dry adiabatically. This orographic effect can be emphasised by the zonal situation and the continentality. The mountain chains on the western part of the American continent act as such a rainfall barrier.

The coastal deserts of Perú–Chile (Figure 10.3) and Namib–Angola owe their arid condition to the existence of *cold ocean currents*, which correspond to the currents of Humboldt and of Benguela, respectively. The evaporation of the sea surface is small and precipitation, fundamentally only fog and dew, are low or almost nonexistent. The air masses condense the water vapour from the cold current, forming coastal fogs and mist, which heat up when they reach the coast, reaching the saturation point.

Figure 10.3.   Marine terraces spaced up to 300 m on the sea level in the Peru–Chile desert, which is the most arid of the world. San Juan of Marcona, south of Peru. Photo. C. Zazo.

Also it is necessary to consider the *albedo effect* or *reflectivity* of the desert surfaces. In warm deserts the heating generates convection currents. Only in deserts of white tones can the reflection favour thermal diurnal losses. On the contrary, in rocky deserts the absorption of the warmth is maximum, especially if a cover of desert varnishes exists on it.

3. Desert climatic features

There are very marked differences between cold and warm deserts. According to Meigs (1953) the limit between both is designated by the mean temperature of the coolest month, which when it is less than 0°C corresponds to a *cold desert*. These are situated fundamentally in the middle latitudes. The most significant characteristic of these deserts is the great thermal annual oscillation that occurs. So, in the Karakum desert of western Asia, temperatures of the soil surface have been registered that vary between 79.4° C and − 40°C (Nicholson, 1993). Precipitation is influenced by the topography and in the high zones is frequently snowfall. Evaporation is low during the cold months and is commonly associated with strong winds.

The *warm deserts* are located in the tropical and subtropical zones. In them the temperatures are high, the precipitations very variable, the evaporation high, and the action of the wind can be very important.

Due to their latitudinal position and to the anticyclonal domain, some deserts suffer a great insolation and the air *temperatures* can reach considerable values, as for example in the Sahara, or Death Valley, as much as 58°C (Nicholson, 1993). As the sky at night stays clear the temperatures diminish considerably by night-time irradiation and the thermal diurnal oscillations can be about 22°C in deserts of high latitudes, although values of as much as 56°C have been registered in Tucson, Arizona (Goudie and Wilkinson, 1977). Annual oscillations are high and seasonal contrast increases with the increase of the latitude. The temperature of the terrestrial surface is higher, having been measured at 83.5°C in the sands of Port-Sudan, at the Red Sea.

In coastal deserts the temperatures are less contrasted with low daily and seasonal intervals, unlike the inland deserts in which oscillations are very marked. In the littoral deserts affected by cold oceanic currents, the daily variations are some 11°C, half of those registered in the inland deserts, and the annual oscillations are about 18°C (Goudie and Wilkinson, 1977).

Precipitation in deserts can be extreme, with dry periods interrupted by convective rains. This great temporal variability is associated with spatial irregularity. These variations increase with the decrease of the annual precipitation. The variability can be given by an index, expressed in a percentage that results from the relationship between the deviation of the mean and the average value of the annual precipitation. In the big deserts (Figure 10.4) the value is up to 30%, which corresponds to the arid and hyper-arid zones (Goudie and Wilkinson, 1977). This space and temporal variability is due to the stormy character of the dominant rainfall, which constitutes a large amount of the total annual precipitation. It is relatively frequent, in most deserts, that some storm rainfall surpasses in 24 h the mean annual precipitation for a 30-year registration period. These events can

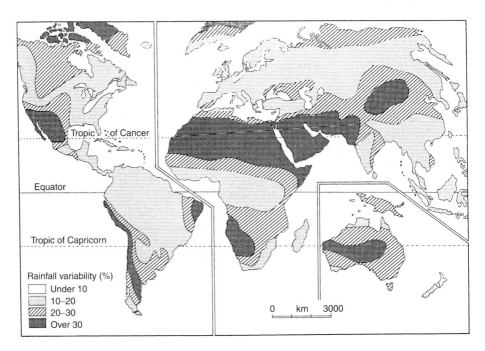

Figure 10.4.   Global variability of precipitation (Goudie and Wilkinson, 1977).

reach extraordinary values, such as a storm in 1925 that discharged 1524 mm in Lima in the Desert of Peru, when in this area the mean annual precipitation is only 46 mm (Nicholson, 1993). In the Atacama, Namib, and Sahara Deserts there can be periods of 10 to 14 years without precipitation. In some places houses are built with salt blocks (Figure 10.5), as in the arid Altiplano or Puna of Bolivia, although is not a hyper-arid zone.

In the coastal deserts, due to the proximity to sea, the air is relatively humid and the precipitation is usually lower than 50 mm. Fogs are very frequent and persist for more than 6 months. They reach tens of kilometres inland, furnishing humidity for the vegetation development. The fogs' discharge constitutes an important part of the total of the precipitation (Demangeot, 1981).

The warm *winds* in the desert regions produce the desiccation of many areas. These winds blow during the day and at night stay calm. They blow up from the desert to the more humid marginal zones and are commonly dust-laden. They are called by many local

Figure 10.5. House built with salt blocks. Uyuni salt flat, Bolivia. Photo. C. Maldonado.

names (*alisio, irifi, sirocco, harmattan,* and other names) (Goudie and Wilkinson, 1977). In particular cases they can evolve into tornado-like, dust-devils, which are hot air masses, whirling and particle loaded. They can reach tens or hundreds of metres in altitude with durations of several minutes. Much more important are the dust storms (*haboob*), that are air walls full of particles. They can produce catastrophic effects on animal life and human activities (Pewé, 1981).

4. Geomorphological differentiation of desert areas

The geomorphic criteria that can be used to establish classifications of arid zones can vary a great deal. The application of rules about dominant processes and general morphostructural types in desert environments allows the establishment of simple differentiations that in general show the main geomorphological features of these morphoclimatic zones.

Two important dominions can be recognised in deserts as a function of the main acting processes (Cooke et al., 1993). The first is the *aeolian dominion* in which predominates the erosive action and sedimentation of the wind. Erosion can create extended groups of yardangs, as those of the Libyan and Arabian Deserts, in which the rock outcrops are clearly lined up by the abrasive action of the wind. Erosive action can also been shown in the blow-out fields or deflation areas present in some desert areas such as the Kalahari Desert. The accumulation of particles transported by the wind gives rise to extensive sand seas that cover approximately a quarter of the deserts. A significant portion of these sand seas are nowadays stabilised by vegetation, as in the southern Sahara, and indicate a major aridity in past periods. The *fluvial dominion* results from the erosive and accumulative activity of run-off waters, which is more evident in the semi-arid zones. Fluvial action remains limited to the valleys, in which erosion and sedimentation are a part of their load, or to the alluvial fan systems, which distribute the water and solid load to the foot of mountainous zones, with a general diminution of the size of the particles towards the distal areas. Both dominions can be superimposed, because the activity of the aeolian and fluvial processes coexists many times in the same zone. Therefore, in the sedimentary record aeolian and fluvial sequences can be interbedded (Figure 10.6). On the other hand, the climatic changes that have taken place during the Quaternary have produced important modifications in the aridity of numerous areas. As a consequence, the activity and intensity of aeolian and fluvial processes seems to be modified considerably. In some regions, such as in the Sahel, the diminution of the aridity emphasised the fluvial activity and it paralyzed the aeolian action, stabilising and partially wasting the systems of dunes by water erosion.

The application of the morphostructural criterion, based on the type of dominant morphologies and in the geologic structure, along with the degree of tectonic stability, permits distinguising between *shield and platform deserts* and *mountain and basin deserts* (Mabbutt, 1977). The first comprises the African, Arabian, Indian and Australian deserts. They are formed fundamentally from cratonic areas basically preformed by Precambrian rocks and platforms of later age. Their main characteristic is the flat relief, interrupted by recent volcanic massifs, like the mountains of the Hoggar and of the Tibesti in the central Sahara, and by important fractures such as that relative to rift-valleys system in Eastern

Figure 10.6.   Aeolian deposits fossilised by Lower Pleistocene detrital material. Escorihuela (province of Teruel), northwards of the Semifosa of Teruel. East-central Iberian chain, Spain.

Africa (Fig. 10.7). These large flats are erosion surfaces of diverse ages (Fig. 10.8) or wide alluvial plans of endorheic basins, such as those of the Lake Chad south of the Sahara; also they are associated with landscapes modeled by the wind, which include regs, fields of

Figure 10.7.   Fault escarpment in the central sector of the Dead Sea rift-valley, Israel.

Figure 10.8.   Surfaces of erosion and, especially of desert pavement. Rann, Great Indian or Thar Desert.

Figure 10.9.   Uyuni salt flat. Altiplano of Bolivia. Photo. C. Maldonado.

yardangs, and ergs. The relief features common to these deserts are caused mainly by the long crustal stability observed in these regions.

The mountain and basin deserts are constituted by long mountainous alignments separated by depressed areas. This relief in a number of places was shaped during the Alpine orogeny in the Mesozoic and Cenozoic ages. As a consequence of the topographical contrast the high zones have suffered a constant denudation and the resultant materials have lodged in the depressions, in the form of coalescent alluvial fans. In the distal areas of the alluvial fans the run-off waters are collected in closed depressions that form ephemeral lakes (Figure 10.9) in which takes place an evaporitic sedimentation (playa-lake). The contact between mountains and basins commonly is due to an existing active tectonic front, along which rise the high zones together with the subsequent subsidence of the deeper parts. Therefore, these are areas of intense subsidence in which there have accumulated an important amount of sediments during the Tertiary and Quaternary. On the other hand, in the depressions aeolian activity of some importance occurs. The deserts of the western United States and those in the Precordillera and Central Cordillera of the Andes present these characteristics. Similarly, the deserts of Central Asia have similar features, although they also develop to a major altitude, for which the climatic contrasts are more marked.

Chapter 11

Weathering processes and resulting forms

1. Introduction

The alteration of the rocks in arid zones is above all the result of the wide daily and seasonal variations of temperature and relative humidity. Moreover, in deserts the rock outcrops are more common than in other climatic zones. The lithologic and structural characteristics of these outcrops influence the action of the processes and the resultant morphologies (Figure 11.1). In these desert environments physical weathering has a more specific importance than the biological and chemical weathering processes. Already it has been pointed out that in cold deserts during the cold months, rock fracture by frost action is an important process and the fundamental characteristics of it have been analysed in the section on periglacial processes. The scarcity of water and high temperatures give rise to weak development in animal life and vegetation and, therefore, biological weathering is much reduced. Equally, the moisture deficit affects the activity and intensity of chemical weathering, because water is necessary in attacking the crystalline structures and for the later mobilization of the ions in dissolution. Some processes like salt weathering, however, are of a physical–chemical type and this implies dissolution and precipitation of highly soluble salts that cause the rock to break later (Cooke et al., 1993). Also in the arid zones can be recognized the formation of iron oxides by capillarity and the mobilization of silica and alumina in alkaline media (Smith, 1994). All this together appears to indicate that the main part of the chemical weathering processes takes place in deserts, but its intensity diminishes considerably in comparison to more humid environments. On the other hand, in arid regions one can find some forms and deposits inherited from different past climatic conditions, such as alteration profiles, laterites, sinkholes, karstic cavities, and so forth. They have all been formed in past climates that were much more humid and in which chemical weathering was more aggressive.

The results of the activity of weathering processes results in granular disintegration, flaking (Figure 11.2), and splitting. The fluvial and aeolian erosive processes mobilize the freed particles, leaving an imprint on the diverse rock morphologies, as a consequence of the unequal activity of the alteration processes. These forms are customarily of small size and are characterized by the presence of shallow hollows in the rock surface (Goudie, 1989a).

In the past few decades there has been considerable progress in the knowledge of the action of weathering processes in desert areas and in the development of the resultant morphologies. The investigations have been carried out in the field or in the laboratory by investigations on durability. The first ones studied detailed temperature and relative humidity variation registers for several years and the influence of this variability on different types of rock weathering. Laboratory studies were designed to investigate

Figure 11.1. Sandstones perforated by weathering controlled by the rock structure. Paleochannels of Miocene sandstone of the Ebro Depression, Alcañiz, Teruel Province, Spain.

Figure 11.2. Lichen-colonized calcareous Oligocene sandstone affected for an intense exfoliation, Ebro Depression, Albelda, Huesca Province, Spain.

the intensity of the processes separately by means of cyclic tests, just as actually happens in the natural situation (Smith, 1994), but did not investigate the processes together or the interaction between them.

2. Weathering processes

2.1. Insolation weathering

Rocks can rupture as a result of temperature changes that cause expansion and contraction leading to breakdown. The heat supplied can come either from solar radiation or occasional fires, relatively frequent in semi-arid environments (Ollier and Ash, 1983). In the first case, the process of physical disintegration is called *insolation weathering*. The rocky surfaces of the deserts are exposed to high temperatures, which can reach values of 82°C in Sudan (Goudie, 1989a), and to diurnal oscillations of 54°C, as in the Atacama desert (Keller, 1946).

Most rocks are formed of different minerals and these present different thermal expansions according to their colour and structure. Dark rocks and minerals absorb more heat than light ones. Moreover, some minerals dilate more easily in one direction than in other, according to their crystalline structure. The expansion produces compressive failure inside the rock and contraction tensile failure. Microcracking and granular disintegration can appear as a result of different expansion and contraction (Birot, 1968a,b). On the other hand, as the rocks are poorly heat conductive, there are important temperature gradients from the surface into the rock (Figure 11.3), which give rise to important expansions and contractions leading to exfoliation (Smith and Warke, 1997). The outcropping stones or blocks broken in parallel fractures (*cleaved boulder*) (Fig. 11.4) are interpreted as produced by insolation weathering (Ollier, 1984), although other investigators estimate that the cause is salt weathering. The abrupt noises or cracking that the explorers heard in the silence of desert nights are thought to be due to breakdown when rocks are cooled during the night.

The classical laboratory experiments of Blackwelder (1925) and Griggs (1936) of heating and cooling cycles over different types of rocks, in the absence of water, show the inexistence of rock cracking after numerous cycles, equivalent to 244 years of daily weathering. When cooling the rock with water instead of dry air microfissures were created after a number of cycles equivalent to two and a half years. It has to be kept in mind that the presence of humidity is general in all the deserts as fogs, night dews and, sometimes, storms that cool the rock rapidly and emphasize contractions. Other circumstances to keep in mind with the early experiments is that rocks go on weakening gradually over long time periods, or are affected by continuous expansion and contraction cycles that bring the rocks to fatigue and finally to cracking (Griggs, 1936; Aires-Barros, 1975). Moreover, it is always necessary to remember that rock breakdown is not only due to a unique mechanism, because in general two or more processes alternate or superimpose.

2.2. Wetting and drying weathering

Rocks in desert environments are exposed to numerous wetting and drying cycles, which produces disruptive effects. The disruptive action caused by water is verified by means of

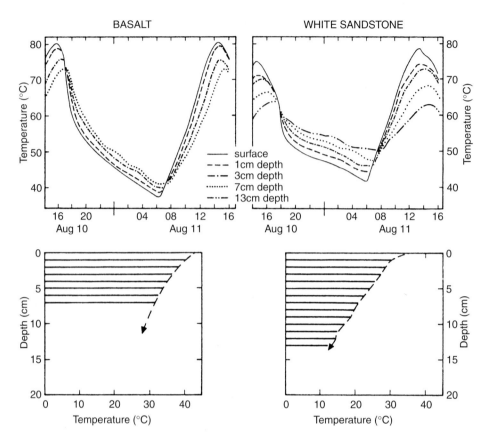

Figure 11.3. Surface temperature variations and at different depths for a basalt and a white sandstone in the Tibesti, central Sahara (Peel, 1974, in Mabbutt (1977), p. 23, Figure 5).

studies of durability carried out on different rocks (Ollier, 1984). These were plunged into water for a day and the next day air-dried. The disintegration was manifest as a superficial desquamation and faulting of the rock, mainly along existent joint planes. The first desquamations were observed after five cycles and cracking began over ten cycles. Complete disintegration took place after 180 cycles. Desquamation is interpreted as resulting from the polarity of water molecules that are attracted by the mineral surfaces, producing an expansive force that separates the confined surfaces. This mechanism is verified in laboratory experiments where rocks were subjected to the activity of liquids of high and low polarity. The high polarity ones produced important disruptive effects, whereas in the low polarity types the disintegrator action was null (Anon, 1966; in Ollier (1984)).

Rainfall, dew, and fog can wet the surfaces of the rocks in the deserts. Dew is common in some deserts and can result in a daily wetting and drying cycle (Cooke et al., 1993). On the other hand, water penetration into the rock depends on pore size and geometry, whereas drying involves evaporation that depends on temperature and wind speed. It is evident that wetting and drying weathering is as much important as the wetting and drying cycles affecting the rock. Therefore, in the North Hemisphere the rock outcrops in

Figure 11.4. Pebbles with parallel splitting. Villicum Sierra, Andean Precordillera, San Juan Province, Argentina.

south-facing slopes suffer more intense desiccation than rocks on north-facing slopes and therefore, the number of wetting and drying cycles will be greater and weathering deeper. These circumstances are verified in the Tertiary sandstones of the Ebro Basin of Spain (Sancho and Benito, 1990) and are corroborated in the ashlars of historical churches and buildings of this region, as in the medieval churches of Torres del Bayo (Gracia, 1985). Both sandstones are intensely altered where they are exposed on south-facing slopes, but weathering is practically nonexistent on north-facing slopes. In the much more arid environments of the Sinai desert, however, observations are totally opposite. A reason should be sought for the production of this contradiction, which may be related to the annual amount, distribution and type of precipitation.

2.3. Salt weathering

The presence of highly soluble salts in extensive accumulations related to lacustrine depressions or as superficial coverings in rocks and sediments is well known in deserts, and is indicated by efflorescences of white tonalities (Mabbutt, 1977). The precipitation of these salts in the porous spaces of rocks, along with the resultant modifications to the activity of physical and chemical processes that affect the salts, gives place to volumetric expansions that generate disruptive forces, which can cause rock disintegration. This group of processes is more effective in desert areas than in zones of more moderate temperatures.

In the last few decades important investigations have been carried out (Doornkamp and Ibrahim, 1990), by means of laboratory experiments directed at investigating the activity and consequences of salt weathering (Evans, 1969 to 1970; Goudie et al., 1970; Goudie, 1974; Cooke, 1979, 1981; Williams and Robinson, 1981; McGreevy, 1982; Sperling and Cooke, 1985; Fahey, 1986; Goudie and Viles, 1995). These works take into account the environmental conditions and the materials affected by salt weathering. Similarly, they have experimented with the activity of different saline solutions in order to know the different salt weathering intensities. They also investigate the effects and the freed particles. Finally, another aspect is involved in the correct knowledge of the processes implied in salt weathering. In Cooke et al. (1993) can be found an excellent synthesis of salt weathering. Also salt weathering has been studied in a qualitative way in order to analyse the origin of diverse micromorphologies. These works will be referred to later, in the section on resultant forms. Similarly, considerable advances have been produced in the field of applications, as long as the activity of the salts causes disturbances and damages to highways, buildings and engineering structures, mainly due to reactions between the cement and gypsum that generate expansions and subsequent cracking (Goudie and Viles, 1997).

2.3.1. Influence of climate

The wide diurnal and seasonal variations of temperature and relative humidity in all deserts, together with the presence of dry and sporadic winds, favour salt weathering. Moreover, these temperature and humidity variations are not ephemeral, because they can also occur over the surface and close to it, so that is why they influence the spatial variability of salt weathering. The high temperatures favour the evaporation of the saline solutions and finally result in their precipitation. Similarly, cooling of saturated solutions can also produce salt precipitation. On the other hand, the humidity increase generates hydrous salts and the temperature increase results in thermal expansion of saline crystals. These crystal growths, hydration, and expansion processes produce important volumetric modifications, which give rise to disruptive forces in the rocks (Cooke et al., 1993).

Climatic variations affect water capillary (Cooke et al., 1982) (Figure 11.5). The capillary fringe can reach a height of 3 m in very arid environments, varying with the size of the constituent particles and the cementation degree of the rock. Their higher limit is conditioned by salt crystal accumulations and when the capillarity front reaches the surface, it produces saline efflorescences (Figure 11.6).

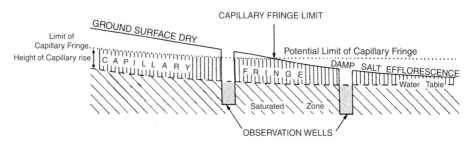

Figure 11.5. Characteristics of the capillary zone (Cooke et al., 1982).

Figure 11.6. Saline efflorescences in Miocene calcareous sandstones of the Ebro Depression, The Tormillo, Huesca Province, Spain.

2.3.2. The affected rocks

The influence of salt weathering on rocks is a function of the greater or lesser capacity of water transmissivity laden with salts. It is therefore necessary to know the physical properties of the rocks, such as effective and trapped porosity, porosimetric distribution, adsorption capacity, penetration capillarity, adsorption, hydric expansion, and resistance to traction, among other factors. Also it is important to know the mineralologic types of the clay minerals.

Studies on durability carried out on different types of rock to discover the degree of resistance to the presence of different saline solutions, reflects the fact that sandstones and limestones are less resistant to salt weathering than igneous and metamorphic rocks. This type of weathering is more effective in porous rocks with great adsorption and water loss capacity, such as poorly cemented sandstones (Mabbutt, 1977).

2.3.3. The salts: types, origin, and concentration zones

The most common salts that take part in salt weathering processes are chlorides, sulphates, carbonates, and nitrates of sodium, calcium, magnesium, and potassium. The carbonate $CaCO_3$ is most common in semi-arid zones, sulphates predominate in arid areas and the chlorides are also quite common in coastal deserts (Cooke et al., 1993). To produce crystallization it is necessary for the alternation of wetting and drying stages, and the higher the number of cycles, the greater the salt weathering intensity. The effectiveness of weathering varies with salt solubility, hydration capacity, and crystallization strength. Some salts, such as mirabilite ($Na_2SO_4 \cdot 10H_2O$) have a disruptive strength much greater than that of freezing water. Laboratory experiments carried out with different types of saline solutions indicate that sodium and magnesium sulphates are the most effective (Figure 11.7) in the disintegration of the rocks by salt weathering (Goudie et al., 1970). On the other hand, the hydration of some salts involves a considerable increase of volume and as a consequence, generates important forces that help to disintegrate the rock. Common hydrations that occur in the nature are the change of

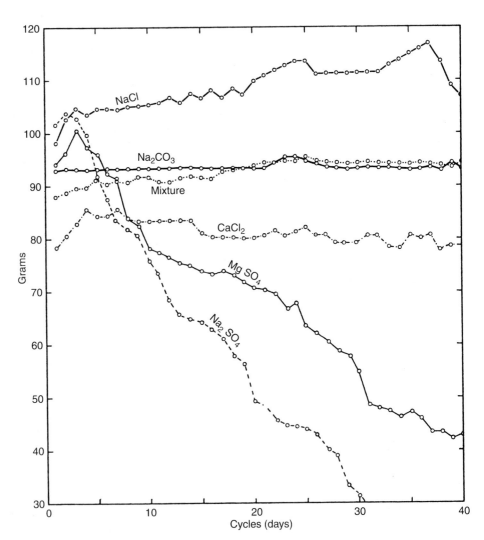

Figure 11.7. Weights variations of samples from the Arden sandstone, after being subdued to salt weathering essays with different types of salts (Goudie et al., 1970, in Fookes and Vaughan (1986), Figure 20.25).

thenardite (Na_2SO_4) to mirabilite ($Na_2SO_4 \cdot 10H_2O$) or that of anhydrite ($CaSO_4$) to gypsum ($SO_4 \cdot 2H_2O$). This last produces a volume increase of about 39% (Jauzein, 1974).

The kind of salts in a region depends upon the lithology of the source area and, therefore, varies from one region to other. In Australia NaCl predominates, in the salt flats of the Atacama Desert the dominant salts are NaCl and sodium and calcium sulphates, in the endorheic area of the Ebro Depression of Spain the most common salts are sodium chlorides and sulphates (Pueyo, 1978 to 1979), and in some salt lakes of mid-eastern Africa the preponderant salt is Na_2CO_3.

Salts come fundamentally from the freed ions in the chemical weathering processes of rocks, which are mobilized in solution by run-off waters and also by subterranean fluxes. In littoral zones, aquifer overpumping can cause saltwater intrusion and salinity increase of the extracted waters. Another salt mobilization source is the water and aeolian erosion of salt deposits. Salts can also come from volcanic dust or gases. In the littoral deserts, marine aerosols coming from the sea commonly constitute an important source of salts (Goudie and Viles, 1997). In these areas there is a clear gradual decrease of salt content towards inland areas (Goudie, 1989a). In Aftout, Mauritania, and in the Senegal delta, where waves strike against the littoral bars they produce water drops that evaporate and generate small salt crystals, which float in suspension giving origin to a whitish fog that the wind pushes toward the continent. Afterwards they can be washed out by the rain or, what is more frequent, by an increase in relative humidity during the night that favours the water fixation in these hydroscopic crystals that fall as a consequence of these enlargements (Tricart, 1969).

Salts coming from these different feeding sources accumulate in a great variety of situations (Cooke et al., 1993). In littoral deserts (Peru–Chile, Namibia and western Australia) salt coming from oceans produced a salt contribution of 150 kg/ha/yr. In tidal flats of seas of high salinity as at the Red Sea, Persian Gulf and Mediterranean Sea an important saline sedimentation takes place in littoral sabkhas. In continental interiors, evaporation of lacustrine waters supplied for run-off, as well as groundwater and aeolian contributions, generate saltlakes. The different solubility of the salts controls their precipitation and results in a zonation that shows the presence of chlorides in the inner part, sulphates in the middle zones, and less soluble carbonates along the external borders (Figure 11.8).

2.3.4. Physical–chemical processes

The mechanisms that involve salt weathering are of chemical and physical character. They include crystallization, hydration, and thermal expansion of high-soluble salts in confined and semiconfined spaces, such as in pores and cracks of the rocks.

Of these three mechanisms, *crystal growth* is the most important (Goudie, 1974) and it happens that as saline solutions become saturated, precipitates and salt crystals are produced in the pore spaces (Winkler and Singer, 1972). These crystals can continue growing between the walls that confine them, if there is a film of saline solution at the crystal/rock interface. Crystallization generates important pressures that are transmitted to the rock (Evans, 1969 to 1970). Precipitation is caused by the evaporation, cooling, or mixing of solutions with a common ion. Some salt solubility (Na_2SO_4, Na_2CO_3, $NaNO_3$ and $MgSO_4$) diminishes with the descent of temperature, which during

Figure 11.8. Zone of sulphate precipitation with selenite domes, El Melah Sebkha, Zarzis, Tunis.

night cooling sometimes produces precipitation (Goudie, 1989a). In general, salt crystal growth is more important in areas in which wetting and drying cycles are common or in zones where salt is concentrated, as in the borders of saline lakes and in fluvial channels.

Hydration involves the adsorption of water by a salt inside of its crystalline structure. This mechanism implies a clear volume increase with its subsequent disruptive effects in confined zones, which gives rise to granular disintegration and cracking in the rocks. In any case, whatever the different types of salts, different pressures of hydration are generated. The salts Na_2CO_3 and the Na_2SO_4 upon hydration increase their volume as much as 300 per cent (Goudie, 1977). The initial precipitation in the interstices of the rocks is in the hydrated salt form, but due to the high existent temperatures during the day, salts change to anhydrous phases. At night, when the temperature is descending and the humidity increasing, they can hydrate. Overall, it can be deduced that hydration is a function of temperature and humidity changes that take place in the deserts (Cooke et al., 1993). These adsorption and dehydration processes of the crystalline structure repeat several times, so that the pressures produced by these volume changes and the disintegration effects in the rocks can be quite important (Goudie and Viles, 1997).

Thermal expansion of crystals of salt, according to laboratory experience, is a less effective mechanism than the previously discussed hydration (Goudie, 1974). The expansion by insolation warming depends upon the thermal expansion of each salt that tends to be higher than that of the minerals of the rocks (Cooke and Smalley, 1968), and on the thermal oscillations that are recorded for the different deserts. Experimental studies have been carried out for temperatures much higher than those existing on the desert rock surfaces (Figure 11.9), but it is supposed that salt volumetric expansion by insolation

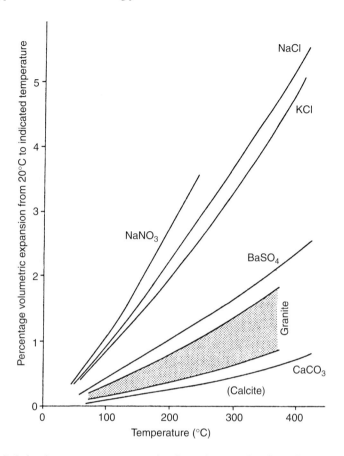

Figure 11.9. Relation between temperature and volumetric expansion for different salts and granite (Cooke and Smalley, 1968, in Cooke et al. (1993), Figure 5.18).

shows similar actions (Cooke and Smalley, 1968). Expansion produced by the salts in the pores and fissures of rocks very close to the surface generates forces that can initiate their cracking. Some studies carried out in the middle latitudes of the northern Hemisphere show that rocks of south-facing slopes show a major weathering degree more than the north-facing slopes. This difference is also attributed to the thermal expansion of NaCl (Johannessen et al., 1982).

2.4. Dissolution

It has been indicated that dissolution plays an important role in salt mobilization in desert environments, but it is also necessary to know the dissolution behaviour of soluble rocks (carbonates and evaporites) in these environments of scarce precipitation. Limestones are affected first of all by biological microkarren due to biokarst activity (Viles, 1984).

Evaporites, such as chlorides and sulphates outcropping in deserts, can be subjected to an intense dissolution that affects the three-dimensions of the rock massif, such as is indicated by the development of exokarstic and endokarstic forms. Despite the scarcity of water, the high solubility of these salts permits a complete development of karst. An excellent example of karstification in evaporites is Mount Sedom in Israel, in the proximity of the Dead Sea, in which the recorded precipitation is only 50 mm/yr (Frumkin, 1995; Frumkin and Ford, 1995). The evaporites are part of a diapir, constituted of salts of marine origin and Late Cenozoic in age. Where they crop out, dissolution of chlorides starts giving rise to a caprock 50 m thick that is constituted of anhydrite, dolomite, and clastic sediments with a dominant nucleus of chloride salts. In the caprock sinkholes have developed, some of which are connected to an important network of subterranean galleries. The outcropping chlorides present spectacular karrens, of the rillenkarren and spitzkarren type (Figure 11.10). In the central Ebro Depression of northern Spain, with a precipitation of about 300 mm, a well-developed formation of Miocene gypsum of continental origin crops out. In it significant fields of sinkholes are developed, many of covered karst, and systems of galleries (Gutiérrez et al., 1985; Benito, 1987). In some areas common phenomena of current subsidence occur that affect buildings and lineal works (Benito and Gutiérrez, 1988; Benito and Pérez del Campo, 1991; Benito et al., 1995).

The lesser solubility of the limestones inhibits in great part the development of karstic morphologies. It is considered that below 250 to 300 mm of precipitation there is no karstic form development (Sweeting, 1972). The studies on actual rates of dissolution carried out on limestone tombs in Haifa (Klein, 1984) and Jerusalem (Danin, 1983) show a weathering rate of 0.005 mm/yr, although this dissolution appears to be influenced by urban pollution and microbiological activity. Therefore, dissolution is very low despite

Figure 11.10.   Karrens (rillenkaren and spitzkarren) developed in chloride rocks, South of the Dead Sea, Mount Sedom, Israel.

Figure 11.11. Structural karren (kluftkarren) with cracks amply widened by dissolution in an area with 100 mm precipitation, Kalia, Midbar Gehuda, Judean Desert, Israel.

that in both cities precipitation is about 600 mm. Further eastwards in the desert of Judea, there are well-developed karren of the kluftkarren type (structural karren) (Figure 11.11), hohlkarren (perforate karren) and kamenitzas in areas with a 100 mm precipitation. Cracks of structural karren appear in many places totally filled with secondary $CaCO_3$ with a leafy structure. These circumstances seems to indicate an ancient stage of higher precipitation than the current, and later more arid periods, with carbonate precipitation between the fractures of the karren.

2.5. Biological activity

Bacteria, fungi, algae, and lichens are abundant in the desert environments, mainly due to the reworking action of the wind. These micro-organisms, whose number of species is very high, produce rock and soil weathering, although their importance in the whole weathering process is not very well known. Environmental conditions, along with the rock or soil characteristics, determine the existent community in the surface and, therefore, the characteristics of biological weathering (Viles, 1995).

These organisms are developed either on the surface of the rocks and soils or inside them. They can grow below the surface developing biocrusts (Figure 11.12), frequently constituted by bands of different colours, reaching up to 4 cm in thickness (Thomas, 1988). Some micro-organisms colonize the existent cracks and others are firmly fixed to the surface (Figure 11.12).

This biological activity, resulting from vital processes or from secreted products, generates substantial changes in the superficial micromorphology. These modifications are

Figure 11.12. Biocrust developed on Miocene clays and superficial colonized by lichens. Observe their major resistance to erosion. Experimental station of Lanaja Ebro Depression, Huesca Province, Spain.

a result of joint physical and chemical processes (Ollier, 1984; Cooke et al., 1993). The expansion and contraction of bacteria and lichens, in relation to water content variations, originate the rock disintegration by granular disintegration and exfoliation in surface as well as in confined spaces. Chemical processes come out by the secretion of organic acids, which attack the rock constituents and generate chelates that help to export cations of low solubility. The carbon dioxide of the microbiological activity favours the dissolution of the limestone surfaces, originating in biological microkarren, characterized by the presence of tiny furrows and hollows (Ollier, 1984). Finally, some of these micro-organisms play a fundamental role in the genesis of the desert varnish.

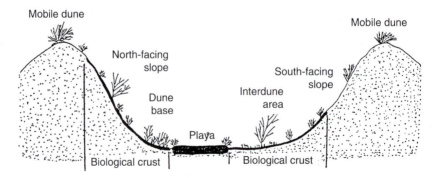

Figure 11.13. Transversal cross section of the longitudinal dunes system in the experimental station of Nizzana, Negev Desert, Israel (Veste, 1995).

Another significant aspect of the biological action is related to hydric and aeolian erosion (Thomas, 1988). It is well known that the vegetation cover affects sediment mobilization, but the existence of a superficial biocrust reduces considerably erosion. Where the aeolian activity holds up, dunes can reach stabilization (Veste, 1995) (Figure 11.13). As the infiltration capacity decreases rilling appears, even though erosion is made difficult by the biocrust cementation (Figure 11.12).

3. Resultant weathering forms

Rock surfaces in deserts can be affected by differential weathering and this can give rise to different types of microforms. These appear at the base of vertical walls forming caves or shelters, in slopes in which develop alveoles and tafonis and in horizontal surfaces or of low inclination in which are recognized gnammas or weathering pits. All these forms can occur in the same rock outcropping and although they are very common in deserts they could occur in other climates (Twidale and Corbin, 1963; Martini, 1978). They mainly develop over igneous isogranular (granite, rhyolite) and sedimentary rocks (sandstones), although they can occur in conglomerates, gneisses, and porphyries.

The water content is higher at the base of the rock outcrops, partly due to capillarity ascent. Therefore, weathering is more important than in the upper zone, where the rock is drier and practically without weathering (Mabbutt, 1977). These circumstances explain the *basal undercutting* of the rock walls, which develops more easily in the south-facing zones of some areas, generating *caves or shelters* (Howard and Selby, 1994). Where undercutting affects isolated hills or inselbergs monoliths of a fungus-like form can be produced. This morphology can also be generated by subsuperficial weathering and later exhumation, with the spectacular flared slopes of the Eyre Peninsula, in south Australia constituting an excellent example of this origin (Twidale, 1962).

On rocky walls of moderate or considerable steepness, rounded hollows can develop that on occasion can completely cover the rock surface. Also they can be observed on the detached blocks of the slope. If the size of these hollows is centimetric they are named *alveoles* that together form the morphology of a honeycomb. Their origin is discussed and is attributed to aeolian erosion, exfoliation, frost action, and salt weathering. They are common in coastal desert environments (Mustoe, 1982).

When dimensions are bigger, decimetric and sometimes metric, they are called *tafonis* (Figure 11.14). They usually appear in groups and have circular or elliptic sections with the bottom covered by detritus that mobilizes with the wind, rain and runoff. They can grow and join others and can also in their growth towards the inside connect with other tafoni. Sometimes, like alveoles, they appear orientated following planes of weakness of the rock, sometimes imperceptibly. Rock surfaces show exfoliation and inside the tafoni saline efflorescences are common. Their origin is discussed. Some authors differ between wall tafonis and basal tafonis, related to basal undercutting processes (Bradley et al., 1978; Smith, 1978). It is in agreement that the dominant processes are those related to water with salt movement close to the surface, which correspond to salt weathering and wetting and drying cycles. Also they adduce the insolation weathering, frost action, dissolution of carbonated cements, and aeolian erosion (Evans, 1969 to 1970; Mustoe, 1983; Young, 1987; Robinson and Williams, 1992). Besides these subaerial processes, weathering under

Figure 11.14. Tafonis on a Oligocene calcareous sandstone, in the Ebro Depression. Tafonis, variable in size, appear orientated, Albelda, Huesca Province, Spain.

the soil surface and a later exhumation can generate some tafonis by erosion (Dragovich, 1969). On the other hand, in environments of marked aridity the development of the tafonis is very rare, because the lack of water impedes the action of many weathering processes.

Another investigated aspect is the morphological analysis of the tafonis (Calvin and Cailleux, 1962; Rodríguez and Navascués, 1982; Sancho and Gutiérrez, 1990; Sancho and Benito, 1990). These last authors note that the tafoni development on Oligocene sandstones of the Ebro Depression of Spain, under semi-arid conditions, depends on the slope surface, textural factors, feldspar weathering, wetting and drying, and salt weathering.

On rock surfaces of horizontal or low slope, outcroppings develop small closed depressions denoted *gnammas* or *weathering pits* (Figure 11.15), which also have numerous local names. They appear in general in groups and their more common diameters are decimetric, although there has also been recognition of centimetric and metric size, where their depth is a function of their diameter. Shapes in plan view are varied, although the circular and elliptic morphologies predominate. Some gnammas have very irregular borders, a consequence of the coalescence of many individuals. Their cross sections occur in a great variety (Gutiérrez and Ibáñez, 1979) (Figure 11.16) and gnammas can be classified as a function of this (Twidale and Corbin, 1963). They are differentiated into *deep gnammas* or *pits* with a depth greater than diameter (Figure 11.15), the *shallow gnammas* or *pans* more shallow than their diameter (Figure 11.17) and in which overhanging sidewalls are common, and finally, *armchair gnammas* developed on gentle

Figure 11.15. Gnammas in or pits, of unequal development, elaborated on Miocenes sandstones. These small closed depressions are filled up with silt-clay material. Region of Alcañiz, Ebro Depression, Teruel Province, Spain.

slopes, with a triangular cross section and constituting an intermediate step towards tafoni morphologies. Their origin is due to differential weathering along joints or especially along intersection of weakness planes, although in some cases no structural control can be observed. They can also develop under a soil or below a mantle of alteration (Figure 11.15),

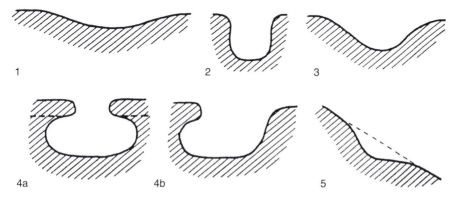

Figure 11.16. Gnammas cross sections of the Alcañiz region, Ebro Depression, Teruel Province, Spain. (1) Gnamma with smooth borders. (2) Gnamma in or pit. (3) Gnamma in or pan. (4a) Gnamma in with overhanging sidewall. (4b) Asymmetrical gnamma. (5) Gnamma in armchair or armchair (Gutiérrez and Ibañez, 1979).

Figure 11.17. Gnamma in or pan partially filled of water, with an overflowing channel, developed on Miocene calcareous sandstone. Region of Alcañiz, Ebro Depression, Teruel Province, Spain.

in which later denudation is necessary to expose them. Exfoliation and granular disintegration and in some cases, abundant lichen colonies can play a certain rule in their generation.

The dominant processes are wetting and drying and salt weathering. In the case of calcareous sandstones the dissolution of the carbonates is very important (Gutiérrez and Ibáñez, 1979), and if sandstones are siliceous, quartz lixiviation takes place if the pH is over 8 (Benito et al., 1993). Once the closed depression is formed, water accumulates inside it and chemical weathering can be predominant. The particles inside the gnamma can be exported by deflation or by water overflowing when the depression fills up. In some outcrops the presence of little channels of overflowing fitted into the rock is common (Figure 11.17). The origin of the overhanging sidewall in some gnammas can be due to a water prevalence in the bottom, for which weathering is major in this zone. When gnammas and tafonis coexist in the same outcropping this can produce the capture of one by the other, leaving the gnamma like an empty background buttonhole. Also on occasion the headward erosion action of the fluvial network can reach in to capture some gnammas.

Chapter 12

Desert surfaces: pavements, patterned ground, varnishes and crusts

1. Introduction

The surface materials of arid zones and their existence at shallow depths are subjected to numerous processes that act together or separately, depending upon lithology, climate, relief, and so forth. The result of this activity leads to the generation of desert pavements, patterned ground, varnishes, and crusts.

2. Desert pavements

2.1. General characteristics

Desert pavements consist of rounded or angular rocky fragments within a matrix of finer sand, silt, and clay-sized material (Mabbutt, 1965b, 1977). They develop not only on low relief surfaces such as alluvial fans, terraces and pediments, but also on slopes. They can occur in other morphoclimatic settings, but it is in arid zones where they reach their maximum expression, going on to occupy significant areas.

Locally they are given many names: *gibber* in Australia, *saï* in central Asia and *hamada*, *reg*, and *serir* in the Sahara, the Near East, and India. Hamadas are tabular relief structures, pedimentation, or glaciplanation surfaces of negligible slope. The surface can be devoid of fragments, in which case it is denoted a *rocky hamada*, or it can be covered in angular blocks (*boulder hamada*) (Figure 12.1) that have not suffered any transport. In some cases the blocks have suffered intense weathering in which a continuous flaking has rounded the fragments (Evenari et al., 1971a,b). Reg or serir can develop, however, on alluvial surfaces with a slight or negligible slope and in these cases the stones are rounded as a consequence of their transport (Figure 12.2). At the present time many reg or serir appear as relict surfaces that are remote or left away from the current drainage network.

A well-developed pavement consists of a stable surface, but when altered by anthropomorphic action (e.g. all-terrain vehicle traffic in the SW USA and Saudi Arabia; Webb and Wilshire, 1983) water and aeolian erosion can lead to substantial modification. Similarly, the famous Nazca Trails, southern Peru (Figure 12.3), which date from thousands of years ago, are desert pavements partially mobilised during their construction. Elsewhere, the pavement's constituent stones can be prehistoric artefacts, and other times the rocks are painted or carved (petroglyphs) (Brakenridge and Shuster, 1986), circumstances by which it is possible to gain information about the relative age of desert pavements.

Figure 12.1. Hamada covered by lag angular boulders. In the central part, there is a small closed hollow (referred to as a "daya" in the Sahara) filled with fine deposits lighter in colour. This deposit may come from eolian material, which has been reworked by run-off. Rann, Thar Desert, India.

Figure 12.2. Typical reg of rounded clasts developed on an alluvial plain. Bhojka, Thar Desert, India.

Figure 12.3. Paths of Nazca built on a desert pavement. Southern desert of Peru. Photo by C. Zazo.

2.2. Processes that participate in their formation

Desert pavements can originate via a single or various processes, which indicates a problem of convergence of forms. Fundamentally the processes can be extrinsic in character, produced by water and aeolian action, and intrinsic, in which the interior of the soil is modified by differentiation processes.

The most universally applied process of pavement genesis is *deflation*, whereby fine particles are exported by the wind, leaving a residue of coarse material that goes on to constitute the pavement. Evidently, exportation is minimal and as a consequence the surface reduction is slight. This is due to the fact that deflation decreases with increasing surface roughness. It has been calculated that with a surface rock coverage of 50%, exportation becomes ineffective (Sharon, 1962). Furthermore, a biocrust film sometimes develops in the spaces occupied by the fine material, which impedes deflation and stabilises the pavement, leaving the surface "sealed" (Campbell et al., 1989).

For some authors (Sharon, 1962) *water erosion* is more efficient in the generation of pavements. In ground plot studies, surfaces with gradients of 5% exhibit a reduction of 5 to 50 cm over a period of 5 years. The exportation of fine material is accomplished via rainsplash and surface wash, although the flow of water is made difficult by the presence of stones. In the case of biocrust occurrence, splash is reduced considerably and the capacity for infiltration diminishes markedly (Poessen, 1986), limiting water penetration to the cracks that usually develop in soils.

Although less important than those mentioned previously, another process which can also give rise to pavements is the upward migration of stones from the soil interior, which then concentrate on the surface (Mabbutt, 1965a; Cooke, 1970a; Mabbutt, 1977). This phenomenon of formation can be produced by freeze pushing in cold deserts. In hot deserts the presence of *expansive clays* is necessary. These swell with an increase in humidity and lead to a rise in ground level. During dry stages the soils contract and crack. Therefore, pavement generation under these circumstances is favoured by enhanced humid/dry cycles.

In general, mechanical surface weathering processes have been proposed for the development of boulder hamadas, which break up the rocks and produce a residual pavement of angular blocks. It is also believed that *sub-surface weathering* can play a part in pavement development (Mabbutt, 1977). This author indicates that the higher sub-surface humidity content, especially in the presence of chlorides and sulphates, leads to a more efficient weathering and, therefore, coarse grains disintegrate gradually. On the contrary, at the surface the water availability is much lower and the pavement stones hardly weather. The end result is that the pavement persists.

2.3. Development

From the analysis of pavement-generating processes it can be deduced that genesis via water or aeolian erosion is rapid, unlike pavements produced by movements in the soil interior. Sub-aerial erosion produces a descent and levelling of the surface after long and continual activity (Mabbutt, 1977). Moreover, as the pavement develops the availability of fine material diminishes. The pavement is subjected to a slow evolution, over thousands of years, during which time the stones weather and the material generated is partially exported (Cooke, 1970a). This slowness in their development was shown in the study of a group of terraces in the Dead Sea region (Hunt and Mabey, 1966), in which the stone-cover percentage increases with terrace age. On the contrary, the oldest pavements in Death Valley, California (Hunt and Mabey, 1966) are characterised by having a higher percentage of fine particles and being more compact than the more recent pavements. In this case the evolution is the inverse of the previous example, perhaps due to a more intense surface weathering.

In addition, it has to remarked that as pavements evolve, their salt content can increase, as the salts get moved by capillary action and stay trapped between the stones. Equally, pavements can experience a continuous contribution of aeolian material. In any case, once formed the pavement is relatively stable, as deflation decreases with an increase in stone coverage, which tends to increase with water erosion. These circumstances mean that the pavement is a fairly stable morphology in which both erosional and depositional processes act, although with low intensities that can also counterbalance each other. This indicates that pavements are resistant to erosion, because they protect the underlying material that is prone to water and aeolian erosion.

3. Patterned ground

We have already indicated in the section related to periglacial morphoclimatic zones that patterned ground (circles, polygons, nets, steps, stripes) reach a great development in both

Figure 12.4.   Desert pavement with clasts veneered with a desert varnish. Note that the clasts are sorted and structured forming circles. Negev Desert, Israel.

high-latitude and alpine areas (Washburn, 1956). The mechanisms invoked in their generation are numerous and complex, practically all related to freeze–thaw processes.

Nevertheless these processes can also develop in arid zones, especially in areas of low slope, (Cooke and Warren, 1973; Washburn, 1979), although in hot deserts the formation mechanisms are very distinct from those in periglacial zones (Figure 12.4). This is, once again, a situation of the convergence of forms or equifinality. These arid morphologies are fundamentally related to humidity variations within the soils and, in an environment of salt precipitation, they occur in relation to periodical inundation and desiccation. The microforms generated can be ephemeral because on occasion they develop and disappear between successive pluvial precipitations.

3.1. Desiccation cracks

Upon drying, humid, fine-grained sediments reach a limit in their contraction and they develop surface cracks. This cracking tends to form polygonal systems (Figure 12.5). The form that the polygons acquire depends on the physical, chemical, and mineralogical characteristics of the material and on the environmental circumstances, as expressed by temperature and humidity percentage. Horizontally, the cracks can be straight or curved and the length, width, depth, and number of cracks, along with how they group, is highly variable (Chico, 1963). The profile of the polygonal cells can be flat, concave, convex, or irregular.

Figure 12.5. Polygonal desiccation cracks on the side of a partially flooded sebkha. Sabknet Bou Jmel. Zarzis region, southern Tunisia.

Most of the studies of crack morphology and genesis are related to experiments concerned with freeze–thaw cycles (Lachenbruch, 1962; Corte and Higashi, 1964 in Cooke and Warren, 1973; Maizels, 1987), and few studies exist related to the origins and characteristics of cracks developed on clay and saline plains in playa environments. Cracks produced in periglacial environments, however, resemble those originating in arid zones, because the expansion and contraction coefficients of ice and salt are similar. The few experiments reveal, amongst other conclusions, that clays develop 40 to 50 times more cracks than sands and gravels (Figure 12.6). Equally, the cracks get shorter with time, due to the development of new cracks which intersect with the older cracks. Furthermore, the space between cracks increases with the rate of desiccation, clay content and percentage of expansive clays.

Stones found on or near the surface act as points of crack division. Also, the polygonal cell form gives information about the origin of these morphologies. Concave surfaces indicate rapid drying of the uppermost material, leading to the production of *mud curls* on the surface lamina (Figure 12.7). A convex morphology reveals the presence of salt in the matrix (Figure 12.8) and, finally, a flat form suggests slow drying in the absence of salt.

On the other hand, Lachenbruch (1962) differentiated between orthogonal systems, in which the cracks intersect at right angles, and nonorthogonal systems where the intersection angles vary around 120° (Figure 12.9). Orthogonal systems are typical of inhomogeneous media where the cracks can display preferred orientations. On the contrary, nonorthogonal systems form in very homogeneous materials which dry uniformly and the cracks develop instantaneously.

(a) Coarse/Medium Sand

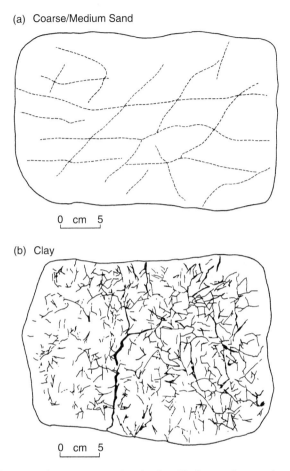

(b) Clay

Figure 12.6. Surface cracking pattern produced after 12 freeze–thaw cycles under experimental conditions. (a) coarse-medium sand. (b) clay (Maizels, 1981, in Cooke et al., 1993).

All of these cracks form in hot deserts with flat topographies, which in the majority of cases correspond to lacustrine depressions. The occurrence of locally developed *giant desiccation fissures* (Figure 12.10) is also possible, with lengths of kilometres and depths greater than 1 m. Their origin is very complex and many causes for their formation have been put forward. They include cracking caused by the subsidence due to extraction of subterranean water, fissurisation by hydrocompaction, contraction from desiccation, seismic activity, and so forth. All of these processes may act alone or in combination.

In playa environments profuse nonorthogonal systems develop in salty clays, which comprise mud plains at the margins of evaporite sedimentation (Figure 12.5). The same system also occurs in gypsum deposits (Tucker, 1978), which on occasion exhibit thrusted borders (Figure 12.11). In Death Valley, California, sorted polygons develop in gypsum-rich material (Hunt and Washburn, 1966) (Figure 12.12), where cracking is generated by processes of salt contraction as a consequence of drying or freezing, and expansion due

Figure 12.7. Mud curls developed on a thin clay layer. Jaisalmer, Thar Desert, India.

Figure 12.8. Nonorthogonal system of convex desiccation cracks with raindrop prints. Closed depression of La Playa, Bujaraloz, Province of Zaragoza, Ebro Basin, Spain.

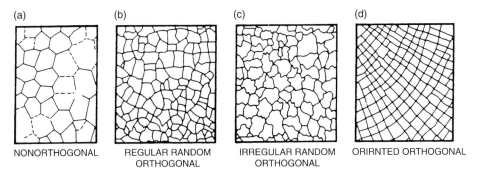

(a) NONORTHOGONAL

(b) REGULAR RANDOM ORTHOGONAL

(c) IRREGULAR RANDOM ORTHOGONAL

(d) ORIRNTED ORTHOGONAL

Figure 12.9. Common systems of desiccation cracks (after Lachenbruch, 1962, and Neal, 1965, in Cooke and Warren, 1973, Figure 2.25).

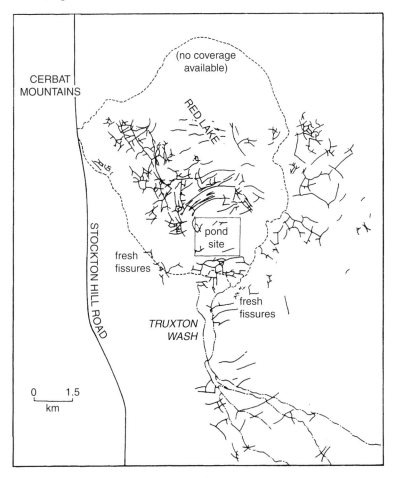

Figure 12.10. A generalised map of giant desiccation polygons, earth fissures and areas of recent fissuring, Red Lake Arizona, based on the interpretation of aerial photographs. The fissures are attributed to long-term desiccation and local subsidence due to water extraction and/or subsurface salt flow (after Lister and Secrest, 1985, in Cooke et al., 1993, Figure 8.6).

Figure 12.11. Gypsum polygons with laterally expanded edges, Sabkha Matti, United Arab Emirates.

to hydration or heating. These processes, together with the mobilisation of salt by capillary action, appear to be the fundamental reasons for the ordering and classification of gypsum-rich soils. Equally, polygons developed over salt surfaces on both littoral and interior sebkhas are very common (Figure 12.13). Such polygons may have suffered a complex evolution.

3.2. Gilgai

Gilgai is an Australian vernacular term meaning pool, which is increasingly used in scientific literature. It is used to describe very small mounds developed on clay-rich soils

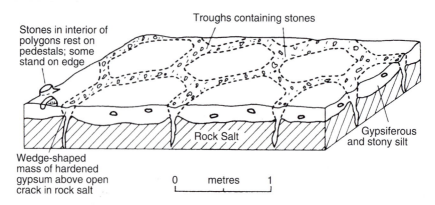

Figure 12.12. Sorted stone polygons related to ridges and cracks in an underlying layer of rock salt, Death Valley, California (after Hunt and Washburn, 1960, in Cooke and Warren, 1973, Figure 2.30).

Figure 12.13. Polygons with laterally expanded edges developed on a salt surface. El Melah Sebkha. Zarzis, southern Tunisia.

and desert pavements (Hallsworth et al., 1955; Verger, 1964; Harris, 1968; Mabbutt, 1977; Hubble et al., 1983; Dixon, 1994a).

Following Harris (1968) a gilgai can be differentiated into three parts on the basis of its cross section (Figure 12.14). A *channel*, which constitutes a depression that can be 2.5 m below the general ground level and can reach 12 m in length. The general level of the microrelief is denoted as the *shelf*, above which small *mounds* of up to 50 m diameter stand out. Some of the authors previously cited differentiate six gilgai types (Dixon, 1994a), which in the main reflect the simple morphological classification of Verger (1964), based on the form and mode of gilgai grouping (Figure 12.14).

Various processes and factors intervene in gilgai formation, which are developed in the monographs of Cooke and Warren (1973) and Mabbutt (1977). Gilgai form in areas of alternating humid and dry conditions. Dry periods, during which the surface cracks, alternate with humid periods in which swelling dominates. They occur mainly in vertisol areas (black expansive clay-rich soils) in tropical and subtropical zones. Gilgai microrelief results from upward movement produced by swelling of the humid subsoil. This expansivity, along with soil contraction, as a consequence of humidifying and drying cycles, explains the genesis of *circular* and *stepped* gilgai (Ollier, 1966) (Figure 12.15). Swelling is due to the presence of expansive clays, such as montmorillonite, and the existence of elevated interchangeable sodium percentages. The gilgai amplitude increases with increasing expansive clay and sodium content. The expansion which gives rise to gilgai can be seen in the soil profile, where small fault surfaces and thrust planes are relatively common. Similarly, stakes placed in the soil are found tilted or even expelled

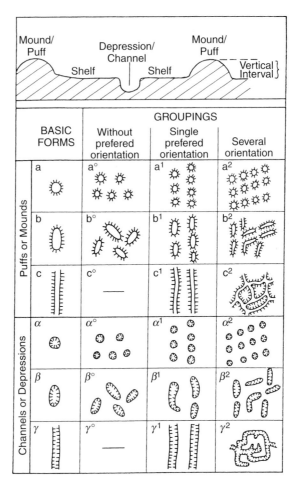

Figure 12.14. In the upper part, cross section of a gilgai (after Harris, 1968). In the middle-lower part, classification of gilgai morphology (Verger, 1964).

after a year. Another way of producing gilgai is given by Hallsworth et al. (1955) and expressed graphically by Verger (1964) (Figure 12.16). During a dry period, particles fall into the interior of the soil cracks and this added material is believed to cause the microrelief during humid periods, because it provokes significant pressure deep within the soil. It is also important to distinguish between the various mechanisms due to the action of distinct forces in diverse situations, in relation to cracking and the expansion–contraction cycles (Knight, 1980), which helps in understanding gilgai genesis more adequately.

4. Desert varnish

A fine, dark film developed on exposed material commonly occurs in arid regions, and this is called *desert varnish* or *rock varnish*. In some deserts, such as those in SW USA, this covering occurs on 75% of the rocky surfaces (Engle and Sharp, 1958).

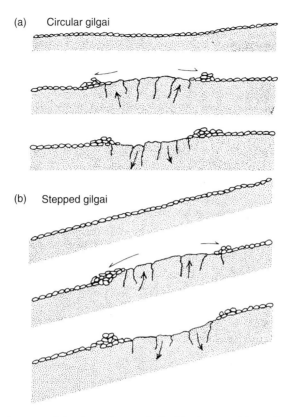

(a)   Circular gilgai

(b)   Stepped gilgai

Figure 12.15. Development of circular and stepped gilgai in the central Australian Stony Desert (after Ollier, 1966).

However, varnishes also occur in alpine, coastal, fluvial, and sub-surface environments, as well as near springs (Dorn and Oberlander, 1982; Dorn, 1998). The dominant color is black or brown, due to enrichment in Mn and Fe.

The origin of varnishes has been the subject of numerous discussions, often lacking precise scientific foundations. With the development of various high-resolution techniques (microscopy and electron microprobe), which permit detailed observations and spot chemical analyses, the understanding of varnishes has progressed considerably. This work started with Engle and Sharp (1958) and since then many advances have been made with regards to composition, structure and origin (Dorn and Oberlander, 1981, 1982), and important results were expected (Oberlander, 1994; Dorn, 1998).

4.1. Composition and age

Varnish covers rocky exposures (Figure 12.17) or surface stones (Figure 12.18). The color tends to be black if it is rich in Mn, more orange if high in Fe and brown if both are present in similar amounts. The varnish film is very thin and varies greatly in thickness (0.002–0.5 mm), although it generally oscillates between 0.01 and 0.03 mm.

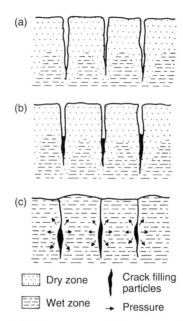

Figure 12.16. The formation of gilgai by wetting and drying (after Verger, 1964).

Figure 12.17. Bedrock surfaces covered by desert varnish. Anti-Atlas, Morocco. Photo by J.L. Peña.

Figure 12.18. Varnish film on clasts of a desert pavement. Negev Desert, Israel.

It develops on all rock types, but mostly on silicates. Varnish is composed mainly of Si, Al, Mn and Fe, with Ca, K, Na, Ba, Ti, Sr, and Cu as minor components. To these components can be added many more in minor proportions, but which can reach significant values locally (Oberlander, 1994). Varnish exhibits microlaminations with different chemical compositions, mainly due to variations in Mn content. Many of these elements are likely to originate from external sources, because they do not occur in the supporting rock. Clay minerals constitute the major part of varnish, some 60 to 80%, followed in importance by Fe, Mn, and Si in the form of amorphous oxihydroxides. Black varnishes with elevated Si percentages also exist which can reach 0.01 mm in thickness and form on siliceous rocks. This type of covering is denoted *desert glaze* or *silica glaze* (Dorn, 1998). It is thought to originate by chemical precipitation of monosilicic acid $Si(OH_4)$ in the form of a gel (Krauskopf, 1956) or via biological activity from a source of opaline phytoliths which dissolve and reprecipitate (Farr and Adams, 1984).

For many decades varnish development or degrees of patination have been used as an indicator of relative age in geomorphological and archaeological investigations (Hunt and Mabey, 1966; Tricart, 1969; Demangeot, 1981). These older studies indicated that varnish did not normally appear on Holocene material, which implies that it needs many years for its formation. Absolute age dating has been carried out using neutron activation analysis, U series, palaeomagnetism, and AMS C-14. For this it is necessary to collect varnish from a surface area of 0.5 m^2 (Krauskopf, 1956), using what little organic material is present. The application of some of these techniques has not been continuous and radiocarbon dating has problems of contamination, through extracting part of the rock substrate along with the varnish. Another way of obtaining relative ages has been suggested, based on the distinct degree of leaching of the varnish constituents. A leaching index has been proposed: Ca + K/Ti, but it presents many treatment and analytical determination problems. All of these methods have been used in studies of fault activity, rate of sea cliff erosion, Quaternary deposit ages and lithic artefacts (Oberlander, 1994).

4.2. Formation processes

Previously, investigators believed that the constituents of varnishes came from the weathering of the supporting rock. Solutions rich in Fe and Mn were thought to rise to the rock surface via capillary action, where they were deposited after humid and rainy periods. Nowadays no one doubts that the constituents derive from external sources, through aeolian dust transport or water transport in solution (Figure 12.19). Electron microscope

Figure 12.19. Rock varnish developed on a sandstone, probably due to the precipitation of waters with silica on the surface. This sandstone is affected by cavernous weathering features (tafoni). Petra, Jordan. Photo by J.L. Peña.

investigations reveal that the varnish is superimposed on, and clearly separated from, the rocky substrate, supporting the theory of an external origin of the constituents.

The most popular hypothesis is based on a biogenic origin, in which Mn enrichment is due to fixing by mixotrophic bacteria which oxidise Mn^{2+} to Mn^{4+}, along with clay cementation which traps other elements. This then repeats itself, producing an accretion and subsequent lamination of varnish, which in turn implies cycles of humidity and aridity. At the present time the bacteria that fixes the Mn is believed to be *Metallogenium personatum*, the oldest known living organism (Oberlander, 1994). Others propose fungi and cyanobacterias as fixing organisms (Krumbein and Jens, 1981). Some advocate purely physico-chemical processes (Engle and Sharp, 1958; Elvidge and Moore, 1979). Mn is more soluble than Fe and in desert environments, Eh values are high, indicating elevated oxidation, and pH is alkaline as a consequence of the low levels of leaching. According to these authors, variations in oxidation–reduction and the basicity of the medium can give rise to varnishes. It is clear that an intense debate exists concerning the origin of varnishes; it is very possible that the conjunction of biotic activity with the physico-chemical variations could be the route to follow in future experiments.

4.3. Environmental implications

The first studies of varnishes emphasised their climatic significance and indicated that a more humid climate than the present was needed for their formation. This implied that existing varnishes were all relict (see Dorn, 1998). The studies also argued for cycles of patination, whereby the varnish is eroded and a new one created.

Currently, with modern high-resolution techniques, the varnish laminations which form microlayers with distinct compositions are interpreted climatically. Periods of reduced Mn accumulation correspond to arid or interpluvial phases. Periods where the Mn/Fe relationship is increased indicate more humid, pluvial or lacustrine phases, where aeolian deflation diminishes because of expansion of the lakes (Figure 12.20). In many places, however, the varnish sections offer an unconvincing register of palaeoclimatic oscillations, and are more likely to indicate that the processes of varnish formation are very complex (Oberlander, 1994).

5. Duricrusts

5.1. Introduction

One of the characteristics of arid zones is the presence of surface and subsurface crusts of differing chemical composition, such as calcareous, siliceous and gypsum crusts. It is also possible to encounter laterites, formed in environments of much greater precipitation, which are mostly found on desert margins. They extend over significant areas in arid zones and some, thanks to their enhanced hardness, are resistant to erosion and are situated in elevated areas. Furthermore, they are good indicators of palaeoenvironmental conditions and an understanding of Quaternary crusts is very useful in the interpretation of past geological stages (Goudie and Pye, 1983). They can also be of economic interest, for

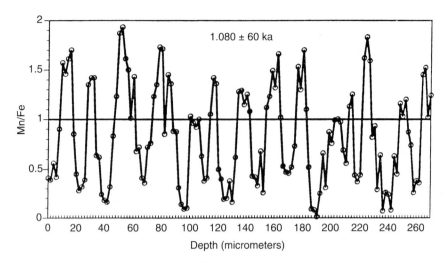

Figure 12.20. Ratios of Mn/Fe through rock varnish on volcanics with K/Ar date of 1080 ± 60 ka, Coso area, California. Peaks and troughs are supported by multiple datapoints, but the record appears too variable to be convincing as an index of climatic oscillations (Oberlander, 1994).

example, calcareous scales may contain uranium and other metals; they can also be used in house construction and as road fill for motorways (Reeves, 1976).

In this section, we will try to analyse the nature, properties and significance of duricrust scales, as well as the geochemical and biochemical implications of their formation, and finally describe the models most commonly utilised in their description.

5.2. Caliches

Caliche is a Spanish word derived from the Latin *calix*, which means lime or limestone, and was used for the first time in 1719 (see Reeves, 1976). It is commonly used in the Unites States and Mexico but not Spain, where the term *costra calcárea* is more common, most probably due to French (croute calcaire) and German (kalkruste) influences. At a global level there still exists a certain confusion because the word has been used for other types of deposits. Moreover, there are many locally-used terms in different areas and countries (Goudie, 1973; Reeves, 1976). The term calcrete, however, used in English literature, is gradually becoming more popular in the scientific community.

Following Goudie (1972a), "Calcrete is a term used for material of continental origin formed dominantly, but not exclusively, of calcium carbonate, found from powdery and nodular to strongly indurated in form and implies cementation of the soil, rock or weathered material within a vadose zone. This definition, however, does not include speleothems, travertines, cemented litoral deposits, or stromatolites in algae lakes."

Caliches are chalky in texture, occupy around 13% of the Earth's surface and occur in areas with an annual precipitation between 400 and 600 mm (Goudie, 1983a,b). Other authors, such as Rutte (1958) in his study of Spanish limestone scale, state that the limits occur between 100 and 500 mm, with no caliches occurring above or below these values. An optimum between 100 and 250 mm is, however, indicated. Although they have a wide

Figure 12.21. Caliche on cemented gravels, resistant to erosion, located on a wadi margin. Jaisalmer, Thar Desert, India.

distribution, caliches are characteristic of arid and semi-arid zones (Figure 12.21) (Reeves, 1976). They develop fundamentally in surface pediment deposits and terraces, and also in aeolian deposits. In the terraces of the River Cinca in the Ebro Depression of Spain, the degree of calcification of the stepped alluvial levels increases with age, and is nonexistent in the more modern levels (Sancho and Meléndez, 1992).

Several volumes have been published concerning caliches (Reeves, 1976; Blümel, 1981; Vogt, 1984; Wright and Tucker, 1991), studying their types, characteristics and genesis. As given by Vogt (1984), the investigation of calcareous crust has to consider different scales or levels. In the first place it is necessary to know the spatial distribution (at the km scale) and its relation to the regional geomorphological context, such as how it may be related to pediment and terrace systems. The next study has to be carried out at the outcrop level, by means of analysing natural cuts and samples that provide sedimentological profiles, thus allowing the caliches to be placed stratigraphically in the sequence. Finally, in the laboratory, thin sections and electron microscopy can be used to analyse texture and micromorphology, which provide data about the formation mechanisms. All of these investigations should be complemented with chemical and mineralogical analyses that help in understanding the generating processes.

5.2.1. Geochemistry and mineralogy

The chemical composition of caliches, calculated from some 300 samples from all over the world (Goudie, 1972b, 1973, 1983a,b), is; $CaCO_3$: 79.28% (CaO – 42.62%),

SiO_2: 12.30%, MgO: 3.05%, Al_2O_3: 2.12% and Fe_2O_3: 2.03%. Those caliches rich in magnesium are called *dolocretes*. The percentages of each distinct component vary enormously from one place to another, as can be seen in the calcretes in southern Australia (Hutton and Dixon, 1981; Dixon, 1994b). In a study of numerous chemical analyses of caliches from Oklahoma and New Mexico (Aristarain, 1970) the calcium carbonate, magnesium, and ferric iron contents were seen to diminish with depth, whereas the silica content increased significantly in the middle part of the profile. Elsewhere, Dixon (1994b) examined the role of topography in the chemistry variations in caliches and demonstrated a higher percentage of calcium lower down in the toposequence and an increase in magnesium in the upper part. In all studies the calcium content fell with depth.

Calcite and dolomite are the dominant carbonate minerals in caliches and are accompanied by quartz, opal, and clay minerals. Caliches can also contain minor amounts of soluble salts (gypsum), glauconite, phosphates, heavy minerals, etc. Mineralogical investigations have centered around the study of the clay fraction and its origin and, as a consequence, improving our understanding of the processes involved in the formation of clay-bearing caliches. The dominant clay minerals are palygorskite and sepiolite, although illite, kaolinite, montmorillonite, inter-layered illite–montmorillonite, and chlorite have all been recognised (Aristarain, 1970, 1971; Gardner, 1972; Reeves, 1976; Hay and Wiggins, 1980). The origins of palygorskite and sepiolite have been attributed to neoformation in an environment containing sufficient magnesium, provided by the weathering of suitable lithologies (Hay and Wiggins, 1980; Hutton and Dixon, 1981; Sancho et al., 1992). Other investigators maintain that they arise through the alteration of montmorillonite and inter-layered illite–montmorillonite (Watts, 1980).

5.2.2. Morphology

Descriptions of caliche morphology are numerous and different classification schemes have been proposed based on distinct criteria. Some are based on sequential evolution (Netterberg, 1969), others on degree of hardness of the carbonates (Reeves, 1976). The degree of calcification also has some bearing on differentiating between soil characteristics (Gile et al., 1966; Machette, 1985). Some distinguish only a few morphological types and others, on the contrary, numerous types (Reeves, 1976).

The fundamental macroscopic morphologies (Sancho and Meléndez, 1992) are *calcified gravels*, which normally appear at the base of a caliche profile, though small levels can be encountered in the interior of the profile. This is constituted of clasts with a vadose carbonate coat and a matrix that may contain root casts. *Chalky caliche* (Figure 12.22) is formed of loose, clay, and sand-sized carbonate particles with practically no host rock material. It can be lightly cemented and have a leafy structure. *Hardpan caliche* occurs in the upper parts of the profile (Figure 12.23). It consists of centimetre and decimetre scale plates that stack laterally. They form ooids, basal speleothemic carbonates, root traces, and dessication cracks. Hardpan caliche can also be found deformed, forming *pseudo-anticlines* (Figure 12.23) generated by expansion and contraction as a consequence of hydration–absorption of water to a minerals of highly soluble salts, periodic wetting and drying, calcite crystallisation and subsequent arching. Deformation may also be due to the presence of expandable clays (Reeves, 1976). Finally, *nodular caliche* can develop in fine-grained material, in which the particles are cemented by carbonate and, in turn,

Figure 12.22. Chalky caliche cemented in the upper most part, and with laminations. Alfés, Province of Lleida, Ebro Basin, Spain.

Figure 12.23. Hardpan and nodular caliche deformed by a pseudo-anticline. La Plana Negra, Province of Zaragoza, Ebro Basin, Spain.

Figure 12.24. Nodular caliche developed on loose fine deposits. Bawri, Thar Desert, India.

generate large quantities of nodules (Figure 12.24). It can also result from the fragmentation of hardpan caliche, targeting loose and cemented nodules (Figure 12.25). A continuous graduation between hardpan and nodular caliches can be distinguished.

The macroscopic caliche characteristics are frequently represented at the microscopic scale: nodular, oolitic and speleothemic textures, carbonate coatings, traces of biologic activity and so forth. All of the investigations directed towards a precise and detailed understanding of these micromorphologies and textures have as an objective an improved understanding of caliche formation processes.

5.2.3. Origin

The analysis of caliches reveals the presence of complex sequences, whereby the superposition of various caliche horizons can be recognised. In addition, the majority of the profiles exhibit diverse stages of precipitation, dissolution and reworking.

For caliche to form there must be a source of calcium carbonate, which may be in the rocky substrate, plant remains, volcanic emissions, aeolian dust, and other factors

Figure 12.25. Nodular caliche with nodule size coarsing upwards. Margalef, Province of Lleida, Ebro Basin, Spain.

(Goudie, 1983b). For many authors, an origin of carbonate transported by wind and dissolved in rainwater (Blümel, 1981, 1982; Machette, 1985) is too generalised, because caliches sometimes develop with great thickness over rocks that are practically devoid of calcium.

In semiarid zones, where a water deficit exists, carbonates do not fully leach and tend to move and precipitate from one place to another. The fundamental process implicated in caliche genesis is the dissolution and precipitation of calcium carbonate in the presence of carbon dioxide:

$$CaCO_3 + CO_2 + H_2O \Leftrightarrow 2HCO_3^- + Ca^{2+}$$

Carbonate precipitation takes place by the loss of CO_2, evaporation, the common ion effect, and biological activity. The biological mechanisms of fixing carbonates in soils can be due to lichens, algae, bacterias and microrrhizae (Klappa, 1979).

Goudie (1973, 1983b) differentiated two kinds of caliche origin; one pedogenic, which seems to be the most generalised, and the other nonpedogenic. In any case, it is very difficult to distinguish between them, because nonpedogenic caliche may be modified by pedogenic processes.

Pedogenic caliches result from the progressive, downward accumulation of carbonates (the *per descensum model* of Goudie; Figure 12.26), through washing of the A horizon and illuviation of the B horizon (petrocalcic horizon). As seen in young profiles, early genesis is manifested by powdery filaments and nodules. In profiles of thousands of years of age, however, the carbonates form a crust over stones and fill the spaces in between particles.

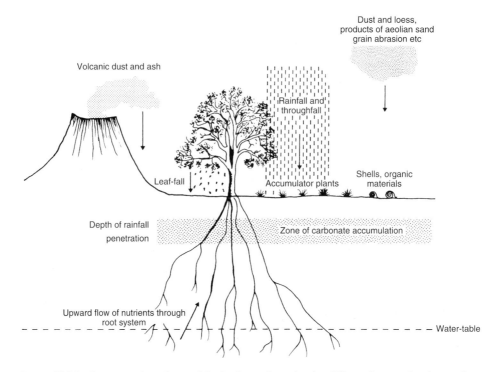

Figure 12.26. Representation of a model of calcrete formation by different inputs of carbonate from above *per descensum* model (in Goudie, 1973b, Figure 44).

The presence of calcified filaments, fibrous calcites and fungus microfossils testify to a pedogenic caliche origin (Vaniman et al., 1994). A problem with this hypothesis is that the caliche thickness is generally much greater than that of the soil; this difficulty can be explained by a slow rate of soil erosion during caliche production.

When carbonate solutions rise through capillary action and precipitate at the surface by evaporation, *nonpedogenic caliches* form, corresponding to the *per ascensum model* of

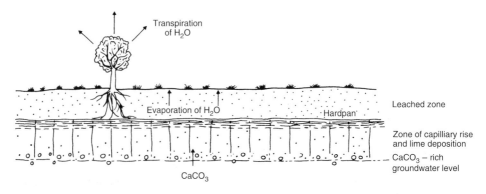

Figure 12.27. A model of calcrete formation by capillarity rise from groundwater *per ascensum model* (after Goudie, 1983b, Figure 4.5, reproduced with permission).

Goudie (Figure 12.27). They can form close to the phreatic level by evaporation in the capillary zone and also beneath the phreatic level as a consequence of a fall in CO_2 content (Land, 1970). This type of caliche can arise through carbonate-rich flood waters, which can reach thicknesses of more than 10 m. They may develop rapidly, because they occur in very young alluvium (Machette, 1985), although this origin constitutes a special case.

5.3. Siliceous crust

Also known as *silcrete*, this is a grey, tan or green coloured, hard and fragile rock, comprised of quartz grains cemented by a microcrystalline and amorphous matrix. Common thicknesses vary between 1 and 3 m. It forms in arid environments, although it can form in more humid zones. It is a product of replacement of surface materials (regolith, sediment, soil) by silica, in which silification takes place at a low temperature unrelated to volcanism, plutonism, or metamorphism (Summerfield, 1983). Siliceous crusts abound in Australia and South Africa, although they have been observed elsewhere. They can occupy the tops of hills due to their resistance to erosion. Precious opals can occur in some silcretes.

In outcrop silcretes commonly exhibit a massive structure accompanied by columnar joints. Plate, botroidal and pillowy structures can also be recognised (Wopfner, 1978). The micromorphology reflects the characteristics of the host rock (grain size and shape, mineralogy and fabric) and the silification processes. This includes micro- and cryptocrystalline quartz with boundaries indented in the mineral grains, which confirms the idea of replacement (Milnes et al., 1991). Chemical analyses indicate that they are usually more than 95% SiO_2 by weight and contain minor amounts of aluminum, iron and titanium. The aluminum is commonly associated with the clay minerals, whereas the titanium appears to be related to the bedrock.

In silcrete formation it is necessary to consider the silica source and its dissolution, transport and precipitation. The primary silica source can be derived from *in situ* dissolution, silicate weathering and aeolian dust. In fluvial and lacustrine environments, diatoms may play an important role. Dissolution and transport are produced by water at alkaline pH, moving vertically and laterally (Summerfield, 1983). The lateral movement model is used in fluvial and lacustrine environments. Vertical transfer has similar aspects to the *per descensum* and *ascensum* models used by Goudie (1973b). Finally, precipitation is produced by evaporation or a fall in pH.

Where siliceous crusts occur in great thicknesses, their age, obtained from diverse places around the world, go back to Mesozoic times, and so are considered relict accumulations, possibly generated in different climatic regimes (Cooke et al., 1993).

5.4. Gypsum crusts

These are found in numerous arid zones, but over much reduced areas (Coque, 1962). They develop in deserts with annual precipitations of less than 250 mm (Watson, 1983). They are defined as accumulations of gypsum found in the uppermost 10 m, having a thickness of 0.1 to 5 m, a gypsum content of greater than 15% by weight and a minimum of 5% gypsum in the underlying substrate (Watson, 1985, 1989a).

Three types can be differentiated on the basis of their structure and stratigraphic position (Watson, 1979). The first type is horizontally stratified crust containing 50 to 80% gypsum. The second is subsurface crust, formed of either large, lenticular crystals 1 to 0.5 m in diameter, commonly known as desert rose, or mesocrystalline material with 0.05 to 1 mm diameter crystals. They can reach up to 5 m in thickness and have a gypsum content of 50 to 70%. The third type is the surface crust, which basically consists of alabaster gypsum with crystals of less than 0.05 mm, occurring in columnar form on a scale of 1 to 2 m, or as loose, dusty deposits. They contain approximately 90% gypsum.

Gypsum crusts can be differentiated on the basis of their generation via capillary rise and evaporation of sulphate-rich waters, or *per descensum* through gypsum-rich aeolian dust (Watson, 1985; Heine and Walter, 1996). These authors state that Tunisian dust originates primarily from the sebkhas, and that dust in the Namib Desert originates from coastal clouds. The nonpedogenic gypsum scale model is based on the precipitation of gypsum in subterranean waters and lacustrine environments.

Chapter 13

The action of water in arid regions

1. Introduction

In the proceeding chapters we have discussed the importance of water in weathering processes, its role in the generation of patterned ground, varnishes, and crusts. In these cases the activity of water occurs in a thin film and its movement is clearly local. We now turn our attention to the action of water activity connected to surface flow, organized in fluvial systems. These are characterized (Schumm, 1977) by a headwater source area, where water and sediments are derived, a transfer zone and, finally, a lower area or sedimentation zone. This fluvial system has a group of independent and dependent variables (Schumm and Lichty, 1965) that have different influences. Time, initial relief, geology (lithological and structural) and climate are the independent variables most significant in influencing the hydrology and erosive activity. The vegetation depends on climate and lithology. The rest of the dependent variables are variation in relief, hydrology (run-off, discharge, sediment production) and the morphology of the drainage network and slopes. In arid regions the first three independent variables vary substantially between different deserts, unlike climate that depends on precipitation and which is, in turn, the cause of run-off and erosion that are conditioned by vegetation cover.

2. Precipitation, vegetation, and evapotranspiration

In the majority of the arid regions of the world precipitation data are rare as there are few meteorological stations and generally, where they do exist, they are very isolated. As a consequence, extrapolations under these conditions are likely to be incorrect. Precipitation is a consequence of different atmospheric processes: storm fronts, tropical storms, orographic effects, and convective cells (Graf, 1988). As most deserts are located in anti-cyclonic belts within the global atmospheric circulation system, precipitation is scarce and limited to the sporadic penetration of storm fronts that are rare and normally produce little rainfall. Tropical storms (cyclones, hurricanes, and typhoons) are rarer than storm fronts, but occasionally they can penetrate the interior of arid regions, such as in the Sonora Desert, between Mexico and Arizona. Orographic effects occur due to a reduction in pressure and a cooling of the rising air, conditions that favour condensation and precipitation. When the air descends on the other side of a mountain range, it warms and, because it has lost the majority of its humidity while rising, it creates a rain-shadow zone. The final type of process is the convective storm that is a few kilometres in diameter and results from warming of the air. When they affect small basins they can produce important floods (such as that of the 28th July, 1990, which developed a delta at the mouth of the lower Huerva River, a tributary of the Ebro in Spain.

Precipitation in arid regions varies significantly (Figure 10.4) and, as a result, greatly influences changes in run-off and sediment production. For example, in the middle of the Ebro Depression, at an experimental station for studying run-off and erosion at Lanaja in Huesca Province, Spain (Figure 13.1), a 4-year record shows that the most important precipitation corresponds to convective storms that occur primarily during spring and autumn, with some in summer. The rest of the rainfall is from convective cells and is of low magnitude (Gutiérrez et al., 1995). For this reason, the highest values of run-off and erosion take place in spring and autumn. Intensities vary considerably within convective cell precipitation. At times, the recorded rainfall from one storm exceeds the total mean annual precipitation. Furthermore, experimental stations with a large quantity of rain gauges show sharp spatial variations and steep gradients (Figure 13.2).

In the literature, a number of cases are cited where precipitation is related to the advection of humid air masses supplying continuous rainfall to a semi-arid region (Inbar, 1972; in Starkel, 1976). The Mediterranean winter rainfall that reaches the mountains of Lebanon, Israel, and Syria is an example. Rainfall during a 4-day period (20th to 23rd January, 1969) varied between 120 and 360 mm, equivalent to a third of the mean annual precipitation. Despite the low mean intensities of 2 to 3 mm/h, occasionally exceeding 8 mm/h, this rainfall corresponded to a return period of 100 years. During this event, 0.04 mm/yr of downcutting, six times the normal rate, was produced in the upper Meshushim basin. This shows that storm precipitation can be of considerable importance in semi-arid regions.

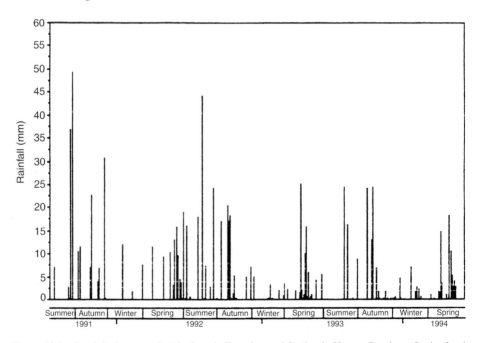

Figure 13.1. Precipitation record at the Lanaja Experimental Station in Huesca Province, Spain, for the period July 1991 to May 1994 (Gutiérrez et al., 1985).

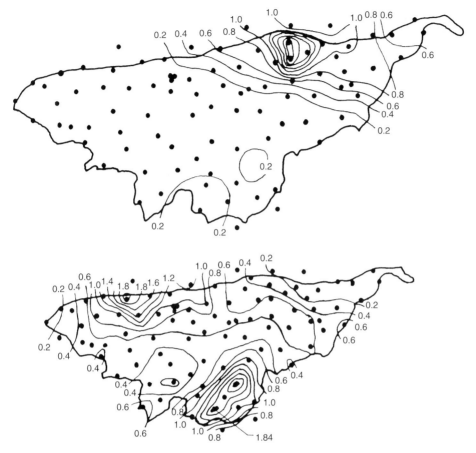

Figure 13.2. The distribution of rainfall from thunderstorm cells over Walnut Gulch, Arizona: Above, a single cell event. Dots represent rain gauges. (after Renard (1970), in Graf (1988), Figure 3.4).

Vegetation in deserts is sparse or even nonexistent. In environments below 100 mm of precipitation the vegetation may only be developed in wadis (Figure 13.3) or closed depressions. If precipitation varies between 100 and 500 mm, the natural vegetation is arboreal. Generally, this has been degraded by human activities to a more open landscape of shrubs and grasses that may also have been affected by overgrazing (Francis, 1994). The scarcity of water and the elevated temperatures mean that plants must adapt to these inhospitable environments by means of physiological or morphological adaptations and procedures that can cope with the stress associated with an increase in aridity (Günster et al., 1993).

We have already seen the role that flora plays in the biological alteration of rocks and the influence that sparse organic material has on the processes of hydrological erosion. There exists, in general, a good correlation between the percentage of vegetation cover and the mean annual precipitation. Interception is a significant factor in hydrological erosion. Návar and Bryan (1990), in a study on the influence of vegetation in a semi-arid region of northeast Mexico, found that after precipitation of 230 mm, divided between 17 storms,

Figure 13.3. Shrub vegetation in the bed of the Nahal Zin in the south of the Negev Desert, Israel.

interception losses were 27%. On the other hand, infiltration is highly variable and considerably greater in vegetated areas than in bare soil. In deserts there are some environments that are more sensitive to vegetation changes and this is manifested in sediment production. It increases on a slope with decreasing vegetation and reaches a maximum on bare soils (Figure 13.4a). Erosion is barely affected, however, where the vegetation cover is low (Figure 13.4b,c) (Schumm, 1977), although for De Ploey et al. (1976) and Morgan et al. (1986) sediment production increased for low values of vegetation cover (Figure 13.4d). Rogers and Schumm (1991) showed experimentally that, on 10% slopes, erosion increases rapidly with a decrease in vegetation cover from 43 to 15% but below this last value erosion decreases. Consequently, with cover below 15%, vegetation does not curb sediment production.

Vegetation affects aeolian erosion by reducing the wind velocity and acting as an obstacle, which generates small dunes, known as *nebkhas*, in the lee of the plant. One observes a decrease in dune movement with an increase in vegetation (Ash and Wasson, 1983) that acts to stabilize them. The same function is carried out by biocrusts that stop dune movement, as observed in the longitudinal dunes of Nizzana in the Negev Desert of Israel (Veste, 1995) (Figure 11.13 and Figure 13.6).

The water that exists on the Earth's surface undergoes *evaporation*, which acts to dry the surface materials. In desert regions drying, brought about by prolonged drought, reaches a depth of various meters. Another type of water loss is due to water uptake through plant roots. *Transpiration* is produced in the stomatal cells, where there is an interchange of water vapor between the atmosphere and the leaves. Transpiration is favoured by low humidity, high temperatures, and wind. In deserts, xerophyte plants adapt

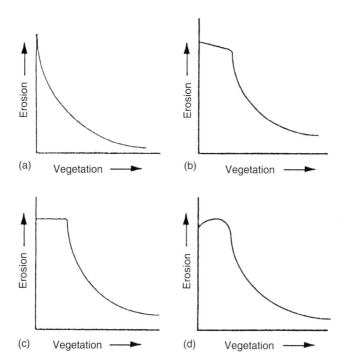

Figure 13.4. Hypothetical relations of erosion and sediment yield to vegetative cover. All show an exponential increase of erosion and sediment yield at higher vegetation density but b, c, and d indicate a decreased influence of vegetation at low values of vegetative cover (after Rogers and Schumm (1991)).

to this water shortage by reducing transpiration through the formation of hidden stomates and by reducing leaf size, or losing them altogether as in the cactus. In addition to xerophytic plants, in the wadis of arid regions, phreatic plants develop. The roots of these plants penetrate deeply to reach the phreatic level, assuring the water source necessary for their development. Evapotranspiration may reach values of 85 to 90% of total precipitation (Renard, 1969), although there are differences between distinct vegetation species. The majority of the evapotranspiration occurs soon after rainfall. The rates of transpiration fluctuate between 2.3 and 10 mm/day and 0.1 mm/day during the driest periods of the year (Thornes, 1994).

3. Characteristics of the fluvial network in deserts

Precipitation in arid regions is scarce and variable. This variability reflects itself in the fluvial systems at the global level. One can distinguish between *endogenous drainage* where precipitation occurs within the desert area and *exogenous drainage* that originates outside the arid zone, but flows through it, such as the rivers Nile, Indus, Colorado in the USA, and the Ebro in Spain. In deserts, an important part of the drainage is *endorheic*, where the water does not reach the sea (Figure 13.7) due to infiltration and above all the prevailing evaporation. Furthermore, there are large closed depressions that receive

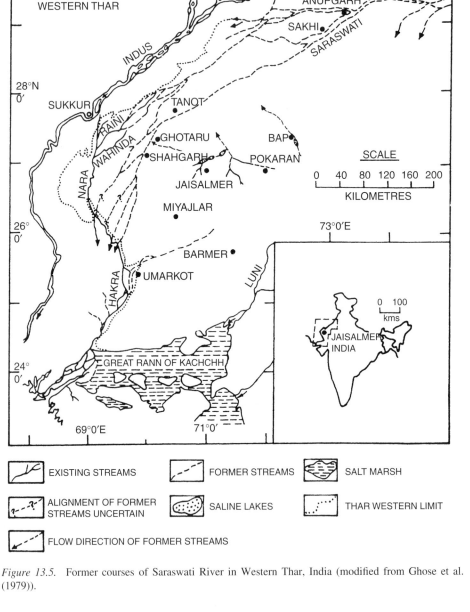

Figure 13.5. Former courses of Saraswati River in Western Thar, India (modified from Ghose et al. (1979)).

superficial waters, such as the endorheic basins of Lake Chad, in the southern Sahara, and Lake Eyre, in the Central Desert of Australia. Guilcher (in Demangeot, 1981) estimated that in the majority of Saharan regions, the largest floods do not travel more than 300 km. Areas with *areic* drainage are those that show no regular circulation of water and occupy

Figure 13.6. Biocrust developed in a system of longitudinal dunes affected by rilling, Nizzana Experimental Station, Negev Desert, Israel.

important extents within desert zones. Some deserts have *exoreic* drainage such as the meridional zones of the southern Atacama Desert of Chile that have a large water source in the Andes.

The river courses of arid regions are usually ephemeral (Figure 13.8) reflecting the scarce supply of water they receive. These rivers (or wadis), therefore, are permanently dry most of the year (Figure 13.9). Perennial rivers also occur, however, if they receive an important water supply from distant sources, usually from surrounding mountain ranges where a contribution of water may come from snowmelt (Figure 13.10). At times, the most distal part of the drainage network may not be currently active, as indicated by greater hydrological activity in the past pluvial periods. This occurs in the central Sahara where complete networks of wadis do not receive water during even the largest floods (Tricart, 1969). Also, the Niger and Senegal rivers that originate in the tropical mountains of Guinea, flow northwards through a very dry region characterized by dunefields that can only be traversed during pluvial periods (Tricart, 1969). The use of aerial photography

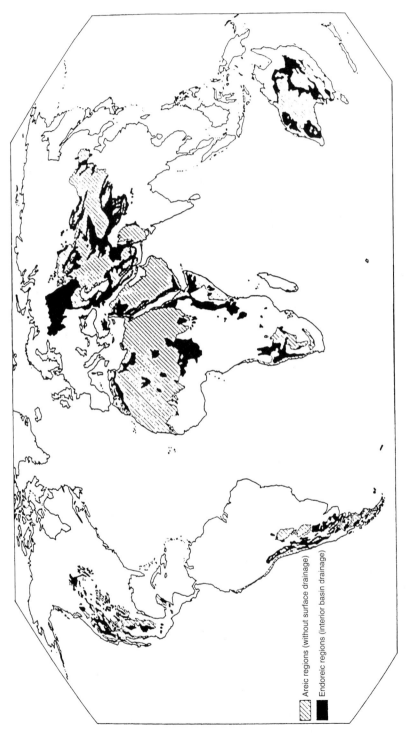

Figure 13.7. Areas of endoreic and areic drainage (after de Martonne and Aufrère, 1928).

Figure 13.8. Network of ephemeral channels in the interior of the large Makhtesh Ramon Depression, that forms from the structural point of view a *combe*, Central Negev Desrt, Israel.

Figure 13.9. Cultivation in the channel of the Sabarmati River, Ahmadabad, northeast India, that indicates the scarcity of superficial run-off that it receives.

Figure 13.10. River course in the piedmont zone of the Atlas Mountains that have significant snow cover, Assif Mellah, southern Morrocco. Photo: J.L. Peña.

and satellite images show that some desert areas had fluvial systems in the past, the courses of which are currently buried beneath the desert plains, as is the case of the River Saraswati in the Thar Desert of India (Ghose et al., 1979; Srivastava et al., 2001) (Figure 13.5). Similarly, at the large endorheic Lake Chad, to the south of the Sahara, an important part of the fluvial network has remained fossilized by aeolian sands (Grove and Warren, 1968). Also, in the piedmont zones of some desert areas, alluvial fan systems develop in which frequent channel shifting occurs. These alluvial fans can be affected by aeolian erosion during periods of water shortage.

Run-off may be initiated in deserts through Hortonian overland flow (Horton, 1933), produced when rainfall intensity is greater than the infiltration capacity of the soil or rock, or when generated by saturation of the upper soil layers. During storm precipitation in arid regions the rainfall exceeds the rate of infiltration and, as a consequence, the run-off is produced by Hortonian flow.

Hydrographs that express the variation of discharge as a function of time show a very sharp rising limb and a declining limb that is initially pronounced before smoothing off (Figure 13.11). These sharp discharge variations over a short time period are typical of arid regions. The curves illustrated in the figure are from the gauging station of Las Tosquillas, a few kilometres downstream of Ugijar, in the Alpujarras of Granada Province (Thornes, 1976). The hydrograph, corresponding to the record of the 24th to 31st August 1969, shows a practically instantaneous flood wave and an equally sudden declining limb, typical of this area where floods are caused by intense summer convective rainfall (Thornes, 1976).

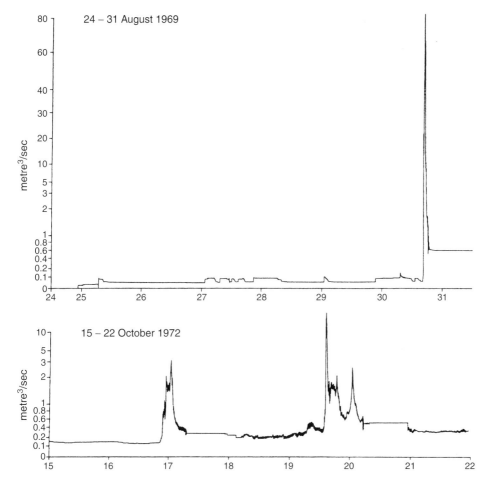

Figure 13.11. Selected traces of stage records from the Las Tosquillas station, Ugijar River, Alpujarras region of Granada (after Thornes, 1976).

The floods in rivers of arid regions can be divided into four types (Graf, 1988): flash floods, events with a single discharge peak, events with multiple peaks and seasonal floods. The first are produced by storm rainfall that covers small areas of around 8 km diameter and are, therefore, limited to basins of 100 km² or less. They are characterized by reaching a peak from zero flow in a matter of minutes or hours and the response is a wave that advances down the channel with a sharp, turbulent front loaded with debris. The example described from Ugijar corresponds to this type of flood. The single peak events have a longer duration than flash floods, from some hours to several days (Ward, 1978). They are produced by tropical storms or storm fronts that affect basins of thousands of square kilometres in size. Those with multiple peaks are caused by precipitation from tropical storms or fronts, like that recorded from 15 to 22 October in the Ugijar River in southern Spain (Figure 13.11). They are produced by successive storms or shower systems that affect a particular region.

In the river courses of arid regions a part of the flow is lost by infiltration and evaporation. Floods in small ephemeral streams are lost over short distances, whereas in larger rivers water is lost gradually, mainly through evaporation, with a clear reduction in discharge and salt precipitation in marginal areas (Figure 13.12). This causes discharge peaks to decrease considerably downstream. These transmission losses are illustrated at two gauging stations located at Walnut Gulch, Tombstone, Arizona, that are 10.9 km apart (Renard and Keppel, 1966). The loss in discharge is on the order of 57% (Figure 13.13).

4. Run-off and sediment transport on slopes

Water from rainfall infiltrates the soil until it is saturated with water, exceeding its *infiltration capacity*. The rate at which the water penetrates is determined by the rate of infiltration that initially decreases rapidly then more slowly later. Infiltration is controlled by a number of factors: type of rainfall (intensity and kinetic energy), texture and structure of the soil, vegetation cover, and the slope of the area. The impact of raindrops causes an increase in compaction that reduces infiltration (Morin and Benyami, 1977). On surfaces with rock fragments studied in the laboratory and in the field with rainfall simulators, infiltration is reduced with an increase in coverage of rocks (Poessen et al., 1990; Abrahams and Parsons, 1991a). Infiltration also diminishes away from shrubs, around which there is a substantial increase in the content of organic material and vegetative detritus (Lyford and Qashu, 1969). The development of biocrusts (Talbot and Williams, 1978) and lichens (Alexander and Calvo, 1990) also reduces infiltration.

Figure 13.12. Evaporation loss in a wadi that has resulted in salt precipitation at the channel margin. At the edge of the Sabkha Mutti, United Arab Emirates.

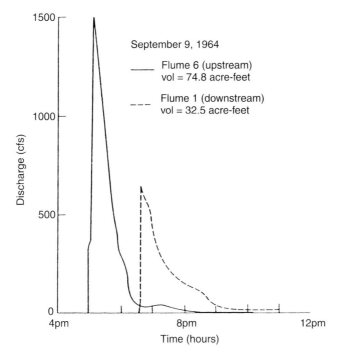

Figure 13.13. Transmission losses for a flood in Walnut Gulch, Arizona, represented by the hydrographs for two flumes 10.9 km apart in the ephemeral-stream channel (after Renard and Keppel (1966), in Cooke et al. (1993), Figure 11.3).

Initial run-off is produced when rainfall intensity exceeds infiltration capacity. Generally in deserts run-off depends on rainfall intensity, as infiltration rate is a more important factor than the state of soil saturation. In arid regions storm events predominate that rapidly exceed infiltration rate. Furthermore, as the vegetation is sparse, interception is low.

On desert slopes practically all the run-off is produced by superficial flow, although in some areas subsurface flow occurs, but only sporadically. Foster and Meyer (1975) differentiated between flow and erosion on slopes produced in rills and inter-rill areas. In both, the hydraulics and mechanisms of erosion are different. In the *inter-rill areas* (Figure 13.14) the water is mobilized, in general, under sheetflow that diverges and converges around objects in its path (for example, rock fragments and vegetation). As a consequence, the velocity and depth of flow vary greatly over small distances, modifying the type of flow (sheet, turbulent, transitional) (Abrahams et al., 1994). In arid regions *rills* are one of the most characteristic features of slopes formed of soft materials (Figure 13.15). Rills have longitudinal and transverse profiles that affect the velocity and type of flow, which is commonly turbulent (Gilley et al., 1990).

In relation to water erosion there are little data regarding slopes in arid regions. Saunders and Young (1983) in their review of rates of slope erosion indicate downcutting exceeding 1 mm/yr in semi-arid climates, whereas in arid regions the values are on the order of 0.01 mm/yr. In semi-arid zones, in areas composed of easily erodable material, the rates of denudation are arguably the highest in the world.

Figure 13.14. Large inter-rill areas separated by small gullies formed in Quaternary clay materials. The slopes are affected by intense cracking and unequal rills development. The Bardenas Reales of Navarra, Spain, in the Ebro Basin.

Figure 13.15. Intense rill formation in Miocene clays similar to badlands. The Bardenas Reales, Navarra, Spain, in the Ebro Basin.

Various factors can be recognized that control the intensity of erosion on desert slopes. The effect of slope was analyzed by Abrahams and Parsons (1991b) in a simulated rainfall study at three experimental stations with distinct lithology and vegetation. At slope angles below 12°, the run-off increased very slowly and sediment production increased with slope due to the gravitational component (Figure 13.16). If the slopes were greater than 12° run-off decreased rapidly, by a value that increased with slope. This reduction compensated for

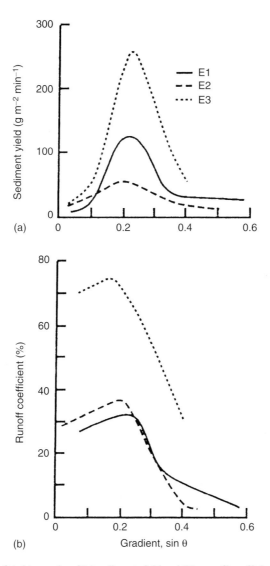

Figure 13.16. Curves fitted to graphs of (a) sediment yield and (b) run-off coefficient against gradient for three sets of experiments denoted by E_1, E_2, and E_3 at Walnut Gulch, Arizona. Experiments E_1 and E_2 were conducted on plots underlain by Quaternary alluvium, with the ground vegetation being clipped for F_1 but not for E_2. Experiment E_3 was performed on plots underlain by the Bisbee Formation (after Abrahams and Parson (1991b)).

the increase due to the gravitational component, therefore, sediment production dropped. The relationship between rock fragment cover and sediment production has been studied in plots by Simanton et al. (1984). For similar slopes, negative correlations were obtained due to a number of factors. In laboratory experiments, however, Poessen and Lavee (1991) found positive correlations when the clasts were greater than 5 cm in diameter. On the other hand, erosion increases after dry periods and decreases during humid ones due to the effect of vegetation that protects the surface from raindrop impact and reduces surface flow.

One of the most important processes that affects inter-rill areas is that of *rainsplash* that provides material that can then be mobilized by sheetflow. When a raindrop reaches the soil (Figure 13.17), the impact generates a small crater by compression and the jolted particles are scattered over the surface by shearing. The greater the kinetic energy of the raindrop, the more important is the effect of rainsplash. If there is a thin film of water on the surface, particle dispersion is increased, reaching a maximum when the thickness of the water lamina is the same as the diameter of the water drop (Palmer, 1963). On horizontal surfaces the distance that the material is ejected decreases exponentially from the point of impact (Savat and Poesen, 1981; Torri et al., 1987). Rainsplash is most effective when there is an increase in slope (Ellison, 1944) and wind speed. This modifies the trajectory of the majority of the mobilized particles. In addition, vegetation cover, the size and percentage of surface particles, organic content, soil chemistry and aggregate stability, amongst other factors, all exert an influence on this process (Scoging, 1989). The rainsplash is an important factor that affects the supply of particles to sheetflow and their subsequent transport in rill and inter-rill areas (Bryan, 1987).

5. Water erosion on slopes

Erosion by surface flow may occur by sheetflow or rilling. According to Leopold et al. (1966), 98% of sediment production during a period of 10 years in a semi-arid area of New Mexico, came from slopes without rills. In environments of greater precipitation in Kenya, Dunne (1980) found no rilling over long slopes, confirming field experiments using rainfall simulation (Dunne and Dietrich, 1980). Similarly, Emmett (1978) indicated that some slopes in Wyoming not affected by rilling suffer uniform degradation through sheetflow erosion. Whereas rill erosion has not received much attention, *inter-rill erosion* has been intensively studied by a number of authors who have focused their investigations on the interaction of splash and sheetflow erosion, both in the field and in the laboratory. These studies have taken into account slope and the physical, chemical and mineralogical properties of soils. They indicate that the processes and rates of inter-rill erosion are extremely complicated and variable, depending on extrinsic factors, such as rainfall intensity, raindrop size, and the presence or absence of wind, and intrinsic factors such as soil texture, the nature of soil aggregates, surface roughness, susceptibility to cracking, and the presence and density of organic detritus (Bryan, 1987).

Rill erosion is predominantly produced by the mobilization of soil particles by concentrated flow and affects a small part of the terrestrial surface. Rills may be developed

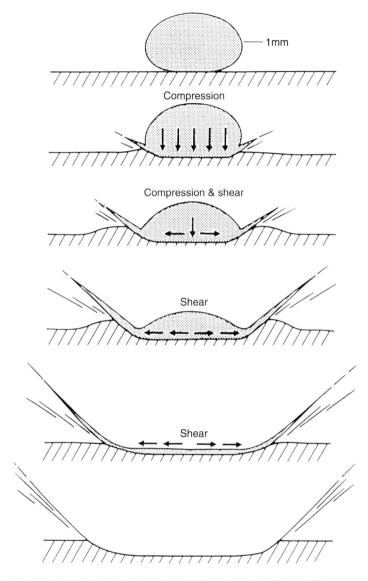

Figure 13.17. Sketch of the effect of a rainsplash (Al-Durrah and Bradford, 1982).

when the run-off is concentrated in small areas of irregular topography. Rill erosion, however, does not occur until the erosive capacity of the flow exceeds the resistance to mobilization of the soil particles, meaning that the run-off may flow a considerable distance over the slope before rills develop (Meyer, 1986). Once the rill is formed, flow increases rapidly and erosion is increased down the slope. The mechanisms that form rills are very complex, as indicated by the fact they may also proceed backwards from the rill head (Bryan and Poesen, 1989).

Dunne and Aubry (1986) believed that the dynamics of rill systems can be explained by the balance between the intensity of rilling and inter-rill processes or the processes of sheetflow and rainsplash. When the latter do not occur, sheetflow becomes unstable, but if it does and sets in motion the impacted particles of the inter-rill areas, the particles tend to fill and eliminate the rills. This indicates various spatial and temporal variations that reflect the changing effectiveness of these processes. At the experimental station of Lanaja in Huesca Province, Spain, investigations of the temporal variations in rill cross sections were carried out using microtopographical survey. Rills were infilled with particles during periods of low precipitation as they were incapable of generating run-off that could transport the loose material in the bottom of the rill (Figure 13.18). Conversely, during periods dominated by convective rainfall of high magnitude and intensity, the run-off generated possessed a greater erosive power. In a 2-year period, some rills, formed in Miocene shales, entrenched 5 cm into the original bed (Sirvent et al., 1997). This indicates the importance of temporal variations and the dependence of rills on inter-rill processes, such as the variation in intensity of rill flow.

In addition, other distinct processes may be recognized in the genesis of rills. In shales, subsurface micro-pipes may collapse and develop into rills (Figure 13.19) (Bryan et al., 1978; Gutiérrez et al., 1988). In other situations rills form due to the intersection of microrill cracks (Figure 13.20) (Haigh, 1978; Gutiérrez et al., 1988). In clay lithologies with alternating clay layers with different expansion capacities, rills may develop in strata with elevated contents of expandable clay. Expansion is accompanied by the development of cracks separated by micro-humps, known as *popcorn* structure, which posses a high

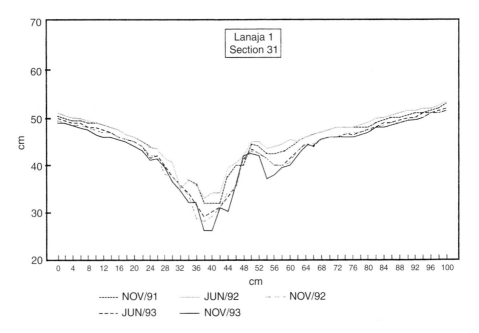

Figure 13.18. Rill cross section taken several times by the profilometer for Lanaja 1 plot (Sirvent et al., 1997).

Figure 13.19. Collapsed subsurface pipes and "bridges" in Holocene shales, that have formed rills or small gullies during their evolution, Lupiñén, Huesca Province, Spain, Ebro Basin.

infiltration capacity (Figure 13.21). In the Dinosaur Badlands, in a semi-arid area of Alberta, Canada, rills form in micro-pediments as a response to the hydraulic conditions of superficial flow (Hodges, 1982). In the same area Bowyer-Bower and Bryan (1986) observed the development of rills by dissolution of resistant siltstones.

Piping or *tunnelling* (Figure 13.19) is common in semi-arid environments and can be defined as the natural development of a subsurface drainage in clastic rocks that consists of the mobilization of solid particles that are transported in suspension by water (Parker, 1963). Piping occurs preferentially in areas whose climates show strong seasonal contrasts, in terms of rainfall variability, such as in Mediterranean climates (Jones, 1981; Bryan and Yair, 1982). Piping results from a complex combination of a number of factors (Stocking, 1976; Bryan and Yair, 1982; Gutiérrez et al., 1997).

Cracking by desiccation is one of the most important factors in the generation of piping (Parker et al., 1990) and reaches a high intensity with the presence of expandable clays (Parker and Jenne, 1967). In summer, the surface materials shrink, storm rainfall fills the cracks, and the clays can be dispersed (Marshall and Workman, 1977) and become mobilized with a sufficient hydraulic gradient (Fletcher and Carrol, 1948; Jones, 1981).

Figure 13.20. Small rills formed in a system of cracks in Holocene shales, Lupiñén, Huesca Province, Spain, Ebro Basin.

Figure 13.21. Formation of rills in shales. At the lower strata rills disappear due to popcorn structures, Custa del Viento, Colola, San Juan Province, Argentina. Andes Precordillera.

In this manner, the cracks enlarge and form *pipes*. These can occur in flat areas incised by the fluvial network and on slopes. Where there is a large contrast in relief, large galleries and pseudo-dolines (or sinkholes) can be formed by collapse, as observed in the area of Lupiñén, Huesca Province, Spain (Figure 13.22 and Figure 13.23), resembling a karstic landscape (Gutiérrez et al., 1988). The continuous collapse of gallery ceilings produced by piping may give rise to the formation of gullies, as testified by the presence of natural arcs in detritic materials (Figure 13.24). These collapses also increase the rate of headcutting in gullies (Crouch, 1983). Piping may also create instability on slopes and contribute to the generation of landslides (Pierson, 1983).

We have already seen how the presence of expansive clays is an important factor in the development of piping, but its presence is not essential for producing expansion (Jones, 1981). In environments rich in sodium, illites and other clay minerals also expand (Imeson et al., 1982), producing an expansion of 10 to 12% (Benito et al., 1993; Gutiérrez et al., 1995). The principal property that controls susceptibility to piping is the content of Na^+ in the exchange complex water with respect to the sum of the Ca^{2+} and Mg^{2+} cations (Sherard and Decker, 1977). These values are obtained by calculating the quantities of these cations in a water extract from saturated paste. Once known one can calculate the exchangeable sodium percentage (ESP) or the sodium absorption relationship (SAR). Clays with an ESP greater than 15 (McIntyre, 1979), or a SAR above 5 (Aitchison and Wood, 1965), contain high pH values (between 8 and 10) and are very susceptible to piping. Furthermore, soils with elevated sodium concentrations facilitate clay dispersion (Heede, 1971). Sherard's diagram (Sherard et al., 1972) establishes the

Figure 13.22. Grouping of pseudo-dolines formed by piping on thick slope deposits. Lupiñén, Huesca Province, Spain, Ebro Basin.

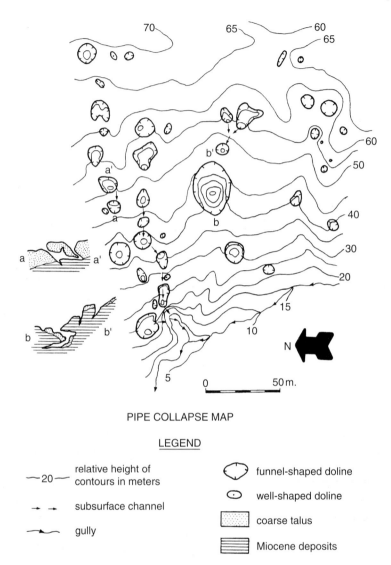

Figure 13.23. Map of pseudo-dolines developed in coarse talus (Gutiérrez et al., 1988).

different dispersive fields in relation to sodium content (Figure 13.25). The points situated in the dispersion field of the graph are susceptible to piping (Benito et al., 1993; Gutiérrez et al., 1997). The study of the physical, chemical, and mineralogical properties of materials is important in civil engineering, especially in the construction of earthen dams as some have collapsed due to piping (Jones, 1981; Parker et al., 1990). If the ESP is elevated, it may be reduced by the addition of calcium, in the form of gypsum or lime (Heede, 1971; Stocking, 1976) that makes the piping processes more difficult.

Figure 13.24. Natural bridge produced by piping processes. The gully has been developed by successive collapses, Barranco del Tormillo or Barranco de la Clamor Vieja, El Tormillo, Huesca Province, Spain, Ebro Basin.

6. Fluvial forms and processes

The principal processes that result from the action of concentrated surface run-off water are channel erosion and the transport and sedimentation of particles. The modifications in channel cross sections during floods are commonly spectacular (Figure 13.26), with important downcutting and rapid accretion (Leopold et al., 1964). Again, semi-arid environments constitute the areas of maximum erosion (Langbein and Schumm, 1958; Fournier, 1960; Corbel, 1964), because these areas present sufficient precipitation and a suitable percentage of vegetation cover that facilitates sediment production.

The energy in a river course is expended, in part, on the erosion of the channel. The processes that act on bedrock river beds are chemical corrosion, or rock dissolution; corrasion, or the mechanical fracturing and abrasion of the rock by transported materials, that may generate rock bedforms; the mobilization of boulders by dragging or hydraulic entrainment and, finally, cavitation, where the rock is broken by bubbles of low pressure water vapour in high velocity currents that hit the rock and explode.

The channels may lengthen by erosion progressing upstream, with headcut erosion of 0.46 to 0.79 m/yr having been measured (Leopold et al., 1964). They may also be widened by erosion at the channel edge, commonly by bank undercutting, and by the detachment or mass movement of saturated material (Figure 13.27). The channels, in addition to lengthening and widening, also deepen their beds, at times producing deep gorges (Figure 13.28). Some desert rivers display breaks in their longitudinal profile (*knickpoints*) (Figure 13.29), where water drops in cascades, giving rise to small closed depressions and

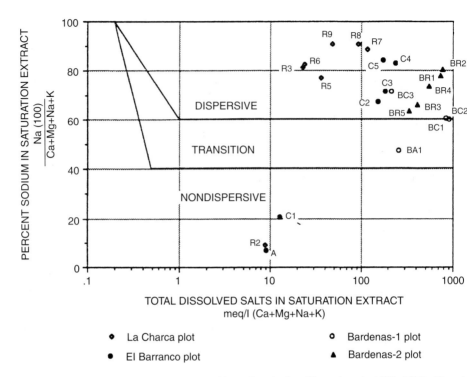

Figure 13.25. Ion composition compared to Sherard's criterion (Sherard et al., 1972, 1976). Note that majority of samples are in the dispersive field with high sodium percentage relative to the total dissolved salts in the saturation extract (Gutiérrez et al., 1997).

undermining at the base that promotes collapse of the walls and the upstream migration of the *knickpoint* (Bull, 1997).

As semi-arid areas have a high sediment production, river courses, where they are active, transport a large sediment load that may reach values of 68% in weight, as shown by Bondurant (1951) for the Puerco River, New Mexico. Sediment transport in fluvial channels is carried out by suspension, saltation and traction processes. The latter two are termed bed load. Perennial rivers transport elevated loads in suspension, whereas bed load predominates in ephemeral channels, which in arid regions, such as the Nahal Yael in southern Israel, reaches up to 87% of the total load.

One of the surprising characteristics of ephemeral channels is that during a flood a certain amount of the bed is eroded and then is filled with approximately the same amount of sediment during the waning of the flood. This occurred in the Arroyo de los Frijoles, in New Mexico (Leopold et al., 1966) (Figure 13.30). The erosion and deposition was produced in both sands and gravels and suggests a state of dynamic equilibrium (Chorley and Beckinsale, 1980). The erosion results in an export of detrital material from the bed that is transported by the run-off. It has been demonstrated that the distribution of the size of the suspended load varies with the velocity of the water or with discharge in desert rivers, contrary to that of perennial rivers (Frostick et al., 1983) (Figure 13.31). The study of bedload is a complex problem. If sediment traps in the river bed are used (Figure 13.32)

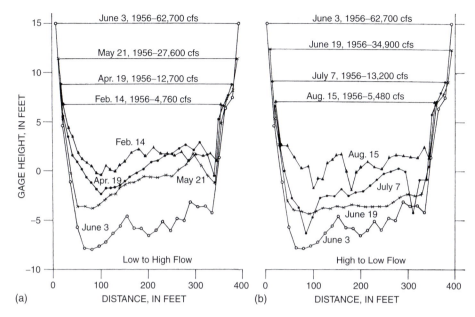

Figure 13.26. Scour and subsequent fill during flood passage, Colorado River at Lees Ferry, Arizona, water year 1965. (a) Low to high flow. (b) High to low flow (after Leopold et al. (1964)).

Figure 13.27. Collapse of large limestone blocks by basal undercutting, Nahal Zin Canyon, Negev Desert, Israel.

Figure 13.28. Todra Gorge, Central High Atlas, Morrocco. Photo: F. Gutiérrez.

they introduce disturbances into the transport processes. The use of painted particles (Leopold et al., 1964) or magnetized pebbles that enable the clasts to be found with a metal detector after the flow, allows the movement of these marked clasts to be studied. Leopold et al. (1966) found movements of clasts up to 3 km downstream during a flood in a predominantly sand-bed channel. Schick et al. (1987) used marked clasts and a detector in a gravel channel to locate where they were buried. Over various floods the importance of excavation and deposition in the transport of buried particles towards the surface and vice versa (Figure 13.33) was apparent.

A typical model for ephemeral rivers was described by Bull (1991, 1997) for *arroyos* of the southeastern USA. This author found a sequence that repeated itself along the arroyo of alternating reaches dominated by processes of aggradation and degradation. The sequence described is the following: a headcut reach (Figure 13.34) that concentrates sheetflow; a channel with vertical walls then directs the flow towards the apex of a channel fan; braided channels that end in an area of divergent sheetflow, and the sequence is closed by convergent sheetflow that again drains towards a headcut reach (Figure 13.35).

Figure 13.29. Break in the longitudinal profile (knickpoint) caused by differential erosion of a thick bank of Cretaceous limestones that are more resistant to erosion. One may also observe the basal undermining caused by the waterfall, Nahel Zin Canyon, Negev Desert, Israel.

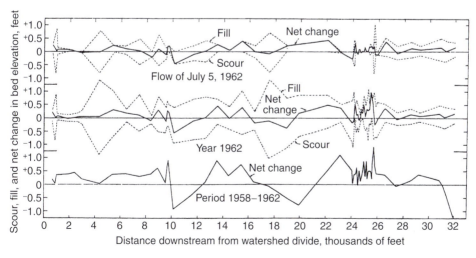

Figure 13.30. Location sketch map of Arroyo de los Frijoles near Santa Fe, New Mexico. Net change of bed elevation during various periods (after Leopold et al. (1966)).

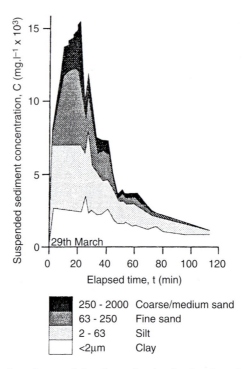

Figure 13.31. Concentration of suspended sediment by size fraction through three flash floods in the II Kimere catchment, Kenya (after Frostick et al. (1983)).

Figure 13.32. Steel mesh trap for the collection of bed load in an ephemeral stream, Nahal Yael Experimental Station, Eilat, Israel.

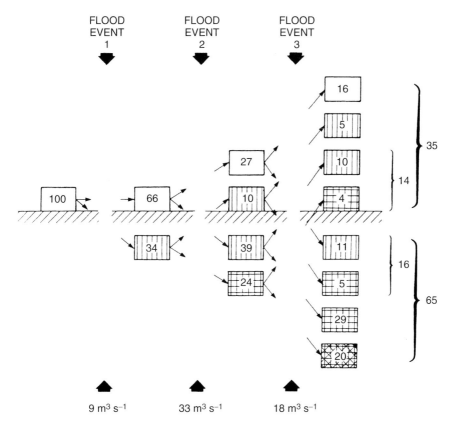

Figure 13.33. Vertical exchange of coarse particles within the scour layer as a result of flood events in Nahal Hebron. Numbers are percentage of the original 282 particles, whose weight ranged between 199 and 3200 g, and which were placed on the channel bed. In this analysis, burial was defined as a state in which at least 50% of the particle was covered (Shick et al., 1987).

The sedimentation of the load transported by fluvial courses in arid regions not only occurs, as in other morphoclimatic zones, in straight, meandering and braided channels, but also by unconfined flows in glacis, pediments and alluvial fans that will be analyzed in another chapter.

7. Valley fills and downcutting channels

One of the most characteristic features of semi-arid zones is the existence of alluvial fills with gullies entrenched within them. This supposes an alternation between stages of accumulation followed by downcutting, interpreted in many cases as a consequence of climate change (Leopold, 1994). When these erosion/accumulation stages are repeated terrace systems are developed (Figure 13.36). Incision takes place where the stream power of the concentrated water flow exceeds the resistance of the materials over which the river

Figure 13.34. The headcut reach of an ephemeral stream eroded in a valley fill over a gypsum bedrock. Barranco de Villafranca, Villafranca de Ebro, Zaragoza Province, Ebro Basin, Spain.

flows (Schumm et al., 1984). This downcutting may form continuous or discontinuous gullies (Figure 13.37). Discontinuous gullies occur in relation to a slope threshold at which the gully does not have sufficient transport capacity to continue forming (Schumm and Hadley, 1957). In the southwestern USA, gullies with vertical walls are termed arroyos (or ramblas in Spain). These morphological features have been the subject of numerous studies in this region (Cooke and Reeves, 1976), with the aim of determining the causes of

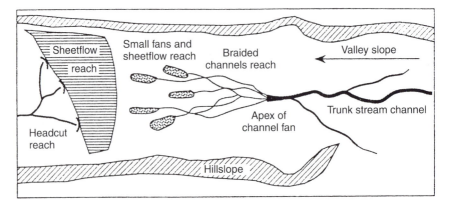

Figure 13.35. Characteristic erosional and depositional reaches of a diagrammatic channel fan of a discontinuous ephemeral stream (Bull, 1997).

Figure 13.36. Different terrace levels in Holocene sediments. El Tormillo, Huesca Province, Ebro Basin, Spain.

Figure 13.37. Small discontinuous gullies on a slope between Rodeos and San José de Jachal, San Juan Province, Argentina, Andes Precordillera.

incision. Climate changes with a tendency towards greater aridity (Antevs, 1952; Leopold, 1976, 1994) and rainfall of elevated intensity and low frequency have been identified as important. The decrease in percentage vegetation cover in semi-arid regions, due to overgrazing and construction work such as the building of communication routes, is one of the fundamental causes of incision. The colonization of the western USA constitutes one of the most significant and most studied examples (Cooke and Reeves, 1976). The decrease in vegetation cover, combined with important storms, was the cause of some gullies incising more than 50 m. This erosive process took place between 1850 and 1920, especially during the period 1870 to 1890, and as a result of this new environmental problem many of the colonists were forced to move on to more favourable areas.

Chapter 14
Slopes in arid zones

In arid regions the production of particles by weathering processes is generally smaller than those mobilized by superficial run-off or by the wind. Evidently, vegetation cover, tightly related to the climate, exerts a manifest influence in this balance. Therefore, it can be said that the degree of aridity controls the balance between supply and evacuation of particles. In the cold mountainous deserts, however, where frost action is the dominant process, the generation of frost shattering is higher than exportation (Figure 8.13), and moreover if aridity is strong and, as a consequence, water erosion is not as important, it is unable to mobilize all the clasts produced.

Although the influence of climate and vegetation is evident in the evolution of the desert slopes, the lithologic composition, along with the frequency of the existing planes of discontinuity, are the most fundamental factors in the development of the modelling of the slopes under these arid environments. These geologic characteristics directly affect the activity and intensity of the weathering processes. These factors, therefore, determine the character and make possible the differentiation of three types of slopes developed on crystalline rocks (Figure 14.1), stratified formations, and highly erodable materials (Mabbutt, 1977). Some authors refer to this first slope type as "massive rocks" and include in it sandstones and conglomerates (Howard and Selby, 1994).

1. Slopes on crystalline rocks

The chemical composition of the magma exerts a direct influence on the morphology of the expelled volcanic materials. The basic products produce extensive basaltic flows, which alternate with pyroclastic deposits. The differential erosion of these formations produces stepped slopes (*trap*), characteristic of the big *plateau basalts* outcroppings, of different ages, existent in the world. Similarly, morphologies of this type occur in the rift valleys of East Africa. The emission of magmas of acid composition, as a consequence of their greater viscosity, generates explosive eruptions that are accompanied by domes and needles, as in the trachytes and phonolites of the Hoggar massif, in the central region of the Sahara.

In plutonic rocks two types of forms are recognized. Some are controlled by *sheeting* and other by cross-jointing systems. Sheeting develops in massive rocks that are fundamentally crystalline, is formed by curved joints parallel to the surface, and is denoted also as topographical jointing. Their origin is attributed to decompression or unloading of rocks that have being subjected to major force inside of the Earth's crust. Later where the rocks ascend to the surface and lose load, they relax and fracture of the rock massif takes

Figure 14.1. Slopes developed on Precambrian granite and on Cretaceous limestones and marls, Negev Desert, Eilat, Israel.

place (Gilbert, 1904). Nevertheless, other origins are also adduced in relation to other strength types (Howard and Selby, 1994). In rocks with numerous discontinuity planes sheeting does not propagate, because the unloading forces dissipate along these discontinuities.

Sheeting occurs in almost all the climatic types and is relatively common in crystalline rocks of the African and Australian Deserts (Twidale, 1981, 1982a,b; Howard and Selby, 1994). Sheeting creates domal or hemispheric forms that were described for the first time by the German geologist Bornhardt (1900). Willis (1936) named these domal inselbergs that stand out on the plains, *bornhardts*, in honour to their discoverer. Many of these bornhardts are the result of the denudation of alteration profiles generated, in past times, in tropical humid climates (Büdel, 1957; Ollier, 1988a,b, 1991; Thomas, 1989a,b, 1994; Twidale, 1990). The climatic change towards more arid conditions unleashes water erosion processes, which brings the rock substrate to outcrop.

Specific types of rock domes are those in form of a mushroom (Figure 14.2) that outcrop in the north of the Eyre Peninsula, in Southern Australia (Twidale, 1962). They are composed of Precambrian granites controlled by curved joints and present in their walls some tafoni and numerous weathering grooves. The concavity is interpreted as originating by subsurface weathering, but in some places in the profile double concavities occur that can be explained as remainders of earlier subsurface weathering levels.

Slopes developed on crystalline rocks affected by jointing systems, seen from afar seem to show rectilinear profiles, but upon closer inspection there can be observed in general a group of rock steps. In his study of the morphology of the igneous rocks in southwestern

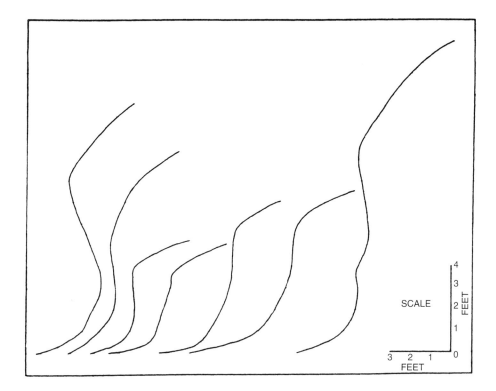

Figure 14.2. Slope profiles from the western extreme of Pidappa Hill, Australia (Twidale, 1962).

Arizona, Bryan (1927) differentiated a steep slope constituted by blocks or *tors* and castle-like forms (*castle kopje* or *koppie*) (Figure 14.3). At the foot was developed a slope covered by detritus, formed by *in situ* weathered blocks; these circumstances can be deduced by the lack of blocks on the lower part of the slope. Nevertheless, on some slopes mobilized blocks do occur. If the joint spacing is appropriate, the slope can show a stepped profile (Carson, 1971); on the other hand the morphology is more rectilinear (Figure 14.4). In the lower parts a wash slope develops that is constituted by fine particles, which are the result of the block weathering and which have been mobilized to the lower zones. Also it is necessary to keep in mind that some slopes have been modelled from weathering profiles in plutonic rocks and, therefore, their morphological and evolutionary characteristics will be different (Oberlander, 1972).

2. Slopes on stratified rocks

The platform deserts (Mabbutt, 1977), which cover cratonic areas of the arid zones of the northern Sahara, eastern Arabia, Thar in the Indian Desert and the Colorado Plateau of the USA, develop slopes on near horizontal stratified formations of different ages. These slopes occur in chains of intermediate type, as those of Israel–Lebanon, and in Tertiary continental basins, such as those existing in arid Spain and in the United Arab Emirates.

Figure 14.3.   Tors and castle-like forms in granite, Phillae, Asswan, Egypt. Photo: J.L. Peña

Figure 14.4.   Precambrian granite with folded pegmatites. The high degree of jointing confers a straight tendency on the slope profile morphologies, Negev Desert, Israel.

Figure 14.5. Simple slope constituted for spectacular massive Neogene conglomerate cliffs, attached to the southern Pyrenean border formations, Mallos of Riglos, Huesca Province, Spain.

Those slopes originating on the morphologies of cuestas or mesas show a great variety. Schumm and Chorley (1966), in their classic study on the Colorado Plateaus, differentiated different types of slopes that result from the combination of lithologic, climatic and vegetation characteristics. *Simple slopes* are fundamentally constituted by one rock type. It can be a clay material affected by water erosion that develops a badland landscape. The other subtype is constituted by predominantly massive rocks, such as sandstone and conglomerates that form important cliffs (Figure 14.5). *Composite slopes* are crowned by resistant caprock below which are developed easily erodable rocks in which can occur badlands or debris slopes that can start from the cornice (Figure 14.6). Finally, *complex slopes* are formed from the alternation of hard and weak layers (Figure 14.7). Simple and composite slopes are be analysed below, whereas complex slopes are considered as a mixed form of the two previous types.

2.1. Simple slopes

In this section we treat the morphology and evolution of the slopes on stratified massive rocks, basically sandstones, and in another section of this chapter badland forms will be analysed, which result fundamentally from the activity of water erosion on easily erodable materials.

Most of the research related to this theme has been carried out in the Colorado Plateaus of the southwestern USA, although it has been also studied in other deserts (Sahara, Australia, Atacama). The evolution of the simple slopes is related to the characteristics of

Figure 14.6. Slope composed of a scarp on limestones and a debris slope that almost totally covers a mainly clayed substratum of Miocene age. The group is partially affected by rilling and gullying, Ballobar, Huesca Province, Ebro Depression, Spain.

Figure 14.7. Complex slope corresponding to a crest front, worked in stratified material of different resistance to the erosion, Dra Valley, south of Morocco. Photo: J.L. Peña.

the rocks and the activity of numerous geomorphological processes, which act or have acted jointly or separately, spatially as well as temporally.

Discontinuities in the sandstone, as with thin clay intercalations or beds with cross stratification, constitute zones more favourable for erosion and generate manifest changes in the profiles of the slopes. Joint systems and vertical fractures constitute weakness zones and along them can form routes of major moisture. As a consequence, the production of weathering particles is increased there and remains trapped in the bottom of the fracture system (Doelling, 1985). Also, in zones of major fault density, slope retreat gives rise more rapidly to successive indentations and salients (Nicholas and Dixon, 1986). It has been already indicated that in massive rocks, such as those that produce these simple slopes, the development of sheeting is common, which has been noted in the arkoses of the famous monolith of Ayers Rock (Uluru), in the central Australian Desert (Twidale, 1978) and also in the sandstones of the Colorado Plateau (Bradley, 1963). This produces a convexity in the slopes and sheets of a few centimetres to a metre in thickness. The dominant *weathering* processes in sandstones are salt weathering and wetting and drying, although in areas with freezing periods, gelifraction can be significant. The composition of the cement also is important. In sandstones with a carbonate cement, this dissolves more easily in cold water and siliceous cements need a pH > 8 to accelerate the dissolution. The loss of cement produces the disintegration of the rock grain by grain. In general, the activity of the weathering processes creates convex forms by rounding of vertexes and edges (Baker, 1936) (Figure 14.8).

Figure 14.8. Rounded forms and tafoni in the Cambrian sandstone of Nubia, Petra, Jordan. Photo: J.L. Peña.

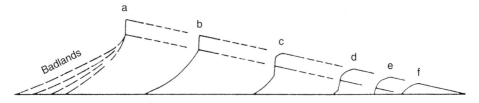

Figure 14.9. Change in slope form during its retreat (Schumm and Chorley, 1966).

In several areas of the Colorado Plateaus (Schumm and Chorley, 1966) modern rock falls can be common. Nevertheless, some of the *landslides* that are recognized in the deserts were produced at other times and, therefore, are relict forms. Some examples have been studied by Grünert and Busche (1980) in the central Sahara and Degraff (1978) in Utah. The analysed landslides are commonly of a rotational type (slump) and were generated in periods of major precipitation during the Pleistocene; moreover, these mass movements can reactivate.

Other erosive process that can be recognized on the simple slopes include *basal undercutting* produced by subterranean water that flows out in the contact of impermeable layers under massive sandstones. Weathering and erosion by water develop shelters (*alcoves*), which in their evolution give rise to rock detachments and slope retreat in sandstones (Higgins, 1984; Laity and Malin, 1985; Baker et al., 1990; Higgins and Osterkamp, 1990). These shelters or cavities are accompanied by alveoles and tafoni.

Simple slopes evolve slowly and at their beginnings can proceed from composite slopes (Figure 14.9), which retreat parallel to themselves. In later stages the outcrop of the underlying clays diminishes, caprocks become rounded and, finally, they convert to simple slopes again (Schumm and Chorley, 1966). An enlargement of these ideas has been carried out by Oberlander (1977a) in his study on the sandstone cliffs of Utah. It also starts in a composite slope (Figure 14.10) with an easily erodable substrate, that when retreating turns into a simple slope. The form of the slope is controlled by the internal structure of the massive sandstones, with thin horizons of interstratificated clays. These produce changes in the slope and the ends of these horizons appear as modifications in the profile. The cliffs

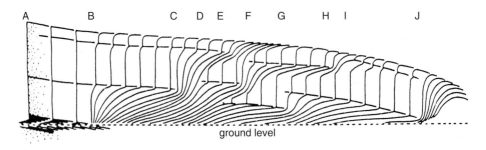

Figure 14.10. Slopes developed on massive sandstones. In A there is a cliff over a fine stratified substratum, which remains hidden in B. Afterwards the intraformational levels control the slope morphology. The segments located above evolve by undercutting, while those located below become rounded (Oberlander, 1977a).

produced evolve by basal undercutting, while the segments situated below the clay strata become rounded.

2.2. Composite slopes

These types of slopes are constituted by a rocky cliff in the upper part and a group of easily erodable layers below (Figure 14.6). The components of those slopes adjust to the ones defined by Wood (1942) and Fair (1947, 1948) in their studies in South Africa, that where extensively analysed and renamed by King (1957a,b). This author considered these slopes as "normal" ones, instead of those indicated otherwise by numerous researchers in climatic geomorphology. For King they are part of a universal model; these slopes from desert environments are the most characteristic. The parts that he differentiated in a slope are (Figure 14.11): (1) the *free slope* that is the upper part, sometimes with a convex profile, as a consequence of weathering and creep; (2) the *free slope* that constitutes the outcrop of the hardest rock and in which general erosive acting processes causes its retreat; (3) the materials resulting from the destruction of the scarp form the *debris slope*, which are basically mobilized by water erosion; and finally, (4) the *piedmont*, which constitutes an important concave element that connects with the alluvial plain.

The *free slope* is rounded by creep, if it exists, weathering and also rainsplash action and sheet erosion. This tendency to rounding increases with the dip of the resistant layer that constitutes the *free slope*, and is emphasized if it is porous or affected by exfoliation (Bradley, 1963). On the contrary, if retreat is caused by faults rounding disappears or decreases.

Free slopes are part of the borders of mesas, structural platforms or cuestas. The thickness of the layer that constitutes the free slopes is important because it controls the cliff height and the length of the debris slope. The rock type is also fundamental, so that the weathering processes are different that disaggregate, for example, sandstones or

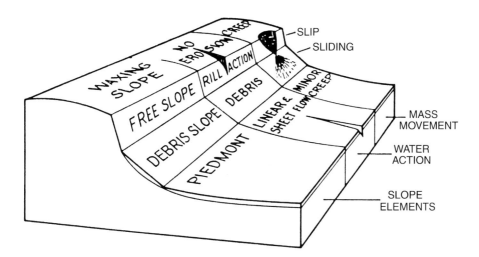

Figure 14.11. Elements of a "normal" slope, water action, and mass movements (King, 1957).

limestones. The cementation degree of the rock is also essential to its resistance. To lithology has to be added the structure. If the free slope is formed by finely stratified or jointed rocks, debris slopes show good development; on the other hand, blocks of a large size will cover the slope.

It has been indicated that debris slopes come from the breaking of the resistant rock that constitutes the free slope. This is what supplies most of the cover of a debris slope, although other parts can come from the underlying substrate. Despite the great development that they can undergo in most deserts, studies of them are scarce compared with the ones carried out in Alpine areas (Oberlander, 1997b).

Schumm and Chorley (1966) indicated that the difference between one slope and others could be explained quantitatively by the relation of weathering (W). The existence or absence of debris at the foot of the scarp depends on the velocity of fragment generation in the free slope (p) before the destruction at the foot of it (d). If this relation is higher than 1 it generates slope accumulation, alluvial fans and a great amount of debris in general. If W is equal to 1, there is a balance between p and d. If W is less than 1, blocks when falling disintegrate upon impact and the remainder is weathered before the next fall. This last is common in the Colorado Plateaus.

This debris can totally or partially cover the slope and, at the same time, be affected by rilling and gullying that gradually work down to a major extension of the rock substrate. The extreme case results from the practical disappearance of the slope cover and, if clays are the outcropping material a badland landform generates. In some cases, hard and weak layers constitute the materials of the substrate; in this case a stepped form originates from composite microslopes (free slope and debris slopes) along the main slope (Figure 14.7). These are the characteristics of the complex slopes.

The fragments of the debris slope show a poor sorting. They are affected by sheet erosion, which exports the fine particles to a lower part of the slope, with the bigger fragments standing out. These also can show an important weathering, with development of alveoles and tafoni in their walls. Finally, in some cases small mass movements can occur that affect to the debris slope.

3. Talus flatirons

These landforms were originally referred to as *talus flatirons* (Koons, 1955), *flatirons* (Everard, 1963), *dreieckshänge* (Wirthmann, 1964; Büdel, 1970), *versants tripartites* (Gossmann, 1976), *tripartites slopes* and *triangular slope facets* (Büdel, 1982), and *talus relicts* (Gerson, 1982). It should be noted, however, that the term "flatiron" has been previously designated only for upfolded, hard sedimentary rock, and "triangular facets" for eroded fault scarps, so considerable care must be utilized to avoid sloppy terminology. These talus flatirons are characteristic forms of arid and semi-arid environments, although they have been also recognized, as seen before, in periglacial areas. They have been studied in the Sahara-Arabian Desert, southwestern United States and in some Mediterranean semi-arid zones.

Their origin initially corresponds to a composite slope, which afterwards is incised by water erosion (rilling, gullies, piping), of the remaining portions of individualized relict slopes (Koons, 1955) (Figure 14.12). These paleoslopes have a triangular form with their

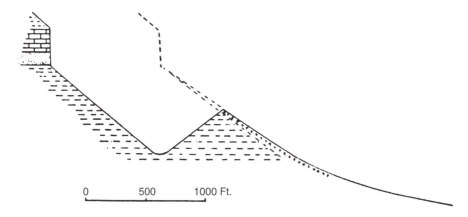

Figure 14.12. Schematic section of a talus flatiron, western Grand Canyon, Arizona (Koons, 1955).

apex pointing to the scarp (Figure 14.13), but trapezoidal morphologies also exist. Talus flatirons are separated from the scarp and are parallel to it. In the distal parts they connect to piedmonts (Blume and Barth, 1972) and, finally, with fluvial or lacustrine terraces. The talus flatirons are concave and are crowned by debris, in general are poorly sorted and generally have a thickness less than 8 m. In the upper part they can reach angles close to 30°, whereas in the distal zones they vary between 2 and 5°. The inner slope of the talus flatiron dips contrary to the first one and it is composed of the substrate materials. It is clear

Figure 14.13. Talus flatiron with triangular apex, developed at the foot of the hill of San Pablo. The area is affected by an intense gullying, Villanueva de Huerva, Zaragoza Province, Ebro Depression, Spain.

that for generating a talus flatiron the alternation of stages is necessary in which accumulation is dominant, followed by another where erosion is dominant. The succession of several alternant stages gives rise to talus flatiron sequences, in which the oldest flatirons are the more distant from the scarp (Figure 14.14 and Figure 14.15).

Talus flatirons are relict slope accumulations and the separation is caused by the incision of the middle slope gullies, where a group of rills converge (Sancho et al., 1988). Once gullies dissect the talus flatiron, they flow to the lower parts where they deposit their sediment load in the form of alluvial fans. Once the talus flatiron is isolated, each one functions in an independent way (Büdel, 1970).

The caprock that forms the free slope above the talus flatiron must not be very thick so that the accumulation does not become excessive (Schmidt, 1987, 1989a), and the slope erosion and scarp retreat will be relatively rapid. For reaching this, the substrate must be constituted of easily erodable rocks. These circumstances appear in composite slopes but not in complex slopes where the alternation of resistant and weak rocks inhibits the development of talus flatirons (Gutiérrez et al., 1998a,b).

One of the formation models of these talus flatirons is due to Koons (1955). A part of a big rock fall covered the slope, followed by a dissection stage that individualized the talus flatiron (Figure 14.16). Nevertheless, the most accepted origin is related to climatic oscillations (Everard, 1963; Gerson, 1982; Gerson and Grossman, 1987; Sancho et al., 1988; Schmidt, 1989b, 1994, 1996; Arauzo et al., 1996a; Gutiérrez et al., 1998b). The stages with more vegetation correspond to a major debris accumulation on slopes. A decrease in vegetation cover can cause slope dissection creating a talus flatiron. Nevertheless, in recent epochs human activity caused destruction of the vegetation cover,

Figure 14.14.   Sequence of talus flatirons in the Chalamera region, Huesca province, Ebro Depression, Spain.

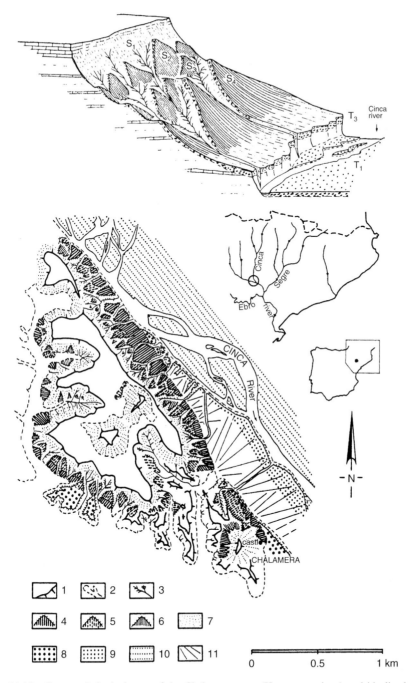

Figure 14.15. Geomorphological map of the Chalamera zone (Huesca province) and idealized block-diagram block of the four slope evolution stages, in which can be observed three phases of talus flatirons. 1 – Mesas and structural scarps. 2 – Infilled valleys. 3 – Gullies. 4, 5, and 6 – Talus flatirons S_4, S_3, and S_2. 7 – Debris slope, S_1. 8, 9, and 10 – T_3, T_2 and T_1 Terraces of the Cinca river. 11 – Alluvial fans (Sancho et al., 1988).

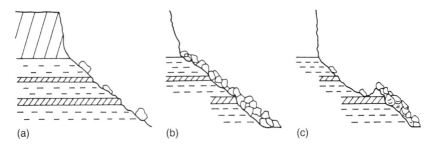

Figure 14.16.   Non-cyclic model of talus flatiron development, according to Koons (1955), in Schmidt (1989a).

which can be the unrestrained factor for debris slope incision (Everard, 1963; Sancho et al., 1988; Gutiérrez and Peña, 1989, 1992, 1998).

4. Free face retreat velocity

To calculate this value it is necessary to know, on the one hand, the lineal dimension that corresponds to the distance between two different moments of the free face, and on the other hand, the time elapsed between these two points. In this way, scarp retreat rates can be obtained (Figure 14.17). There is an important difficulty in obtaining these data, which implies a large imaginative effort and the adoption of numerous and varied assumptions to reach retreat values. As a consequence, the methodologies applied for the

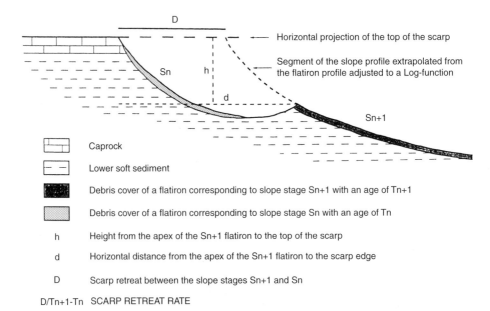

Figure 14.17.   Scheme of a talus flatiron in which is indicated the different parameters used to calculate the scarp retreat rates (Gutiérrez et al., 1998a).

obtaining of these data are very variable (Oberlander, 1997a). On the South Sinai Peninsula, Yair and Gerson (1974) calculated rates of 0.1 to 2 mm/yr from fault free face retreats. The presence of a thin lava cover over a retreating free face in Northern Arizona, dated by K/Ar, let Lucchitta (1975) obtain values of 6.7 mm/yr. Schmidt (1980, 1989b), analysing the geometry of the captured consequent valleys of the Colorado Plateaus, obtained values of 0.3 mm/yr and 0.5 to 6.7 mm/yr. Other methodologies, much criticized by several authors, are based on the pack-rat accumulations of *Neotoma* existing in the mouths of Arizona caves and dated by C-14. The values obtained are 0.45 mm/yr (Cole and Mayer, 1982). Young (1985) studied the free face retreat from the Lower Eocene erosion surface in northeastern Arizona, from a deduced position of the scarp based on the location of the straight valleys and obtained values of 0.16 to 0.17 mm/yr. Schmidt (1987, 1988, 1996) calculated the retreat by graphical constructions starting from talus flatiron profiles. In his 1996 publication he correlated the talus flatirons with Illinoisian and Wisconsinan glaciations and the resulting values were 0.2 to 0.35 mm/yr. Most of these works were based upon supposed ages. Sancho et al. (1988) fixed the date from archaeological remains in slopes and talus flatirons, obtaining free face retreat rates of 0.3 mm/yr in the Tertiary formations from the Ebro Depression of Spain. Temporal values are more accurate when different slope talus flatirons are dated with C-14, in the central Ebro Depression, which have ash and carbon remains in their contents (Arauzo et al., 1996b; Gutiérrez et al., 1998b). The rates obtained by these authors are 0.9 to 1 mm/yr for the last 35,000 years. All these data reflect that the retreat rates in arid zones of different world parts are about 0.1 to 7 mm/yr.

It has already been indicated that most of these values were obtained with simplified indirect methods and with supposed ages. All together these constitute an important but inconvenient method for correlation of these values, as long as there does not exist a clear reliability of the methods and, therefore, in the results. Another problem is related to the sinuosity of the free face. Generally, it supposes a straight front, which is more common as much of the scarp thickness is; despite that it is not a universal rule. When using the method of extrapolation of talus flatiron segments (Sancho et al., 1988) until their intersection with the prolongation of the top of the free face, intersections are obtained at different distances, probably due to the free face sinuosity (Figure 14.18). Nevertheless, while having many intersection points, an arithmetic mean can be obtained that is much closer to the actual retreat rate.

One of the most important characteristics, relative to the major or minor retreat velocity, is related to the lithologic and structural features of the constituent material of the free face, which at the same time controls the resistance degree to the erosive processes (Schumm and Chorley, 1966; Nicholas and Dixon, 1986; Schmidt, 1989a). The mineralogical composition of the rock is fundamental as it determines the reaction to the weathering processes dominant in the area. The erosion rate is also influenced by the dip of the layers (Howard and Selby, 1994), so that the higher the dip, the smaller the volume of eroded rock, and vice versa. Equally, the thickness of the rock of the free face is in inverse ratio to the retreat rate (Schumm and Chorley, 1966; Schmidt, 1987, 1989b). The presence of important thicknesses of constituent rocks of the free face facilitates the application of the acyclic model of the origin of talus flatirons of Koons (1955), supported by Schipull (1980) and Schmidt (1987). With this origin, free face retreat takes place starting from strong impulses, unconstrained by big rock falls, between which alternate the

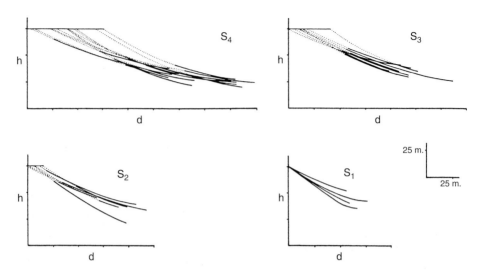

Figure 14.18. Profiles of talus flatirons corresponding to different stages (S_1–S_4) drawn in unbroken line and extrapolated curves to points. The altitude of the apex of the talus flatiron with respect to the current scarp is h and d is the distance of the apex to the scarp, measured on aerial photography (Sancho et al., 1988).

production of particles by the weathering processes present in the area. On the contrary, if the free face materials are a little thick, free face retreat is much faster. The degree of fracturing of the materials of the free face constitutes one of the most significant factors with respect to the retreat rate and mainly controls the degree of resistance to erosion. In this way, the rocks affected by a highly significant fracture density show higher values of retreat rate. Evidently, the erosion intensity of the substrate also affects free face retreat, because where it is composed of highly erodable rocks such as clays and marls, rills progress toward the base of the scarp producing basal undercutting and rock falls (Schipull, 1980; Gerson and Grossman, 1987).

It has been already indicated that there exists a balance between the supply rate of fragments from the free face to the slope (p) and the erosion rate (d) of these accumulations, which constitutes the weathering relation (W) of Schumm and Chorley (1966). In some cases, p clearly surpasses d and the slope accumulation surpasses the free face, fossilizing it. In these conditions free face retreat is paralysed, although chemical weathering can continue to act (Figure 14.19). This balance of production–destruction can occur without big oscillations because the changes do not significantly affect the system. It can lead to major erosion, however, so that the debris slopes gradually remain suspended or isolated. This breakdown of the balance is due to an environmental change and implies the overcoming of a geomorphic threshold (Schumm, 1979). Most researchers estimate

Figure 14.19. Slope partially covered with debris that fossilizes the scarp. Afterwards it is affected by an intense rilling that affects the red clay Miocene substratum, Hill of the Alfambra castle, Teruel Province, Iberian range, Spain.

that the slopes have been subjected to numerous climatic changes in which in the humid/cold stages accumulation dominates and in the dry/warm climate the erosion processes prevail. Free face retreat takes place in both climatic types, but it is possible that it occurs more in one climate that in another due to the weathering-intensity variation, generating mass movements, increasing basal undercutting, and so forth.

Finally, it is necessary to keep in mind the speed of base-level decline (Howard and Selby, 1994) that affects the drainage system, and which can also be due to climatic or tectonic causes. Similarly, the proximity of the slope to a river with great erosive power can break the slope profile in its lower part and unleash erosion that affects the free face retreat, just as it appears in the Cinca river in the Chalamera area of Huesca Province in Spain.

5. Badland slopes

These are areas of very intense water erosion, with high-drainage density (313–820 km/km^2), lack vegetation, and having steep slopes that make transit across them difficult. Badland slopes basically develop on labile materials, in arid and semi-arid environments, commonly at the foot of the scarps (Figure 14.20). The morphology of the divides can be rounded (Figure 13.14), sharp (Figure 13.15), and in their retrogression they can generate turreted hoodoo forms (Figure 14.21) (Scheidegger et al., 1968). Some badland areas are the consequence of the anthropic activities, such as mining, deforestation, construction, exploitations, dumps, and so forth (Aghassy, 1973). As erosion rates are very fast, they offer

Figure 14.20. Badland zone with slashed divides that affect clays and lenses of Miocene sandstones. The top of the formation is eroded by a cemented accumulation glacis that is Pleistocene in age, Bardenas Reales, Navarra, Ebro Depression, Spain.

a small laboratory with a complex microrelief in which the processes and the modifications of the forms can be analysed in short periods of time. Nevertheless, one must be careful when extrapolating these results to areas of major surfaces (Campbell and Honsaker, 1982; Campbell, 1989).

Numerous processes affect badlands that control its development. Therefore, it is necessary to know the physical, chemical and mineralogical properties of the materials (Gutiérrez et al., 1995; Gallart et al., 2002) that constitute these zones of high erosion, to try to understand the activity of the implied processes and their influence on the morphology of the area. So, the dispersion index and swelling capacity are physical properties of great importance in several processes. Chemical analysis permits the obtaining of the sodium absorption ratio (SAR) or the sodium exchangeable percentage (ESP), which are directly related to the susceptibility to piping of the materials. X-ray diffraction analysis of the clay fraction, carried out on oriented aggregates, shows the mineral species existent. The presence of swelling minerals, such as montmorillonite, affects considerably the weathering and erosion processes while also producing successive shrinking and swelling in the superficial materials.

In general, badlands develop under arid and semi-arid climates, where the droughts are prolonged and the dominant rainfall is storm-like. Nevertheless, they can develop in areas with 1075 mm rainfall, such as in Perth Amboy, New Jersey, although these badlands are of anthropic origin (Schumm, 1956a). The role of rainfall is important from the chemical point of view (Bryan and Yair, 1982), because in areas of scarce rainfall salts of high solubility are not reached to dissolve. On the other hand, if water

Figure 14.21. Turreted morphology developed on Miocene clays and sandstone from the Tudela formation, in an area of intensifying water erosion. Cabezo de Castildetierra, Bardenas Reales, Navara, Ebro Depression, Spain.

availability is very abundant, calcium and magnesium will lixiviate. With intermediate rainfall calcium predominates. These variations reflect upon swelling, the dispersion rate of the clays in the regolith and, therefore, upon the rates of infiltration. At more detailed scales, slope exposition influences water availability and, therefore, the activity degree of several geomorphic processes (Yair et al., 1980). In zones subjected to frost action, as in western Colorado, a variation in permeability is produced, converting a less permeable rill surface, into a very permeable one covered with aggregates (Schumm and Lusby, 1963; Schumm, 1964). In Alberta, Canada, the freeze–thaw cycles affect the superficial material, preparing it for a quick denudation by subsequent run-off (Campbell, 1974).

The surface outcropping materials are subjected to numerous fundamental physical and chemical processes that destroy the primitive rock by making a formation of a *regolith*. That of the arid regions is of limited thickness (20 to 30 cm) and the superficial layer presents a polygonal cracking when the clays have a little swelling capacity. If it is high, the shrinking and swelling cycles produce the breakdown of this superficial layer

Figure 14.22. Popcorn structure in which can be observed the fragmentation of the superficial layer of the Tertiary clays. Cuesta del Viento-Colola, San Juan Province, Argentina, Andean Precordillera.

and convert it in fragments with big voids and popcorn structure (Figure 14.22). This layer also exhibits an intense lixiviation of the soluble ions. Below occurs a 5 to 10 cm thick layer that can be an amorphous dense crust (Gerits et al., 1987) or a granular aggregate (Schumm and Lusby, 1963). The transition to fresh rock is represented by a slightly weathered and fragmented layer. There is, however, a great diversity of regoliths in terms of rock composition, the steepness, and the exposure of the slope. After rainfall, if the regolith is dry, water infiltrates following the cracks, pipes and pores until it saturates, so that run-off starts after a few minutes. The regolith swells when wetted, the cracks begin to close, and the internal flux is restricted to the big cracks and to micropipes (Figure 14.23) (Hodges and Bryan, 1982). Afterwards, the wetting front descends gradually; infiltrating the water until it finally generates superficial flow and the almost complete closing of the cracks.

The badland surface is affected by different water-erosion processes, such as the impact and splash of rain drops (rainsplash), rilling, gullying, and subsurface erosion (piping), which has already been described before and which is a most important erosive agent. Nevertheless, these areas are also subjected, on occasion, to mass movements.

Mass movements figure enormously in badland morphology. If regolith easily absorbs water, as occurs with the Chadron Formation in South Dakota (Schumm, 1956a,b), slopes evolve by creep and by small pellicular slides, which fundamentally affect the regolith (Figure 14.24). It leads to the development of rounded divides. On the other hand, when the absorption of water is small, as in the Brule Formation, rilling is important and a high

STAGES IN THE INITIATION AND DEVELOPMENT
OF RUN–OFF FROM SHALE SLOPES

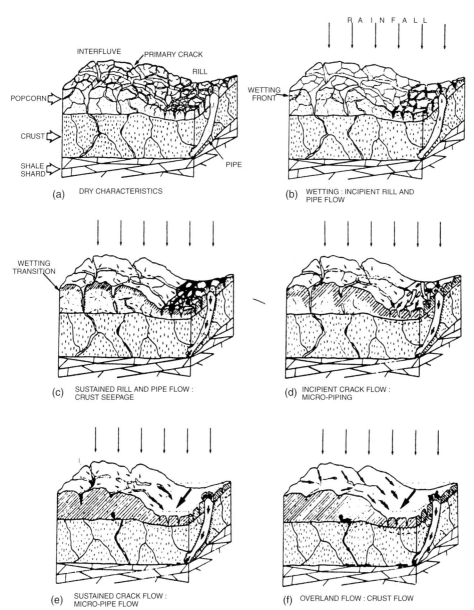

Figure 14.23. Stages in the initiation and development of the run-off on a shale slope (Hodges and Bryan, 1982).

Figure 14.24. Badland zone with rounded divides, in which can be recognized numerous pellicular slides and cicatrices with a half-moon beginning shape. The substratum corresponds to Miocene marine marls of the Guadix-Baza depression. Experimental station of Las Dehesas, Granada, Betic Range, Spain.

drainage density, steep slopes and sharp divides, characterize the landform. Nevertheless, on these slopes mudflows can be produced that are channelled along the existent rills (Figure 14.25). In the Ebro Depression of Spain, the generation of mudflows originates during the wintertime, during which rainstorms are the norm. Rainfall and the lower evaporation, as a consequence of the lower temperatures, cause regolith saturation that can reach liquid states and when flowing gives origin to the mud flows that fill the rills. During convective rainfall these mud flows are re-eroded, starting a new cycle (Gutiérrez et al., 1995).

These badland areas lacking vegetation show very high *erosion rates*. As a consequence, microforms develop very fast and allow erosion monitoring over short time intervals. To measure this soil loss, dynamic and volumetric methods can be used (De Ploey and Gabriels, 1980). The first method measures the amount of transported sediment by water erosion with collectors (Figure 14.26) or Gerlach troughs. The second measures the denudation with erosion pins (Figure 14.27) and bidimensional (Figure 14.28) and three-dimensional microprofilometers (Campbell, 1974, 1982). It is evident that the values of soil loss obtained depend on the utilized technique (Yair et al., 1980) and the erosion rates registered with volumetric techniques are, for some authors (Takei et al., 1981; Rogowski et al., 1985; Hall, 1988), higher than those obtained with dynamic methods. On the other hand, Sirvent et al., (1997) reached an opposite conclusion, which indicated that the rates obtained with collectors were bigger than those recorded with erosion pins. Nevertheless, one of the big problems is the conversion of the volumetric data, because it

Figure 14.25. Mud flow channelled in a previous rill. Miocene clay of the Hoya de Ayerbe, Los Corrales, Huesca Province, Ebro Depression, Spain.

is necessary to know the bulk density of the regolith and this has varied through time (Regües et al., 1992).

Water erosion produces a lowering that has varied between 2 and 20 mm/yr for different experiments carried out in different deserts (Campbell, 1989). The maximum known rates of erosion are recorded in China where they reach values of 38,000 t/km^2/yr. Hadley and Schumm (1961) obtained values of about 13,400 t/km^2/yr in the South Dakota badlands. The values fluctuate considerably as a function of the material type, magnitude and rainfall intensity, time of recording of the experimental stations, technique utilized, and so forth.

As the erosion velocity is very rapid, we can at least have a partial knowledge of the *development* and *evolution* of this landform. Many aspects of the major morphologies are reproduced in miniature in the badlands (Figure 15.8), which furnishes information in understanding their development (Scheidegger et al., 1968). Works carried out in Perth Amboy by Schumm (1956a) on the drainage system developed in these badlands, reflect

Figure 14.26. Experimental Station Bardenas-2 for the study of water erosion, in which has been installed fences and a collector device in a badland area over Miocene clays, Bardenas Reales, Navarra, Ebro Depression, Spain.

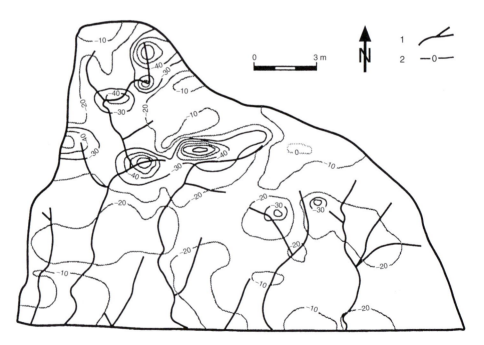

Figure 14.27. Ground lowering contours recorded by erosion pins between November 1991 and November 1993 in the experimental plot Lanaja-1. Huesca Province, Ebro Depression, Spain. Legend: 1 – rills; and 2 – ground lowering contours (mm).

Figure 14.28. Microtopographic profilometer made up of 50 pins, that lets us obtain a profile between two fixed points. Profiles are made every 6 months in order to study microtopographic variations.

that their evolution is similar to those of large fluvial systems. Although a dendritic network can be randomly developed, the channel spacing and their distribution is intimately related to the erodibility of the material and the run-off generated per surface unit. The channels incised rapidly into the materials and generated steep slopes. As slopes retreated they formed micropediments at the foot, which increased in size with the retreat of the slopes. These piedmonts also finish eroding and slopes evolve (Schumm, 1962).

The investigations carried out in regions with different aridity degree show erosion rates of 0.45 mm/yr in the northern Negev of Israel with a mean precipitation of 90 mm (Yair et al., 1980). In the badlands of Alberta, Canada, rates of 4 mm/yr with a precipitation of 330 mm are indicated (Campbell, 1982). All of this leads one to think that in a certain badland zone, when analysing for long time intervals, erosion rates might have been quite variable, because these arid and semi-arid regions have been subjected to numerous climatic changes. In them, rainfall varies substantially from the more dry (interpluvial) to more humid (pluvial) periods and, therefore, the erosion rates will be lower or higher, respectively.

Chapter 15

The arid region piedmonts: glacis and alluvial fans

Piedmonts constitute the transition between the mountain areas, where erosion is the predominant process, and low-lying areas where transport and sedimentation are the dominant processes. Aridity controls weathering on the hillslopes and in this context being the particle production from rock fragmentation, it is normally small. Nevertheless, within mountain and basin deserts (Mabbutt, 1977), the climatic conditions are usually variable in the uplands, but the contribution of rainfall or snow melting is important. In this way, as a consequence of such external supply the piedmont zones will receive water flows with a higher sediment load than those occurring in typical arid areas. The mobilized materials within the mountain catchments are transported by confined streams, which emerge into piedmont zones where the water flow is abruptly unconfined and then transported sediments spread out radiating downslope. The unconfined water flow decreases its velocity and transport capacity abruptly, giving rise to progressive sedimentation. On the other hand, in the backing hillslopes the development of erosive-accumulative ramps is typical, which constitute another characteristic landform of arid zone mountain fronts.

The landform assemblage generated within the piedmonts, therefore, can hold either an erosive-nature pediment (glacis) or a depositional one (alluvial fans). Pediments and alluvial fans can be intimately related, and sometimes may be difficult to differentiate (Figure 15.1). Bull (1977) suggested that fans can be distinguished from pediments when the thickness of the associated deposits is up to 1/100 times the axial length of the depositional area. In addition, a fan can be levelled in response to an external base-level fall, giving place to pediment-like surfaces, which in turn can be buried by renewed alluvial-fan deposition. On the other hand, pediments tend to develop in low-relief areas (low and narrow ranges and modest escarpments), whereas alluvial fans are usually linked to more energetic topographic configurations (mountain fronts, valley junctions, and so forth).

1. Glacis and pediments: definition, terminology and geographic distribution

According to López-Bermúdez (1973) who carried out a good updated revision on *glacis*; this term (which means the gentle sloping ramp of the ancient Roman fortifications) was first used by Dresch (1938), to designate specific landforms with a gently concave longitudinal profile and with slopes down to 6°. Tricart (1969) considered that glacis constitute vast piedmont plains with a low gradient, and coated by a thin sedimentary veneer, which eventually connected with the desert axial fluvial systems constituted by the ephemeral stream channels of *ouadis* (wadis) (Figure 15.2). Important terminological problems arise with the progressive incorporation of this term into different languages,

Figure 15.1.   Piedmont deposits in which the discrimination between alluvial fan and pediment surfaces proposed by Bull (1977) are not applicable. This is due to the absence of channel incisions which impede the direct observation of the underlying bedrock. This picture can illustrate a mantled pediment, and also coalescent alluvial fans. La Sierra del Cerro del Coronel, Province of San Juan, Argentina. Andean Cordillera.

Figure 15.2.   Pediment surfaces merging with the upper terrace of the Ouad Ziz. Central Atlas, Morocco. Photo by F. Gutiérrez.

which give rise to an extensive terminological and conceptual misunderstanding (Tator, 1952, 1953; Dumas, 1967; Tricart, 1969; Pécsi and Sailárd, 1970; Cooke and Warren, 1973; Whitaker, 1973, 1979; Zuidam, 1976; Demangeot, 1981; Cooke et al., 1993). The English writers usually use the word *pediment*, which was first used by Gilbert (1877), and latter defined by Mabbutt (1977), as a flat piedmont surface cut in the bedrock and separated from the backing hillslope by an abrupt knick-point or piedmont junction (Figure 15.1). Where a thin veneer of debris coats the erosive surface of the pediment, it is called *mantled pediment*. In the German literature the equivalent term is the so-called *fussfläche* (Mensching, 1958).

French authors make an extensive use of the term glacis, but either for clarification or descriptive purposes, they label it with different adjectives generating a large list of terms (Tricart, 1969). López-Bermúdez (1973) compiled 28 different denotations, but Whitaker (1973) managed a larger list of terms. Those glacis directly carved in the bedrock without associated deposits (equivalent to pediments) have been denoted as *bare glacis* (Birot and Dresch, 1966), *ablation glacis* (Dresch, 1970), *erosion glacis* (Dresch, 1938), or *denudation glacis* (Tricart, 1969; Tricart et al., 1972). Different terms are also used depending on the thickness of the associated detrital deposits. In consequence, Tricart (1969) discriminated among *denudation glacis* (*glacis de denudation*) with a thin and discontinuous sedimentary cover, *sheet-flood glacis* (*glacis d'épandage*) with a continuous and apparent (1 to 3 m thick) veneer of sediments and finally *accumulation glacis* (*glacis d'ennoyage*) where the sedimentary cover is over 5 to 10 m thick. This terminological overabundance joins with a significant conceptual confusion. For some French authors the erosion glacis are only developed in soft rocks, like some limestones and marls; Dresch (1957), for example, set aside the term "pediment" to characterize only erosion surfaces in crystalline bedrock (Birot and Dresch, 1966). Birot (1960) considered the definition of glacis as the main problem of arid zone geomorphology. Therefore, here we propose that, to avoid any kind of terminological confusion, the term *pediment* should be used for pure erosive surfaces carved in the bedrock, and *mantled or covered pediment* for erosive surfaces fossilized by a thin veneer of alluvial sediments.

Pediments develop in a wide variety of climatic, lithologic, and tectonic contexts (Dohrenwend, 1994). They occur from subpolar latitudes to arid or humid-tropical climates (Whitaker, 1973, 1979) and therefore they can be considered plurizonal landforms (Figure 15.4). The dictionary of Whitaker (1973) even quotes the term *cryopediments* in relation to nivation processes occurring in periglacial areas. Pediments, however, develop extensively throughout arid and semi-arid regions where they are characteristic landforms (Figure 15.2). In arid zones, vegetation is scarce, thus reducing the impediments to the erosive work of overland flows that generate the pediments. We understand that pediment is synonymous for erosion glacis, as proposed by Dresch (1957). In consequence, we think that most of the previously- listed terms are usually referred to arid and semi-arid climatic conditions, or they can constitute ancient arid palaeo-landforms.

1.1. Pediment morphology

Pediments are commonly fed by small mountain catchments with limited relief. They can emerge from mountain fronts, which constitute the junction between the catchment areas

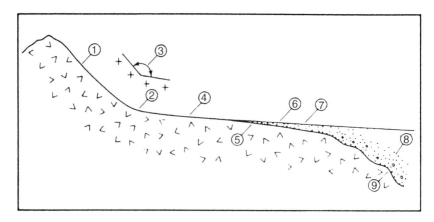

Figure 15.3. Terminology of pediments and related features of the desert piedmont zone. (1) Backing hillslope. (2) Piedmont junction. (3) Piedmont angle. (4) Pediment. (5) Mantled pediment. (6) Mantle. (7) Alluvial plain. (8) Alluvial fill. (9) Suballuvial floor (after Mabutt, 1977).

and the piedmont plains. Mountain fronts may have a linear or sinuous geometry depending on the associated tectonic activity. In any case, they display spaced indentations (irregularities) corresponding to the different valley outlets from the mountain catchment. Valley-outlets are called "quebradas" in South America, where they usually correspond to narrow rocky valleys with steep slopes. In the absence of tectonic activity, mountain-front sinuosity increases with time (Bull and McFadden, 1977). On the other hand, linear fronts develop in relation to active faulting or to lateral erosion of longitudinal stream flows along the front-toe. Linear fronts develop well- defined, and sharp, *knick-points*. Conversely, in other areas the piedmont junction is developed in a progressive way, giving rise to a large concave hillslopes where it is difficult to identify the precise beginning of the pediment surface (Figure 15.2 and Figure 15.3).

   Few publications have been focused on the morphometric analysis of pediments (Corbel, 1963; Mammerickx, 1964; Cooke, 1970b), consequently it is basic to list the variables considered in these studies (pediment area and slope). Pediments present a variable surface area, ranging between 2.5 and 650 km^2 in the Desert of Sonora in Arizona. On the other hand, the fraction of the total surface area occupied by pediments in a specific desert zone also may have great variability. In the Mojave Desert of California the portion of area occupied by pediments is only the 6.7% of the total desert area, but in the Sonora Desert, this proportion increases to 30%. Pediments surfaces usually develop from the lower segments of the hillslopes down to an axial gully or sedimentation plain. The longitudinal profiles of pediments may vary from concave to linear, and they everywhere show a consistent downslope gradient reduction. When pediments border isolated relief, or *inselbergs*, long-profiles display a characteristic radial form (Figure 15.5 and Figure 15.7). These may be of two types: *position inselberg* (*fernlinge*, from the German authors), which preservation is due to its watershed location, and *resistant inselberg* (*härtlinge*) developed by the occurrence of differential erosion.

   The pediment slopes range between 0.5 and 11° (Tator, 1952), and they may vary along their concave longitudinal profiles (Figure 15.6). Where pediment development occurs

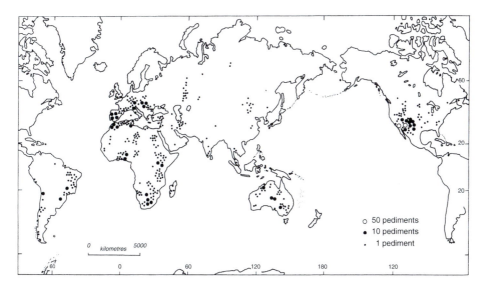

Figure 15.4. Map of the pediments mentioned in the literature (after Whitaker, 1973).

along active tectonic fronts their gradient is steeper than in areas of tectonic quiescence. There are also relationships between slopegradient and clastsize, in such a way that when the steeper the slope, larger clastsizes occur. For some authors lithology is not a relevant factor (Mammerickx, 1964), while others point out its impact on the

Figure 15.5. Inselberg carved in Cambrian sandstones with the development of gently slope radial pediments. Jodhpur–Jaisalmer region. Semi-arid zone of the Thar Desert (India).

Figure 15.6. Segmented hillslope with a succession of convex and concave segments. The relief terminates in a pediment developed from a concave knick-zone. Machakos, Kenya.

slopegradient (Cooke et al., 1993). Climate seems also to affect glacis gradients, with smaller glacis slopes in the more arid regions (Mabbutt, 1977).

The channel network of pediments is usually entrenched in the proximal zone, but channels radiate downslope towards the distal areas, where commonly overland flows

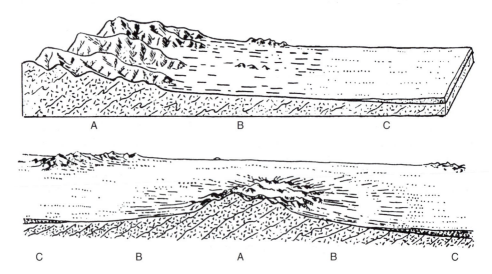

Figure 15.7. Different elements of the pediment. (A) Dissected zone (up) and inselbergs (down). (B) Pediment with residual reliefs. (C) Mantled pediment (Johnson, 1932a).

gradually diminish by infiltration. Most typical pediment channels have a braided pattern. They are shallow channels subject to significant switching during extreme floods. During periods of base-level fall, channel entrenchment can also propagate to distal locations giving rise to subparallel drainage patterns (Cooke and Warren, 1973). In most cases, pediments are planation surfaces directly carved in the bedrock, but mantled by a thin and discontinuous alluvial cover. In particular cases some small relief emerges from the glacis surface as residual landforms denoted by the French authors as "chicots" (Figure 15.7).

1.2. The role of geology and climate in pediment development

As aforementioned, some French authors differentiate between glacis developed on soft or in hard rocks. This division attempts to emphasize the great influence exerted by *lithology* and *geological structure* on glacis location and evolution. When pediments develop on granites, or crystalline bedrock, they usually present a sharp knick-point attributed by some Australian authors (Mabbutt, 1966; Twidale, 1967) to the occurrence of a more intense weathering in the piedmont junction zone. As chemical weathering processes are very limited in arid zones by water availability, the occurrence of sharp knick-points is attributed to previous humid periods occurring during Quaternary, or even Tertiary times. In the south of Arizona, where schists and basalts are the dominant bedrock lithology, knick-point angles are significant (Kirkby and Kirkby, 1974). On the contrary, in more erodible substrates constituted by several types of weak rocks, the transition between the backing hillslope and the pediment surface is gradual, but marked by a large concave knick-zone (Dohrenwend, 1994; Oberlander, 1997b). This latter is the case of the mantled pediments developed in the Ebro Depression of Spain and throughout the Mediterranean littoral.

The recent tectonic history of a region (i.e. *neotectonic* activity) exerts a primary influence on pediment development. Tectonically stable areas are appropriate zones for the initiation and development of glacis, which in addition are characterized by sinuous mountain fronts. On the other hand, where tectonic activity is relevant, mountain fronts are linear and alluvial-fan development prevails (Bull, 1977; Bull and McFadden, 1977). All the published observations indicate that tectonic quiescence (crustal stability) is a very important factor for pediment development.

In arid zones, pediment development is influenced by water availability for surface run-off. In hyperarid areas (i.e. Atacama, Central Sahara) the deficient rainfall gives rise to the nearly absent overland flow erosion (erosion paralysis of Oberlander, 1997b), and therefore the existing pediments should derive from previous wet (pluvial) periods. Climatic changes also affect the size and extent of the vegetation cover on the pediment surface, and consequently to the relative interaction of the working erosion and depositional processes. The role of climatic changes and their impact in pediment development is analysed in detail in Section 15.1.5 (pediment evolution).

1.3. Dominant processes in pediment development

The processes involved in the generation of overland flows are may be the more important ones in desert areas. Because the arid regions are specifically defined by a permanent deficit of water (Tricart, 1969), however, the previous statement may be considered a contradiction. Nevertheless, the typical desert rainfall events are rare high-magnitude storm events during which energetic surface run-off processes (*arroyada*) are generated. Overland flows can be unconfined generating sheet-floods, but eventually can be confined in gullies.

In regions of tectonic stability, the pediment surface can be divided into a proximal erosive area, an intermediate transport area, and a distal depositional area (Johnson, 1932b). During episodes of increasing sediment supply these different areas shift upward to the proximal zone, and conversely when it diminishes, these different pediment segments move downwards to distal areas (Cooke and Mason, 1973). The erosive and transport areas are characterized by the occurrence of mobile shallow braided channels but also by sheet-flow processes. In these two zones of the glacis, we can observe former pediment surfaces slightly elevated over the functional ones. These are relict landforms in which the dominant geomorphic processes are surface and subsurface weathering, eolian reworking, development of desert pavements with varnish or special soils such as *pedocals*.

For some authors (mainly Australian ones), the subsurface weathering plays an important role in pediment development, mainly on those generated on crystalline bedrock (Ruxton, 1958; Mabbutt, 1966, 1977; Twidale, 1967, 1983). The upper regolith cover can be easily removed by overland flows. This fact facilitates the exposure of the basal surface of weathering, generating an *etched plain* or *etch surface*. This kind of process is common in savannah areas, but they also occur in the semi-arid environment. The zones subjected to deeper weathering are localized at the piedmont junction along the contact between the bedrock and the regolith, which constitutes a linear weakness zone.

In softer bedrock the overland flows have higher erosive power. In the case of calcareous rocks, planation is favoured and assisted by karstic corrosion (Nicod, 1992). In the Iberian Range of Central Spain, mainly constituted of calcareous rocks, pediment surfaces have an excellent development, giving rise to extensive pediplains.

The study of micro-topographic changes, carried out in small experimental parcels by Schumm (1962), provides very valuable data about the variations of the micro-pediments over an 8 year time-series. They were monitored with erosion pins in the gentle hillslopes developed on the siltstones of the Badland National Monument, South Dakota. The micro-pediment surfaces underwent a progressive lowering that increased with their slope gradient, and the pediment junction experienced parallel retreat. On the other hand, in some of the cases, active deposition took place in the distal areas (Figure 15.8). In that paper, Schumm pointed out that the formation of such miniature pediments did not have to be necessarily similar to the real case. Chorley et al. (1984), however, indicated that this process may occur in many places at larger scales, suggesting that the analysis of badlands can improve our understanding about the development of some type of pediments.

1.4. Hypothesis on the origin of pediments

Pediment surfaces occur preferentially in zones of low-relief mountain catchments. When the catchments are more modest, the bigger will be the surface area occupied by pediments. The main proposed hypothesis for pediment origin is based on the explanation of those processes governing slope degradation and piedmont generation. The different

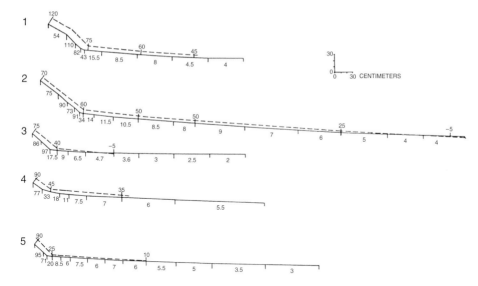

Figure 15.8. Micro-pediment profiles from the Badland National Monument, South Dakota. Lower numbers indicate pediment slope (%) and upper numbers show the location of the different erosion pins indicating the value of ground lowering (mm) produced by the erosion. Solid lines show pediment profiles obtained on May 1961 and dotted lines those obtained on July 1953 (Schumm, 1962).

proposals can consider either: (a) the type of overland flows working at the piedmont zone, or (b) the set of previous degradation processes preparing the bedrock for its later planation. We have to consider both working hypotheses as complementary ones.

The most classic hypothesis is that of the *sheet flood* of McGee (1897), where in the confined streams coming from the mountainous uplands, spread and merge downslope from the piedmont angle, giving rise to a 0.5 m thick sheet of water and silts that spills out in an unconfined turbulent flow. These water flows correspond to flood events during which the erosive effect on the bedrock incorporates particles of very diverse size into the running waters. Sheet-flood processes have been also recognized by Blair (1987) as active processes in alluvial fans based on the analysis of aerial photos of the Rocky Mountains. Tricart (1969) pointed out that a significant relief is necessary to concentrate the surface waters into streams that later spill their sedimentary load on to the piedmont. This indicates that with a very low relief the sheet-flood hypothesis does not seem applicable. On the other hand, because the pediment feeding channels emerging from the mountain front have some spacing, the intervening (inter-channel) areas are not commonly affected by these sheet floods. Tricart (1969) also suggested that it is necessary for the occurrence of a flat or gently topography, prior to the development of sheet floods. Lawson (1915) introduced weathering processes and mass movements to explain the parallel retreat of slopes during pediment development. This author also recognized that sheet flood is the ultimate process in pediment development.

The hypothesis of *lateral planation* for pediment development was first proposed by Gilbert (1875, 1877), later accepted by Bryan (1922) and improved by Johnson (1932a,b). In the piedmont, the confined channel becomes a group of small channels with a braided disposition. Each individual channel has decimetric widths and depths. These little channels can find obstacles in their trajectory, such as bushes and large blocks that have to be bypassed. The braided pattern of these channels is the same one as those developed in the big rivers, which implies that lateral planation occurs in response to the continuum divergence of these small channels. The erosive work exerted by the lateral shifting of the channels gradually removes all the small existent reliefs and, eventually it develops the pediment. The radial systems of glacis developed at the foot of the inselbergs can be explained in this way, otherwise it is difficult to understand from the sheet-flood hypothesis alone (Tricart, 1969). Some authors consider that lateral planation is more effective when it affects the less resistant rocks (Sharp, 1940; Denny, 1967), whereas in the granites or crystalline rocks the denudation is smaller (Howard, 1942). Although the activity of the diffuse overland flows is verified in field studies, this hypothesis should not be considered as a general explanation (Mabbutt, 1977).

Twidale (1962) and Corbel (1963) considered that the processes of chemical weathering have some impact on pediment development. In shield and platform deserts, the denudation of ancient weathering profiles can generate etchplains that later are transformed pediment in response to planation processes (Mabbutt, 1977). On the other hand, weathering is especially intense at piedmont angle locations (Twidale, 1962, 1967, 1976a,b), which collects all the overland flows generated on the adjacent hillslopes. This author relates the origin of both pediment and flared slopes with this kind of selective weathering. During humid periods the weathering front progresses downslope, while during the dry periods active denudation removes the previously weathered material (Figure 15.9).

1.5. Pediment evolution. Pediplains

The overland flows running downslope throughout the surface of the glacis eventually pour into endoreic depressions, seas, oceans or, more frequently to a desert stream. If the initial base level of the pediment changes, then different geomorphological processes reshape the primitive pediment surface. In the first case, the endoreic basins undergo a continuous aggradation, in response to the erosion of the basin borders which give rise to a progressive reduction of the pediment slope. When sea-level falls, just as happens in glacial periods, headward erosion propagates upslope dissecting, separating and isolating the previous pediment surfaces, as is evidenced in numerous examples on the Mediterranean fringe. On the other hand, when sea-level rises the area occupied by the pediments decreases, being able to even be completely covered by the marine waters.

The most common cases of pediment evolution are those related to the *wadis* (desert streams). The materials transported by the overland flows generate an erosion pediment in the proximal areas, and accumulation ones in the mid and distal areas, where deposits left by the desert streams may merge with the alluvial plain. In these zones of coalescence an interfingering between both kinds of deposits occurs. Pediment deposits are generally formed by angular clasts of generally the same composition, due to the shared sediment source of the glacis channels, whereas the alluvial plain deposits are polymictic and more rounded due to their longer longitudinal transport (Gutiérrez and Peña, 1976). When base-level falls in response to climatic or tectonic causes, the lateral alluvial systems incise

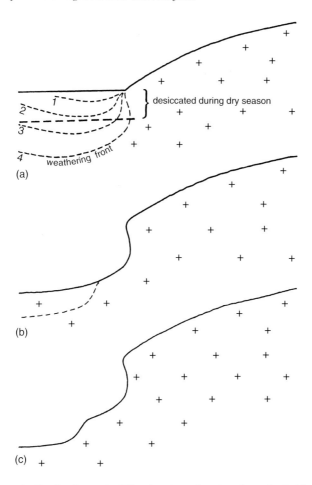

desiccated during dry season

weathering front

(a)

(b)

(c)

Figure 15.9. Stages in the development of flared and overhanging slopes by knick-point (scarpfoot) weathering and subsequent differential erosion: (a) subsurface infiltration of moisture in the knick-point area and lowering of the weathering front; (b) base-level fall and stripping of weathered debris resulting in the exposure of the weathering front as a flared slope; (c) repetition of process and development of double flare (Twidale, 1976a,b).

gradually into the previous pediment, resulting in the dissection and compartmentation of the previous pediment surface and the generation of a terrace in the margins of the desert channel (Figure 15.10). The development of similar cycles of base-level fall originates a group of staircased pediments (Figure 15.11). These correspond to the *glacis embôités* (Joly, 1950; Dresch, 1957) of the French authors and to the *stepped pediments* or *staged pediments* (Mabbutt, 1977) of the English writers. This system of staircased pediments reveals the occurrence of successive phases of incision of the desert channels, with intervening periods of glaciplanation or pedimentation. This model characterizes glacis evolution, but mainly for soft rocks that evolve quickly in semi-arid climates (for example, the Ebro Depression and inter mountainous basins of the Iberian ranges of Spain) (Figure 15.12). On the other hand, in arid climates *sensu stricto*, where precipitation is

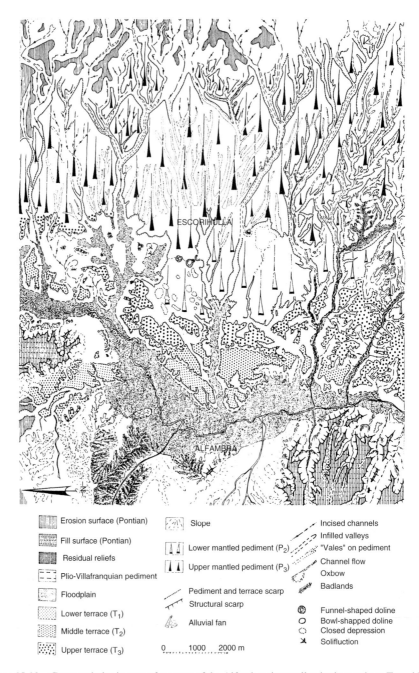

Legend:

Erosion surface (Pontian)

Fill surface (Pontian)

Residual reliefs

Plio-Villafranquian pediment

Floodplain

Lower terrace (T$_1$)

Middle terrace (T$_2$)

Upper terrace (T$_3$)

Slope

Lower mantled pediment (P$_2$)

Upper mantled pediment (P$_3$)

Pediment and terrace scarp

Structural scarp

Alluvial fan

0   1000   2000 m

Incised channels

Infilled valleys

"Vales" on pediment

Channel flow

Oxbow

Badlands

Funnel-shaped doline

Bowl-shapped doline

Closed depression

Solifluction

Figure 15.10. Geomorphologic map of a sector of the Alfambra river valley in the northern Teruel Basin, Iberian Range, Spain. The upper mantled pediment (P$_3$) interfingers with the upper fluvial terrace (T$_3$) and the lower mantled pediment (P$_2$) with the corresponding lower terrace (T$_2$). Interbedding of mantled pediment and fluvial deposits occur at the connection zone. (Gutiérrez and Peña, 1976).

Figure 15.11. Staircased pediments in which are evident the concave knick-zones with the backing slopes of the Sierras Exteriores in the Pyrenees. Hoya de Ayerbe, Province of Huesca, Spain, Ebro Depression.

Figure 15.12. Staircased pediments and fluvial terraces in the Sierra del Pobo, Escorihuela region, Province of Teruel, Northern Teruel Basin. This picture covers a small zone of the area mapped in Figure 15.10.

Figure 15.13. Mantled pediments that evolved to a desert pavement with varnish. Negev desert, Israel.

very scarce, the alternation between periods of planation and incision takes place within large spans of time (Figure 15.10), because climatic oscillations of great magnitude are needed to alter the dominant geomorphological processes. Finally, in hyper-arid areas fluvial processes are almost completely inactive. In hard rocks, as the granitic ones, the incision of the lateral tributaries is impeded by the own resistance of the rock, but a previous weathering for pediment development seems necessary.

In areas of tectonic stability, when arid conditions persist over long periods of time, the pediments continue their enlargement, as a consequence of the continuous parallel slope retreat of the adjacent mountain catchment. Afterwards the pediment surface can develop isolated inselbergs that eventually are also denuded. The final relief corresponds to an extensive planation developed in arid climates, corresponding to those labelled by King (1953) as pediplains in opposition to the concept of peneplains of Davis (1899), which develop in humid climates. The *pediplain* is the final result of a *pediplanation* cycle that is reactivated by surface uplift (Figure 15.14). King (1953) thought that pediments were the fundamental landforms for all the worldwide landscapes. Pediplanation gives rise to several consecutive pediplains within the continents. These are normally assembled incised one into another generating an staircasement (*piedmonttreppen*) of pediplains (King, 1962, 1976). These pediplains can be fossilized or exhumed, as the pre-Triassic and intra-Miocene surfaces of the Iberian range, or may outcrop, as the different erosion surfaces later developed in this same mountain system (Gutiérrez and Gracia, 1997) (Figure 15.15). By means of the correlation of the erosion and sedimentation records, King (1976) differentiated six global planation cycles from the Jurassic to the Quaternary.

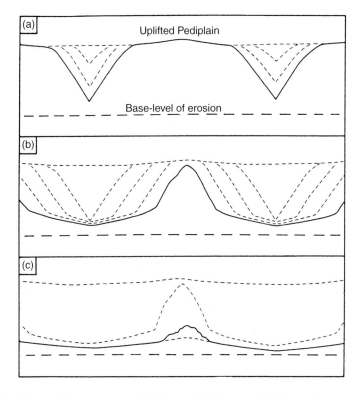

Figure 15.14. The pediplanation cycle. (A) Youth. (B) Maturity. (C) Old age (Interpreted by Small, 1970).

2. Alluvial fans

2.1. Definition, antecedents, and terminology

Numerous known definitions are based on the geometry, sedimentation area, and processes involved in fan development (Rachocki, 1981). The definition of Bull (1968), one of the most excellent investigators in this field, points out that an alluvial fan is a fluvial depositional body and the surface of which approaches the segment of a cone radiating downslope from a point where a stream channel emerges from a mountain area (Figure 15.16).

The history of the knowledge of alluvial fans, also called dejection cones (Gómez Villar, 1996), goes back to the 19th century when these landforms were initially brought to modest attention. Only in the late 19th century were some excellent papers published (Gilbert, 1875; McGee, 1897) in relation to the exploration of the arid territories of the southwestern USA. In the first half of the 20th century the investigators started to inquire about the origin, sedimentation processes, reworking, and tectonic disturbance of fan surfaces (Johnson, 1932a,b; Blackwelder, 1928, 1931). The leading impulse to the analysis of these depositional alluvial-fan landforms,

Figure 15.15.   Planation surface levelling an old folded cretaceous sedimentary sequence. This surface corresponds to the pediplain elaborated during the middle and upper Miocene in central Spain, where due to its outstanding outcropping area is denominated as the "Fundamental Erosion Surface of the Iberian Range." The picture illustrates the Aliaga zone, of Teruel Province. northeastern Iberian Range.

however, took place during the early second half of the last century with the publication of the key works of Blissenbach (1952, 1954), Bull (1963, 1964a,b), and Denny (1967).

After that, alluvial fan analysis was focused on different topics, and their study also extended to their identification and characterization in the ancient stratigraphical record. The study of both ancient and modern fan deposits, took into account their geometry, sedimentation and erosive processes, as well as the factors involved in fan development and evolution. The specific works resulting from these studies will be cited and commented in the following sections. Nevertheless, it is convenient to highlight some papers and thematic monographs in which the bulk of knowledge about alluvial fans is synthesized (Yazawa et al., 1971; Bull, 1972, 1977; Schumm, 1977; Colombo, 1979, 1989; Rachocki, 1981; Nilsen and Moore, 1984; Ashida, 1985; Lecce, 1990; Rachocki and Church, 1990; Cooke et al., 1993; Blair and McPherson, 1994a,b; Gómez Villar, 1996; Harvey, 1997a).

Alluvial fans have also been studied by means of simulations (Rachocki, 1981) and experimental laboratory models. In spite of the considerable amount of problems generated by simulations, very good results have been obtained (Blair and McPherson, 1994a). In this way, laboratory models have reproduced the development of sieve lobes in sand-box analogues (Hooke, 1967). Moreover, using such sand and mud analogues, it has been concluded that debris flow deposits predominate in the fan proximal zone, those interfingered with channel deposits characterize middle fan locations, whereas at distal

Figure 15.16. Alluvial fans in the Sierra de Alhamilla piedmont, Almería Province. Betic Cordillera. The distal zones are subject of intensive greenhouse agriculture.

areas fluvial deposits dominate (Hooke, 1967, 1987). Other laboratory analogues were specifically projected for the establishment of the relationships among drainage basin evolution, fan development and fan stratigraphy (Schumm, 1977; Hooke and Rohrer, 1979; Schumm et al., 1987).

Once the antecedents and trends on alluvial fan research are examined, it is necessary to emphasize some relevant fan features, as well as to specify those terms commonly used in the literature. Alluvial fans are characteristic landforms at mountain front locations (Figure 15.16), but they also occur in intermontane basins and valley junctions (Figure 15.17) throughout all the different morphoclimatic zones, and with very variable sizes. An alluvial fan results from the sedimentary output of single or multiple drainage basins and its morphology and stratigraphy record the process of basin excavation. The water and sediment fluxes concentrate in the lowermost zone of the basin where a mainstream channel issues from the mountain catchment. These mixed fluxes expand in an unconfined flow at this kind of inland basin outlet where, as a consequence of the abrupt slope and stream-power reduction, the transported load is deposited to build up the alluvial-fan bodies.

In these settings individual fans frequently coalesce away from the mountain front locations. In arid and semi-arid zones the alluvial fans constitute the most characteristic landform on the piedmonts of basin and range desert zones, (Figure 15.18). Distally, the coarse detrital fraction that constitutes the main fan-bodies, may grade to the thinner sediments of playa-lake environments.

Figure 15.17. Alluvial fans dissected by the Mendoza River. Uspallata, Mendoza Province, Argentina. Andean Cordillera.

Blissenbach (1954) divided the alluvial-fan bodies into three different parts (Figure 15.19): the proximal area, or fanhead, located at the apex; the middle area, between the fanhead and the lower parts of the fan; and the distal area constituted by the zone more distant from the apex. Finally, it also included a lowermost area where fan coalescence occurs.

The climatic classification of alluvial fans into "humid" and "dry" fans, or in other words *fans of arid zones* formed by intermittent fluvial flows, and *fans of temperate zones* generated by perennial flows (Schumm, 1977) has a positive acceptance. Nevertheless, this classification is still controversial and it was enlarged by McGowen (1979) when considering that in arid zones debris flows are more common than in humid areas, where fluvial sedimentation prevails. Recently, Nilsen (1993) and Stanistreet and McCarthy (1993) related the form and processes of the fan with the climate, concluding that their semi-conical shape is built by debris flows in desert environments ("dry type"), whereas other authors consider that this geometry is also developed in humid temperate regions (Harvey, 1984a; Blair, 1987).

In dry regions fan development is favoured by different reasons (Harvey, 1997). The vegetation cover is sparse or absent, intense storm rainfalls are frequent, and overland flow processes predominate on the hillslopes. In consequence, there is a high rate of sediment production during the storm conditions. The steep mountain streams with flashy run-off regimes give high rates of sediment transport and delivery to mountain front locations. Finally, the sediment transfer from the drainage basin to the fan body is sporadic and basically linked to the aforementioned intense storm conditions.

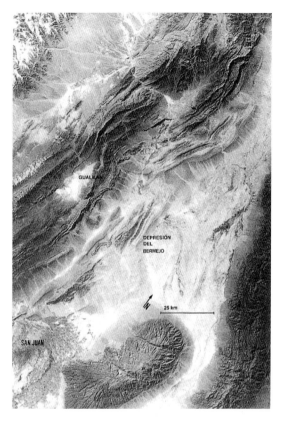

Figure 15.18. Landsat image of the San Juan region of Argentina. Hills located to the northeast are the Sierras Pampeanas, and those located to the northwest belong to the Andean foreland, which are separated by the Bermejo Depression. The piedmont zone of these reliefs are constituted by coalescent alluvial fan systems, which occasionally merge with playa-lake systems, locally called "barreales." San Juan City is located to the south of the picture.

2.2. Alluvial fan morphology

The main morphological features of alluvial fan systems (Blair and McPherson, 1994a) (Figure 15.20) are the following ones. The *drainage basin* constitutes the upland area, generally with steep slopes, from which water and sediments are derived. The *feeder channel* transfers the main fluid and solid fluxes to the *fan apex* and it corresponds to the outlet point of the drainage basin in the mountain front, this apex is obvious if the front is sharp. The *incised channel* is the downslope prolongation of the feeder channel on the fan, and it may be divided into several channels. The incised channel usually terminates in proximal or mid-fan areas, corresponding in its location to the *intersection point*. From this point flows depart, become unconfined and expand downslope giving place to the active depositional lobes. The *headward-eroding gullies* are common features in the distal part of fans. These channels may progress upslope by headward erosion, giving eventual rise to

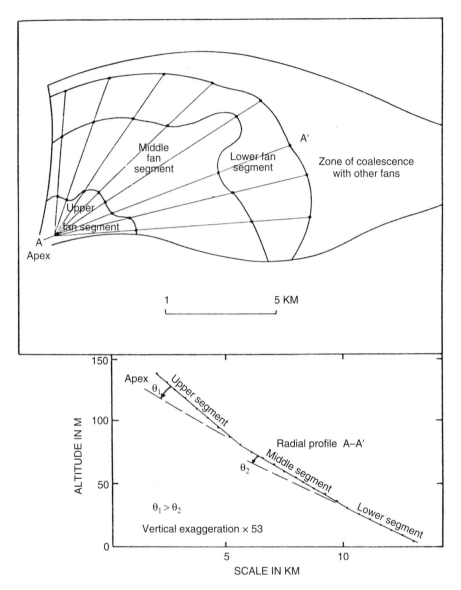

Figure 15.19. Typical segmentation and longitudinal profile of an alluvial fan. Example from the Tumey Gulch fan, western Fresno County, California (Bull, 1964a).

the capture of the incised channel and promoting the downslope shift of the active depositional lobes (Denny, 1967).

Because the depositional processes are strongly controlled by the sediment supply on to the fan surface (Harvey, 1997), the bulk morphology of an individual fan should reflect the main characteristics of its source area. Many of the studies on the morphology of alluvial fans are based on the analyses of morphometric variables, such as fan shape, fan area and fan gradient (slope).

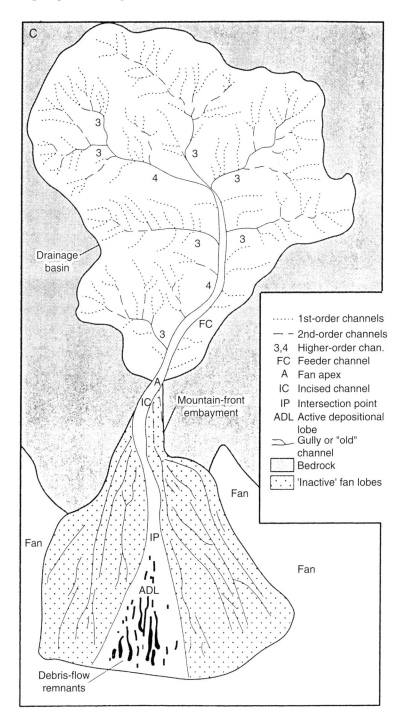

Figure 15.20. Main geomorphological features of an alluvial fan system. Example from the Trail Canyon fan, Death Valley, California (Blair and McPherson, 1994a).

The *fan shape* is semicircular, or a cone segment, with concave longitudinal profiles (Figure 15.19) and convex traverse ones. Data extracted from detailed topographical maps allow development of the following mathematical relationships (Troeh, 1965):

$$Z = P + SR + LR^2$$

where Z is the height of a particular point on the fan surface, P the height of the fan apex, S the fan slope in point P, R the radial distance from P to Z, and L the average slope along its longitudinal profile.

The *fan area* is the more studied variable and it has been demonstrated that there is a simple relationship between fan area and drainage basin area that may by expressed by the equation:

$$A_f = pA_b^q$$

Where A_f and A_b are fan area and drainage area (measured in km^2), respectively. Studies of different groups of fans have demonstrated similar values for the exponent q in different zones ranging from 0.7 to 1.1, but the values for p show a wider range between 0.1 and 2.1. These variations are interpreted as a consequence of differing fan age and history between different regions, but also to the different rock resistance within the drainage basins (Harvey, 1997). Figure 15.21 displays the relationships among the different morphometric fan variables for different areas. Fans of the California Coastal Ranges have larger surface area per unit drainage area than fans in the Death Valley region. The regression line for the fans Spain of southeast has been obtained from 68 fans (Harvey, 1987a).

The *fan gradient* is usually taken as the axial fan surface slope in the upper part of the fan and the most characteristic values range from 2 to 12°. Its relation to the fan is:

$$G_f = aA_c^b$$

Where G_f is the fan gradient. Figure 15.19 illustrates that the values of the exponent b range between -0.35 and -0.15, but values for the constant a show a greater range from 0.03 to 0.17, which is interpreted as linked to different sedimentary processes. Fans in the Death Valley region are much steeper than the California Coastal Ranges, whereas the SE Spain fans show an intermediate position. On the other hand, in the SE Spain fan group of southeast Spain, those dominated by debris flows have steeper slopes than those dominated by fluvial and/or sheet-flow processes (Figure 15.22), which reveals the relevant impact of sedimentary processes on fan gradient (Harvey, 1984a).

2.3. Factors influencing alluvial-fan development

Fan deposition takes place if the sediment supply overcomes the transport capacity of the feeder channel. This indicates that within the channel the *critical stream power threshold* defined by Bull (1979) is surpassed. However, fan sedimentation is also affected by different factors, the impact of which is variable depending on the different climatic and topographic scenarios.

Topographic factors have a strong influence on sediment supply. In that way erosion rates tend to be more important on steep slopes, favouring in addition the generation of debris flows. These predominate in fans supplied by small steep drainage basins.

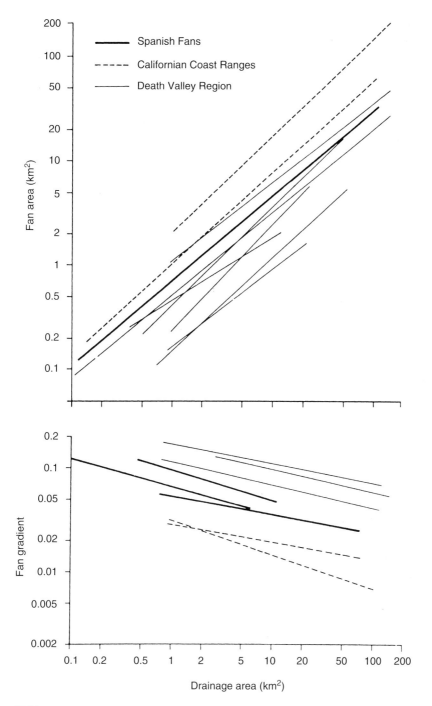

Figure 15.21. Regression relationships of drainage area to fan area and fan gradient for three different regions (Harvey, 1984a).

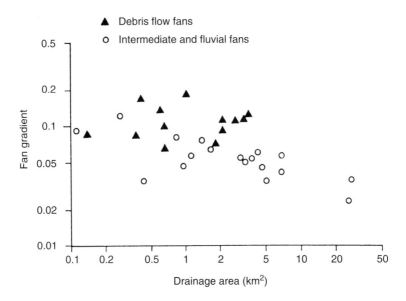

Figure 15.22.   Drainage are to fan gradient relationship for selected alluvial fans in SE Spain. Contrast plotting position of steep slope debris flow fans with the gentle slope intermediate and fluvial fans (Harvey, 1987b).

On the other hand, fluvial processes tend to characterize the fans supplied by large and less steep basins.

The clasts mobilized within the drainage basin of the fans came from the weathering, collapse or removing of the outcropping bedrock substrate in their slopes. This *lithologic factor* exerts a relevant influence on alluvial-fan development. The weathering processes, mainly mechanical in arid zones, work- conditioned by the rock type, and consequently the resultant sediments can be quite different. The resultant clast sizes mainly depend on the granulometry (in detrital rocks) and on the spacing of discontinuity planes affecting the bedrock. Therefore, when there is a wide spacing in the source rocks, they yield blocks or coarse clasts. Because in arid zones chemical weathering is practically absent, the silt and clay particle sizes are commonly scarce. Their abundance increases considerably, however, where source rocks are fine grained, such as marls, mudstones, siltstones and tephras. These fine materials are necessary to generate debris flows within the drainage basins, just as Harvey (1984a, 1987) pointed out in different examples of alluvial fans of SE Spain.

The *climatic factor* has both a direct and indirect influence on the weathering processes conditioning the sediment production within the fan basins. Water availability directly affects rock weathering, sediment transport, and also indirectly the vegetal cover development. The intensity, duration, and frequency of strong rainfall events are the most influencing factors in sediment production. The lack or scarcity of rainfall events promotes a drastic reduction on the weathering and transport processes. Nevertheless, in areas such as Death Valley, with a mean annual precipitation of 43 mm, the sediment suite accumulated within the fan bodies is only derived from rainstorm events (Blair and McPherson, 1994b).

The *temperature* can also affect fan development because it may increase the rates of chemical weathering. On the other hand, it is necessary to consider the common occurrence of altitudinal coolness gradients, which may give place to congelifraction in the headwater zones. Also, the amount of precipitation increases with altitude. These *orographic gradients* may affect the *vegetal cover*, which in turn has a relevant impact on soil erosion and slope stability. This is mainly driven by the development and penetration of the roots in the ground, which is different for different types of plants. Consequently, when climatic changes, or human activity (overgrazing, deforestation, fires, etc.), impact the expansion and type of the vegetal cover, the affected hillslopes may became unstable.

Fans can develop in areas of tectonic quiescence, but also in active tectonic zones. The latter are characterized by the occurrence of basin and range deserts such as those developed in the semi-arid southwestern USA and central Asia. The tectonic activity (*tectonic factor*) may condition the appropriate topographic settings for fan development, such as active mountain fronts (Figure 15.23). At these locations, thick sequences of alluvial fans may develop, recording an important depositional history all over the recent

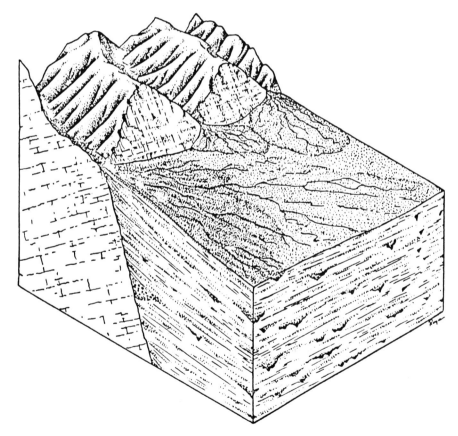

Figure 15.23. Schematic block diagram of an active mountain front showing the typical internal structure of the thick alluvial sequence of the piedmont. The occurrence of triangular facets indicate recent faulting along the mountain front (Bull, 1977).

Figure 15.24. Debris coming from the erosion of the Andean Cordillera develop alluvial fan systems from active range-front faults. Cobija, north Chile. Photo by J. Rodríguez-Vidal.

Figure 15.25. Tectonic tilting and erosion of Pleistocene alluvial deposits in the Andean piedmont close to the City of Mendoza, Argentina.

geological periods. Neotectonics constitutes the main control in fan location and development in these areas. Many of the old sedimentary sequences of alluvial fans are commonly interpreted from a tectonic point of view (Harvey, 1997). On the other hand, the tectonic activity is not limited to the main range-front fault (Figure 15.24), but ground deformations and faulting may also affect the entire fan system. These recent tectonic ground dislocations may affect the feeder channels of the fans giving rise to relevant channel offsets and displacement with the consequent shifting on the locus of the active sedimentation (Figure 15.25).

Tectonic activity can also affect fan development by means of direct modification of the fan slope. When the uplift rate of the mountainous front is greater than the incision rate of the fan channel (Figure 15.26), fan slope increases and sedimentation occurs in the proximal zone. On the other hand, if the channel incision rate exceeds the uplift rate the active deposition shifts to the mid- or distal zones of the fan. Figure 15.26 illustrates how during limited aggradation conditions the development of fanhead trench in the proximal zone transfers the active deposition to distal locations (Bull, 1964a, 1977). This process gives rise to segmented fan surfaces constituted by fan segments of different ages located at different topographic positions. These different fan segments can display a different degree of incision, as well as soil development (calcretes, desert pavements, and varnish, etc.). These *segmented fans* (Figure 15.27) still have a very controversial origin, and different hypothesis, based on recent extreme events, and complex responses to climatic and tectonic changes, have been proposed (see Cooke et al., 1993).

Finally, Blair and McPherson (1994b) consider a set of *extrinsic factors* related to the effect of neighbouring environments on fans. In this way, fans can be affected by aeolian sedimentation, because windblown deposits can modify or even obstruct the fan channels. The development of sandsheets or dunes in distal fan locations may occasionally promote channel ponding. The fluvial channels developed in downfan locations can capture the main fan channel, triggering an abrupt fall in the fan base-level and its subsequent dissection. If the alluvial fans are bounded by either lacustrine or marine environments, the subaerial fan flows may transform into fandeltas. In these cases, sea- or lake- level changes directly affect the fan base-level influencing its erosional or sedimentary behaviour. Finally, volcanic activity can generate lava flows that can interfere and modify the internal drainage of fans. The occurrence of ashfall within the drainage basins favours the generation of debris flows from the steep slopes during storm events.

2.4. Sedimentary processes

The sedimentary processes involved in fan development are of two types, primary and secondary (Blair and McPherson, 1994a,b). The primary processes, or *sediment supply processes*, are those responsible for the transportation of sediment from the drainage basin to the mountain front locations where alluvial fans develop. The dominant transport mechanism will be strongly conditioned by the water to sediment ratio in the fan flows, as well as by the availability of fine, mud-sized materials within the drainage basin (Wells and Harvey, 1987; Harvey, 1997). A low concentration of fines and high water/sediment ratios favour the occurrence of fluvial transport of coarse clasts by traction. When the content of fines increases the clasts can be matrix-supported and mobilized in a debris flow

Figure 15.26. Development of segmented alluvial fan in response to the different uplift rate of the mountain front. In the upper sketch uplift and aggradation rates are bigger than channel incision. In the lower sketch channel incision is bigger than uplift. In this case active depositional zone has shifted downfan as consequence of channel entrenchment an intersection point development giving place to a telescopic fan (Bull, 1977).

Figure 15.27. Segmented fan surface. The upper segments are highly eroded by gulling, but also by the dissection of the Jachal River. The picture was taken from a place located between Rodeos and San José de Jachal, San Juan Province, Argentina.

movement by plastic flow, in which the solid and liquid phases move in a cohesive way. The secondary processes, or *reworking processes*, lead to fan degradation and erosion through the remobilisation of the sediments previously deposited by the primary processes. Many of these processes work during prolonged spans of time on to the fan surfaces, whereas the sediment supply events are generally triggered by low- frequency and high- intensity storms.

2.4.1. Sediment supply processes

These may be divided into two classes: fluid-gravity and sediment-gravity flows (Blair and McPherson, 1994a,b). The fluid-gravity flows can be described as typical fluvial flows, whereas the sediment-gravity flows are mass movements on the colluvial slopes or on fragmented bedrock within the drainage basins. These mass movements can be rockfalls (i.e., gelifraction), rock avalanches (i.e., overturns), gravity slides (i.e., landslides, mud flows) and debris flows.

The rocky materials coming from the upper parts of the drainage basin can reach the proximal area of the fans. *Rock avalanches* take place by the abrupt fall of a large bedrock mass that can disintegrate into boulders and clasts of diverse size, which move downslope as a high-velocity granular flow. In the rock avalanche of the Monte Huascarán (Cordillera Blanca, Peruvian Andes), that took place in May 1970, a mean speed of 280 km/h was reached. These granular flows may feed the fans (Nicoletti and Sorriso-Valvo, 1991) and can reach run-out distances of several kilometres. Their resultant deposits, denoted as "megabreccias" by some authors, display a characteristic coarsening-upwards bedding with blocks up to 20 m at the surface, and matrix-rich lower layers. In plan view they present a lobate shape, as in the case of the rock avalanches of the Sierras Pampeanas in NW Argentina, studied by Fauque and Strecker (1988). They were 3 to 4 km long, 0.5 to 1.5 km wide, and displayed a front rim of 8 to 10 m high. Inside these lobes older similar deposits were recognized.

Gravity slides (landslides) correspond to rapid downslope movements of the colluvial slopes and/or the rocky substrate along a preferential shear surface. These include rotational and translational slides. In these cases the mobilized material is a coherent and homogeneous block, opposite to the granular flows involved in rock avalanches. These processes are important because they directly or indirectly feed the fans. Also, many times the generated landslide scars work as the nucleation zones for gully or stream channel generation (Sorriso-Valvo, 1988). As in the case of rock avalanches, seismic shaking can trigger most of these slope movements, whereas the debris flows are commonly generated by the progressive addition of water to the colluvial slopes that eventually fail by plastic flow.

Debris flows are one of the most characteristic deposits of arid and semi-arid zones in both modern and ancient alluvial records. They constitute flows of sediments formed by a mixture of coarse clasts embedded in a fine-grained matrix with a variable content of water and air (Figure 15.28) (Johnson, 1970; Innes, 1983; Costa, 1984, 1988a,b; Johnson and Rodine, 1984; Takahashi, 1991; Coussot and Meunier, 1996). Their source areas are normally constituted by the colluvial slopes or high weathered zones of the drainage basins. In the mountainous deserts, a complementary source area is constituted by alluvial materials deposited during previous stages of fan activity. The water required to generate the debris flows usually comes from high intensity storms, but sometimes also from the rapid thaw or rainfall events on the snow covers located in the upper zones of the drainage basins. These facts produce rapid overland flows that may generate debris flows where they incorporate the heterometric fragments and silt/clay-size particles mantling the hillslopes. Where the flow velocity increases and the internal shear resistance reduces, they work as a viscous fluid (Johnson, 1970). The viscosity of debris flows may vary considerably, and when they present a high viscosity and volume, the channel-floor erosion can give rise to sharp levees along the channel banks. On the other hand, with a low viscosity, their downslope movement over grassland slopes produces a very limited erosion and smaller levees (Selby, 1993). All these processes are governed by a flow of wet sediments moving downslope within the stream channels. The morphology of debris flows is variable, but generally displays a set of common features. Their most frequent morphologies are constituted by hummocky surfaces with lobate

Figure 15.28. Deposits faulting affecting to fan debris flow deposits (left) and bedded fluvial deposits (right). Pleistocene alluvial fans of the Blanco river valley. Andean Precordillera, Mendoza Province, Argentina.

frontal areas (Figure 15.29). In some cases the lateral and central area of the debris flow lobes terminate in a frontal talus (Johnson and Rodine, 1984). Debris flows may exert a relevant impact on the main fan channel, even overflowing it, and leaving deposits on both banks. The internal structure is massive with large boulders randomly distributed and embedded in a fine-grained matrix, but locally they may exhibit thin sand or silt lenses deposited by intervening "fluvial" flows. These bedded deposits, denoted as transitional deposits, can be generated during transitional stages between debris to fluvial flows, apparently at the end of debris flow events (Wells and Harvey, 1987). These debris flows are more frequent in the proximal areas of alluvial fans. *Mud flows* are mainly constituted by silt-clay materials with occasional clasts. These clasts can display imbricate fabrics parallel to the depositional flow, especially when clasts are elongated. Mud flows, however, are rare deposits in the alluvial fans developed in arid and semi-arid regions.

Two mechanisms have been traditionally considered for the origin of debris flows. The more accepted idea considers the generation of debris flows from the downslope failure of steep colluvial slopes (27 to 56°) triggered by the progressive wetting and saturation of the ground during rainfall or snow-melt events. The water saturation of loose detrital materials promotes the reduction of their shear stress assists their downslope movement as debris flows (Campbell, 1975; Costa, 1988a,b). Other proposals suggest that debris flows are generated by the mobilization of the loose materials mantling the hillslopes by single

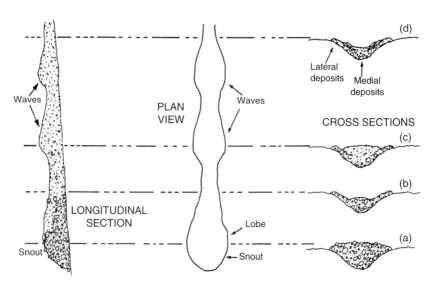

Figure 15.29. Idealized sketch of a debris flow lobe, showing the surface debris waves (Johnson and Rodine, 1984).

overland flows (Johnson, 1970; Johnson and Rodine, 1984). These authors denote this type of starting process, as the "hosepipe effect." Whatever the case, the production of debris flows needs the occurrence of *flash floods* originated by high intensity storm events. These energetic rainfall events are unusual, and the recurrence interval for debris flow generation has been estimated to range from 300 to 10,000 years (Costa, 1988a,b), but locally return periods may be shorter.

In *fluvial processes* sediments and water remain as separate phases during transport, and the detrital particles are mobilized by means of turbulent flows as bed or suspended load. When the solid load constitutes an important fraction (20 to 47%) hyper-concentrated flows are generated (Costa, 1988a,b). Two types of fluvial processes can be differentiated; *sheet floods* and incised *channel flows* (Blair and McPherson, 1994a,b).

Sheet floods are the most common process in the sediment input of alluvial fans. These are broad unconfined flows that expand downslope over the piedmont surface. From aerial photo analysis Blair (1987) described the catastrophic sheet flood event occurred in the Roaring River fan (Rocky Mountain National Park, USA), as a continuous sheet of water of 0.5 m depth flowing over the 320-mlong, active depositional lobe of this fan. Sheet flooding is produced by strong storms events of low frequency and high intensity that can even promote catastrophic flooding (Gutiérrez et al., 1998). These sporadic events are separated by periods of rare fan activity, during which the fan surface is reworked by secondary processes. The deposits can be constituted by a varied suite of blocks, gravels, sands, and silts. These usually are arranged in interbedded laminar couplets 10 to 30 cm thick, with depositional slopes of 2 to 8° parallel to the fan surface (Blair and McPherson, 1994a). The longitudinal axes of the bigger blocks may have an orientation perpendicular to the slope, but sometimes they display imbricated fabrics. Wavy gravel ribbons transverse (transverse ribs) to the fan slope have been also described as products of sheet flooding.

The *incised channels* dissect the fan apex and can be several metres deep. They constitute paths for the transfer of sediments between the proximal and distal zones of the fan, and consequently the feeding channels of the active depositional lobes. These channels also contribute to fan sediment reworking by means of erosive work on their incised channel beds. The resulting sedimentary facies are characteristic of confined flows including fluvial sand and gravels mobilized during flooding events.

A distinctive sedimentation variety of alluvial fans are the so-called *sieve lobes* or *sieve deposits* which were described by (Hooke, 1967) after laboratory studies with sand and granule features resembling fans. Characteristic lobate deposits build up from the points where the water flow loses its transport capacity due to its rapid infiltration into the permeable fan substrate, commonly in the mid-fan intersection points. The coarsest material is accumulated in the frontal sector of the depositional lobe, but upslope finer sand-sized sediments prevail. Sieve deposits commonly develop during flooding events and are mainly deposited in mid- and distal fan zones. They can also originate from previously deposited debris flow lobes where overland flows sieve the finer material. Commonly the result is a matrix-free, clast-supported fabric. In this last case, the origin of the sieve lobes is secondary.

The three-dimensional analysis of alluvial fans provides evidence for the occurrence of numerous and complex graded bedding features, which can be interpreted as climatic or tectonic responses. Typically particle size decreases downslope with the coarser material in the proximal areas and the finer material in the distal ones. The channelized fluvial blocks and gravels, and debris flow deposits prevail in the proximal areas (Figure 15.30c,d) (Harvey, 1997). According to Miall (1978) if the dominant sedimentary facies are fluvial ones we have "Scott"- type sedimentary sequences. On the other hand, the profusion of debris flow deposits distinguishes the so-called "Trollheim-type" sequences (Figure 15.30a,b). In mid-fan areas sieve deposits can be developed, but sheet-flood deposits commonly extend towards the distal parts of the fans. On the other hand, strong changes in the distribution of sedimentary facies can be occur between untrenched and proximally trenched fans; in the latter the more recent deposits are shifted towards the distal areas of the fans (Figure 15.30e,f).

2.4.2. Reworking processes

As previously mentioned, alluvial fans mainly develop in response to high intensity storm events, but the return period of these energetic events is important. In consequence, the fan surface is exposed to a variety of reworking processes during the long periods of sedimentary inactivity. During these periods the depositional surfaces created by the primary sedimentary processes are largely modified. In addition, in active tectonic areas, fan surfaces can be affected by ductile and brittle deformations, such as ground folding, faulting (Figure 15.30f) and warping. These reworking processes are diverse and they have been analysed by different authors (see Blair and McPherson, 1994b).

The most relevant reworking processes are the *overland flows* derived from ordinary rainfall episodes that dominate the long periods of time between strong storm events.

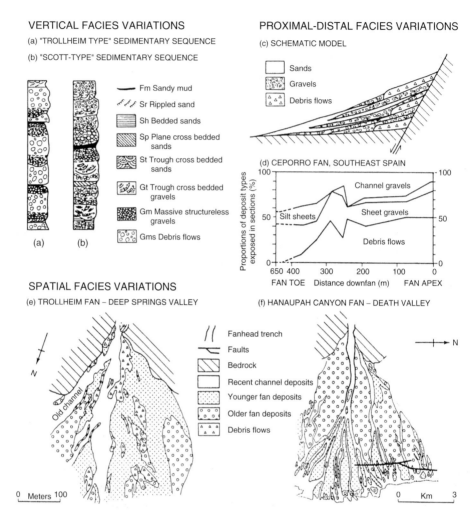

VERTICAL FACIES VARIATIONS

(a) "TROLLHEIM TYPE" SEDIMENTARY SEQUENCE

(b) "SCOTT-TYPE" SEDIMENTARY SEQUENCE

— Fm Sandy mud

Sr Rippled sand

Sh Bedded sands

Sp Plane cross bedded sands

St Trough cross bedded sands

Gt Trough cross bedded gravels

Gm Massive structureless gravels

Gms Debris flows

(a)   (b)

PROXIMAL-DISTAL FACIES VARIATIONS

(c) SCHEMATIC MODEL

Sands

Gravels

Debris flows

(d) CEPORRO FAN, SOUTHEAST SPAIN

Proportions of deposit types exposed in sections (%)

Channel gravels

Silt sheets   Sheet gravels

Debris flows

650 400   300   200   100   0
FAN TOE   Distance downfan (m)   FAN APEX

SPATIAL FACIES VARIATIONS

(e) TROLLHEIM FAN – DEEP SPRINGS VALLEY

(f) HANAUPAH CANYON FAN – DEATH VALLEY

Fanhead trench

Faults

Bedrock

Recent channel deposits

Younger fan deposits

Older fan deposits

Debris flows

Old channel

0 Meters 100

N

0   Km   3

Figure 15.30. Variation of sedimentary facies in alluvial fans: Vertical facies variations: (a) "Trollheim" type and (b) "Scott"- type sedimentary sequences (Miall, 1978). Proximal-distal facies relationships: (c) Schematic model (Rust, 1979). (d) Downfan variations in proportion of facies exposed in fan section on Ceporro fan, southeast Spain (Harvey, 1984b). Spatial facies variations: (e) Apex area of a proximally aggrading fan, Trollheim fan, Deep Springs valley, California (Hooke, 1967). (f) Proximally entrenched and distally aggrading fan, Hanaupah Canyon fan, Death Valley, California (Hunt and Mabey, 1966). (After Harvey, 1997).

These processes result in the winnowing and transference of the existing fine material towards the distal fan areas. Erosive flows may result in the generation of a net of rills and distributary channels radiating downslope from the fan apex to distal areas where wash channels dominate (Denny, 1967). It is also necessary to consider the activity of the subsurface groundwater within the fan bodies, which work as important aquifers. Groundwater availability impacts plant growth rates and, consequently on subsequent processes of overland flow erosion. On the other hand, water infiltration can favour the calcium carbonate cementation of fan surfaces. In rich-carbonate semi-arid zones these processes can give rise to the occurrence of petrocalcic soils.

Wind action is an important reworking process in arid regions. Windflow produces the pick-up and winnowing of fine particles from the fan surface that can eventually give place to the development of desert pavements. The exposed gravels undergo aeolian abrasion resulting in the generation of ventifacts sometimes coated with desert varnish. The sand-sized particles transported by windflows can also be deposited on to the fan surface giving place to continuous sand-sheets or generate dune- type accumulations around obstacles, such as plants. These aeolian deposits can attain an important thickness in favourable "leeside-slopes" of the fans giving place even to relevant disturbances on fan channels.

The *weathering* of fan surfaces includes the action of mechanical, chemical and biological processes. Mechanical weathering, such as salt crystal growth, mineral hydration–dissecation cycles and gelifraction, disintegrate and fragment the clasts and they are conditioned by the abundance in water and their salt content. Biological weathering, or bioturbation, is generated by the burrowing and material removal produced by the activity of plants and fauna. Calcium carbonate precipitation near the fan surface produces the development of petrocalcic soils (Gile and Hawley, 1966; Machette, 1985; Harvey, 1987a). The cementation of fan sediments generates resistant calcrete crusts, which prevent the erosion and reduce the infiltration of water on fan surfaces (Harvey, 1990, 1997).

2.5. Dynamics of alluvial-fan development

The alluvial fans are the landforms of faster development rates within arid areas (Mabbutt, 1977). Average accumulation rates in alluvial fans of the Death Valley (USA) are of 1 m every 1000 years (Hooke, 1972). Alluvial fans are dynamic geomorphic systems showing progressive morphologic changes during their development (Harvey, 1997). This author proposed a model for fan evolution in which the relationships between the fan- surface and fan- channel profiles control the balance of aggradation an incision in mid and distal fan areas (Harvey, 1987a, 1997) (Figure 15.31). Types A and B are characterized by the occurrence of fanhead entrenchment and distal aggradation from the mid-fan intersection point. In Type C, scour rather than deposition takes place at the intersection point, and in Types D and E this scour is sufficiently effective to create a series of headcuts, which may shift to distal areas as a consequence of a progressive downfan migration of successive intersection points. These successive headcuts can eventually link together as a mid-distal fan trench, dissecting the fan throughout its length (Type F). Fan dissection can also occur by base level-induced headward erosion from the fan-toe (Type G). This can be considered a key model for the understanding of fan aggradation and dissection dynamics, in which tectonic disturbances, climatic changes, human activity, and extreme storm events play a relevant role in the modification of the dominant erosive or depositional processes.

There are other models of fan development described in the classic literature, such as that proposed by Davis (1905) based on their cyclical conception of landscape evolution. This author, proposed a juvenile fan stage in which the higher relief continuously provides

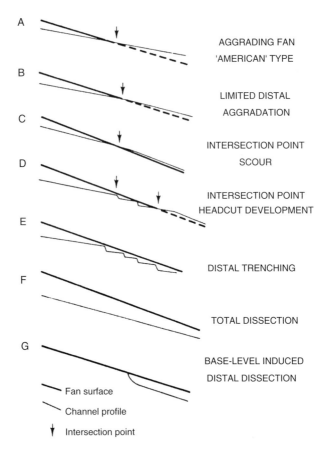

Figure 15.31. Alluvial-fan development model based on the relationships between fan- surface and fan-channel longitudinal profiles and their impact on the dissectional or aggradational behaviour of the fan (Harvey, 1987a).

sediments to the piedmont. The progressive relief reduction gives rise to a dominance of sheet flooding and laminar erosion during the final stages of the cycle, where pediments are generated.

Sorriso-Valvo (1988) proposed an evolutionary model for fans based on mass movement occurrence in the Calabria region of Italy. This author proposed four main stages for fan development (Figure 15.32). The first stage, labelled as creep-stage, is characterized by shallow landslides located downslope, whereas upslope creep processes are dominant during long periods of time. In stage B, or collapse-stage, landslides and debris avalanches dominate upslope where important gravitational scars are produced, whereas in the piedmont the collapsed material is accumulated as colluvial cones. The collapsed material can be removed throughout the main channel during high intensity and low frequency rainfall events. Both the weakened bedrock and the debris remaining in the scar, are mobilized by subsequent mass movements (reactivation of the previous

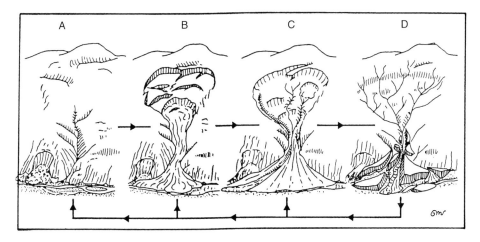

Figure 15.32. Cyclic evolution of an hypothetical landslide-related fan. Arrows indicate the possible successive stages. The shape is typical of bedrock with ductile behaviour. In brittle and jointed bedrocks the main flow stage B may take the form of a high-speed debris avalanche. (Sorriso-Valvo, 1988).

ones) and/or debris flows. These processes can build a fan in a relatively short term. This stage C corresponds to the alluvial fan construction. Finally, the fan destruction, or stage D, takes place when the slided material has been removed and stream flowing from the scar attain a higher erosional power generating the erosion of the fan, which progress from the apex. For this author this model is cyclic and considers the succession of stages A–B–C–D–B–C–D... as the most common evolutionary trend on fan evolution.

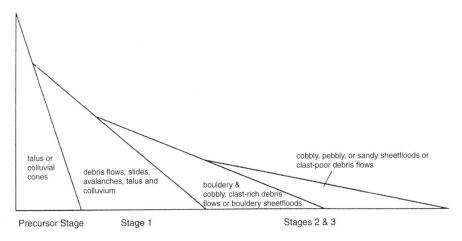

Figure 15.33. Schematic diagram of the increasing slopes during the different evolutionary stages of fan development (vertical exaggeration × 2). The different stages present various associations of dominant sedimentary processes. During fan development depositional slopes diminish, fan radii lengthen, and knick-points decrease in elevation (Blair and McPherson, 1994a,b).

However, the model of Sorriso-Valvo (1988) basically considers the contribution of mass movements to the generation of alluvial fans, but undervalue the contribution of true fluvial processes.

Blair and McPherson (1994a,b) proposed a similar model with a four-stage evolutionary scenario (Figure 15.33), initiated by mass movements and development of talus cones at mountain front locations (Figure 15.34). In stage 1 the incipient drainage basins hold an abrupt relief, where first order channels prevail and landslides and debris flows are the dominant processes. During stage 2 the gravitational mobilization of these deposits buildsup the alluvial fan and enlarge the drainage basin, which acquires an amphitheater shape. The area of the fan is also increased during this stage and fan radial length may reach values of 1 to 5 km. On the other hand, fan slope reduces during this stage displaying characteristic values between 5 and 15°. During stage 3 the expansion of the fan system continues, whereas in the drainage area the number of first-order channels increases and 4 or 5 channel orders develop. Mass movements are still characteristic within the drainage basin, but fan sedimentation is driven by debris flow and sheet-flood processes (Figure 15.35). In this stage the fan

Figure 15.34.   Debris cones in Punta Vacas, Mendoza Province, Argentina. Andean Cordillera.

Figure 15.35. Alluvial fan and main fan channel (right). Debris cone with incipient drainage basin modelled in amphitheatre shape. East of Punta Vacas, Mendoza Province, Argentina. Andean Cordillera.

radial length may enlarge from 3 to 10 km and the fan slope diminishes to 2 to 8°. Finally, this last stage can be maintained during long periods of time until rivers or glaciers will eventually dissect the fans, nevertheless the fan system can be reactivated by tectonic activity.

Chapter 16
Desert lakes: playa lakes and sebkhas

1. Main features

Large areas of desert regions are characterized by endorheic drainage and, consequently, waters collect in the lower topographic areas creating lakes of different sizes and origins. Lakes can last if water input (surface run-off, ground waters, rivers) is higher than output (mostly evaporation). If solute discharge increases, the lake becomes saline. Lakes from arid zones can dry out during long periods of time, although in most cases, they show a seasonal regime, drying out during the summer season. Only the Aral, Caspian, and Dead Sea (Figure 16.1) are permanent water bodies, as their main water input is derived from nearby mountains or far-away areas. The Dead Sea, fed by the Jordan River, outlet of the Sea of Galilee, has a very high salinity (300,000 ppm TDS), which impedes the development of life in its waters. Saline lakes with a salinity of about 5000 mg/l are relatively common in arid and semi-arid areas (Shaw and Thomas, 1997). This figure, mainly based on biological tolerance, is considered the boundary between fresh and saline water bodies (Euster and Hardie, 1978).

There is a very wide terminology referring to closed depressions developed in arid environments, largely due to the use of vernacular names (Tricart, 1969; Cooke and Warren, 1973; Neal, 1975; Cooke et al., 1993; Currey, 1994; Rosen, 1994; Shaw and Thomas, 1997; Briere, 2000; Gutiérrez, 2001). The term *playa* has a Spanish origin and was introduced in the English language from the exploration of the North American southwest by the Spaniards (Gutiérrez, 2001; Gutiérrez et al., 2002) (Figure 16.2). Recently, Briere (2000) presented a thorough compilation of names used to designate these depressions. This author proposes the use of the term playa for intracontinental depressions with a negative water balance that remain dry more than 75% of the year, and playa lake, a transition between playa and lake, for depressions that host water from 25 to 75% of the year. In many cases these definitions may be difficult to apply since the hydrologic regime of the playas may change substantially in successive years. Tchakerian (1999a), in the Encyclopaedia of the Deserts, defined playas as depressions typical of arid and semi-arid zones, affected by flooding by ephemeral surficial waters or fluctuations in the water table. The definition of *pan* given by Dregne (1999) is very similar to the previous one. Several authors consider both terms as synonyms (Shaw and Thomas, 1997; Goudie, 2004b). On the other hand, the word *sabkha* has clear geographic meaning restricted to depressions located in coastal plains (Tchakerian, 1999b; Briere, 2000).

Other synonyms are *sebkha*, an Arabic term used in northern Africa and *sabkha*, common in the Arabian Peninsula (Figure 16.3). In Iran, *kavir* denotes depressions rich in salts. Another Arabic word, *chott*, means a saline basin with groundwater input, and is

Figure 16.1.   The northerncentral sector of the Dead Sea from the western margin, Israel.

mainly used in the Sahara region (Chott El Jerid, Tunisia). *Takir* is a clay-rich depression in Central Asia, somewhat similar to the *barreal* in the Argentinean Andes (Figure 16.4). In the South American deserts, *salar*, *salina*, and *bolsón* are used for depressions with clay-rich or saline sedimentary deposits.

Figure 16.2.   Saline closed depression with old salt exploitation facilities in the foreground, Laguna La Playa, Central Sector of the Ebro Depression, Bujaraloz, Zaragoza Province, Spain.

Figure 16.3. Sebkha surrounded by dune fields. Al Liwa, Rub Al Khali Desert, United Arab Emirates.

Figure 16.4. Barreal Blanco (whitish lacustrine mud) composed of carbonate-rich silts transported from the calcareous Cambrian–Ordovician formation outcrops in the watershed, Andean Precordillera, San Juan Province, Argentina.

Figure 16.5. Large halite accumulations in the central areas of the littoral sebkha, El Melah, Zarzis, Tunisia.

   Glennie (1970) distinguished between interior and *littoral sebkhas* (Figure 16.5). The latter are saline plains located in littoral arid areas, affected by marine water penetration and with tidal-related deposits. High-soluble salts are commonly precipitated in these littoral sebkhas because they are located in areas with high evaporation rates. Some modern examples occur in the Persian Gulf and south of Tunisia.
   Most of the closed desert depressions are small. The largest one is the Salar de Uyuni located in the Bolivian Altiplano at 3653 m a.s.l (Figure 10.9 and Figure 16.6) with a surface area up to 9000 km^2 (Rettig et al., 1980). During the Pleistocene period, the Salar was a lake (Minchin Lake) that reached more than 43,000 km^2 (Blair, 1986). Lacustrine depressions represent less than 1% of the total desert area (Shaw and Thomas, 1997), although in some regions of South Africa they cover up to 20% of the surface (Goudie and Wells, 1995). In NW Texas there are between 30,000 and 37,000 lacustrine basins (Osterkamp and Wood, 1987). With only the exception of tectonic lakes, most basins located in arid zones do not exceed 15 m water depth at the maximum lake level (Mabbutt, 1977). Tectonic lakes are commonly elongated, parallel to the regional tectonic faults and structures. Deflation, or wind-erosion, can also originate elongated lakes following the direction of the prevalent winds. In most cases, however, closed desert depressions are approximately circular in shape. If water and sediment input is low, the bottom of the lake is usually irregular. With increasing clastic deposition, small topographic bottom irregularities disappear. Waves, currents, and other dynamic agents produce bars, barrier islands, spits, and other geomorphic features in the saline lakes, as in the Laguna de Gallocanta in Zaragoza and Teruel provinces in Spain (Figure 16.7, Gracia, 1995).

Figure 16.6. Salar de Uyuni, from the volcanic outcrops known as the Isla del Pescado, Bolivia. Photo: C. Maldonado.

Figure 16.7. Depositional littoral morphologies during the dry season (September 1996) in the Laguna de Gallocanta, Zaragoza and Teruel provinces, Spain, Central Iberian Range. Photo: B. Leránoz.

From an ecological point of view, lacustrine habitats are limited to water bodies deeper than 2 m; water bodies less than 2 m deep are considered as palustrine or paludal (swamp) habitats (Cowardin et al., 1979, in Currey (1994)). Lacustrine environments contain numerous plant- (blue green algae and reds) and animal-salinity-tolerant species (Eugster and Hardie, 1978). Lakes with high salinity are mined for chlorides, nitrates, sodium carbonates, borax, and other minerals of lithium and potassium. Those lakes with large flat areas can be used as airports, or for sport and military activities. Archaeological sites have been found in the littoral areas of some saline lakes where ancient people settled during pluvial periods (Demangeot, 1981).

2. Origin of closed depressions in arid environments

Desert lakes have a variety of origins. Hutchinson (1957) distinguished two main types (structural and geomorphic) and enumerated 13 processes that led to 75 different types of closed depressions. Mabbutt (1977) identified four main lake basins: tectonic, erosive, depositional, and crater lakes (volcanic and meteoritic). Each one of these main types is subdivided into more categories. Shaw and Thomas (1997) followed this classification and added some new cases.

Tectonic basins originate as a result of fault-bounded depressions or large sag basins. Examples in the East African Rift Valley system are the Dead Sea in Israel and Jordan, Lake Turkana or Rudolph in Kenya, and Lake Natron in Tanzania. Lake Victoria and Lake Chad in Africa, and Lake Eyre in Australia are related to large, tectonic intracontinental sag basins. Examples of landscapes composed of mountains and depressions due to tectonics are the Basin and Range province in the western US deserts (Figure 16.8) and in

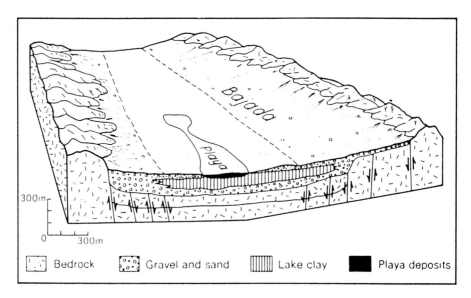

Figure 16.8. Block diagram of a tectonic basin with playa, typical of the southwest United States (from Motts (1970) in Mabbutt (1977), Figure 55).

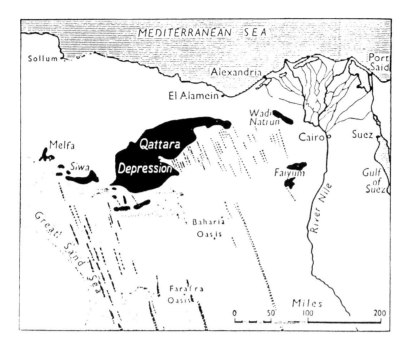

Figure 16.9. Map showing depressions, sand seas, and lines of dunes in the Egyptian Desert (Holmes, 1944).

the Argentinean Andean Pre-cordillera. Small depressions fed by ground waters typically occur along strike-slip faults.

Erosive basins in arid regions are essentially originated by deflation and wind erosion and they are very common in most world deserts. The largest deflation basin is the Qattara Depression (Figure 16.9) in NW Egypt with a surface area of 18,000 km^2 and the lowest point located 134 m below sea level. The sediments eroded form a large dune system located SW of this depression (Glennie, 1970). Erosive basins of this size are rare, but basins of hundreds of metres of diameter are quite common, and excellent examples occur in the central areas of the Duero River Basin of Spain (Gutiérrez et. al., 2005). Absence of vegetation is a requirement for formation of deflation basins.

Evolution of deflation basins can be halted if particle size of the exposed beds is too coarse to be transported by the wind (Tricart, 1969). A similar effect is caused by a thick salt layer. A groundwater table close to the surface also limits the depth of deflation (Shaw and Thomas, 1997). Deflation can also form sebkhas between the dunes (Figure 16.3). The *hamadas*, large plains in the Sahara and India, contain some closed shallow depressions, denoted as *dayas*, that have been interpreted as a result of mechanical disaggregation and wind erosion (Tricart, 1969), or karstic activity according to other authors (see Shaw and Thomas, 1997, Gutiérrez et. al., 2002). The deflated sediment particles may be deposited along the edge of the closed depressions and form aeolian build-ups shaped as half-moons (*lunette* (Hills, 1940) in English literature and *bourrelet* in French literature) (Figure 16.10). These aeolian deposits are composed of clay to sand sized particles. They occur in areas with annual rainfall between 100 and 700 mm. They may be a few kilometres long and

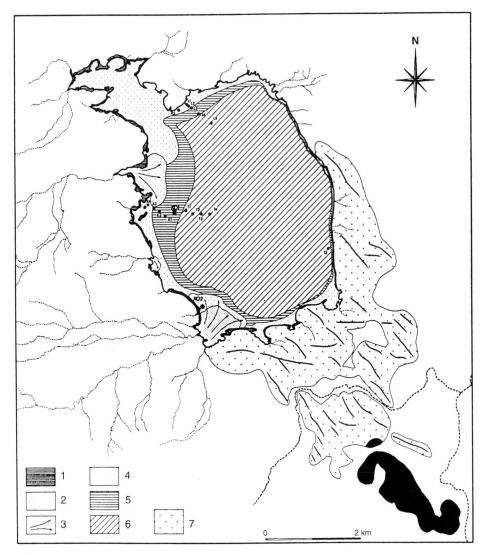

Figure 16.10. El Kourzia sebkha and its associated lunette. 1 – Small cliff surrounding the sebkha's edge. 2 – Carbonate-rich silt area. 3 – Alluvial fans from the main tributaries. 4 – Area of micropolyhedra. 5 – Carbonate-rich silt area with thin gypsum lenses. 6 – Area of black mud with summer halite crust. 7 – Lunette of el Kourzia Sebkha. In black: older lunette (modified from Perthuisot and Jauzein (1975)).

more than 60 m height. Lawson and Thomas (2002) suggested that groundwater fluctuations may be important in controlling sediment supply to lunettes in South Africa.

Within this category of erosive basins, we also include the closed depressions caused by karstic dissolution of surface and subsurface rocks, particularly evaporites. One example is the Bujaralóz region, in the central Ebro Basin of Spain, which contains a large number of karstic depressions, seasonally flooded with water (Figure 16.11 and Figure 16.2).

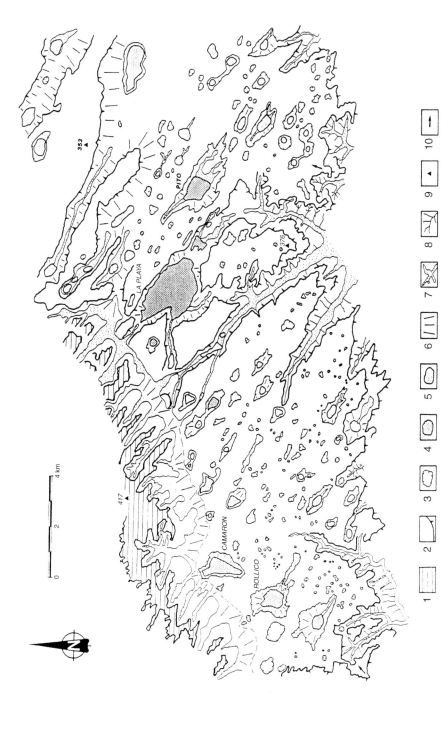

Figure 16.11. Geomorphological map of the Bujaraloz salt lakes. 1 – Structural landforms on gypsum. 2 – Platforms and cuestas on limestones. 3 – Karst depressions and dolines. 4 – Salt lakes. 5 – Areas of temporal swamp. 6 – Recent glacis and slope deposits. 7 – Infilled valleys. 8 – Gullies. 9 – Topographic point. 10 – Dip.

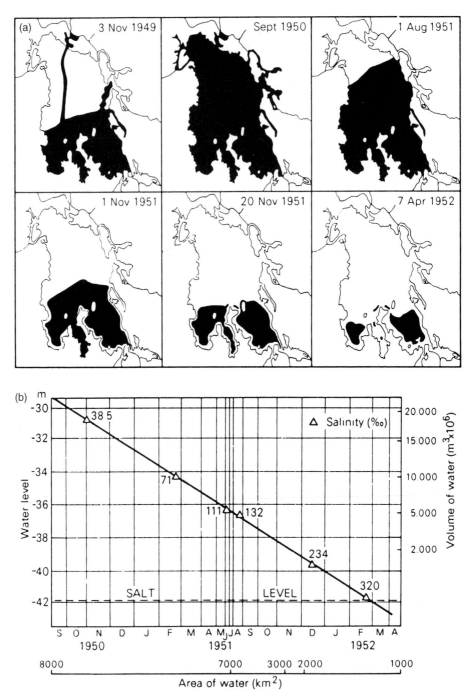

Figure 16.12. Drying cycle of Lake Eyre North, Australia, 1949 to 1952. (a) Extent of surface water. (b) Relationship between level, volume, area and salinity of water (from Boython and Mason, 1953, in Mabbutt (1977), Figure 54).

The largest one is called La Playa, and it is more than 4 km in length. The dissolution of the underlying Miocene formations composed of alternating beds of limestone and gypsum is responsible for the genesis of many small depressions. The basins are elongated with the main axis following a NW direction, which corresponds to the prevailing winds ("cierzo") and the main tectonic lineaments and fractures (Quirantes, 1965; Pueyo, 1978 to 1979, Sancho and Gutiérrez, 1993).

Piping is another process leading to the appearance of closed depressions in semi-arid regions (Figure 13.23). In Lupiñen in the Huesca Province of Spain in the central sector of the Ebro Depression, piping occurs along the contact between the loose, very-porous, thick (up to 15 m) slope deposits and the Miocene Formations. The enlargement of the tunnels and pipes decreases the mechanical resistance of the ceilings and the cavities finally collapse. On the surface, the collapsed areas are indicated as pseudolines up to 35 m wide and 20 m deep and with funnel or cylinder shapes (Figure 13.22 and Figure 13.23) (Gutiérrez et al., 1988).

Animal activity provides some examples of erosive basins. Erosive processes may be accelerated because of increased grazing by large herbivores and by soil compaction caused by large herds. Elephants excavate with their legs and trunks closed depressions up to 23 m in diameter and 1.5 m deep (Thomas, 1988).

Sedimentary depressions can be formed after a drainage system is abandoned or changed by another fluvial network. Maizels (1987) described a complex series of paleochannels in Oman that during the Plio–Pleistocene alternation of humid and dry periods caused closed depressions. Also in floodplains some lacustrine depressions can be formed. Dunes can impound creeks and rivers, as it happens south and east of Timbuktu in Mali. Lava flows may produce the same effects. Littoral evolution causes the appearance of beaches and bars, and the formation of littoral sebkhas (Glennie, 1970). Some of the *albuferas* in the Spanish Mediterranean coast, such as the Pego and Elche Albuferas, have the same origin (Sanjaume, 1985).

Finally, volcanic activity builds *craters* that could be partially filled up with water as with Zuni Salt Lake in New Mexico. Meteoritic impacts in arid zones form *impact craters*; Meteor Crater in Arizona is the most outstanding example of this type of closed depression.

3. General features of hydrology of lakes from arid zones

Lake level fluctuations depend upon water input and output balance. Inputs are surface run-off, groundwater, and secondarily, direct rainfall over the lake. Outputs are evaporation, infiltration and, when the lake basin overflows, the flow of the outlet. When input and output are not equal, the lake level fluctuates and shorelines retreat and advance. In arid regions these fluctuations are very frequent, except in very deep lakes such as the East Africa rift valley lakes. Desert climates are characterized by scarce rainfall and intense evaporation, and consequently, lakes dry out very often.

Lake water volume depends on the relative altitude of the watershed boundaries in relationship to the lake bottom and the lake bathymetry. With time, watershed highs are eroded and bottom irregularities are covered with sediments, and, consequently, water storage capacity of the basins decreases. Height and width can be modified by aeolian deflation, which usually tends to widen and deepen the closed depression. The shape and the storage capacity of the lake basin can be modified by tectonics in active regions. Volcanism plays a similar role. This is the case of the numerous volcanoes located in the East Africa rift valley, which in more active periods altered the morphology of the lakes located in these large tectonic structures.

Some complex lake basins are composed of connected sub-basins, and water, solutes and sediments are transferred and mixed from one sub-basin to another. Sediment and solute transport occurs along the surface channels or during flooding episodes when one sub-basin overflows (Currey, 1994).

Lake level decrease in desert lakes is mostly controlled by evaporation which can easily account for 2 to 3 m loss per year. Near Timbuktu in Mali, 2.5 to 3 m of water evaporated from the diffluent Niger lakes was recorded during an 8-month period (Tricart, 1969). As precipitation in arid zones is very irregular, after a long drying-out period, intense rains may flood the lakes for several years. These large flooding episodes are characteristic of large basins, because small basins commonly dry out in less than a year.

Lake Eyre in the Simpson Desert of Australia has a surface area of 8000 km^2, a watershed about 100 times larger, and an annual rainfall between 100 and 500 mm (Bonython and Mason, 1953). During the mid-20th century, the lake was flooded after a 200 to 400 mm rainfall in the southern area of the lake during February and March of 1949 (Figure 16.12a). After more intense rainfall during the next year, particularly about 500 mm during the January to April period, the lake filled up in September 1950. The lake level desiccation phase started the following month, with a salinity increase and lake level, water volume, and lake surface area decreasing (Figure 16.12b). In 1953 the lake had completely dried out. The average salt composition is NaCl (96.14%) and CaSO$_4$ (2.38%). Lake Eyre fills up completely about 12 times per century.

From a geomorphological point of view, lacustrine processes are more active in the shallow, littoral environments, where the water column is mixed by turbulent flow, and low-energy waves are active transport agents towards the coast and along the littoral zone. Beaches, bars, and other morphologies are common littoral features.

Detailed studies of littoral morphology and evolution, and analyses of stratigraphic records have allowed reconstructions of raising and lowering lake levels of many lacustrine systems during the Pleistocene and Holocene ages. These reconstructions are the hydrograph of the lake during those periods (Figure 16.13).

4. Sedimentation in desert lakes

Sediment deposition in closed depressions located in arid zones is mostly controlled by the hydrological regime. In dry playa lakes only affected by flooding, sediments are mostly silts and clays transported by the water inflows. During higher energy floods, sand can be transported into the playa and deposited in small channels. This kind of clastic sedimentation occurs mostly along the margins of the basin; alluvial fans may also develop in these areas. Waters reaching the central areas of the basins are subjected to strong evaporation, and consequently, salinity increases and salts precipitate after chemical saturation is reached. In wet playa lakes, saline ground waters are dominant and sediment deposits comprise salt-imbedded clastics and evaporites (Mabbutt, 1977).

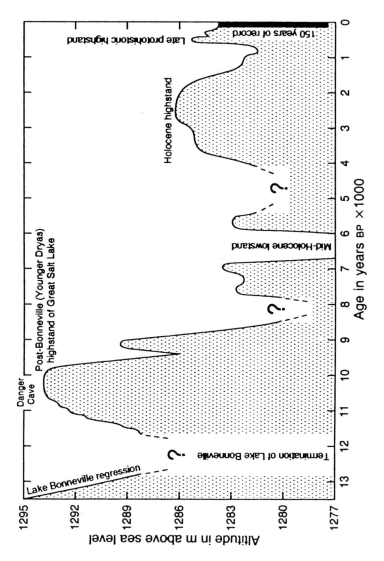

Figure 16.13. Hydrograph of the Great Salt Lake Basin, Utah during the last 13,000 years (Murchinson, 1989, in Currey (1994)). First records of human activity have been dated between 10 and 11 ka in Danger Cave.

In mountain deserts and tectonic basins, playa lake models are commonly characterized by a topographically high margin, which corresponds to the raised block (Figure 16.8). At the foothills, alluvial fans develop, with coarse gravel and sand proximal facies grading into silty and clayey distal facies, commonly associated with salts. Finally, in the topographic lows, a shallow lake develops and evaporites form (Eugster and Hardie, 1975; Hardie et al., 1978; Anadón, 1989). Most of these depressions are located in tectonically-active areas, with high subsidence rates, and consequently, thick sediment series accumulate. Many examples occur in the western United States deserts.

Ephemeral lakes located in platform and shield deserts commonly have very low topographic gradients and thin sediment sequences composed of sand, silt, and clay with some evaporites. In many cases, the rocky substrate crops out in the middle of the depressions. These lakes show large surface area expansions during the recharge periods. During dry periods, aeolian activity is the dominant process and, if the basin is large enough, dune systems can develop in different areas, as has happened in Lake Chad (Central Africa).

4.1. Clayey closed depressions

These are flat areas dominated by clay sediments with minor amounts of sands and silts. They occur at the bottom of the lake or surrounding the inner evaporite-dominated areas of the lake (Figure 16.8). Finer particles remain in suspension and deposit more slowly, whereas coarser silt and clay particles only reach the inner part of the depression during flooding episodes. Surface waters have commonly low solute concentrations and dissolved solutes are transported into the saline core of the lake or are mixed with the finer clay fraction. Desiccation of the muddy sediments form polygon systems characterized by concave cells if they lack salts, or with convex cells if salts precipitate with the fine clastic sediments.

An interesting example of formation small clayey closed depressions has been described by Osterkamp and Wood (1987) in the Texas and New Mexico High Plains. First, small depressions form by deflation and are partially filled with water. Waters percolate and slowly dissolve underlying calcrete horizons, which causes subsidence and a widening of the depressions. These small basins are filled with fine clastic material without salts.

4.2. Saline lakes

Saline lakes are natural laboratories where many active processes can be studied. They also provide useful modern analogue environments to interpret fossil sedimentary records (Reeves, 1968). From a geological point of view, they have been intensively studied during the last two decades, mostly because of their economic significance through their association with valuable oil and evaporite minerals. They also contain a detailed archive of past climatic, hydrological, sedimentological, geochemical, and biochemical conditions (Eugster and Kelts, 1983).

One of the first problems to solve in saline lakes studies is the origin of the salts. In the Senegal Delta, the marine spray caused by waves is the main *salt source* for some of the *sebkhas*. Sprayed waters evaporate and small salt crystals precipitate and remain floating in suspension forming a whitish fog in the littoral areas. Winds transport this fog inland and the salt crystals are finally deposited during rainy events. Daily temperature changes provide another mechanism for crystal deposition. At night, increased relative humidity causes water molecules to adhere to the hygroscopic salt crystals, increasing their size and forcing them to settle. During stronger rain events, the salt crystals disperse in the surface dissolve and are transported to the interior of the *sebkhas*. Finally, waters evaporate and salts are precipitated (Tricart, 1956). Marine spray is a very important solute source in littoral arid regions with strong wave action. The amount of salt of marine origin rapidly decreases at distances farther away than 100 km from the coast (Mabbutt, 1977).

Another significant solute source is chemical weathering. One example of such types of chemical reactions is the dissolution of salt-bearing rocks, such as the Triassic and Tertiary formations of some Spanish basins. Salt recycling has been common through geological time, as it has been shown using isotopic signatures of some of the elements. The final water compositions after chemical weathering reactions take place depends on the existing rock minerals, their relative abundance, the degree of weathering, the type of chemical reactions, and the specific drainage conditions (Eugster and Hardie, 1978).

Finally, another salt source is saline crusts from beaches and littoral *sebkhas* subjected to deflation. Wind transport salt mineral particles considerable distances.

Chemical composition and concentration of lake waters from desert lakes are very variable. The actual mineral composition of the precipitated salt depends on the brine composition, but other factors are also important. Such factors include for example, brine concentration, gas diffusion, particularly CO_2, mixing of water of different composition, and temperature changes. The interplay of all these processes on different brine types is responsible for the precipitation of different mineral assemblages (Table 16.1) (Eugster and Hardie, 1978).

With increasing chemical concentration in water, the precipitation sequence commonly starts with Ca-carbonates, and is followed by Ca–Mg-carbonates with increasing Mg^{2+} content. Sulphates are the next to precipitate, and as a result, the Mg/Ca ratio increases, and the brine becomes more conducive to the formation of protodolomite and magnesite. After carbonate and sulphate precipitation, water concentrations increase by evaporation until they reach saturation in the most soluble salt minerals (chlorides). As chemical concentration increases, the different solubility of the salt minerals controls the precipitation and, as a result, a concentric pattern is common in many saline deposits (Figure 16.14) (Hunt et al., 1966). Calcium and magnesium carbonates occur in the outermost areas, sulphates the intermediate, and the inner core is occupied by chlorides (Figure 11.8).

Saline deposits alternate with fine-grained sediments, as fluvial and aeolian sand, silt, clay, and organic-rich beds (kerogens) (Figure 16.15). Saline lacustrine sequences reflect the complex sedimentary evolution of the basins, with flooding and desiccation periods and a large spatial and temporal variability of the depositional processes (aeolian, fluvial).

The changes in water content in playa lakes define a saline cycle, composed of a flooding stage, evaporative concentration, and desiccation phases (Bryant et al., 1994). Other consequences of flooding events are the reworking of the playa lake surfaces, re-dissolution of salts, and mobilization of fine particles.

The micromorphology of the saline lake surface comprises a large number of *microforms*, mostly controlled by the chemical composition and the depth of the water

Table 16.1. The major saline minerals of the different brine types (Eugster and Hardie, 1978).

Brine type	Saline minerals	
Ca–Mg–Na–(K)–Cl	Antarcticite	$CaCl_2 \cdot 6H_2O$
	Bischofite	$MgCl_2 \cdot 6H_2O$
	Carnallite	$KCl \cdot MgCl_2 \cdot 6H_2O$
	Halite	$NaCl$
	Sylvite	KCl
	Tachyhydrite	$CaCl_2 \cdot 2MgCl_2 \cdot 12H_2O$
Na–(Ca)–SO$_4$–Cl	Gypsum	$CaSO_4 \cdot 2H_2O$
	Glauberite	$CaSO_4 \cdot Na_2SO_4$
	Halite	$NaCl$
	Mirabilite	$Na_2SO_4 \cdot 10H_2O$
	Thenardite	Na_2SO_4
Mg–Na–(Ca)–SO$_4$–Cl	Bischofite	$MgCl_2 \cdot 6H_2O$
	Bloedite	$Na_2SO_4 \cdot MgSO_4 \cdot 4H_2O$
	Epsomite	$MgSO_4 \cdot 7H_2O$
	Glauberite	$CaSO_4 \cdot Na_2SO_4$
	Gypsum	$CaSO_4 \cdot 2H_2O$
	Halite	$NaCl$
	Hexahydrite	$MgSO_4 \cdot 6H_2O$
	Kieserite	$MgSO_4 \cdot H_2O$
	Mirabilite	$Na_2SO_4 \cdot 10H_2O$
	Thenardite	Na_2SO_4
Na–CO$_3$–Cl	Halite	$NaCl$
	Nahcolite	$NaHCO_3$
	Natron	$Na_2CO_3 \cdot 10H_2O$
	Thermonatrite	$Na_2CO_3 \cdot H_2O$
	Trona	$NaHCO_3 \cdot Na_2CO_3 \cdot 2H_2O$
Na–CO$_3$–SO$_4$–Cl	Burkeite	$Na_2CO_3 \cdot 2Na_2SO_4$
	Halite	$NaCl$
	Mirabilite	$Na_2SO_4 \cdot 10H_2O$
	Nahcolite	$NaHCO_3$
	Natron	$Na_2CO_3 \cdot 10H_2O$
	Thenardite	Na_2SO_4
	Thermonatrite	$Na_2CO_3 \cdot H_2O$

table. Depositional processes and deflation, both influenced by the water table depth, also play also a significant role in the genesis of micromorphological features (Neal, 1965).

Microforms very common in the sulphate-rich zones of saline lakes are hemispherical mounds, named *domes* or *tumuli* (Figure 11.8). These morphologies also appear in other environments in Spain, such as the gypsum-rich Tertiary formations in the Ebro Depression (Artieda, 1993) and in the Sorbas region of Almeria (Pulido-Bosch, 1986). The mounds are hemispherical, with diameters about 1 m long, and heights less than 0.5 m. The central part of the domes is commonly collapsed, which allows assessment of the thickness of the deformed bed (about 20 cm) and the hollow inside of the mound (Figure 16.16). Several hypotheses have been proposed to explain the origin of these structures. Artieda (1993) considered that *in situ* dissolution and precipitation of surface

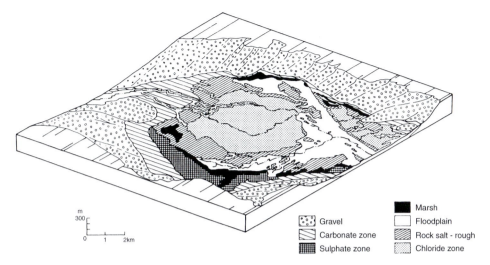

Figure 16.14. Salinity zones in a playa surface. Cottonball Basin, Death Valley, California (from Hunt et al. (1966)).

gypsum are the main factors. Secondary gypsum crystals filling the existing pores increase the volume and the crystallization pressure is responsible for lateral gypsum expansion and the appearance of dome morphologies. Calaforra (1996) described three hypotheses: (1) the existence of compressive tectonic forces; (2) the volume increase during the transformation of anhydrite to gypsum; and (3) the crystallization of a gypsum-rich matrix within the interstitial space between the macro-crystalline gypsum that would provoke

Figure 16.15. Section showing the halite surface crust and the underlying organic-rich mud, Sebkha el Melah, Zarzis, Tunisia.

Figure 16.16. Tumulus developed on the sulphate zone of a saline lake. The top has collapsed and the inside of the tumulus is hollow, Sebkha el Melah, Zarzis, Tunisia.

stress, volume increase, some uplifting and, consequently, the formation of hemispheric structures. The latter hypothesis is somewhat similar to Artieda's (1993).

Similar structures (*spring mounds*) can also be generated in relationship to springs or artesian flows; some mounds related to deep faults can reach up to 100 m in diameter (Reeves, 1968; Mabbutt, 1977; Roberts and Mitchell, 1987). The mound shape is due to the mechanical or chemical sedimentation around the flowing spring. Inorganic travertines and organic tufas (algal pinnacles up to 10 m high) (Neal, 1965) also develop around the springs. Some other small closed depressions appear after methane or air accumulated in the sediments escapes and the overlying salt bed collapses (Reeves, 1964).

Millington et al. (1995) described in playas from Tunisia *saline ramps* formed by salt crystals on top of salt crusts. Commonly they occur in groups and are shaped as half moons, orientated in the direction of the prevailing wind. Wind concentrates the small salt particles on both flanks of the ramp.

In many playas, desiccation after flooding leaves a thin salt film that covers the lake surface. High water tables or capillarity pumping maintain this thin bed as wet and stable (Eugster and Kelts, 1983). Desiccation of this thin layer forms *saline blisters* (Figure 16.17) and *saline polygons* up to 10 m in diameter (Figure 16.18). These are the two more common surface morphologies in saline playa lakes. Cracks limiting the polygons are caused by changes in volume due to wetting and desiccation of the sediments (Mabbutt, 1977). Krinsley (1970), in his study of the Great Kavir in Iran, proposed a number of phases in the evolution of the lake surface micromorphology (Figure 16.19). The polygon edges are sites of enhanced evaporation and precipitation

Figure 16.17. Salt blisters in a thin saline mud layer from an interdune sebkha, Al Liwa, United Arab Emirates.

Figure 16.18. Salt polygon system with thrust edges, Salar Pastos Grandes, Puna, Salta Province, Argentina. Photo: C. Sancho.

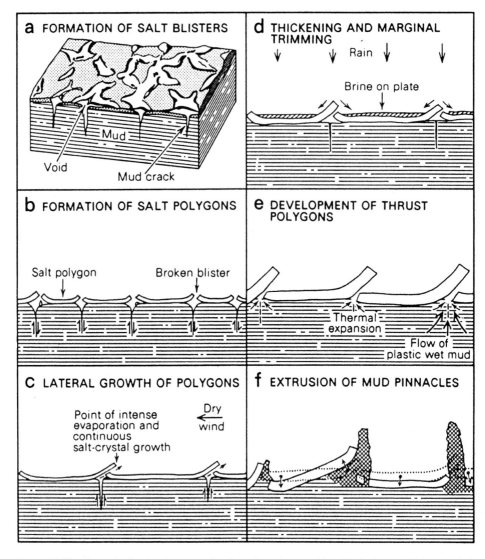

Figure 16.19. Stages in the development of an irregular salt crust, Gran Kavir, Iran (Krinsley, 1970, in Mabbutt (1977), Figure 58).

from brines rising along the mud cracks, particularly those on the upwind side of the prevalent dry winds. The growth of the edge of the saline microplate produces small thrusting surfaces over the adjacent plates to leeward, creating an illusion of overthrusting (Reeves, 1968) (Figure 16.20). Concave-shaped plates fill up with brine and consequent salt precipitation causes their thickening. Mud pinnacles are produced by the extrusion of the underlying saline mud. This process occurs because of either the thermal expansion of the black mud under the light-coloured saline crust or the load effect of the thick salt crust (up to 30 cm) over the underlying plastic wet mud.

Figure 16.20. Thrust polygons developed on a halite bed. The halite crust is mostly covered by aeolian sand, Sabkha Matti, United Arab Emirates.

The combination of saline ramparts and mud pinnacles makes the surface of most saline playa lakes very irregular and hard to walk on (Figure 16.21).

In areas with saline sedimentation, patterned ground occurs, which is similar to those in periglacial environments. This is a clear example of convergence of morphologies. Hunt and Washburn (1966) identified them in saline playas close to alluvial fans in Death Valley, California. They describe polygons, nets, steps, and bands. In the carbonate zone of the playa lake, classified and nonclassified nets occur, with detrital sediments filling up the cracks and causing a clastic classification. Polygons develop when a gypsum or salt layer underlies a clastic layer; desiccation of the salt produces cracks that are filled up with the overlying clastic material. The filled cracks provoke a lateral expansion of the polygons when they are wet, which causes the uplifting of the central part of the cell.

5. Littoral sebkhas

Littoral sebkhas are supra-tidal plains where a number of different depositional processes concur. During the last two decades, several sebkhas have been intensively studied because of their economic interest, among them, the littoral sebkhas of the Persian Gulf, particularly in Abu Dhabi (Ortí, 1989). Modern and recent depositional processes in these littoral depressions are used as depositional models to interpret pre-Quaternary rock formations, as with the Permo–Triassic in Europe.

Figure 16.21.   Irregular salt crust in a playa lake, mostly composed of thrust plaques and pinnacles, Salar de Atacama, Chile. Photo: J.L. Peña.

Figure 16.22.   Dome-shaped stromatolites, Sebkha el Melah, Zarzis, Tunisia.

Figure 16.23. Sandy-silt, organic-rich tidal plain, Sabkha Matti, United Arab Emirates.

The sediments accumulated in littoral sebkhas are composed of shell fragments, ooids, algal mats, stromatolites (Figure 16.22), evaporites, and organic matter (Figure 16.23). Some of the large littoral sebkhas show lateral changes in depositional environments. One example is Sabkha Matti in the United Arab Emirates, which inland becomes an interior sebkha dominated by alluvial deposition (Glennie et al., 1994). Deflation is also intense and causes small nebkhas, microyardangs, and dunes. Water sources for littoral sebkhas are large tides, storm waves, and sometimes, continental sources. Towards the sea, intratidal plains, lagoons, barriers, and islands develop (Cooke et al., 1993). The evaporite surface of the sebkhas present similar micromorphologies to those described for saline lakes.

FIFTH PART
Aeolian geomorphology

Chapter 17

Aeolian processes and erosive landforms

1. Introduction

Aeolian processes are those derived from wind activity, and they comprise erosion, transport, and sedimentation. Small-sized particles are mobilized and, finally, deposited in different environments, sometimes at a great distance from their source area. Desert winds are able to build large dune fields, such as those of the Rub al Khali desert (Figure 17.1), in the Arabian peninsula, that covers about 560,000 km^2 (Allison, 1997). Also, as a consequence of aeolian processes, more than 300 m of loess were deposited in China (Derbyshire, 1983a). Aeolian deposits are studied by different scientific disciplines such as physics, geology, biology, and agronomy (Pye and Tsoar, 1990). This multidisciplinary scientific research has provided significant advances in the knowledge of aeolian activity and its effects. Recent research on aeolian geomorphology deals mainly with quantification of processes, analyses of forms and internal structures of the aeolian deposits, and also with the information that those supply in order to understand the climates of the past.

From an historical perspective, during the 19th century, wind was considered as a transporting agent of limited importance, in comparison with water. At the end of that century, and at the beginning of the 20th century, the scientific trend reversed, and some researchers (especially German ones such as Passarge and Walther) pointed out that aeolian erosion was predominant in deserts, and inselbergs with their surrounding plains were a consequence of wind activity (Goudie and Wilkinson, 1977). Research activities focused on exploration of new territories and the description of landforms, and were not particularly concerned with the connections between forms and processes. Bagnold's (1941) work was a milestone in this subject because he quantified wind characteristics and transport processes. Since then conventional aerial photographs have become an important tool but the decisive boost started with research using satellite images, clearly showing the geomorphic work of the wind (McKee, 1979a,b). With these techniques it is possible to analyse yardang systems, that can extend up to 100,000 km^2 (Lybian desert, Tibesti massif, desert of Lut in Eastern Iran, and others). Recent studies on orbital images from Mars and Venus show clearly the importance of wind processes on these planets (Greeley and Iversen, 1985). The discovery of oil and hydrocarbons in Mesozoic aeolian sandstones from Europe and USA was also a stimulus to work on these subjects. Knowledge of the aeolian domain has increased with the large number of absolute dates, carried out since 1980, using C-14 and luminescence techniques.

Figure 17.1.   Dune field. Al Liwa, desert of Rub al Khali, United Arab Emirates.

2. Origin and characteristics of aeolian particles

Materials transported by the wind are commonly of diverse composition; mineral and rock fragments, and organic particles, such as shell remains. The grain sizes of the mobilized particles are usually silt and clay, but sometimes clay aggregates can be transported if salts are present (Bowler, 1973; Wasson, 1983). Quartz is the most commonly occurring material, because it is abundant in the Earth's surface and, moreover, it is very resistant to chemical weathering (Bagnold, 1941). Feldspars and heavy minerals are also present, in lower percentages, and, in addition, they undergo a progressive weathering.

The main physical characteristics relative to aeolian particles that influence erodibility and wind transport capacity are the specific weight, size, and shape of the particles (Watson, 1989b). Particles have different specific weights as a consequence of mineralogical differences. Commonly, aeolian particles show bimodal or polymodal grain-size distributions, showing the influence of the mineralogy in mobility (Willets, 1983). Aeolian sand accumulations are usually well sorted, although they have significant variations, both in size and sorting, in different deserts, in the same erg, and even in individual dunes (Watson, 1989). Grain shape also affects mobility. Therefore, flat-shaped particles are less prone to be mobilized than rounded grains (Willets, 1983). Generally quartz sand grains from desert dunes present subrounded shapes and, considering an aeolian accumulation as a whole, the biggest particles are more rounded than the smaller ones. It is also necessary to point out that abrasion during transport has an influence on the degree of roundness (Krinsley and Wellendorf, 1980). Aeolian particles present a higher degree of roundness than those of glacial, fluvial, and littoral environments.

Figure 17.2. Aerial view of a group of coastal dune ridges in the Guadalquivir river mouth, Doñana National Park, Huelva Province, Spain.

The origin of particles from desert aeolian accumulations is very variable (Demangeot, 1981; Pye and Tsoar, 1990). Wind-transported particles can be weathered grains entrained from the surface of a rock, but most of the particles come from loose materials. These are localized in continental areas in the bottom of desert rivers and its terraces, pediment covers and alluvial fans, lacustrine terraces, and in the bottom of sebkhas or playas. Marine beaches are also an important source area and they originate coastal dune system ridges. (Figure 17.2). During the periods of gradual sea-level lowering as a consequence of glaciation, large submarine surfaces became subaerial regions, and some were affected by aeolian activity (Glennie, 1972).

On the other hand, besides the transport of sand-sized particles, mobilization of aeolian silts also occurs. Aeolian silts commonly travel as dust storms and can reach distances of thousands of kilometres.

3. Particle mobilization by the wind

The entrainment of the aeolian sands is carried out in different ways, including complex processes. Loose particles, subjected to wind action, are affected by opposing forces, like gravity, friction, and cohesion (Iversen et al., 1987). Wind speed increases over any ground protuberance, as pressure decreases, according to Bernouilli's equation. As a consequence, a loose particle can undergo *uplifting*, and this force is much more intense in the contact with the ground, where wind-speed is maximum (Cooke et al., 1993). The lifting is much more important in dust turbulences (vortex and whirl-winds). *Creeping* includes friction between the particle and the air, called surface creeping, that causes rolling, and shape creeping, due to the difference of pressure between the stoss side and the lee side of the particle (Livingstone and Warren, 1996). When grains are airborne, *bombardment* against the substrate and collisions between grains begin. These processes cause the possible creeping and lifting of other loose particles.

One of the fundamental parameters is the minimum speed required to mobilize particles of a particular size. Entrainment is a function of the average size of the particles and the shear-wind speed. Bagnold (1941) studied this relationship and obtained values for the critical thresholds of shear speed (U^*) considering different particle sizes. This relationship is represented in Figure 17.3, where it is possible to observe that the bigger particles need a higher speed to be mobilized, but the smallest particles (<0.06 mm) also need high wind speeds to be entrained. This is due to the electrostatic cohesive forces. Sand size particles (0.04 to 0.40 mm) are the most easily mobilized. As lifting and dragging energy increases, there is a critical value of wind-speed shear to start particle

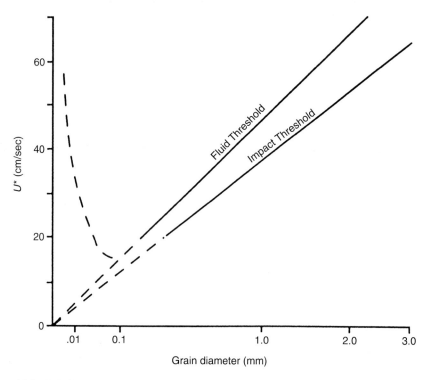

Figure 17.3. Relations between fluid and impact threshold shear velocity (U^*) and particle size (Bagnold, 1941).

mobilization. This is the *fluid threshold* (or *static*). When entrainment of grains has begun, ballistic impacts reduce required energy to keep the movement of particles. This is the *impact threshold* (or *dynamic*), where slower speeds are required to move a stationary grain by fluid pushing, and it is equivalent to 80% of the static threshold (Andersson and Haff, 1988). These equations adjust correctly on flat, homogeneous, dry surfaces, with isolated, loose clasts (Williams et al., 1994). The values of the movement thresholds, however, are also affected by the variability of sediments, surface crusts, slope inclination, humidity, and roughness.

4. Types of aeolian transport

When particles pass the mobilization threshold they can be transported in four different ways; creep, reptation, saltation, and suspension (Figure 17.4). The transportation type is mainly dependent on the grain size of available particles (Bagnold, 1941). The boundaries between these different transport types are not sharp and the transition from one to another is not very clear.

Creep involves translation or turning of coarse particles (0.5 to 2 mm) on the surface, pushed by the wind force or by the ballistic impact of grains during saltation. Creep can begin some moments before the mobilization of the particle by saltation (Nickling, 1983). It is difficult to establish the amount of material that moves by creeping because is very difficult to distinguish creep from reptation. We can estimate, however, that creep represents one quarter of the mobilized bed load.

Recently, a transitional type between creep and saltation has been defined. It is called *reptation*, in which particles move in small jumps caused by the impacts of high-energy

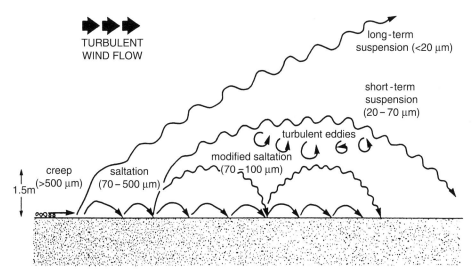

Figure 17.4. Types of particle transport by wind. Indicated particle size-ranges in different transport modes. (Reproduced with permission of Pye (1987)).

Figure 17.5.   Sand grains mobilized by saltation. Al Liwa, desert of Rub al Khali, United Arab Emirates.

grains (Andersson and Haff, 1988). As a result of the impact, grains can loose 40% of their speed, and the transmitted energy can mobilize other grains from the bed. The main difference between reptation and creeping is that, in the former, particles change continuously from reptation to saltation transport.

Saltation (Figure 17.5) consists of the uplifting of a particle, that covers a distance in a ballistic trajectory (Bagnold, 1941) and falls down, impacting on the surface. The energy of this impact is used in the uplifting of other grains. Therefore, saltation is a very important transport mechanism with cascade effects (Nickling, 1988). Uplifting angles fluctuate between 30 and 50°, and speeds are higher in the stoss side facing the wind. Most particles, according with different models and experiments, descend with angles between 10 and 15° (Anderson, 1989). Jump distances reach 12 to 15 times its height (Livingstone and Warren, 1996), and the smaller particles, entrained at a higher speed, can reach more height and so the distance of jump is longer (Anderson and Bunas, 1993). Height reached during jumps can be 3 m, but the average is 0.2 m (Pye and Tsoar, 1990). Bagnold (1941) recognized that jumps are higher over a rocky surface or a surface with pebbles than over a bed of loose sand (Figure 17.6).

Pure saltation develops when turbulent vertical speed has no significant effect on the particle trajectories. Between pure saltation and pure suspension the trajectory of the particles is modified due to speed settlement and inertia. This type of transport has been called "*modified saltation*" (Figure 17.7) (Nalpanis, 1985). This indicates that there is not a sharp transition between saltation and suspension.

The smallest particles (< 0.06 mm) are transported by suspension within a turbulent airflow (Figure 17.7) that can keep this fine sediment at a considerable height during many days, and finally deposited it as dust or loess. This constitutes a deposit generally between 30 and 60 m, occupying large surface areas, and being able to travel along more than

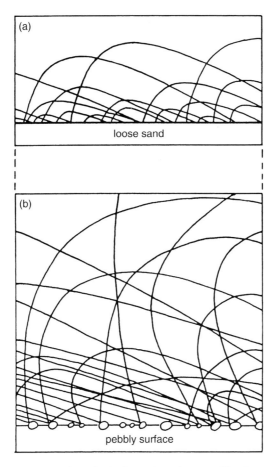

Figure 17.6. Schematic representation of saltation trajectories over (A) a loose sand and (B) a pebbly surface (Bagnold, 1941).

300 km. Within the suspension transport we can distinguish the airborne grains for long periods of time (*long-term suspension*) from those that fall quickly on the surface (*short-term suspension*) (Tsoar and Pye, 1987) (Figure 17.4 and Figure 17.7).

5. Ripples

Ripples are the smallest aeolian accumulations, and they represent the initial response of sand surfaces to wind transport (Figure 17.8). Although they are depositional forms, they will be analysed in this chapter, because they are closely related to the types of transport previously described. Ripples are oriented perpendicularly to the wind transport direction, even though on slopes they flow obliquely, because they are affected by gravity. Most of them have a short-term existence and they move much more quickly than dunes (Livingstone and Warren, 1996). Usually ripples are asymmetrical in a transverse section.

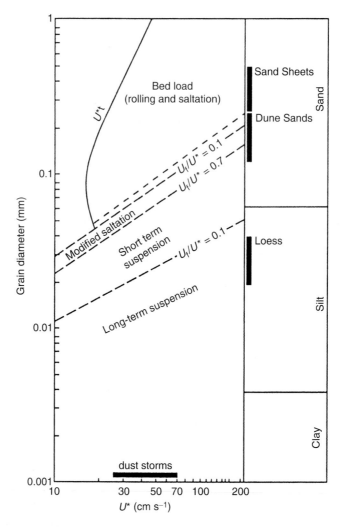

Figure 17.7.   The relationships between grain diameter, shear velocity, and mode of sediment transport. U_f is the particle fall velocity (Tsoar and Pye, 1987; modified in Thomas (1997d), Figure 16.2).

The stoss side is convex with angles of 8 to 10°, and the lee side has a slope angle between 20 and 30° (Sharp, 1963). Some ripples have a sharp crest; some others present a rounded one. Wavelengths vary between a few centimeters up to tens of meters, and the heights range from 1 cm to about 30 cm. Wavelength and height are the result of the particle sizes; ripples of coarse sand present a longer distance between crests than those of fine sand (Seppälä and Lindé, 1978).

Megaripples are remarkable landforms, with a wavelength between 1 and 25 m. They are originated by strong winds and are composed of coarse sand. In cross section they are generally more symmetrical than the smaller ripples, and present a higher contrast in grain size between crest and trough areas. According to Bagnold (1941) several centuries are

Figure 17.8. Ripples in Al Liwa, desert of Rub al Khali, United Arab Emirates.

necessary to build megaripples, however, Sharp (1963) estimated that during a wind storm a few hours are enough. *Chiflones* are wavy ribbons of coarse and fine sand, 100 to 150 m long and 1 to 1.5 m wide (Simons and Eriksen, 1953). They form when coarse sands are channelized between obstacles such as shrubs.

Several models have been proposed to explain ripple formation. The first one (Bagnold, 1941) established a relationship between ripple wavelength, observed in experimental tunnels, and the length of sand paths during saltation. Irregularities in the surface act as sites for erosion and sedimentation. The stoss side is bombarded by particles transported by saltation that transfer energy when they collide, as a consequence new grains are mobilized and the process continues. Trajectories have the same length as ripples if sand size and wind speed do not change. This hypothesis is very attractive, because it explains a large number of characteristics and variables related to ripple formation. Sharp (1963) estimated that particles in ripples move essentially by reptation. Irregularities on the bottom surface and interactions between moving grains form local elevations on the former surface, originating incipient ripples of short wavelength and low height, which progressively grow bigger. A third model, proposed by Anderson (1987), was developed using experimental data and numerical simulations. The model establishes differences between low energy impact reptations and long trajectory high-energy impact saltations. It identifies different saltation paths, some of them longer than the wavelength of ripples. This model disagrees with Bagnold's model, because saltation impact does not contribute directly to ripple formation. According to Anderson's work, reptation processes are dominant. Because reptation distances increase with wind shear strength, the ripple wavelength also increases. This agrees with data obtained in experimental tunnels (Seppälä and Lindé, 1978).

In conclusion, aeolian ripples are closely related to sand saltation, although new models suggest that reptation provides the initial conditions for ripple formation, and they quickly evolve by saltation impacts and, as a consequence, by grains transported by reptation (Lancaster, 1995).

6. Aeolian erosion and resulting landforms

When particles are lifted by the wind from the surface and impact on the rocks at significant speeds, they produce aeolian erosion landforms. They are more widely spread in areas of scarce or absent vegetation, especially in arid zones, but are also present in periglacial and marine beaches environments.

Wind erodes by *deflation*, that is, the result of entrainment and transport of loose material of the ground. Average deflation rate is about 1 mm/year (Cooke et al., 1993). *Abrasion* is the mechanical wearing out of cohesive material. Ballistic impacts of grains cause surface rock fracture, until the surface becomes stabilized. This suggests that abrasion speed reaches a maximum and later decays. Most authors estimate that abrasion is produced only by sand grains, however others think also that the aeolian dust originates abrasion, although secondarily (Greeley and Iversen, 1985). Abrasion speed rates are very variable, about several millimeters per year, and they depend on the type of rock, wind speed and amount of grains provided by the wind. The height where maximum abrasion occurs fluctuates between 0.1 and 0.4 m. These figures have been obtained from field experiments during intense storms (Sakamoto-Arnold, 1981) and from theoretical models (Anderson, 1986).

6.1. Ventifacts

Ventifacts are pebbles with smooth surfaces, polished and faceted by the impact of particles carried by the wind (Figure 17.9). Pits, flutes, grooves, and helicoid patterns are developed both in boulders and in rock outcrops. These micro-morphologies appear only in weathering-resistant rocks.

Facets are oriented perpendicularly to wind direction. The number of facets is very variable, and up to 20 have been described (Higgins, 1956). The most common, however, is the presence of two or three facets that intersect forming edges. The German words *einkanter*, *zweikanter*, and *dreikanter* are commonly used to indicate that they have one, two, or three facets, respectively. *Pits* are small closed depressions that can be originated by the action of wind over a nonhomogeneous rock or on pre-existent holes. They are inclined 55 to 90° from wind direction and, in consequence, they indicate the windward side. Where the angle decreases considerably we find the transition to *flutes* (Sharp, 1949). These have a spoon-shaped form, with dead ends and U-shaped cross sections. Commonly they are open to the windward. Their length, width, and height vary greatly, from millimetric sizes up to centimetric ones (Cooke et al., 1993). They appear in horizontal surfaces or in gentle slopes, in very different kinds of rocks, and they are not related to the structure (Breed et al., 1989; Laity, 1994). *Grooves* are opened in both extremes and are longer than flutes. They indicate, as those described before, the predominant wind

Figure 17.9. Ventifacts (*zweikanter* = two facets; *dreikanter* = three facets). Terrace pebbles from the central sector of Duero depression.

direction in the area. Sometimes rock composition is nonhomogeneous, and the wind erodes softer areas more intensely. These forms, generated because of the presence of inhomogeneities are called *etching*. Finally, other microforms are the *helicoid forms* developed on marbles from the Mojave Desert of California. They are shallow grooves with spiral morphology in relation to the direction of the wind. They fluctuate from several millimeters up to some centimeters wide and deep. They occur in areas of very high wind speed (Laity, 1995).

The origin of ventifacts has been studied both in field research (Sharp, 1964, 1980) and in the laboratory (Kuenen, 1960; Whitney and Dietrich, 1973), for different particle sizes. They are the result of the impacts of particles transported by saltation and suspension, the role of sand being the most important in their configuration (Laity, 1994). It is unusual, however, to find ventifacts in dune fields or in its surrounding areas, but they are common in desert pavements with scarce sand.

Former interpretations pointed out that facets were inherited forms, or at least they conditioned its posterior abrasion (Higgins, 1956; Kuenen, 1960). Other explanations establish an origin related to the existence of winds coming from different directions (Kuenen, 1960). This hypothesis can explain some ventifacts, but in mountain zones winds blow essentially from a single direction (Sharp, 1949). A well-accepted theory is based in the rotation of pebbles around a vertical axis, with rare overturning caused by excavation beneath due to deflation. The capsizing can be also caused by expansion and drying of clays, and by animal activity (Livingstone and Warren, 1996). This explains that the smaller the ventifacts, the larger the number of facets they present, and vice versa. Moreover, overturning justifies the existence of ventifacts with facets in the upper part and in the bottom. Some authors have pointed out that these micro-morphologies (polished surfaces, pits, flutes, grooves, and spiral forms) are probably originated by a vortex of aeolian dust (wind rotating around an axis), more than by sand impact (Maxson, 1940; Higgins, 1956). These micro-morphologies can be reproduced in wind-tunnel experiments (Dietrich, 1977; Whitney, 1979). In any case, most of the shapes related to ventifacts are explained by sand grain impact, and aeolian dust erosion (Breed et al., 1989, 1997).

Wind direction can be determined from ventifacts and their associated forms, and, as a consequence, fossil ventifacts provide good records of wind circulation during past geological periods. If, in addition, they are dated using desert-varnish techniques (Dorn, 1995), it is possible to know the paleoenvironmental conditions and the intensity of aeolian activity.

6.2. Yardangs

The word *yardang* has its origin from the eastern limit of the Taklimakan desert in China. The shape of the yardang looks like a capsized ship hull (Figure 17.10) even though in many cases yardangs are flat topped and lack a keel. The windward face is wider, whereas the leeside becomes narrower and looses height. Yardangs show varied forms (Figure 17.11). Halimov and Fezer (1989) described eight types of yardangs in the depression of Quaidam in China, according to its morphology, and they suggest a temporal succession. The ratio between longitude and width varies from 3:1 up to 10:1 (McCauley et al., 1977). Yardangs appear usually grouped, and their long axes are parallel to the direction of the more intense winds. They occur in every large desert all over the world, except in the Australian desert. They have been also identified on Mars (Greeley and Iversen, 1985). Some authors (Cooke et al., 1993) distinguish yardangs according to their size: microyardangs (1 m) (Figure 17.12), mesoyardangs (10 to 100 m) and megayardangs (1 km). The more useful classification by Livingstone and Warren (1996) of landforms up to 100 m long, called yardangs, and those of longer size, that form ridges (Mabbutt, 1977, Laity, 1994) or megayardangs (Figure 17.13). Cross sections of megayardangs present a U shape, with the bottom as the trough corridor. They are commonly covered by sand, pavements, and sometimes, by barchan dunes, parallel to the ridges. Megayardangs are well developed in the southern part of the Tibesti Massif (Central–Eastern Sahara), where the ridges can reach up to 200 m in height, up to 4 km in length and a separation between 0.5 and 2 km; they cover a surface area greater than 600,000 km^2 (Hagedorn, 1968; Mainguet, 1972). Very large groups appear also in the hyperarid deserts of Egypt, Iran, and Perú. Yardangs are carved generally in soft materials, such as aeolian sands and fluvial sediments and Cenozoic lacustrine deposits, although many of the Saharan ridges are excavated on Cambrian sandstones. They can be developed in a wide variety of lithologies, from semi-consolidated sediments to granites or even harder rocks

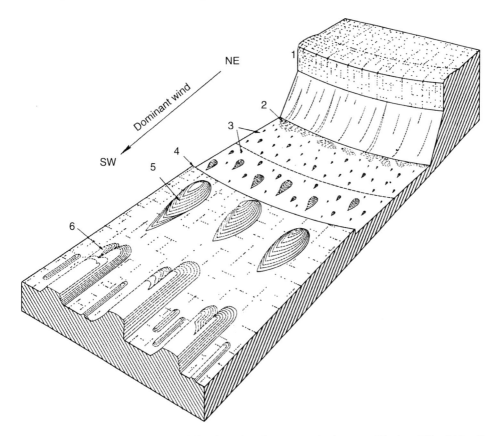

Figure 17.10. Scheme of the general situation of yardang groups in the area of Borkou, Tibesti Massif (Republic of Chad). 1: Cliff that separates the basin from a higher surface. 2: Lacustrine area, sometimes subactual. 3: Small yardangs of the depression. 4: Scarp windwards over the lower surface. 5: Yardangs of platform margin. 6: Sandstone crests and corridors (after Mainguet, 1972).

(Goudie, 1989b). Moreover, direction, dip and cleavage of the rocks affect the shape and development of yardangs.

Yardangs are probably shaped by abrasion and deflation processes (McCauley et al., 1979; Ward and Greeley, 1984), with slight modifications caused by water erosion (Laity, 1994). Abrasion contributes to polish and to basal excavation that can reach 1 or 2 m in depth (Grolier et al., 1980). Deflation is important in the development of yardangs in poorly consolidated lithologies. As happens with ventifacts, a vortex of dust particles seems to be an important process in shaping the morphology of the yardangs (Whitney, 1985). Many yardangs have previously undergone fluvial erosion, and they appear more elongated if the drainage network is parallel to the prevailing winds. Periods with fluvial–lacustrine activity, and others with dominant aeolian action occur and deposition and transport of sediment in the troughs alternate (Mainguet, 1972; Gutiérrez et al., 2002). Finally, wind-tunnel experiments (McCauley et al., 1979; Ward and Greeley, 1984), performed with synthetic sediments, show that it is possible to identify positive main

Figure 17.11. Flat-topped yardangs. The top is a cemented sandstone. Kar-Al-Ajban, United Arab Emirates.

Figure 17.12. Microyardangs on a cemented dune. Sebkha Matti, United Arab Emirates.

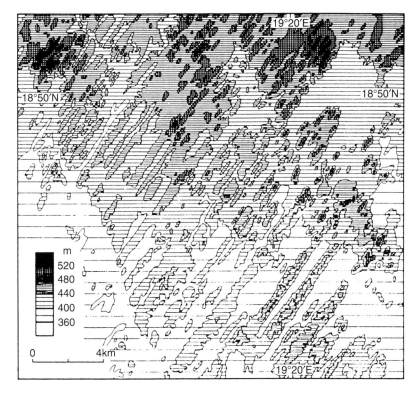

Figure 17.13. Wind-sculpted sandstone landscape, originating a morphology of crests and corridors. Borkou lowland, south of Tibesti Massif, Chad Republic (Hagedorn, 1968, in Mabbut (1977), Figure 40).

fluxes and secondary negative ones returning along the flanks. They conclude that abrasion and deflation are the fundamental processes in yardang formation.

6.3. Deflation basins

In the previous chapter we dealt with the origin of closed depressions, and we analysed some of the fundamental features that characterize the so-called erosive basins. In this section we study some aspects more deeply that are essentially related to their geometry, distribution and origin.

Deflation basins or blowouts are closed depressions that can reach 12% of large surfaces (Cooke et al., 1993). They usually have a rounded, elliptical, or kidney shape, with rounded margins. They are more common in the semi-arid zones of the world, and have a great development in central and southern Africa, the Pampas and the Pantanal of South America, southern and western Australia, and Manchuria and western Siberia (Goudie, 1991; Goudie and Wells, 1995). Their size can vary considerably; with the surface area of the basins in southwest Australia ranging between 0.004 and 100 km^2, with an average size of 0.05 km^2 (Killigrew and Gilkes, 1974), whereas in the south of Africa they fluctuate between 0.05 and 30 km², with an average of 0.2 km² (Goudie and Thomas, 1985). These authors calculate densities of 100 basins every 100 km².

Deflation basins are more commonly developed in poorly consolidated sediments. Some lithologies, such as the sandstones from the south of Africa, are the most suitable for the development of these closed depressions. Some rocks quickly disintegrate by saline weathering, such as some claystones with a high salt content, making deflation easier. The presence of swelling clays implicates that during dry periods clayey regolith cracks deeply, and the product of this disintegration can be easily exported by the wind (Price, 1963).

During the Late Pleistocene period desiccation of the great pluvial lakes in the western USA, a period of intense deflation occurred that caused excavation of the lake bottoms, mobilization of dune fields and generation of lunettes (Goudie and Wells, 1995). Similar effects are identified on other continents.

Ancient drainage systems developed in large areas of low relief are suitable places to generate deflation basins. Excavation of some areas of the creeks and small rivers and generation of small deflation basins may disconnect the drainage networks, although paleochannel paths are still identifiable (Bowler, 1986). Dunes can also cut off the channels, producing deflation basins.

In sandy deserts, there are large surface areas with interdune deflation basins. They are formed as a consequence of the growth of dune crest fields derived from parabolic dunes. The development of these dunes implies the formation of deflation basins of an elliptical shape (Lancaster, 1978).

Also in littoral areas, with more humid conditions, is possible to recognize large fields of deflation basins. In the Carolina Bays, between Maryland and north of Florida, they cover a strip parallel to the coast of about 1100 km, and about 500,000 basins have been identified (Prouty, 1952). They were apparently formed during the Wisconsin glaciation as the elliptical depressions of giant parabolic dunes, built by northwestern winds that blew along the sandy surfaces of the coastal plain, when the phreatic level was low. When the water table rose, depressions were stabilized. After that, during dryer periods, many lunettes were built (Price, 1968; Goudie and Wells, 1995).

In the leeward margins of the deflation basins sand–clay dunes with a half-moon shape developed commonly. They are called *lunettes*. They often reach a height between 10 and 50 m, however in the Sebkha Kelbia, in the centre of Tunisia, their maximum height is 165 m, with a 17-km length and a 9-km width, being the highest lunette in the world. The surface of these dunes is often colonized by vegetation, which facilitates accretion. Particles from the bottom of the deflation basin become fragmented into micro-polyhedrons because of desiccation, and they are mobilized by winds with a speed higher than 3 m/s, being able to transport these particles and deposit them in the outer rim of the closed depression (Perthuisot and Jauzein, 1975). The presence of salt, which blocks vegetation development in the bottom of the depression, causes the aggregation of clay particles that will be also exported out of the basin. *Pellets* of salt and clay that form the lunette are lixiviated by rain, and salt returns in part to the sebkha bottom; so a cycle of salt is established (Tricart, 1969).

In some cases in the leeward area several lunettes can occur (Figure 17.14). The most external dune has a much higher content of quartz, and the rate of clay increases toward the internal dunes, becoming clayey dunes (Bowler, 1973, 1986). This has been explained

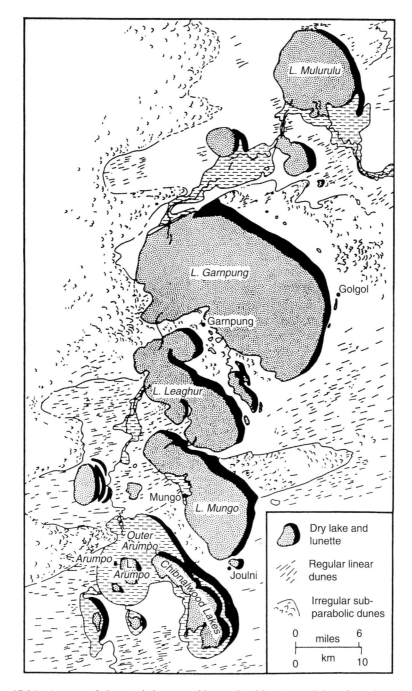

Figure 17.14. A group of characteristic pans, with associated lunettes and dune forms in northwest of New South Wales, southeastern Australia (Bowler and McGee, 1978, in Goudie and Wells (1995), p. 13, Figure 9).

by deflation of fluvial sediments to build the external dune. On the other hand, the internal lunettes are developed by deflation from playa deposits, accumulated after the initial deflation stage (Goudie and Wells, 1995) (Figure 17.15).

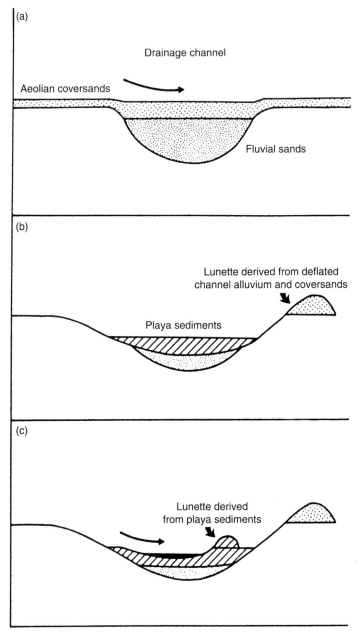

Figure 17.15. A model of lunette development caused by deflation of channel alluvium, aeolian sheets and playa deposits (after Goudie and Wells (1995)).

Great deflation basins (Figure 16.9), such as the Quattara depression, commonly have a polygenic origin, although deflation plays the fundamental role. Albritton et al. (1990) show that this depression was initially a fluvial valley that was later split by karstic processes during Upper Miocene, and later it was deepened by deflation. Moreover, mass movements and fluvial processes modified it somewhat. In addition, saline weathering prepared materials for subsequent deflation. We should also consider the role that Quaternary climatic oscillations played, modifying the geomorphological processes and, as a consequence the genesis of depressions. We can conclude that great deflation basins have a long and complex history. In smaller deflation basins deflation is dominant, although other processes are also present, as we previously indicated.

Chapter 18

Aeolian accumulations

1. Introduction

Aeolian accumulations are the result of sedimentation of particles carried by the wind, producing deposits of sand and dust. We shall begin with the study of the aeolian sands, analysing their distribution, processes and types of accumulations.

Dunes are, undoubtedly the most outstanding of desert forms (Figure 18.1). Research on these landforms has experienced a boost during the last two decades with the use of satellite images. Aeolian sands cover about 5% of the Earth's surface (Thomas, 1997d), and 20% of the arid zones of the world are occupied by sand accumulations. The proportion of this cover varies according different deserts, less than 1% in North and South America (Lancaster, 1995) up to 50% in Australia (Mabbutt, 1977). The word *erg* or *sand sea* means a large surface covered by sand, at least 125 km^2 (Fryberger and Ahlbrandt, 1979; Thomas, 1997d). The erg is called *koum* in Central Asia and *nafud* in Arabia. We can see the global distribution of great ergs in Figure 18.2, where active ergs are distinguished from marginal ergs, stabilized by vegetal colonization. Most ergs are active with an average annual precipitation of 150 mm (Wilson, 1973). The largest number of sand seas is concentrated in the Northern Hemisphere. Wilson (1973) identified 58 ergs with a surface extent greater than 12,000 km^2 in the different deserts all over the world, most of them in the Sahara desert, where the erg Chech in southern Argelia (312,000 km^2) is the largest in this desert. The erg of Rub al Khali, which occupies one-third of the Arabian Peninsula (Figure 18.3), has a surface area of 560,000 km^2, and the Great Sandy-Gibson Desert, in western Australia, covers a surface area of 630,000 km^2.

In some areas it is difficult to distinguish between coastal and desertic dunes. Examples dealing with this problem occur in the coastal zones of Namibia (Lancaster, 1982; Lancaster and Ollier, 1983), Perú (Finkel, 1959) and Oman (Goudie et al., 1987). In some zones, continental dunes obtain their sand supply in the coastal regions, whereas in other areas ergs have prograded towards the sea (Pye and Tsoar, 1990).

A hierarchic classification of ergs (Wilson, 1972; Lancaster, 1994) is based on the identification of three classes with distinct spatial patterns: (1) aeolian ripples (0.1 to 1 m); (2) isolated individual dunes (50 to 500 m); and (3) compound and complex dunes (>500 m). Each element in this hierarchy responds to the dynamics of the wind pattern of the region.

Other large bodies of aeolian sands do not have dune forms and are called *sand sheets*. Their surface area ranges from a few square kilometers up to 100,000 km^2 in the eastern Sahara (Breed et al., 1987). Their development is controlled by vegetation, grain size, shallow phreatic levels, seasonal floods, and the occurrence of superficial crusts (Kocurek and Nielson, 1986).

Figure 18.1.   Dune field in Jaisalmer, Thar Desert, India.

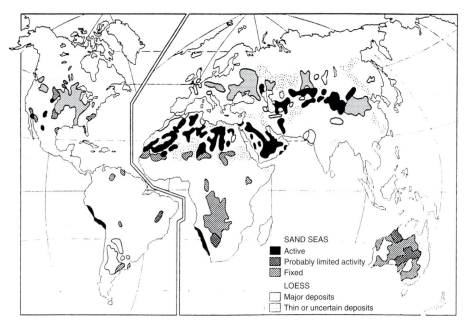

Figure 18.2.   The global distribution of aeolian deposits (Thomas, 1997).

Figure 18.3. Barchanoid forms in the southern erg of Rub al Khali, United Arab Emirates.

2. Factors controlling erg development

Three requirements are necessary in order to build ergs: (1) large supplies of sand; (2) suitable climatic and topographic conditions; and (3) enough wind energy (Pye and Tsoar, 1990). About 90% of particles from active ergs are sand-sized because silt is characteristically transported far away from deserts, and clays are mobilized by the wind only as *pellets*. Most of the sand comes from fluvial, littoral, and lacustrine deposits. In the Namib desert (Lancaster and Ollier, 1983) and in the Sinai desert, dunes have been originated by aeolian transport in fluvial and littoral, environments. Sands can be mobilized and flow by fixed paths, covering large distances, and reaching the distal zones of the ergs (Wilson, 1973). Maximum speeds have been obtained in aeolian sheets from Mauritania, fluctuating between 62.5 and 162.5 m^3/m width/yr (Sarnthein and Walger, 1974), whereas the typical speed is about 3.49 m^3/m width/yr in barchans. Where sand supply speed is high, large dune fields can grow in only decades or centuries. Nevertheless, the time required for the development of ergs is much higher. The average thickness of the Sahara's ergs is about 21 to 43 m, although in the Simpson desert Australia it is 1 m (Wilson, 1973). These values are very low by comparison with the thickness of several hundred meters of Paleozoic and Mesozoic aeolian formations. This implies a subsidence in the sedimentary basins, and possibly a greater aeolian activity during these periods, especially before the development of many terrestrial plants (Kocurek, 1988).

Aeolian processes are usually more important when vegetal cover is scarce, with the result that extensive ergs occur in zones with <250 mm of rain, although in the south of Africa the limits of active dunes vary between 100 and 150 mm (Lancaster, 1981). In some ergs, distinctive accumulative phases and hiati can be identified as an evidence of past climatic changes, as we can see in the Gulf of Oman (Cooke et al., 1993).

According to their mobility, ergs can be dynamic or static. First, they occur in flat areas where the migration of the erg, in the direction of the predominant wind, does not encounter significant obstacles (Pye and Tsoar, 1990). On the contrary, static ergs are localized in topographical depressions, limited by steep slopes (Figure 18.4a) (Wilson, 1973). A clear example appears in the Great Eastern Erg, located to the south of the cliffs of Tademait plateau in Algeria (Fryberger and Ahlbrandt, 1979). These authors proposed different models of topographic influences in the development of ergs (Figure 18.4b). They occur: (1) in the shadow of topographic barriers; (2) in shallow desert depressions; and (3)

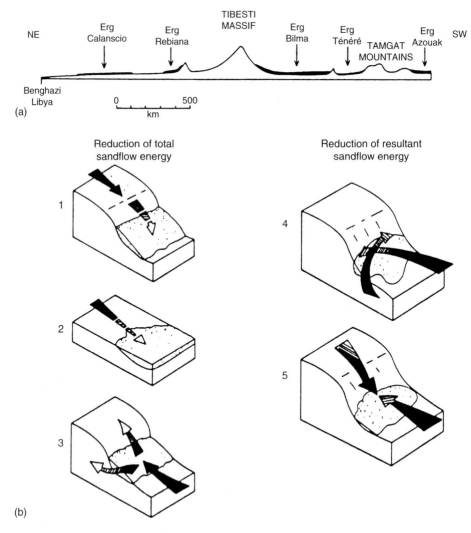

Figure 18.4.   (a) The relationships of ergs in North Africa to topography (after Wilson, 1973). (b) Models of topographic influences on ergs development (Fryberger and Ahlbrandt, 1979).

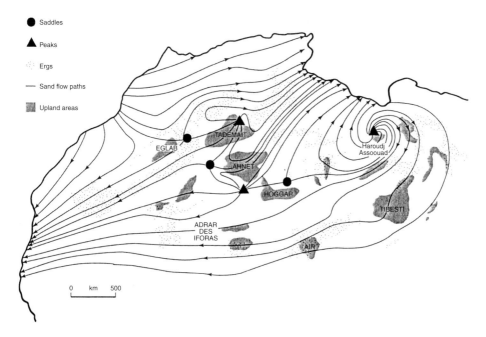

Figure 18.5. Sand flow across the Sahara showing major source and deposition areas (after Wilson, 1971; Mainguet, 1978).

in zones with a reduction in sand-transporting energy. On the other hand resultant energy can dissipate; (4) when surface winds are deflected leading to sites of favourable accumulation; and (5) when winds blow from opposite directions.

When sand transport speeds are high, flow is higher than sedimentation, and aeolian sheets and barchans can develop. On the other hand, if sand flow is saturated and accumulation exceeds transport, ergs are generated (Wilson, 1971). This author used the sand path flow obtained by weather stations to draw sand flow maps in the Sahara desert (Livingstone and Warren, 1996) (Figure 18.5). In this way he obtained the source areas and the sedimentation areas of Saharan ergs.

In tropical deserts, the environments with high aeolian energy are located in the Trade Wind belts that border the anticyclonic areas, whereas the zones of low energy are situated in the proximities of the high-pressure centres (Livingstone and Warren, 1996). The Sahel region, in the south of the Sahara desert, is a low energy area where particles transported by the Trade Winds from the northeastern areas of higher energy are accumulated.

3. Dominant processes in dunes

The onset, development, and achievement of a dune-equilibrium morphology reflects the balance between erosion and sedimentation, both controlling the dune's shape (Lancaster, 1995). A decrease in sand transport results in an affluence that exceeds the evacuation, with resulting sedimentation and increasing accumulation. On the other hand, ground irregularities have considerable impact on transport speeds, because of the interaction with the flow and by the generation of secondary flows (Lancaster, 1994).

The onset of *dune formation* is not a well-known process. The starting of a dune implies a local sedimentation, due to the descent of the transport speed. Kocurek et al. (1992), in their studies about the development of dunes in Padre Island (Texas), indicated that accumulation begins in small irregularities of the terrain surface, or small vegetation obstacles or boulders, originating shadow dunes. In this work they recognized five steps of dune evolution, with progressive changes on the lee-side slope, due to the separation and expansion of the flow (Figure 18.6): (1) small irregular patches of dry sand, a few centimeters high; (2) protodunes of 0.1 to 0.35 m high, with ripples all over its surface; (3) protodunes 0.25 to 0.40 m high with sand grains falling down on the lee side slope; (4) grainfall protodunes 1 to 1.5 m; and (5) dunes with grainflow, 1 to 2 m high. The three first stages can be considered as generating embryonic dunes.

Bagnold (1941) pointed out that an intense sand transport implies an instability transverse to the flow, originating sand deposition in longitudinal strips 1 to 3 m wide. He pointed out that these strips can be the core of further longitudinal dunes. In small sand accumulations, erosion prevails with low and intermediate wind speeds, but when these are higher, accretion processes dominate (Greeley and Iversen, 1985). Dunes can be also built by strong wind gusts able to sweep the sand, depositing it when wind intensity decreases; small aeolian accumulations can be the beginning of dune formations (Warren and Knott, 1983).

Many hypotheses have been considered to explain the regular repetition of successive dunes, with comparable shapes and similar separation. Models are based in dunes developed under water, and their evolution is founded in the kinematic instability and in turbulence physics (see Cooke et al., 1993; Livingstone and Warren, 1996). On the other hand, as dunes grow there is a streamline convergence of wind towards the crest, and as a

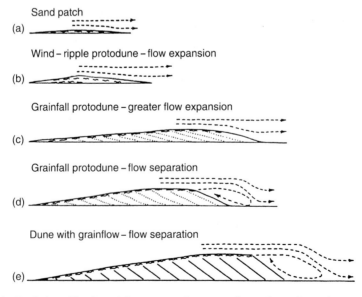

Figure 18.6. Evolution of lee-face airflow patterns for stages of protodunes. Internal structure indicated by wavy dashes for wind-ripple laminae, dotted lines for grainfall laminae, and solid lines for grainflow cross strata (after Kocurek et al., 1992).

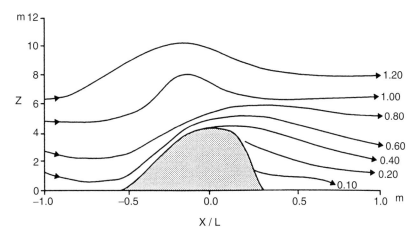

Figure 18.7. Wind flow pattern over an isolated barchan, in Salton Sea, California. A convergence of stream lines, with compression, is observed on the windward slope, and an expansion in the lee-side slope (after Lancaster, 1995).

consequence, wind speed increases, but conversely an expansion and divergence on the lee side slope, originating varied secondary flow systems (Lancaster, 1995) (Figure 18.7)

Windward slopes of dunes are zones of important changes, becoming apparent by modifications of slope inclination, roughness, shear speed and sand discharge (Livingstone and Warren, 1996). The entire slope is a zone of maximum erosion (Bagnold, 1941). Wind speed increases towards the highest point. Differences of wind speed between slope and crest, measured on barchans, can reach a ratio of 1:42 or even more. These conditions occur on straight slopes with an inclination of 5 to 10°. High wind speeds, however, have been pointed out on very steep slopes (Lancaster, 1985), but other authors consider that in zones of strong winds the inclination of slopes is low (Gaylord and Dawson, 1987). A clear disagreement exists, indicating that further research is needed.

Crests can have a straight form or occupy the highest point, separated from the brink (at the top of the leeward face) (Figure 18.8). These variations are explained considering first

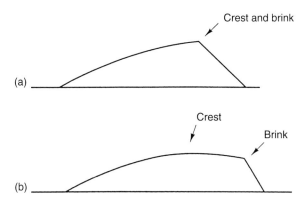

Figure 18.8. Dune crest (the highest point) and its brink (the top of the slip face). (a) Crest and brink are together. (b) They are separated (Livingstone and Warren, 1996).

Figure 18.9.   Bottleneck-shaped grain flows in the lee face of a dune, Ythan Estuary and Sands of Forvie, Aberdeenshire, Scotland. Photo: F. Gutiérrez.

small dome-shaped dunes, with a clear separation between crest and brink, becoming bigger dunes where the separation is much smaller, evolving finally to a straight slope with a crest (Lancaster, 1987). Inverse evolution has also been suggested (Capot-Rey, 1963). Another interpretation, indicated by Bagnold (1941) points out that the separation between crest and brink is due to variations in shear speed and sand transport speed. Finally, this author indicates that in dome-shaped dunes the crest can be mobile whereas the bottom remains static. These variations have been documented by Lancaster (1985), who pointed out speed relations between the bottom speed of the dune with regard to the crest of 158:1 and 13:1, for strong and gentle winds, respectively. These observations indicate that sand transport occurs only in the crest areas of the dunes during low wind speed periods, whereas during strong winds mobilization is present all over the dune. Crest height decreases during periods of low speed winds, and increases with high speed winds.

In *lee-side slopes*, wind speed and sand transport decrease quickly because of the flow expansion between crest and brink and their separation in the lee-side face. Flow is very irregular, gusty and with flow inversions, which can be detected with smoke. Wind speeds in the bottom of this slope are low and variable; in crescentic dunes they fluctuate between

Figure 18.10. Slid sand plates on the lee side slope, Beach of Trabucador, Ebro delta, Spain. Photo: F. Gutiérrez.

0.4 and 0.8 respective to the crest values in the desert of Namib, and up to 0.04 in the dunes of Padre Island, Texas (Sweet and Kocurek, 1990). In longitudinal dunes and in star dunes flow is complex, especially if the crest line is sinuous. Flow separation takes place on the crest, but when it is oblique to the crest, whirls appear on the leeward slope and sand is transported along the dune (Tsoar, 1983a). Sand grains can fall down from the crest or cross it by saltation. They are also affected by mass movements, from fracture places with scarps of 5 to 10 mm, flowing as avalanches, causing bottleneck morphologies (Figure 18.9). These avalanche tongues are a few centimetres deep, about 0.5 m wide, and move at speeds around 0.2 m/s (Hunter, 1985). With morning dew or in saline environments, increased superficial cohesion may allow mobilization of sand plates (Figure 18.10) (McKee, 1979a,b).

4. Classification of dunes

A large number of dune types have been differentiated, even though the main processes, previously described, are always present. Besides this, a large number of terms have been applied to the same form (Breed and Grow, 1979), giving a certain degree of confusion to the attempts for classifying dunes and their associations.

Existent classifications belong to one of the following two groups (Lancaster, 1995). First are the morphodynamic classifications that relate the dune type to the generating winds or the sediment supply. The second group, morphological classifications, only consider the external shape of the dune. A large number of morphodynamic classifications have been proposed (Aufrère, 1928; Clos-Arceduc, 1969; Hunter et al., 1983). These latter

Table 18.1. A classification of aeolian dunes (Livingstone and Warren, 1996).

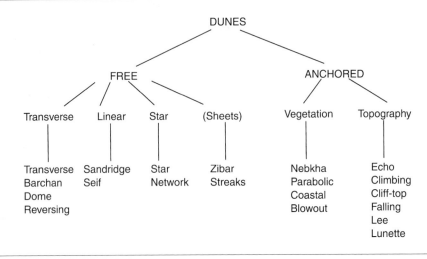

authors, relating dune courses and wind directions, distinguished between transverse, oblique, and longitudinal dunes. Research about dune dynamics demonstrates, however, that different parts of the same dune can be at once transverse, oblique, and longitudinal (Tsoar, 1983a; Lancaster, 1989b). Mainguet (1983, 1984) differentiated between erosive forms (parabolic dunes and sand ridges) and depositional morphologies (barchans, transverse chains, longitudinal dunes, and pyramidal dunes). The morphological classification elaborated by McKee and Bigarella (1979) is based upon the shape and number of faces. Taking into account this classification, Lancaster (1995) and Thomas (1997d) proposed other classifications of dunes. Moreover, yet another type of dune classification is based on practical finalities, as to their limitations for traffic.

We shall adopt, because of its simplicity, the classification proposed by Livingstone and Warren (1996), resulting from the modification of a previous one (Cooke et al., 1993), which is based on the form and the movement characteristics of the dunes, related in some cases to the presence of vegetation and topographic obstacles (Table 18.1).

4.1. Transverse dunes

The main flow direction is normal to the crest, during almost all the year. These dunes grow and keep an equilibrium in shape and size. They are longer in the direction perpendicular to the predominant wind and the windward slopes present less inclination than those on the the the lee side.

The word *barchan* has a Turkish origin, and is used for isolated, half-moon-shaped dunes, developed over a coherent substratum, like a pediment, a desert pavement, or a sebkha. They commonly occur on the margins of ergs and in sand transport corridors associated with depositional zones. As a general rule, they are small sized, and their height is one tenth of their width. The smaller ones are quickly mobilized, whereas those that reach several decameters high migrate very slowly. The stoss-side slope is clearly convex,

Figure 18.11. Small barchan in the foreground, and nebkhas in the background, Jaisalmer, Thar Desert, India.

with an average angle of 12°, whereas the lee-side slope presents an inclination of 33 to 34° (Pye and Tsoar, 1990). The horns point out in wind direction (Figure 18.11). Barchan advance speed is directly related to sand transport speed over the crest, and presents an inverse relation with the height of the crest (Bagnold, 1941). Speeds are very variable, and can reach up to 63 m/yr (see Thomas, 1997d). As sand supply increases, barchans join together laterally to form barchanoid ridges (Kocurek et al., 1992), also called megabarchans (Cooke et al., 1993; Lancaster, 1995; Thomas, 1997d). If the amount of sand increases even more barchanoid complex ridges are produced (Figure 18.3) (Thomas, 1997d), also called compound crescent dunes (Breed and Grow, 1979; Lancaster, 1989a). These dunes characteristically present a principal ridge, from 20 m up to 80 m over the interdune area, with a large number of barchanoids on the stoss-side slopes and in the crest zones. The separation between the bigger ridges ranges from 700 m up to 2000 m. Single barchans, and barchan associations occupy about 40% of sand deserts (Fryberger and Goudie, 1981).

The main characteristics of erosion and sedimentation in barchans are the presence of erosion in the stoss-side slope and sedimentation in the lee-side slope. Howard et al. (1978) measured, during a period of 2 weeks, the changes of height of the barchan surface. They essentially found maximum erosion in the centre of the stoss side, and sedimentation in almost all the lee-side areas (Figure 18.12).

Dome dunes are included in the group of transverse dunes, because they have a similar orientation and the same type of sand transport. Usually they are low, with a flat crest, without a slope more abrupt than the others, and with a circular or elliptical form. They are not common in most deserts, but they are very abundant in the Chinese desert of Taklimakan (Zhende, 1984). In contrast to other sandy areas where their size is very small, here they

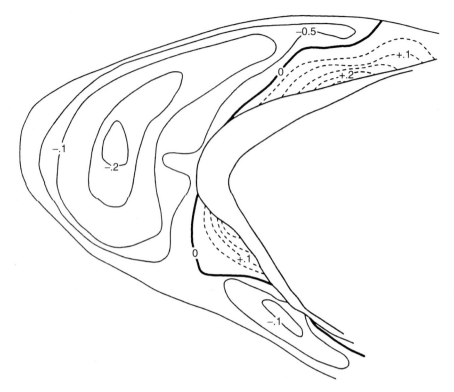

Figure 18.12. Changes in surface elevation of a barchan dune during a period of 2 weeks (after Howard et al., 1978).

reach 40 to 60 m high and 500 to 1000 m in diameter. Dome-shaped dunes are common in the margins of dune fields. Glennie (1970) pointed out that if wind speeds are low, oval hills appear and with high wind speeds longitudinal forms are built. Other authors indicate that the action of strong unidirectional winds is the cause of the origin of dome-shaped dunes, as it impedes crest forming (Breed and Grow, 1979; McKee and Bigarella, 1979).

Reversing dunes are included in the group of transverse dunes because sand transport is perpendicular to the crests. They are originated by totally opposite winds, corresponding to two different seasons. So, the reversing dunes of Oman (Cooke et al., 1993), are formed by the strong summer monsoons (the bigger ones), and by the opposite weaker winter winds (the smaller ones).

4.2. Linear dunes

These are the most common dunes and can have a great length, up to more than 200 km. They are quite straight-edged, parallel, with a regular separation between ridges. The transverse section is more symmetrical than in the transverse dunes. A large number of crests and lows can be observed in the longitudinal profile. These are typical for the great dunes and the middle-sized ones. The smaller ones, however, are quite straight. These dunes usually move in the direction of the dominant wind, but in some cases they

Figure 18.13. Seif "The Dragon," on the coast of Iquique, north of Chile. Photo: J. Rodríguez Vidal.

are mobilized in an oblique direction. Linear dunes are also called *seif*, *sif*, *silk*, etc., and the corridors between sand ridges are known as *gassi*, *goud*, etc. (see Cooke et al., 1993).

Among linear dunes it is possible to differentiate simple, composite, and complex varieties (Pye and Tsoar, 1990). The simple ones are narrow ridges (Figure 18.13) that can be longer than 200 km with a straight or wavy crest. The transverse profile is straight or rounded, with a regular interdune spacing and a high dune to interdune area ratio. They are widely developed in the Kalahari and Simpson deserts (Lancaster, 1995). In these regions they are from 2 m up to 35 m high, with widths of 150 to 250 m, an interdune spacing of 200 to 450 m, and they can reach up to 200 km long. Sometimes these dunes appear partly covered by vegetation, as in Nizzana, on the border between Israel and Egypt (Figure 18.14). A characteristic of these dune fields is the coalescence of crests forming a "Y" or in a diapason shape (Figure 18.16), which are common in areas where dunes are very close. Interdune zones can be covered by sand or can be desert pavements, partly covered by vegetation.

Composite linear dunes are formed by two or more dune ridges, very close or superimposed on the crest. They have been well studied in the Namib desert (Lancaster, 1983), where ridges are asymmetrical, from 25 m up to 40 m high, with a spacing of 1200 to 2000 m, and the interdune areas are covered by sand. Respective to the linear complex dunes, the big dunes of the Namib, 50 to 150 m high and with a spacing of 1 to 2 km, have a wavy main crest, and crescentic dunes on the flanks. The interdune areas are covered by sand with ripples subparallel or slightly oblique to the main crest.

Wind patterns in longitudinal dunes are variable. The most common are unimodal or bi-directional but, sometimes, several directions have been identified. They are often essentially associated with permanent winds from a single direction, with superimposed oblique winds.

Figure 18.14.   Simple linear dunes partly covered by vegetation, Experimental station of Nizzana, Negev Desert, Israel.

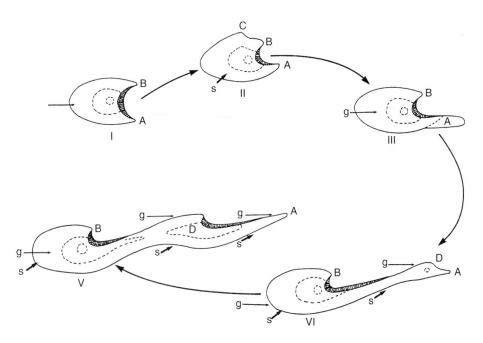

Figure 18.15.   A model of the development of linear dunes (seifs) from barchans; "s" refers to strong winds and "g" to gentle winds (Bagnold, 1941).

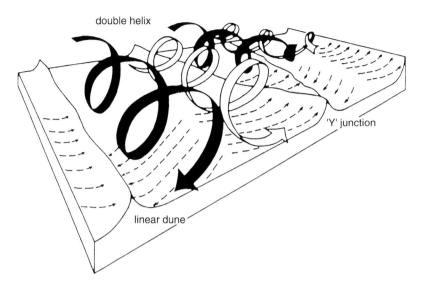

Figure 18.16. Bagnold's hypothesis for linear dune formation by induced roll-vortices (interpreted by Livingstone and Warren, 1996).

The origin and dynamics of linear dunes have been the aim of a large number of studies, reflected in the syntheses of Pye and Tsoar (1990), Cooke et al. (1993), and Livingstone and Warren (1996). Bagnold (1941) proposed that seifs are derived from barchans by the lengthening of one of the horns in an area with winds from two directions of different intensities (Figure 18.15). Others consider that lengthening is produced in the front of a parabolic dune (Verstappen, 1968). The hypothesis of the origin of linear dunes from helicoidal flows (Bagnold, 1953) is still considered at present (Figure 18.16), supported by field and wind-tunnel research (Tseo, 1993). Dunes develop in convergence zones and corridors occur between them. This model seems to explain also the spatial regular pattern.

4.3. Pyramidal and network dunes

The common characteristic to both types of pyramidal and network dunes is the occurrence of a certain number of slip faces developed as a consequence of the coexistence of different wind directions. Pyramidal dunes are also called star dunes, cones, sand mountains, ghourds, and other local names (Pye and Tsoar, 1990; Cooke et al., 1993). They are the biggest dunes present in ergs, and in Algeria they can be more than 400 m high (Wilson, 1973). As a consequence, they support a large amount of sand and they are placed in zones with a high sedimentation rate. Besides their large size and pyramidal morphology, they also present three or four radial arms (Figure 18.17). About 11 per cent of all dunes belong to the pyramidal type (Fryberger and Goudie, 1981) and in the Great Eastern Erg of Algeria they occupy 40% of the region (Breed and Grow, 1979). The arms coalesce on the top of the dune, where the slope inclination ranges between 15 and 30°, whereas near the bottom of the dune, slope inclinations are lower (5 to 10°) and it is possible to see small crescentic and reversing dunes there. Spacing between dunes varies, according to Lancaster's compilation (1995) between 150 and 6700 m, and dune widths

Figure 18.17.   Pyramidal dune, Erg Chebbi, Morocco. Photo: J.L. Peña.

between 180 and 6000 m. Generally there is a clear direct relationship between height of the dune and spacing in pyramidal dunes (Lancaster, 1989c). These dunes can grow several centimeters every year, and it seems that they migrate slowly, instead of keeping a compensatory balance (Nielson and Kocurek, 1987).

The wind studies done over pyramidal dunes in different deserts indicate that winds are multidirectional with variable energy (Lancaster, 1995), especially during the months when most of the sand transport takes place. As a consequence, net transport is usually low. The interactions between the seasonal wind changes concentrate sand sedimentation in the central parts of the dune. The main arms appear approximately transverse or slightly oblique to the directions of the local prevalent wind. Minor arms of the pyramidal dune are placed parallel and transverse to the bearing strike of secondary winds (Figure 18.18) (Lancaster, 1989c). Besides this, topographic barriers have exerted a clear influence in the existence of pyramidal dunes (Breed and Grow, 1979) because they modify the regional winds, increasing their variability or creating traps affecting sand transport.

Network dunes are developed under a continuous cover of sand, and are formed by dunes a few meters high. They are interlaced, alveolar, or rhombic-shaped networks of several hundred meters. They have been interpreted as groups of transverse dunes overlapped as a consequence of the different directions of the seasonal winds (Aufrère, 1935). Their complexity is variable, depending on the intensity and persistence of the seasonal winds.

4.4. Sand sheets and zibars

Sand sheets are accumulations of small dunes without slip faces that are developed on a flat, smooth terrain (Figure 18.19). *Sand strips* and *sand ribbons* are elongated form

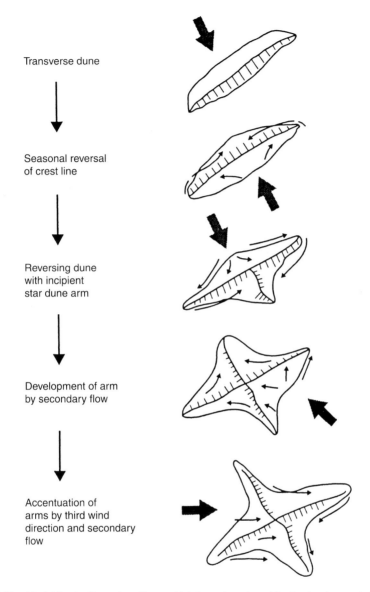

Transverse dune

Seasonal reversal
of crest line

Reversing dune
with incipient
star dune arm

Development of arm
by secondary flow

Accentuation of
arms by third wind
direction and secondary
flow

Figure 18.18. Model for the formation of pyramidal dunes (star dunes) by the development of secondary wind flows, while the dune is mobilized in a multidirectional wind regime Modified from Lancaster, 1989b.

varieties of the sand sheets (Breed and Grow, 1979). The surfaces can be irregular, undulated with or without *ripples*, and so forth. Many sand sheets are the basis or support of other mobile dunes. Globally, the extent of the sand sheets is more than one and a half million km^2 and, at the beginning of this chapter we have pointed out that the maximum surface known is $\sim 100,000$ km^2, in the borders between Egypt, Sudan, and Libya.

Figure 18.19. Wavy sand sheet, with some ripples. Matagusanos valley, Andean Precordillera, San Juan Province, Argentina.

The thickness of the deposits in sand sheets varies from a few centimetres up to 10 m (Breed et al., 1987). Grain size fluctuates between fine sand, sometimes with a significant content of silt, and poorly classified coarse sand (Pye and Tsoar, 1990). In Egypt we deal with sandy plains with residual deposits of fine gravel and pebbles that form the surface levels. Generally, with rare exceptions, sand sheets are poorly sorted deposits. Underlying levels are usually formed by subhorizontal layered sand beds (Lancaster, 1995).

Sand sheets are affected by several factors conditioning their formation (Kocurek and Nielson, 1986). Vegetation reduces the movement and dune growth, and consequently, accretion of horizontal sand laminas occurs. These circumstances of sparse vegetation are common on the margins of deserts. Where a coarse sand layer occurs, it is likely that the fine fraction has been exported by deflation. If there is a shallow phreatic level or periodic or seasonal floods occur, formation of dunes is made difficult or impeded. Finally, the presence of surface crusts or algal mats makes impossible the transport of sand in order to build dunes.

As happens in sand sheets, *zibars* are flattened areas with dunes without slip faces, with thin deposits. They form ripples and megaripples, sediment is coarse sand, and they are settled on a hard substrate. This explains that the term zibar, derived from the Arabic word zibara, means a hard surface passable by vehicle. They are common in areas where fine materials have been deflated. They can be found in different environments, from the centre of the Sahara Desert to the zones with scarce vegetation in southern California (Cooke et al., 1993).

4.5. Dunes anchored by vegetation

The areas that present dunes with low mobility are suitable places for plant growth, because water retention by sand occurs (Tsoar and Moller, 1986). This vegetation cover modifies the wind flow and favours aeolian sand trapping.

Dunes anchored by vegetation, also called phytogenetic dunes (Cooke et al., 1993), are originated by the stabilization of aeolian sand around the plants (*nebkhas* and coastal dunes) and by the erosion of surfaces covered by vegetation (*blowouts* and parabolic dunes). Even though they are almost fixed dunes, many of them can undergo changes in their shape and size, during seasonal or even longer cycles.

The most common type of dune anchored by vegetation are the *nebkhas*, also called shrub dunes, shadow dunes, hummock dunes, rebdou, and other terms (Pye and Tsoar, 1990; Cooke et al., 1993). They occupy enormous areas in the lowlands of the semiarid areas. They look like small ridges trapped by the plant, elongated in the sense of the sand transport, and located on the lee side (Figure 18.20 and Figure 18.11). The particles forming the nebkhas are sand, silt, and clay aggregates. The shape of a nebkha is a consequence of its size, density, and the growth of plants (Capot-Rey, 1957). Plants need to be at least 10 to 15 cm high for an effective trapping of sand. The size of the nebkhas varies considerably; they can reach up to 10 m high and 1 km long when they get trapped by a group of trees (Cooke et al., 1993). The dune reaches its maximum height near to the plant in the lee-side zone, because here the inverse flows are higher. Plant longevity is also important, because only long-living species can retain a large quantity of sand. On the contrary, many nebkhas are ephemeral, because they are placed on annual plants. In some cases perennial plants suffer changes in the phreatic level, and rain and sediment supply

Figure 18.20. Groups of nebkhas in the margins of the Thar Desert, Jaisalmer, India.

produce sudden modifications in the aeolian dynamics (Gile, 1975). Some nebkhas are affected by processes other than the aeolian ones, such as impact and splash by raindrops, runoff and animal activity. Therefore, there is a sum of processes, and it is difficult to evaluate the input of each one (Cooke et al., 1993).

Blowouts or deflation basins are morphologies produced by deflation, as closed depressions on unconsolidated aeolian sands, up to the limit of plant roots (Livingstone and Warren, 1996). Some authors use the term blowout in a restricted way, only for those generated in vegetated areas, whereas others such as Lancaster (1989c) extend the terminology to those shaped on bare dunes (Figure 18.21). Blowouts occur in continental dunes, essentially in the desert margins, and in vegetated coastal dunes. One of the most studied groups of blowouts is the Meijendel one, in the coastal dunes of Holland (Figure 18.22) (Jungerius, 1984). Blowouts elongate in the direction of the wind, and the major axis is 10 to 30 m long, exceeding 100 m in some cases. Vegetal cover is very important in the development of these forms. Therefore, loss of vegetation reduces the roughness, increasing, in consequence the scour surface and the export of material. This loss can be produced by overgrazing, rodent activity, and fire. Also the plants of the upper parts of the dunes are the first to become dry and so aeolian erosion takes place preferentially in these areas. Sometimes vegetation can be pulled out by hurricanes and unusual storms. Deflation deepens in the area without vegetation. Winds reach their maximum speed in the bottom of the depression, originating vertical vortices that spill the sand towards the edges of the hollow. In areas where rain precipitation is considerable, water erosion helps to widen the blowout (De Ploey, 1980). Exported sand is settled on the plants at the edges of the closed depression, and the aeolian erosion becomes maximum during the dry season, where the depression widens and it is able to migrate. On the other

Figure 18.21.   Blowout developed on the dune chains of Al Liwa, Desert of Rub al Khali, United Arab Emirates.

Figure 18.22. Blowout with sandy bottom, and vegetation of shrubs and trees. Small blowouts can be seen in the foreground, Meijendel, Coastal dunes of Holland.

hand, most blowouts have a short existence because they get quickly covered by vegetation, and so, half of the blowouts of Holland disappeared in about 9 years (Jungerius and Van der Meulen, 1989). Entrained sand takes the form of a fine layer with a plume shape and, when erosion is intense, a dune appears in the leeward zone, which advances over the existing vegetation. As a consequence, the area with floating sand is enlarged and a parabolic dune can be generated.

Parabolic dunes are U- or V-shaped, with the two arms partly covered by vegetation in the stoss-side area. They are localized in cold climates, coastal areas, and desert margins. The largest zone of parabolic dunes is found in the Thar desert of India, covering an area of about 100,000 km^2 (Verstappen, 1970) (Figure 18.23). Arms are 1 to 2 km long, and the front is 10 to 70 m high: they essentially advance by means of avalanches. Many of these dunes coalesce, because they have different movement speeds, giving rise to imbricated or multilobated parabolic dunes, with multiple crests and slip faces (Wasson et al., 1983). Few data are known about advance speeds of parabolic dunes. They fluctuate between 0.05 m/yr in zones of dense vegetation and 13 m/yr in some coastal regions. These dunes seem basically associated with the existence of a vegetal cover with a moderate development and with a unidirectional wind regime. The role of vegetation in the genesis of the parabolic dune is fundamental, because it protects the less mobile arms against aeolian action, allowing the advance of the central part. Parabolic dunes seem to derive from blowouts and they grow using the sand supplied from the underlying sediments. The sand supply can stop if a cohesive bedrock crops out or if the capillarity front, where sand is moist and unable to be entrained, is reached. In other cases the phreatic level can appear, originating swamps and

Figure 18.23.   Parabolic dune front slightly covered by vegetation, Tena-Jaisalmer, Thar Desert, India.

lakes. After this, the arms of the parabolic dune converge, decreasing their height and finally a sand tongue passes over the apex (Figure 18.24) (Pye, 1982).

Coastal dunes are the result of a combination of aeolian processes interacting with lithological, marine, fluvial, slope, edaphic, and cultural processes (Livingstone and Warren, 1996). The type and extent of these dunes depend on the quantity of available sand and the supply rate. This is a function of the existence of a sediment source and the strong littoral drift towards land. Their origin is related to rivers, beaches, coastal erosion, and estuarine sands.

Coastal dunes can be classified, in a simplified way into primary dunes, with their development directly related to coastal processes, and secondary dunes, where the sand has a littoral origin, but other processes also take place. These last dunes include aeolian sheets, blowouts, and parabolic dunes.

If sand supply to coastal dunes is poor, the sand spreads inland forming a thin layer. On the other hand, if sand supply is abundant, ridges parallel to the coast line appear (Figure 17.2), that move slowly towards inland, and they can be more than 100 m high, keeping their slip faces straight (Goldsmith, 1985). Where sand supply is very abundant, vegetation can be buried, and dunes very similar to the desert ones appear. These dunes are known as transgressive, and are usually of a transverse type. If vegetation is very high, a continuous ridge is developed or, sometimes, a system of parallel ridges in a prograding coast appears. These ridges can grow at high speeds. Where vegetation is sparse as in the arid coasts, ridges are discontinuous and are formed by sand mounds lined up with vegetation.

The evolution of coastal dune ridges can originate very complex morphologies, with blowouts, transverse dunes and parabolic dunes, by successive building of parallel ridges, which development can last for hundreds of years.

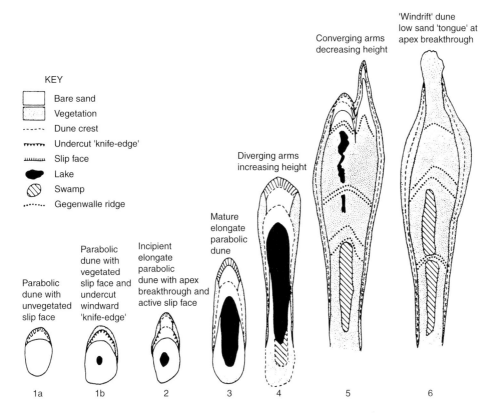

KEY

☐ Bare sand

▦ Vegetation

------ Dune crest

⌢⌢⌢ Undercut 'knife-edge'

⌣⌣⌣ Slip face

● Lake

◉ Swamp

········ Gegenwalle ridge

'Windrift' dune low sand 'tongue' at apex breakthrough

Converging arms decreasing height

Diverging arms increasing height

Mature elongate parabolic dune

Parabolic dune with vegetated slip face and undercut windward 'knife-edge'

Incipient elongate parabolic dune with apex breakthrough and active slip face

Parabolic dune with unvegetated slip face

1a 1b 2 3 4 5 6

Figure 18.24. Stages in the growth and eventual dissipation of elongate parabolic dunes in the Cape Flattery area, Australia (Pye, 1982).

4.6. Topographically anchored dunes

When moving aeolian sands encounter any topographical obstruction, like hills, cliffs, blocks, or buildings, accelerations and expansions of the wind flow around the obstacle occur. These circumstances provoke erosion or sand accumulation, and static dunes occur.

Experiments carried out in wind tunnels (Tsoar, 1983b) have provided a large number of data about the relationship between topographic obstacles and the generation of dunes. If the wind blows perpendicularly into a slope presenting more than 60° of inclination, *echo dunes* appear, separated from the obstacle a distance equivalent to three times the height of the obstacle. At the bottom of the cliff an inverse flow appears, sweeping the sand and generating a corridor between the obstacle and the sand ridge. The small initial echo dune increases its height by this mechanism, until it reaches an equilibrium condition, corresponding to 0.3 to 0.4 times the height of the obstacle. If the cliff is sinuous, the dune shape changes along its length.

Where the slope of the obstacle presents an inclination of less than 30°, sand is transported over it, but with higher inclinations sand is deposited and *climbing dunes* are

Figure 18.25. Climbing siefs, Aswan, Egypt. Photo: J.L. Peña.

formed (Figure 18.25). They are very common on desert hills and cliffs. The sands that reach the crest of the cliff settle down because they enter a zone of reduced speed. These aeolian accumulations are called *cliff-top dunes*. Leeward of wide obstacles the wind calms down and *falling dunes* are deposited. When the obstacle is narrow, *lee dunes* appear, sand ridges are commonly straight, which because they are sheltered by the obstacle, can extend for long distances. In the Draa of Malichigdane in Mauritania they get up to more than 100 km long (Breed and Grow, 1979). Several dunes present around rounded craters of Mars have been interpreted as lee dunes (Greeley and Iversen, 1985). *Lunettes*, half-moon-shaped dunes, have been described in the previous chapter.

5. Desert dust

During the last two decades of the 20th century the study of aeolian dust deposits as paleoenvironmental records has steadily increased. They include terrestrial sequences, deep-sea sediments, and cores from the polar ice caps. Research provides information about changes on the extent and intensity of aridity, glacial periods and localization of glaciers, and changes in the global atmospheric circulation (Pye, 1995).

The entrainment, transport, and sedimentation of aeolian dust are very important for the geomorphologist, but also present a relevant interest for meteorologists, climatologists, ecologists, and environmental scientists. Besides the geomorphic implications that open a whole range of research possibilities (loess, varnish, crusts, erosion, and so forth), there is also the interest of applied studies, which are more and more important because of the increasing number of inhabitants in arid zones (Péwé, 1981; Middleton, 1997).

5.1. General characteristics

Desert dust consists of small particles carried away by the wind of arid environments in suspension and deposited afterwards. Atmospheric dust can come from different environments than those of arid zones (cosmic dust, volcanic dust, marine salts, etc.) and the source areas can be distinguished by their composition and size.

In western Europe desert dust is a familiar feature, because almost every 7 years or so occur the so called "red rains" of aeolian dust originating from the Sahara desert, that can be observed as a thin film on top of cars. This dust settlement rate is on average 0.25 kg/m^2/yr (Goosens and Offer, 1990). Globally it is possible to estimate that dust carried out from deserts is about 1800 to 2000 million tons per year (D'Almeida, 1989). A dust storm is recognized when visibility is less than 1000 m. In Egypt some ten storms are observed every year, about 30 in China, and in the Mexican capital and in Kazakhstan an average of 60 storms occur every year (Goudie, 1978).

According to grain size, there are two main types of desert dust that reflect the distance they have travelled. If dust has been transported less than 100 km, sizes are between 0.005 and 0.05 mm; if transport distances are longer, the diameter is smaller than 0.002 mm (Péwé, 1981). The largest particles are transported by wind whirls and dust storms and produce extensive loess deposits, whereas the smaller ones move like an aerosol in the troposphere, and remain suspended until they are entrained by the rain and deposited.

Desert dust is composed of inorganic and organic materials. The most abundant mineral is quartz, accompanied by feldspars, calcite, dolomite, mica, clay minerals, oxi–hydroxides, heavy minerals, gypsum, halite, and opal. The organic part is composed of diatoms, phytoliths, spores, and pollen. As we could expect, composition changes according to the source area and the distance.

5.2. Source area and generating processes

The main currently active areas providing aeolian dust are the subtropical deserts, which form a wide belt from western Africa to Central Asia, and the arid and subhumid regions where dry soils are exposed to strong winds during several periods through the year (Péwé, 1981; Middleton et al., 1986)(Figure 18.26).

The areas covered with loose sediments that contain significant quantities of sand and silt, but are poor in clay, are the most favourable for generating aeolian dust. One of these environments is the poorly-sorted and unvegetated sediments of the *ouads*. They are usually braided channels with a great lateral migration.

The playas, chotts, and littoral sebkhas, although they vary enormously in their composition, contain zones characterized by the presence of fine grained material, which are another source of aeolian dust. Alluvial fans are supply zones, especially their middle and distal areas, where the percentage of fine particles is important. Ancient dry lakes also can supply diatom dust, salt, gypsum, and clay minerals. The large regions covered with loessic sediments are important dust source areas, particularly where vegetation is nonexistent. An important source area of dust comes from volcanoes, as with the emissions from Mount St. Helens in USA in 1980, and Pinatubo in the Philippines in 1991, where large quantities of dust were emitted into the atmosphere, and transported afterwards (Pye, 1987; Cooke et al., 1993; Livingstone and Warren, 1996). Mechanisms

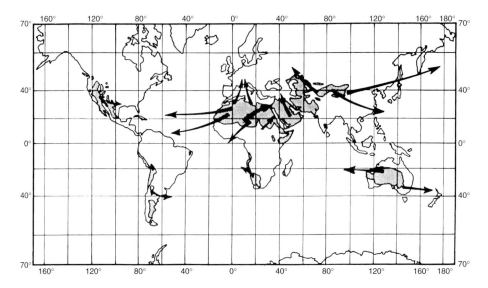

Figure 18.26. Major deserts, directions, and distances of dust transport (after Péwé, 1981).

that form fine particles are quite varied. Clay-sized particles are produced essentially by chemical weathering, but the origin of silt-size particles, that basically form the loess, is controversial. Weathering by frost can play a fundamental role in the generation of fine particles in cold deserts. Processes of thermoclastism, wetting and drying, besides the haloclastism or salt weathering, are the basic mechanisms of fine-grained material production in deserts. Besides these, we can add aeolian abrasion. In beach environments agglomerates are produced from silt and clay particles that are exported by the wind. Aeolian dust can incorporate particles from chemically-weathered profiles, developed in tropical humid environments. Finally, another source is of organic origin; phytoliths, diatoms, radiolarians, spicules of echinoderms, pollen, spores, and so forth.

5.3. Entrainment, transport, and sedimentation

The entrainment of dust particles from the soil surface is a function of wind speed, and it is controlled by the type of wind, sediment characteristics, and the presence of obstacles (vegetation cover, topography, and so forth) (Middleton, 1997). Strong winds able to mobilise and transport dust particles occur under very different conditions, and at different scales (Pye, 1987). These dust masses are identified and studied using satellite images and the classical meteorological observations.

Wind types transporting dust vary from simple gusts to dust storms with advance fronts up to 1000 km long (Figure 18.27). *Dust devils* are small convective vortices that appear when an intense warm up of the ground surface has taken place, causing the movement of air and dust upwards. These vortices are visible from a long distance and present an inverted cone shape. They commonly reach heights between 3 and 100 m, although they can exceed 1000 m (Idso, 1974). Diameter on the ground is 0.5 to 3 m. They move laterally and can cause significant damage.

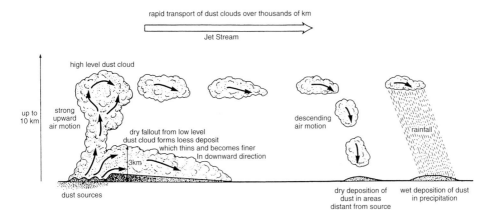

Figure 18.27. Schematic representation of modes of aeolian dust transport and deposition (Pye, 1995).

Dust storms, also called *haboob* (an Arabic word that means violent wind) are probably the result of descending air currents from large cumulonimbus clouds. They look like a dust wall of 300 m up to 3000 m high, which can advance at speeds up to 200 m/s. As we can see on satellite images they measure 500 to 600 km wide and extend up to 2500 km with a plume shape (Péwé, 1981).

Other winds that also transport very fine particles are the *dust plumes* that move parallel to the soil surface as a horizontal vortex. The *shamal* is a wind that blows towards the north during summer in Iraq and the Iranian Gulf. Most of these winds loaded with dust do not reach 1 km in height and their ground speed is less than 40 km/h. Another transport of aeolian dust is the *katabatic wind*, that sweeps the sediments from the foothills of large mountain ranges, such as with the *zonda* wind in Argentina and the *Santa Ana*, in California. Similar winds occur in the Andean altiplano and on the plateau of Tibet (Pye, 1987; Livingstone and Warren, 1996; Middleton, 1997).

In semi-arid zones of higher latitudes, the passing of cold fronts is probably one of the most important causes in the mobilization of aeolian dust. Huge walls of dust appear, as in the haboobs, up to 3 km high, and move at a maximum speed of 40 m/s. They have been described in the north of China, Israel, Egypt, in the High Plains from Texas and New Mexico, and in the south of Australia (Pye, 1987).

Sometimes dust reaches considerable heights, and it is transported at a great speed by the jet streams, covering distances of thousands of kilometers (Figure 18.26). Thus Saharan dust travels to central and northern Europe, and crosses the Atlantic Ocean, reaching the Caribbean Islands (Prospero et al., 1970).

Sedimentation of aeolian dust takes place in different ways: (1) wind speed and turbulence reduction; (2) aggregation of particles with electrostatic charges; (3) dragging of dust suspended in the atmosphere by rain (Pye, 1987). Although sedimentation also takes place in oceans and wet areas, we shall deal essentially with dust deposited in arid zones. The quantities deposited are variable, although the average is about 100 t/km^2/yr near the source areas (Goudie, 1995). This amount decreases quickly with the distance from the area of origin. Desert dust with a coarser size travels within the lower parts of the atmosphere, and often is deposited when it encounters obstacles perpendicular to the wind flow. Nevertheless the finest particles that travel in the higher parts of the atmosphere are not affected by the roughness of the terrain, and can remain suspended for long periods of time (Coudé-Gaussen and Rognon, 1988; Coudé-Gaussen, 1991).

In hyperarid zones only small quantities of dust occur, because precipitation is rare and, generally, obstacles are not common. On the margin of deserts, however, corresponding to semiarid zones, there are great quantities of aeolian dust deposited. These particles are easily trapped by the disperse vegetation. In the north of Nigeria, up to 3 m of desert dust have been deposited during the last 40,000 years (McTainsh, 1987) and, at present, sedimentation rates vary between 0.1 and 0.7 mm/yr in different zones of Israel and the Sahel.

5.4. Geomorphological implications

Desert dust has a considerable importance to the better understanding of certain geomorphic processes that have strong implications for human and economic activities. Some superficial formations appear because dust deposited in the vicinity contributes to their genesis in semi-arid and arid areas. So, the supply of calcium carbonate to the soils produces caliches in areas without calcareous rocks, as we can interpret from the studies on calcretes in Namibia and southeast of Spain (Blümel, 1982). Also dust rich in silica can be important in the formation of silcretes in deserts (Summerfield, 1983). Likewise, sedimentation of gypsiferous dust is the main cause of the formation of gypsum crusts in southern Tunisia (Coque, 1955). Desert varnish, that forms a thin film rich in manganese and iron, covers rocks in arid zones. As was pointed out in a previous chapter, particles that form the varnish are provided by the sedimentation of aeolian dust. Laminations that constitute the varnish are analysed carefully and paleoclimatical interpretations are carried out, according to the fluctuations in manganese and iron contents (Oberlander, 1994). Desert dust can also affect the weathering developed on buildings. In dune zones, sedimentation of dust and salts can form a crust that is enough to stabilise the dunes, allowing the establishment of biocrusts and plants (Pye and Tsoar, 1987). Aeolian dust is also collected in playas, sebkhas, alluvial fans, and ouads causing modifications of their sediment composition. These sediments are quickly mobilized by water within their depositional environments. Many soils from desertic and peridesertic regions contain significant quantities of aeolian dust, and some soils are virtually formed by these allochtonous supplies. According to Rapp (1984), the terra rossa from Spain, Italy, and other Mediterranean countries are formed essentially by reworked dust coming from the north of Africa. This desert dust fertilizes many soils, affecting even some of those developed in the Caribbean Islands.

Desert dust deposited in oceans and seas constitutes a significant percentage of oceanic sediments. The study of these accumulations provides very useful information on modifications of wind systems, and about environmental changes that have taken place on the continents during the Quaternary period (Middleton, 1997). Therefore, this research provides information on the arid phases and their intensity in terrestrial environments. Ice accumulations, especially in the polar ice caps, preserve a record of aeolian dust that is known because test drillings have been done. In Greenland, during the Last Glacial Maximum (18,000 years ago), dust concentration was 40 times greater than now. Phases with high dust content are followed by others with small or nonexistent quantities of dust, indicating the alternation of periods with high and low aridity.

5.5. Loess

Loess, a term derived from the German *löss*, is a clastic terrestrial sediment, composed essentially of particles of a silt size and formed by the accumulation of aeolian dust (Pye, 1995). It presents a variety of colours; gray, white, yellow, brown, and red. It is possible to differentiate between primary loess, with an aeolian origin, and secondary loess, redeposited or originated by nonaeolian processes. In loess outcrops, a certain degree of chemical and biological weathering occurs, and also edaphogenic processes.

Loess accumulations (Figure 18.28) cover about 10% of the Earth's surface above sea level (Pécsi, 1968), being mainly developed in Eurasia, USA, and Argentina (Figure 18.2). In China loesses occupy more than 1 million km^2. The layer of loess has an irregular thickness and covers the irregularities of the terrain. Thickness is usually less than 30 m, although in Lanzhou, China, it exceeds 300 m (Derbyshire, 1983b).

Figure 18.28. Loess accumulations in the region between Jaipur and Agra, India.

Loess mineralogy is very variable, reflecting the composition and evolution of the source area. Usually quartz predominates (45 to 55%) accompanied by feldspars, carbonates, heavy minerals, volcanic flint, and clay minerals. Loess has more than 50% of silt-sized particles, but the contents of clay and sand are very variable. Where loess contains more than 20% of sand, it is called sandy loess, and if the clay content exceeds 20%, clayey loess (Pye, 1987).

Loess has a mechanical behaviour with a tendency to collapse. It supports vertical loads when it is dry, but its shear strength decreases considerably when wet, and the underlying material flows and slides (Middleton, 1997). Where loess is dry it presents vertical fracture systems in scarped outcrops.

The formation of a loess deposit can be simple or quite complex. Loess can be generated by weathering and further deflation. Usually there is an intermediate phase of fluvial transport, previous to the aeolian action. Initial particles can also be the result of glacial weathering and/or erosion, or can be transported and deposited by fluvioglacial action, and finally mobilized by deflation (Smalley, 1972).

There are many processes producing particles of a silt size; particles from bed rock; glacial crushing; gelivation; fluvial abrasion; aeolian abrasion; haloclastism; chemical weathering; clay-pellet aggregation, and biological processes (see Pye, 1995). According to the generating processes some authors denote the extensive loess deposits related to Quaternary continental glaciations, periglacial, or glacial loess. Others refer to "peri-desert" loess in connection to those on the margins of arid zones.

In the deserts of central Asia particles are weathered in high mountain environments and later are transported and deposited by the rivers in lower zones, where they are entrained by deflation. In those areas with active tectonics, large quantities of sediments, including silts, were produced during the Quaternary era (Smalley, 1990). Due to the environment where they were generated, they are called "peri-mountain" loess. In deserts located on shields and platforms, extensive deposits of loess do not occur, because those areas have a low relief, are tectonically stable and, as a consequence, have low sediment production, in contrast to the deserts of central Asia (Smalley, 1995).

An accumulation of thick and extensive loess deposits needs an important supply of dust during tens or even hundreds of thousands of years in order to be produced, and the existence of topographical and vegetation barriers to stop the dust. In the deserts of low relief the production of sediments is poor (peri-desert loess) whereas in deserts of mountains with neotectonic activity sediments are generated at a very high rate (peri-mountain loess).

SIXTH PART

Applied geomorphology and arid regions

Chapter 19

Applied geomorphology and arid regions

1. Introduction

The applied geomorphology of desert areas is not very different from that of other morphoclimatic areas, although it does have some peculiarities. The lack of water, scant or zero plant cover and much eroded soils leave the regolith with no efficient protection against the action of atmospheric agents (Tricart, 1969). The most important processes in arid areas are wetting and drying, salt weathering, water erosion, channel changes, flooding, and wind erosion.

Practical studies on arid regions are important because of the extraordinary extension of these environments; the world's two largest morphogenetic systems are those made up by the arid and cold regions. It is difficult, however, to generalize about the effects of geomorphology in these areas because of their great climatic, geological and tectonic differences. Further, these regions are usually underdeveloped and scantily populated, and data on even their basic features, such as their climatic and hydrological characteristics, are few and geographically disperse. In most cases, the only data available come from a few rare and rather specific studies (Gutiérrez and Sancho, 1993).

In 1979, the population of the world's arid areas – 651 million – was 15 per cent of the planetary total. Between 1960 and 1974, however, an increase of 63.5 per cent was experienced. Some 1200 million people inhabited these regions by the year 2000. This population is distributed mainly as large concentrations; there are 355 cities over the world's arid regions with more than 100,000 inhabitants, although most of them are in Asia. The geomorphological and environmental problems caused by the expansion of these cities are numerous (Cooke et al., 1982; Goudie, 1990a,b; O'Hara, 1997).

Many of the landforms of these dry regions were generated under climatic conditions different to those of the present day. A detailed study is therefore needed to differentiate between those shaped by modern processes and those that are relicts of earlier times (Gutiérrez, 1986). It is also important to decide on the basis and goals of any practical geomorphological study to be performed in a desert region. This has been clearly stated by Cooke (1977):

> Applied geomorphology in deserts is primarily concerned with the survey and evaluation of landforms, superficial materials and processes, and with managing, monitoring and predicting landform and process changes. It relates to practical problems in which geomorphological information normally only forms a part of the required body of knowledge, and its philosophical basis (in so far as it exists independently of "pure" geomorphology) is inevitably dictated by the demands of specific problems. Axiomatically, the applied desert geomorphologist must be able to talk to and work with those in other cognate disciplines, and

he (or she) must be able to communicate ideas comprehensibly to others who may have no geomorphological knowledge.

2. Weathering

The weathering of desert areas is mainly driven by strong variations in temperature and humidity. Fluctuations in temperature give rise to insolation weathering, and at high altitude to freeze–thaw phenomena. Variations in humidity encourage wetting and drying processes as well as haloclasty. The lack of water impedes chemical modification, and biotic weathering is mainly limited to the physical and chemical degradation caused by algae and lichens. Weathering in these areas is therefore rather peculiar and gives rise to a set of micro-landforms such as honeycomb, tafoni, caverns and gnammas, which can be very abundant in medium-grained rocks (Mabbutt, 1977).

2.1. Weathering by wetting and drying cycles and salt weathering

Weathering by *wetting and drying* cycles leads to the disintegration of rocks by flaking and cracking (Ollier, 1984). It is an important agent of erosion in desert areas, causing considerable breakage of rocks (Ollier, 1977). The water circulating in the pores of rocks carries very soluble salts, especially chlorides and sulphates, and when these precipitate because of evaporation they can have an even greater disruptive influence than freeze–thaw. This process, known as *haloclasty*, acts alongside wetting and drying. Of all the weathering processes at work in desert areas, haloclasty, which involves the growth of crystals, volumetric expansion, and salt hydration (Cooke, 1986; Cooke et al., 1993), is the most important (Doornkamp and Ibrahim, 1990). It can disintegrate rocks into particles the size of silt. Evidence of this is mainly seen along the edges of salt lakes, in ephemeral channels, and where the area of capillarity intersects the soil surface. The builders of ancient civilizations knew the disruptive effects of salts on rock very well (Ollier, 1977). The Great Pyramid of Egypt (Fig. 19.1), for example, was originally covered in a protective layer of resistant limestone slabs. Unfortunately, about 1000 years ago part of this covering was destroyed and the weathering of the pyramid's core began (Emery, 1960). Although natural materials are still used in construction today, many structures are made of concrete, but the weathering behaviour of this material can be studied in the same way.

The effectiveness of haloclasty (Cooke et al., 1982) depends on environmental conditions such as those produced by daily fluctuations in temperature and humidity (which give rise to the crystallization or hydration of salts), and on fluctuations of the capillary front. Equally important are the properties of the materials affected, especially those related to porosity and pore distribution. Finally, different salts have different effects: the most aggressive are sodium and magnesium sulphate (Fig. 11.7) (Goudie et al., 1970; Goudie, 1974 and Cooke, 1979).

The most common and most serious problems related to engineering structures (Cooke, 1986; Goudie, 1994a; Goudie and Viles, 1997; Goudie, 1998) concern fluctuations in the capillary front. These can affect the foundations of buildings, roads, canals, oil pipelines

Figure 19.1. The upper part of the Great Pyramid of Egypt is covered by unweathered limestone slabs. The underlying stone, however, shows marked weathering. Photo: J.L. Peña.

and other structures. Strict control over the materials used is therefore vital, a good idea being to use only those with low salt content and which are highly resistant to salt weathering. Laboratory analysis of their durability is recommended. Some buildings clearly show the height reached by the capillary front. In the arid areas of Bahrain, Dubai and Egypt, this is never higher than 4 m, and usually between 2 and 3 m (Cooke et al., 1982). In the semi-arid area of Alcañiz, in the Ebro Basin of Spain, however, heights of 5 m have been reached (Fig. 19.2). The chlorides and sulphates of sodium, calcium and magnesium are the salts that most affect concrete, in which they can cause expansion and disintegration. The use of sulphate-resistant cements is recommended (Mehta, 1983).

In the construction of roads, areas made wet by capillary rise (Fig. 19.3) should be avoided. In such places, salt weathering causes cracking, hollowing, crumbling and disintegration (Cooke et al., 1982). A way of assessing the risk of haloclasty is to measure the electrical conductivity of water in the vadose zone and to make salinity maps (Fig. 19.4).

2.2. Dissolution

Sweeting (1972) indicated that karst landforms in carbonate formations do not develop when rainfall is below 250–300 mm. Therefore, the karstification seen in the calcareous rocks of arid areas can only have been produced during times that were wetter than the present. Chlorated and sulphated materials, however, are more soluble than limestone and are susceptible to karstification. According to Jakcus (1977), gypsum and halite are, respectively, 183 and 25,000 times more soluble than calcite in distilled water at 20°C. These figures show the intensity and velocity of karst development in salt and gypsum.

Figure 19.2. An undulating capillary front in the sandstone slabs of the Calatravas Castle (18th century), reaching 5 m above ground level. The most important weathering is seen in the lower slabs. Alcañiz, Province of Teruel, Ebro Basin, Spain.

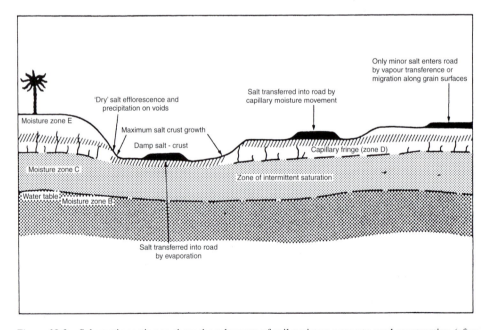

Figure 19.3. Schematic section to show the relevance of soil moisture zones to road construction (after Fookes and French, 1977).

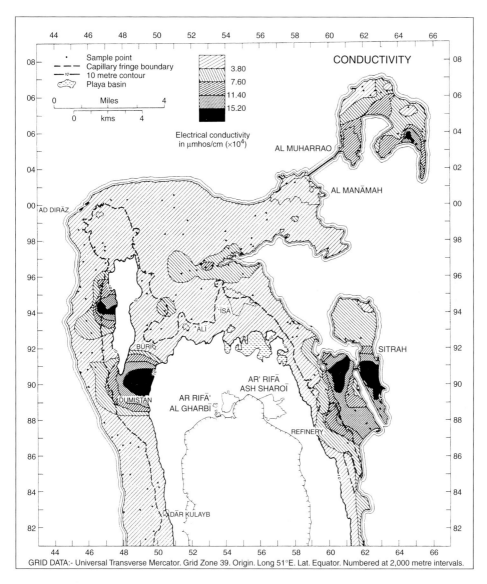

Figure 19.4. Aggressive ground conditions in northern Bahrain: the capillary fringe limit, and depth to groundwater beneath the surface (after Power and Water, Unpub. data; in Cooke et al., 1982).

The outcrops of these evaporites around the world occupy a surface area of 60 million km^2 (Ford and Williams, 1989); although karst is found mainly in the northern hemisphere, it appears in all kinds of climates (Klimchouk et al., 1996).

The high solubility of the gypsum in the Tertiary formations of the Ebro Basin, where conditions are basically semi-arid, has given rise to extensive karstification and a series of associated dangers (Gutiérrez and Gutiérrez, 1998). Dissolution is much more marked where the gypsum is covered by alluvial deposits (glacis, terraces, infilled valleys).

This gives rise to the risk of subsidence, which can have important socio-economic consequences (Gutiérrez and Peña, 1994). The alluvial deposits overlaying the evaporitic formations, which can be over 100 m thick, have been explained by sedimentary karstic subsidence (see Gutiérrez and Gutiérrez, 1998). The alluvial cover subsides because of ductile flexure or fragile collapse.

Around Zaragoza, Spain, subsidence affects roads (Figs 19.5 and 19.6) and railways (Benito and Gutiérrez, 1988; Pérez del Campo, 1989; Benito and Pérez del Campo, 1991; Benito et al., 1995), houses, warehouses and even the water supply to the west of the city (Casetas-Utebo) (Zuidam, 1976a,b; Gutiérrez et al., 1985; Soriano et al., 1994; Soriano and Simón, 1995; Gutiérrez and Gutiérrez, 1998). The most spectacular case, however, is that of the new town of Puilatos (about 30 km north of Zaragoza), whose construction had to be abandoned. What had been built had to be destroyed because of the numerous cracks caused by subsidence through karstification (Benito and Gutiérrez, 1988).

The area's lined irrigation network has also suffered the effects of subsidence. In 1954, the *La Violada* concrete canal gave way because of a collapsed doline 25 m in diameter (Llamas, 1962; Riba and Llamas, 1962). Non-lined canals, such as the *Canal Imperial de Aragón*, supplied water to the underlying gypsum, starting a karstification process and the generation of dolines. A relationship exists between the density of dolines and their proximity to irrigation ditches in the area of Villamayor in Zaragoza. The highest concentration of dolines is close to the irrigation ditches, especially at their confluence (Fig. 19.7) (Benito and Gutiérrez, 1988), showing that water leakage from the ditches is the main cause of dissolution in the area and therefore of the ensuing subsidence.

Figure 19.5. Flexure between the Pikolín factory and the Zaragoza-Aragón highway in Spain caused by karstic subsidence of the underlying gypsum (June, 1996). Photo: F. Gutiérrez.

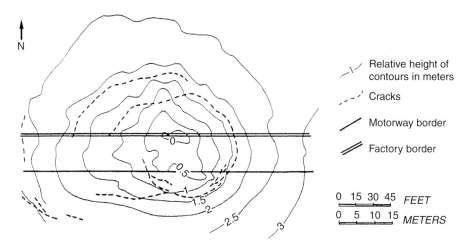

Figure 19.6. Karstic subsidence affecting the Zaragoza-Alagón highway and a factory. The area has been filled with 3 m of asphalt agglomerate over the highway's 25-year existence. Current subsidence is about 25 cm/yr and is affected by traffic and the interannual variability in rainfall (according to González, Unpub. data; in Benito and Gutiérrez, 1988).

The irrigation of crops can play a role in doline formation through the pumping out of water (Benito, 1987). Finally, some dolines are used as rubbish dumps, which contaminates the groundwater.

The semi-arid area around Calatayud in the Province of Zaragoza, which lies in a tectonic *graben* filled with Neogene and Quaternary sediments, experiences very similar problems to those of the area around Zaragoza. The Quaternary alluvial vary greatly in their thickness (they can be 100 m thick, e.g. at the confluence of the Jalón and Jiloca rivers) because of karstic subsidence. This process is very active on the alluvial plains and affects drainage canals, communications, irrigation ditches and the buildings of Calatayud. The conservation of buildings of cultural interest in the city is faced with serious problems (Fig. 19.8) (Gutiérrez, 1995, 1996).

Other environmental problems of gypsum terrains include the poor quality of water for consumption, which can be polluted by fertilizer, oil products, and even radioactive materials (such as those deposited by the Chernobyl disaster). Gypsum mining can also cause collapses and flexures, and the removal of groundwater can intensify karstification (Klimchouk and Andrejchuk, 1996).

3. Changes in volume

Soil volume changes can be caused by high sodium concentrations in the cationic exchange complex of clays, and by the presence of shrink–swell clays. High sodium concentrations cause clay dispersion and tensional stress to manifest themselves at the surface as a system of cracks with domal blocks. In the Miocene shales of the Ebro Basin

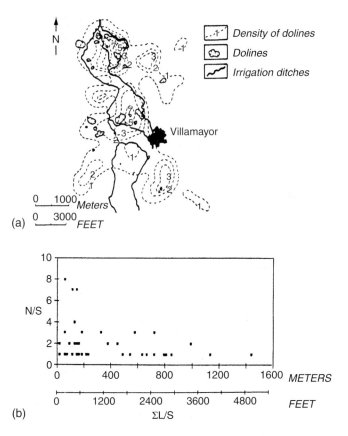

Figure 19.7. (A) Relationship between doline density and the network of non-lined irrigation ditches. (B): Relationship between doline density and distance from the nearest irrigation ditch. N = number of dolines. S = unit of surface area, ΣL = sum of the distances of the dolines per unit area with respect to the nearest irrigation ditch (Benito and Gutiérrez, 1988).

in Spain, where shrink–swell clays are absent, swell values of around 12 per cent have been recorded (Gutiérrez et al., 1995).

More important is the expansion caused by wetting when the soil contains significant percentages of shrink–swell or expansive clays (smectites, vermiculite and halloysite). In semi-arid areas, this process is much more common in the lowest lying plains (Cooke et al., 1993). Variations in the water content of the soil cause its expansion and contraction, and pressures of up to 10 kp/cm^2 can be generated (Salinas, 1988). This swelling and shrinking leads to gilgai soils that are characterized by a gently rolling landscape. In this micro-relief, only a small percentage of the land is suitable for cultivation. Most fertile land has to be used for grazing because, if machinery is used to level the land, the characteristic micro-relief returns within 3 years. Even stakes driven into this type of soil are pushed out by movements generated by vertical forces (Harris, 1968).

In the USA, more than 250,000 houses are built every year on expansive soils – and about 10 per cent of them suffer significant damage (Jones and Holtz, 1973). In Spain,

Figure 19.8. Cracks caused by karstic subsidence in the Colegiata de Santa María la Mayor, a mudejar church of the 12th century. Calatayud, Province of Zaragoza, Spain.

about 32 per cent of all geological formations contain expansive clays, and about 67 per cent of the country experiences the type of climate which can lead to swelling (by about 10%) and shrinkage (Salinas, 1988). An increase in the soil volume of about 3 per cent can cause damage, so special foundations are required. If these are to be made properly, it is important to know the depth to which such expansive behaviour occurs. This zone given to variation in volume is known to geotechnologists as the "active layer." According to Salinas (1988), volume changes cause the deterioration of earth cuttings (slopes and road cuttings), the breakage of pipes, the destruction of walls, the deformation of paved surfaces, the breakage of ditch bottoms and pavements, the distortion of foundations, and cracks in buildings (Fig. 19.9). Some damage can be made worse by the breakage of drains.

Another manifestation of this activity is the production of giant desiccation fissures resulting from the shrinkage of *playa* materials belonging to ancient desert lakes. These

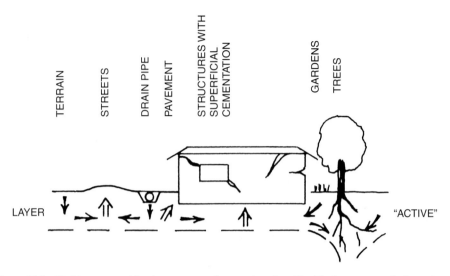

Figure 19.9. Problems caused by the presence of expansive clay. The black arrows mark the moisture transfer lines. The open arrows indicate possible damage (Salinas, 1988).

are dangerous and can form suddenly, causing serious structural damage to linear constructions and introducing trafficability problems (Cooke et al., 1993).

4. Fluvial systems

Despite the low rainfall of arid regions, they are significantly shaped by drainage processes (Tricart, 1969). Rainfall is often local and short-lasting, as are the ensuing runoff and drainage processes. Areas with scant vegetation and gentle topography can cause unconfined flow. During times of flood, sediment transport is discontinuous and channel morphology can change rapidly from the meandering to braided type.

Within a drainage system, surface runoff and sediment production are related to rainfall and temperature. These factors also influence the development of plant cover, and a set of complex relationships between these variables can be established. For example, when rainfall increases, so too does runoff. If the temperature goes up and there is more constant rainfall, runoff is reduced because of evaporation. These two variables influence the production of sediment in drainage basins.

For a given amount of rainfall, the sediment concentration increases with temperature; and for a given mean annual temperature, the concentration falls with an increase in annual precipitation. This means that in arid regions, more sediment is moved per unit of water than in other regions, but because runoff is not very common, the total amount of sediment transported per unit time is less.

Temperature and precipitation have a clear relationship with plant cover. Plant cover increases with rainfall, and this leads to a considerable reduction in erosion. Knox (1984) established a critical plant cover threshold of around 70 per cent (corresponding to a mean annual rainfall of 400–500 mm) for the prevention of serious erosion. Where plant cover is below 70 per cent, the amount of soil vulnerable to erosion increases rapidly. Francis and Thornes (1990) lowered this threshold for the conditions of the semi-arid Mediterranean, although they did differentiate between low and high density rainfall.

4.1. Water erosion

The erosion of slopes by water involves the detachment of particles from the soil and their consequent transport. The energy required for this is supplied by the impact and the splashing of raindrops, by surface flow, and by interactions between these. The size distribution of soil particles, structural stability and crusting are the main properties of the soil, determining its vulnerability to water erosion in arid environments. Land use can also influence this (Lal, 1990). About nine per cent of the world's dry regions are affected by water erosion induced by human activity (Thomas and Middleton, 1994). The countries of the Mediterranean have some of the highest human activity-induced erosion rates in the world (Woodward, 1995).

The techniques for calculating water erosion rates are numerous and include the use of experimental stations, measurement of the sediment load in rivers, the use of radioisotopes, satellite imagery and map analysis, and the revised universal soil loss equation (RUSLE), and so forth. The results obtained with these techniques can be used in soil conservation, the prediction of reservoir longevity, and to decide on the best places to store waste, and other factors.

The influence of human activity on water erosion was studied by Iverson (1980) in the Mojave Desert. The intense use of four-wheel drive vehicles has caused much surface disturbance (Fig. 19.10) that has led to greatly increased runoff and erosion. The responses to this activation were studied with a rainfall simulator, and it was found that damaged areas had a reduced infiltration capacity, that there was more production of sediment (Fig. 19.11), and that runoff became channelled into wheel ruts.

Evenari et al. (1971a,b), who worked in the Negev, reported another case related to the water erosion of slopes. Some 2500 years ago, water was collected on slopes in this desert and channelled towards the valleys for irrigation purposes. The runoff increased considerably during this time, stones accumulated on the surface, and the underlying loess became encrusted through the action of the rain, reducing its infiltration capacity.

Human activity was also involved in the rapid reduction of the suspended load carried by the Colorado River (in its Grand Canyon stretch) between 1940 and 1946. This reduction (of around 50–100 × 10^6 metric t/yr), was attributed to conservation practices, flood control, a reduction in stock raising, and drought (Schumm, 1977).

Badlands are areas of intense water erosion that can affect man-made structures. Rilling and gullying destroyed a large part of the town of Pastici in the Province of Materna, Italy, in 1688, and a new town had to be built (Zachar, 1982). On the Yellow River in China, intensive gully erosion produced a rapid gully head retreat affecting houses in its basin (Zachar, 1982). In a 5-year study, Crouch (1983, 1990) used a microtopographic profiler to obtain gully head retreat values of 0.5–1 m/yr and of 2.5 m/yr in areas with intense subsurface flow.

Figure 19.10. Erosion of slope by vehicles. Kalia, Judean Desert, Israel.

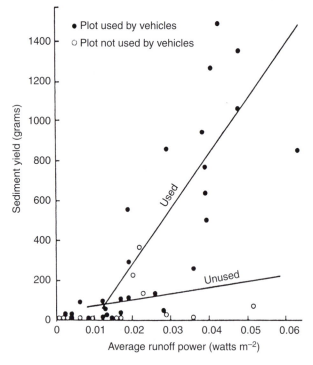

Figure 19.11. Computed regression lines for runoff power versus sediment yield from meter-square erosion plots used and unused by off-road vehicles. Mojave Desert (Iverson, 1980).

Formations with expansive clays or high exchangeable sodium percentages (ESP) are potentially capable of causing piping (i.e. of making subsurface conduits). In detrital rocks such as shales, the effects of piping look somewhat similar to those seen in soluble rocks, with the production of tunnels, caves, dolines, natural bridges, and subterranean drainage systems similar to karst systems (Parker et al., 1990). Piping is a serious problem for roads, and can cause the undermining of bridge abutments, the breakage of lateral concrete drainage systems, the undermining of concrete drains, and even the breakage of the road itself (Fig. 19.12) (Parker and Jenne, 1967).

Many earthen dams are subject to piping, a process that has caused more than a few to break. In Australia, fracturing in dams is due to the presence of montmorillonite group clays and the high percentage of exchangeable sodium (Ingles and Aitchinson, 1970). Studies performed in the USA and other countries report the same conclusions (Sherard et al., 1977; Sherard and Decker, 1977).

In some cases, pipes can cause landslides and subsidence and vice versa. Pipes can make slopes susceptible to these processes, which end up causing sheet and rotational slides. The undercutting of piping helps to generate mass movements on slopes (see Jones, 1981).

Piping has important effects on agriculture (Fig. 19.13). In some parts of Australia, deep ploughing is used to temporarily destroy existing pipes (Hudson, 1981). Generally, intense piping occurs in gullied terrain of little agricultural value. Piping often occurs because of inadequate deforestation practices or overgrazing. The soil conservation agencies of several countries, therefore, try to ensure the rational use of agricultural and forestry resources.

The prevention and control of piping is an important civil engineering and forestry problem. The solution used in earthen dams is to install sand filters to seal the escape of dispersive clays (Sherard et al., 1977), or to intensely compact the material to reduce hydraulic conductivity. The remedies to alleviate piping in linear constructions are based on surface drainage (Parker and Jenne, 1967). Some authors suggest the reforestation of the affected areas as a long-term solution. Reforestation would diminish the rapid drying that occurs during summer and prevent subsequent cracking. It would also increase the organic matter of the soil (Newman and Philips, 1957). Unfortunately, it is difficult to get vegetation to take root properly in arid areas. Another way to alleviate the effects of piping is to add gypsum to the soil to reduce the ESP, and therefore reduce clay dispersivity (Stocking, 1976).

4.2. Fluvial channels

The types of channel that exist in desert regions are straight, meandering and braided, but their topography can change rapidly. Research into hydraulic geometry and the behaviour of drainage channels in these areas has led to important advances in fluvial geomorphology. Though the types of channel are the same as those occurring in wet regions, some aspects of fluvial theory, such as plain flooding, cannot be extrapolated to arid areas because the feed mechanisms are different (Graf, 1988). The instability of these channels is quite obvious, and human activity commonly causes fluvial changes that entail considerable risk.

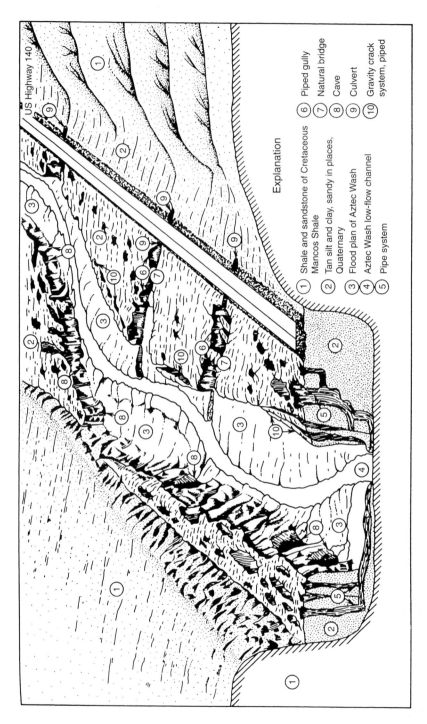

Figure 19.12. Idealized block diagram of Aztec Wash, southwestern Colorado, showing damage to US Highway 140 where it traverses a dissected and extensively piped valley fill. 1. Cretaceous shale and sandstone. 2. Quaternary silt; clay and sand. 3. Floodplain. 4. Low-flow channel. 5. Pipe system. 6. Piped gully. 7. Natural bridge. 8. Cave. 9. Culvert. 10. Gravity crack system, piped (from Parker and Jenne (1967), in Costa and Baker (1981), Fig. 10.15).

Figure 19.13. Pipe in a field of cereals. Azlor, Province of Huesca, Ebro Basin, Spain.

Changes in channel type are relatively frequent, and have been recorded on several rivers of the western USA (Table 19.1), where meandering channels have become braided (Graff, 1988).

One of the most surprising examples is the metamorphosis of the Cimarron River in Kansas, which was analysed by Schumm and Lichty (1963). This river was a deeply cut

Table 19.1. Changes in channel width in rivers of the western USA (Graf, 1988).

River	Change	Period
Canadian River, Oklahoma	0.8–3.2 km	1906 flood
Río Salado, New Mexico	15–168 m	1882–1918
Red River, Texas–Oklahoma	No change	1874–1937
Red River, Texas–Oklahoma	1.2–0.8 km	1937–1953
Cimarron River, Kansas	15–366 m	1874–1942
Cimarron River, Kansas	366–168 m	1942–1954
Platte River, Nebraska	1161–111 m	1860–1979
S. Platte River, Colorado	790–60 m	1897–1959
N. Platte River, Wyoming	1200–60 m	1890–1977
Gila River, Arizona	45–90 m	1875–1903
	90–610 m	1903–1917
	610–61 m	1917–1964
Salt River, Arizona	No change	1868–1980
Fremont River, Utah	30–400 m	1896 flood

meandering channel about 15 m wide which, after the floods of 1914, became a straight bedload channel about 336 m wide. This dramatic change was interpreted as being due to climatic fluctuations, agricultural activities, and the destruction of the natural vegetation.

Another no less surprising example is that of the Gila River that runs through Arizona and New Mexico (Graf, 1988). Before 1890, the channel was narrow and meandering, but in 1905 flooding modified it to become braided, and to have a width of 1 km in some places. Later, the growth of vegetation and sedimentation narrowed the channel, and by 1980 – almost a century later – it had returned to its old meandering nature. These results show that the change from braided to meandering channels occurs much more slowly than in the opposite direction.

The effects of human activity have been very clear in changes to the South Platte River of Colorado, which now has a braided flow. This river went from being 800 m wide in 1897 to only 60 m in 1959, becoming a narrow channel with a tendency to form meanders (Fig. 19.14). This profound modification was related to the regulation of the river and the loss of depth due to the use of water for irrigation. This caused an invasion of vegetation on the floodplain and sandbars (Nadler and Schumm, 1981).

The erosion of the concave banks of meanders and sedimentation on their convex parts creates lateral instability. It also causes the migration of meanders downstream that can interfere with man-made constructions. Such is the case of the Cimarron River near

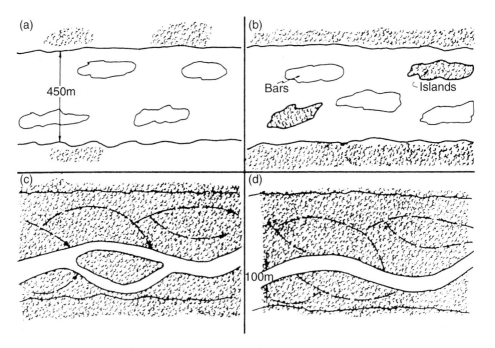

Figure 19.14. Model of South Platte River metamorphosis: (a) early 1800's, discharge is intermittent, bars are transient; (b) late 1800s, discharge is perennial, vegetation is thicker on floodplain and islands; (c) early 1900s, drought allows vegetation to establish itself below mean annual high-water level, bars become islands, single thalweg is dominant; (d) modern channel, islands attached to floodplain, braided patterns on floodplain are vestiges of historic channels (from Nadler and Schumm, 1981).

Perkins, Oklahoma. When it was built, the river was straight and braided in the area of the bridge, but there was a large meander one mile upstream. The position of this meander drifted, and in 1968 it affected the structure of the bridge (Chorley et al., 1984). Gravel extraction from river beds can also cause headcut erosion and the undermining of bridge pillars, as occurred on the Salt River in Phoenix, Arizona (Chorley et al., 1984).

Dams are perhaps the clearest form of human interference in river systems. They cause the deposition of sediment in the reservoir while downstream channels become degraded and rapids can form. This occurred after the construction of the dams on the Colorado River (Graf, 1980).

The effects of the Aswan dam on the Nile have been truly important. The downstream delivery of silt has been reduced and the delta has retreated because of a lack of water in its channels. It is estimated that this retreat might be as much as 30 km in some places by 2100: this would have dire consequences for the area's 1,000,000 plus inhabitants. There has also been a generalized fall in the water table and an increase in the salinity of nearly all the delta lands (Mainguet, 1991; Beaumont, 1993).

The River Ebro in northeast Spain also suffered an important fall in its bedload during the 20th century due to the construction of numerous dams. During the first third of that century, the sediment load was estimated at 1,000,000 metric t/yr. During the period 1950–1970, however, it had dropped to some 730,000 metric t/yr, and between 1970 and 1990 it was down to 300,000 metric t/yr. For 1990–2000, the mean was below 100,000 metric t/yr. When these results are compared with the transport capacity of the coastal part of the Ebro Delta (320,000 metric t/yr), a negative sedimentation balance can be appreciated which, since about 1970, has caused the acceleration of the delta's erosion (Guillén et al., 1992). Further, the horizontal and longitudinal instability of its channels – many of which are found in arid regions – are frequently accompanied by quite serious vertical instability. This is shown in successive stages of accumulation and incision, which are reflected in the morphology of the terraces. Human activity in semi-arid areas has also caused the formation of gullies. Several examples are analysed in the chapter on climatic change in arid areas.

As a consequence of the spectacular increase in the settlement of arid areas, river systems have suffered important changes. During construction, natural vegetation is removed and the surface is strongly modified by earth-moving machinery. This increases soil erosion. When construction is over, soils are impermeable and consequently there is an increase in runoff during times of rainfall and an increase in the depth of the water carried in the channels (Leopold, 1968, 1994). Flash floods can also occur.

4.3. Alluvial fans, debris flows and pediments

Alluvial fans, characteristic features of piedmonts in arid and semi-arid areas, are formed by the discharge of sediments transported by rivers from mountainous areas with gentle slopes. Their morphology is triangular and coalescence is very frequent.

Commonly, alluvial fans are controlled by tectonic, climatic or vegetation changes that lead to modifications in the production of sediments and runoff (Harvey, 1997). Deposition occurs locally when the supply of sediments is greater than the transport capacity, i.e. when the transport potential of the fan is below that necessary to move the sediment through the system. This is a critical energy threshold of the stream flow (Bull, 1979). Temporal variations can cause modifications to the sedimentation velocity and even the aggradation or dissection of the fan surface.

Alluvial fans present serious difficulties to human activity (Beaty, 1974; French, 1987). Flows are rare and unpredictable, and there can be wide variation within and between successive flows. Further, they can be dominated by water or by detritus which can even block channels. They may also undergo important migration and avulsions are common (Cooke, 1984).

The risk of flooding is evident, but not all parts of fans are subject to the same degree. The risk is extreme along the active channel, moderate at the head and distal areas of the fan, and low in the mid sector (Kesseli and Beaty, 1959). A common planning and engineering solution is to design and maintain a single channel for the flow (Graf, 1988). But the best defence against flooding is to identify the most dangerous areas and make flood risk maps such as those produced for the city of Suez and its environs (Cooke et al., 1982). These maps help to determine construction densities and to situate projects away from the channels (Fig. 19.15). Sediment deposition can also be fought using check dams and by dredging the channels in the catchment area. Flood alert systems can also be installed (Porath and Shick, 1974).

The channels running across fans are unstable and frequently overflow because of boulder sized sediment on their bottoms. This creates new channels which on occasion form part of abandoned channels. Important floods occurred in the south of Jordan in 1966 that partially destroyed the town of Maan, killing 70 people. The cause was mainly the re-use of an abandoned channel by the rising water (Graf, 1988).

Debris flow is one of the most common types of mass movement in semi-arid areas. It can be caused by intense rainfall moving abundant colluvial material through channels and over alluvial fans (Johnson and Rodine, 1984) and depositing material in areas with gentle slopes – commonly at the head of the fan. The distinction between the deposits made by debris flows and fluvial processes is not always easy to make because there is a continuous gradient between them (Hooke, 1987). Both are generated by concentrated flows of large transport and erosion capacities, but with different viscosities.

In their movement, debris flows can uproot vegetation, cover roads, channels and railways, drag along cars, destroy constructions and cover fields (Hooke, 1987). In semi-arid areas with clay soils, dangerous mud flows can occur that become channelled down rills and slopes.

Debris flows can be started by landslides, along with materials supplied from other sources and by a sudden saturation in detritus (the fire-hose effect) (Johnson and Rodine, 1984). Debris flows are very common in the semi-arid regions of the world and have been frequently studied, especially in California. In the La Canada Valley, Los Angeles County, debris flows occurred in 1934, 1938 and again between 1962 and 1971, causing the deaths of 63 people and important material losses (see Johnson and Rodine, 1984).

The area around Gissar in Tajikistan, close to the Afghanistan border, suffered an earthquake in 1988 that caused landslides on loess hills. The deposits were saturated by irrigation water and in the initial moments reached a liquid state producing a debris flow that moved some 20 million m^3 of material a distance of 2 km. Houses on the outskirts of the town were buried (Ishihara, 1989).

In some areas, urban development reaches up to the apex of alluvial fans, allowing detritus and mud flows to extend beyond their source area. Urbanized substrates facilitate

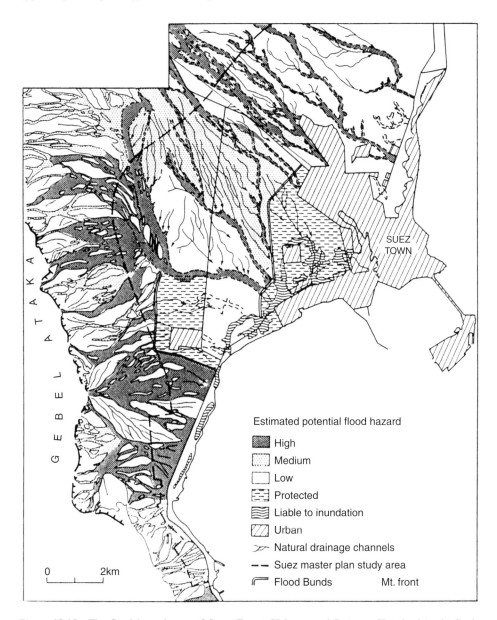

Figure 19.15. The flood hazard map of Suez, Egypt (Halcrow and Partners, Unpub. data; in Cooke et al., 1982).

the propagation of the flow and, because they are impermeable, prevent the loss of water (Rantz, 1970).

Perhaps the most important problem in determining the risks associated with these fans are that they remain dry for long periods of time. In southern California, for example, the flood recurrence interval is 70 years. Disasters only occur, therefore, from time to time, and only the really important ones become the object of public note (Burton et al., 1978).

Pediments have high drainage densities, but in contrast to alluvial fans they have no radial network. Pediments can be very problematical areas for human activity because they are at risk of flooding, channel instability and sedimentation (Rhoads, 1986). Because of their many small channels, linear constructions are often flooded and eroded. To solve this problem, dams are constructed on the upper ramp of the pediment to deviate water for flood control. The sediments transported also influence human activity by blocking drains, sewers – even narrow bridges – and thus hold water back and increase the flow. The small channels can overflow into one another. All this makes it more difficult for engineers to stabilize the channels. Where there are pediments, the best solution is to stabilize the largest channels and to be prepared for rising water (Graf, 1988).

4.4. Floods

On alluvial fans, flood waters are mainly sheet flows running across the fan surface. Floodwater poses important risks to human activity, and on occasion can be the cause of catastrophe. Buildings can be damaged or destroyed, and railways (Fig. 19.16), roads (Fig. 19.17), canals, bridges (Fig. 19.18) and agricultural land damaged by erosion or the accretion of the solids carried by the water.

Figure 19.16. The Arroyo del Medio fan, over which a railway passes, has been cut by the Río Grande. The tracks have been disturbed and left hanging in the air. Cordillera de los Andes, Province of Jujuy, Argentina. Photo: J.L. Peña.

Figure 19.17. Large accretion of sediments on the River León near a road bridge. Andes Cordillera, Jujuy Province, Argentina. Photo: J.L. Peña.

Figure 19.18. Bridge broken by flood water. Be'er Sheva, The Negev, Israel.

The inherent risk of flooding is increased by constructing close to rivers and by the agricultural use of alluvial soils; this increases the population close to rivers and their floodplains.

One of the most catastrophic flood events of recent history occurred in the spring of 1887 when the Yellow River or Huang Ho (China) broke its banks, killing between 1.5 and 7 million people (Davis, 1992). It affected 2000 towns and villages and material losses were huge. Similarly, in September of 1911, the Yangtze flooded, affecting four Chinese provinces, killing over 200,000 people and leaving half a million homeless. On the 17th September 1954, in Farahzad, Iran, 2000 pilgrims who were camped in a narrow gorge were killed when a flash flood hit them. In semi-arid Spain, the worst ever flooding occurred in Valencia in November of 1957 with 86 deaths, and in Tarrasa–Sabadell, in September 1962, where 267 people lost their lives (Martínez et al., 1987). López-Bermúdez et al. (2002) developed an historical analysis of the floods which occurred in the Spanish Mediterranean zone.

The fight against flooding has gone on throughout history in arid areas. The town of Calatayud in the Province of Zaragoza, Spain, was founded by the Arabs at the beginning of the eighth century, and it partially occupied an alluvial fan exposed to frequent flooding during storms (Gutiérrez, 1998). With the aim of reducing this risk, a weir and channel were built that deviated the water from the Barranco de la Rúa and the Rambla de Ribota (Fig. 19.19). In the Barranco de la Rúa basin, the runoff over a surface area of 6.3 km^2 (total basin area 8.2 km^2) was diverted by Arab engineers. As the population grew and settled closer to the River Jalón, mainly in the 13th and 14th centuries, the risk of flooding increased. During this time, or perhaps a little later, the Balsa de Valparaiso reservoir was constructed close to the fan apex to alleviate the problem. After the 16th century, a tunnel nearly 1 km long was excavated that drained the water towards the Barranco de Longia, and in so doing some 98 per cent of the runoff of the Barranco de la Rúa was diverted from its drainage basin, avoiding the risk of flooding. The Barranco de las Pozas, with its smaller drainage area, still causes problems for Calatayud (Gutiérrez, 1998). In the neighbouring town of Daroca in the Province of Zaragoza, the Mina tunnel, also about 1 km long, was excavated in 1555, to divert water from the main throughway, which passed through what is now the Calle Mayor (Main Street) where it caused serious flooding.

To avoid or at least reduce the chance of flood disaster, many and costly projects are undertaken in the world's arid regions, although recent studies suggest that the engineering solutions adopted are not always very adequate (Cooke et al., 1982). Dikes, dams and ditches are all part of these projects. Small dams at the headwaters of channels reduce erosion and control the runoff over small areas, whereas large dams downstream allow the attenuation of large rises in water levels (Costa and Baker, 1981). Dams, however, do not provide control over flooding, but rather "partial or specific protection" (Leopold and Maddock, 1954). It is also very important to control human use of floodplains with adequate legislation. Each local government should have detailed information concerning areas at risk of flooding. Flood alert systems have been improved by the installation of automatic hydrological information systems (AHIS), the aim being to establish emergency alerts that minimise flood damage (Ortiz, 1993).

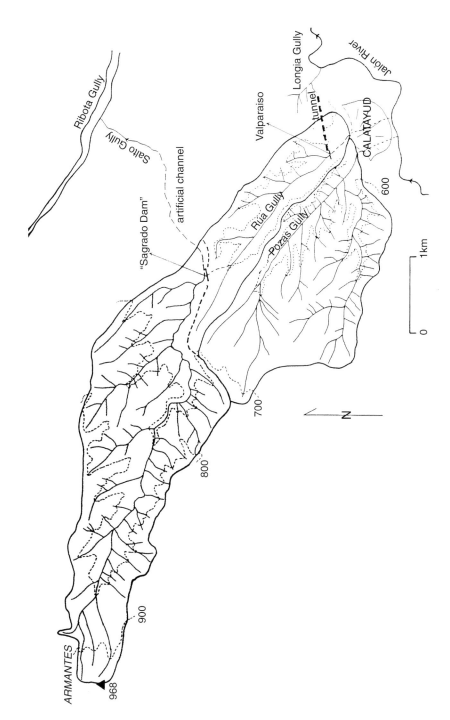

Figure 19.19. The fight against flooding in Calatayud, Province of Zaragoza, Spain (Gutiérrez, 1998).

5. Lake systems

Closed depressions, or undrained lakes (Langbein, 1961), are relatively common in arid areas, and are often found grouped together. They are not usually very deep and are ephemeral in nature. Because evaporation commonly exceeds water input, they are often dry and saline, and the water level fluctuates because of the high variability in available water.

These lakes can, however, overflow as a consequence of excess inflow. In September and October 1968, it rained for 38 days over the greater part of Tunisia. Rivers burst their banks and lakes overflowed, and the excess water ran down overflow channels giving rise to flooding. Some 542 people were drowned and hundreds of thousands made homeless. Damage to construction and agriculture was severe (Davis, 1992).

The salinity of these desert lakes can be related to the groundwater flow toward the lake, as occurs in the closed Bujalaroz basin in the Province of Zaragoza, Spain. The continuous supply of water provides salts to the system – and the only way to get rid of them is the wind. This makes the use of such lands for irrigated agriculture inadvisable; the water supplied to the irrigation system increases the level of these lakes, perhaps making them permanent. This can affect the flora and fauna of the region. Additional irrigation water can cause changes that may lead to serious environmental problems (Sánchez et al., 1993).

Human activity associated with the lakes of arid regions can be very varied, but often they lead to their slow eutrophication or degradation (Goudie, 1986). Sometimes, in semi-arid environments, lakes are partially dried to make way for crops. This is quite a normal fate for small lakes. In the Caspian Sea, the biggest lake in the world, man-made activities have caused a fall in the water level of over 3 m since 1929 (Goudie, 1992). In Israel, the largest hydrological project ever undertaken was the National Water Carrier, which transports water from the Sea of Galilee to the whole of the country's Mediterranean coast where it is required for irrigation (Beaumont, 1993). Consequently, the salinity of the Dead Sea has increased considerably as the quantity of fresh water flowing into it from the River Jordan, rising just north of the Sea of Galilee, has decreased. This increase in salinity has killed off the Dead Sea's animal and plant life, although it has favoured the salt extraction industry at the lake's southern tip.

Over the last few decades, the Aral Sea, the fourth largest lake in the world, has been the focus of one of the greatest ecological disasters ever recorded. This lake lies in a classic arid region and is fed mainly by the Rivers Amu Daria and Syr Daria, which arise in the mountains of Pamir and Tian Shan, respectively. Since 1960 this lake has suffered a serious reduction of its perimeter because of huge irrigation projects (Létolle and Mainguet, 1993) (Fig. 19.20). It once had a mean depth of 16 m and a maximum of 68 m close to the eastern shore. Table 19.2 shows the main hydrological data for the period 1960–1990.

The removal of water has left ports high and dry, fishing has fallen from 48,000 metric t/yr in 1957 to zero, fishing villages have been reclaimed by the desert, the water is polluted by pesticides and its salinity has increased considerably, the water level in the area's wells has dropped by 10 m and is polluted, and the fauna and flora are degraded, etc. (Létolle and Mainguet, 1993). Moreover, many of the irrigated soils have become saline and have been abandoned (Beaumont, 1993).

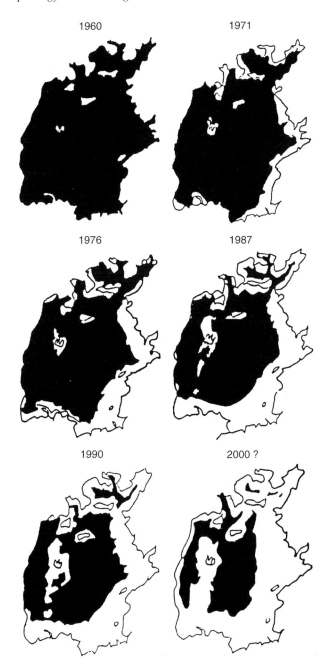

Figure 19.20. Evolution of the Aral Sea during the period 1960–2000 (Létolle and Mainguet, 1993).

Table 19.2. Hydrology of the Aral Sea from 1960 to 1990 (Létolle and Mainguet, 1993).

Year	Discharge of rivers (km^3)	Water level (m)	Area (km^2)	Volume (km^3)	Mineralization (g/l)	Fishing (t)
1960	40	53.5	67,900	1090	10	43,430
1965	31	52.5	63,900	1030	10.5	31,040
1970	33	51.6	60,400	970	11.1	17,460
1975	11	49.4	57,200	840	13.7	2940
1980	0.5	46.2	52,500	670	16.5	0
1985	0	42	44,200	470	23.5	0
1990	0	39	41,000	330	26.5	0

6. Subsidence caused by human activity

Human-induced subsidence has different causes including the extraction of fluids (water, oil, gas and geothermic fluids), hydrocompaction, compaction and later expansion due to the reduction of the load, the drainage of peat bogs, the drying of clay, and underground mining (Costa and Baker, 1981; Waltham, 1989). In arid areas, the most common type of human-induced subsidence is that caused by the exploitation of groundwater. The lack of water in these regions obliges people to extract it from the aquifer so that agriculture can be practiced. Abusive extraction can induce subsidence when the hydrostatic pressure in the storage rock falls. This subsidence is more frequent in non-consolidated Tertiary and Quaternary detrital deposits, formations in which sand and clay alternate. With the extraction of water from the sands, the pressure in the pores falls and the grains of sand have to support the overlying pressure. Sands do not compact much, but the compressibility of clays is high, and this allows surface subsidence (Marsden and Davis, 1967). This kind of subsidence generally affects large areas, producing shallow depressions that cause damage to buildings and other features. These endorheic areas of anthropic origin are sensitive to flooding, and if such subsidence occurs on the coast the sea may invade the area. This happened in Galveston Bay in Texas (Fig. 19.21), where residential areas were flooded (Coates, 1983, 1987).

In this same area, the extraction of groundwater has caused a wide sink hole and more than 160 faults with a total length of some 500 km (Fig. 19.21) (Holzer, 1984). Of these, 86 have been historically active, and are probably the result of the reactivation of Tertiary faults in the overlying Quaternary sediments. The groundwater is extracted mainly from sand and clay beds of the Chicot Aquifer that form part of a thick sequence of non-consolidated river and marine sediments. Pumping began at the end of the 19th century but increased during the 1930's. The result was the generation of an area of subcircular subsidence of about 12,200 km^2; maximum sinkage was 2.5 m. Although the Goose Creek oil field lies in the same area, its contribution to this subsidence appears to have been very small.

The subsidence of this area is accompanied by faults at the centre and the flanks of the depression, the density of faults being variable. Historically these faults have become

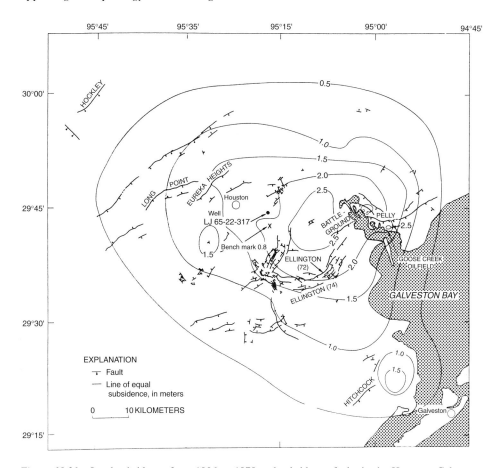

Figure 19.21. Land subsidence from 1906 to 1978 and subsidence faults in the Houston–Galveston, Texas, region (in Holzer (1984) with data from several authors).

reactivated and escarpments of 1.12 m have been observed. These faults are easily identified where they cross roads, airstrips, and other features.

In terms of the magnitude and variety of subsidence caused by water extraction, however, the record goes to Central Valley, California. The San Joaquín Valley is the most important area with 13,500 km^2 affected by settling and 9 m of maximum subsidence. This began back in the 1920's, and gradually increased until the mid 1950's when the sinking rate reached 0.55 m/yr (Poland and Davis, 1969). It then fell somewhat to 0.33 m/yr between 1963 and 1966, and in 1973 was imperceptible, all due to the bringing in of water from northern California. In 1976 and 1977 the subsidence again accelerated due to heavy drought that provoked the need for groundwater.

In central-southern Arizona, groundwater was heavily exploited in the mid 1940's, and this again caused subsidence. More than 50 areas were identified, with a total area of 8000 km^2 affected that became subject to subsidence, cracking and faults (Holzer et al., 1979). The cracks are usually about 1 m wide and several meters deep. The lengths of the faults vary from 1–16.7 km, and their escarpments are 1 m high. Serious damage has been caused to roads and irrigation systems, and the trajectories of some aqueducts have had to be modified (Coates, 1983).

In the San Joaquín Valley and central-southern Arizona, a slow uprising has followed the pumping out of groundwater. According to the US Geological Survey, this reached 6 cm between 1948 and 1967 in the area close to Casa Grande, Arizona. It is thought this was caused by isostatic rebound following the extraction of the water, which eventually reduced the load on the underlying materials (Costa and Baker, 1981).

7. Wind action: problems and control

The mobilization of sand and dust by the wind occurs in many types of climate, but in arid areas it is at its greatest. Given the scant or absent vegetation in these areas, much of the soil is subject to wind action. Further, the limited development of soils leads to high erosion vulnerability with some 39 per cent of the world's arid land affected (UNEP, 1992). Locally, tornadoes can have intensely destructive effects on construction and agriculture, as occurred on the 8th August 1992 at Ejea de los Caballeros, Province of Zaragoza, Spain (Fig. 19.22). The movement of wind-driven particles has a considerable effect on human settlements, and can cause enormous damage to buildings, crops, and transport and communications networks. The accumulation of wind deposits can destroy vegetation and increase the problem of desertification (UNESCO–FAO, 1977).

Figure 19.22. Destruction of grain silos by a tornado on August 8th 1992. Ejea de los Caballeros, Province of Zaragoza, Spain. Photo: C. Sancho.

Problems arise because of deflation, transport and sedimentation. Deflation can reduce the quantity of fine particles and nutrients in the soil, causing it to lose its fertility (Middleton, 1990). It can also undermine structures (e.g. lampposts or telephone posts), causing them to collapse. During transport, the particles cause abrasion that produces hollows and furrows. Dust in the wind can also reduce visibility at airports and choke people and animals (Pewé, 1981; Pye, 1987; Middleton, 1997). For these reasons, a dust storm alert system has been set up in Arizona, with electronic signboards on roads and announcements on public radio warning of any threats. Dust can also enter houses, causing health problems and contaminating food and drinking water. If the dust is saline, it could affect buildings. The positive effect of wind-blown dust is that it can provide soil nutrients to new areas.

The quantities of dust mobilized are of the same order of magnitude as those transported by rivers (Livingstone and Warren, 1996). A significant part of windblown dust comes from human activities. Ploughing activities break up soil aggregates, and in dry periods plumes of dust can be seen behind tractors (Lee et al., 1993). Another source is traffic driving over unsurfaced roads. The removal of vegetation by grazing or fire can also increase wind erosion (O'Hara, 1997). The drying of the Aral Sea has increased the amount of windblown dust (Goudie, 1994b). Wind sedimentation, on the other hand, can bury houses, crops, canals (Fig. 19.23) and roads (Fig. 19.24). Digging them out can be very costly if large areas are affected (Livingstone and Warren, 1996).

Control measures should be aimed at reducing the wind's transport capacity, reducing the supply of sand and, if required, deviating windblown sand (Watson, 1990; Pye and

Figure 19.23. An irrigation ditch, partially buried by sand, that was part of the infrastructure for irrigating an area of the Thar Desert using water from the Indira Gandhi Grand Canal. Jaisalmer, Province of Rajasthan, India.

Figure 19.24.   Dunes advancing across a road. Maharaga, United Arab Emirates.

Tsoar, 1983a; Mainguet, 1991). The techniques employed (Cooke et al., 1982; Watson, 1990) involve stabilizing areas by recovering the natural vegetation or the planting of other species, depending upon how arid the area is. In hyper-arid areas, the extreme lack of water may render this impossible. It is important to plant fast-growing species (e.g. *tamarix* sp., acacias and eucalyptus) (García Salmerón, 1967; Pye and Tsoar, 1990). Around the railway from Bagoton to Lanchou in China, vegetation corridors 300–500 m wide have been planted to slow down erosion (Watson, 1990).

   Another control method lies in the stabilization of the surface. This could involve covering it with stones, oiling it, or covering it with asphalt, synthetic latex, polyvinyl, gelatin or cellulose fibres. These can be effective but for periods of less than 5 years. Chemical sprays have been widely used to produce a chemical crust a few millimetres thick which, combined with the reestablishment of vegetation, has provided some good results (García Salmerón, 1967; Cooke et al., 1982; Livingstone and Warren, 1996).

   Windbreaks are commonly used in desert areas to slow down the wind and trap the sand it carries (Fig. 19.25). They are commonly made from palm leaves, but also from low-cost wood produced by local vegetation. Such structures reduce the transport capacity of the wind both in front of them and behind them. The volume of sand that accumulates is proportional to the height of the windbreaks used. These should be installed some distance from the object to be protected and perpendicular to the prevailing wind. It is a good idea to set up a series of windbreaks at different distances from the object. The furthest away will become covered rapidly, but it can increase the lifespan of those in the intermediate distance by a factor of four, and of the closest by a factor of nine (Fig. 19.26) (Kerr and Nigra, 1952). Some windbreaks are set up in a zigzag fashion to restrain winds from several directions (Fig. 19.27) (Watson, 1990).

Figure 19.25. Windbreak system for protecting a palm grove. Nefta, Tunisia. Photo: J.L. Peña.

Dune stabilization is the objective of several techniques, but their effectiveness is very variable. Sometimes, however, it is only necessary to deviate the direction of a dune. If dunes are small, it is sometimes decided to simply move the sand somewhere else. This procedure is cheaper if it can be used in construction. The destruction of a dune by digging ditches parallel to its axis is an expensive solution – and only temporary. Stabilization with vegetation can be very expensive and the use of windbreaks or covering dunes with

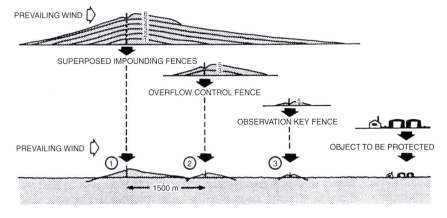

Figure 19.26. Three fence system to protect extensive areas, such as villages, shops, yards, industrial plants, and other features (after Kerr and Nigra, 1952).

Figure 19.27. Windbreaks arranged in a zigzag for trapping sand blown by winds coming from different directions. Erfoud, Morocco. Photo: J.L. Peña.

stones is more useful. The surface treatment of the arms of crescent-shaped dunes with oils allows the deflation of the central part, and an accumulation of sand (very much greater than that associated with the original dune) occurs to windward (Fig. 19.28) (Kerr and Nigra, 1952).

8. Degradation of arid areas. Desertification

The degradation of soil results in a reduction of its quality and, therefore, of its potential productiveness. Such degradation can be physical – manifested by the movement of the smallest particles, biological – related to the loss of organic material, or chemical – linked to the concentration of highly soluble salts. This degradation of the soil is the main factor causing desertification (Mainguet, 1991). Desertification affects some 65 million hectares that were previously productive and on which some 800 million people live (Fig. 19.29). Desertification and salinization are two of the nine main problems cited by the World Bank (Goudie, 1990a,b).

In Spain, some 27 million hectares, 53.4 per cent of the country, are affected by important soil degradation. Annual soil losses are about 1,000,000 t (MOPU, 1987). In south-eastern Spain, desertification is a severe problem with erosion rates of up to 300 metric t/ha/yr (López Bermúdez, 1986). Data on other processes such as salinization, wind erosion and the physical and biological degradation of the soil are not available for the country as a whole (López Bermúdez, 1988). Recently, the study of water erosion in Spain has intensified with the Lucdeme Project. The Network of Experimental Stations for

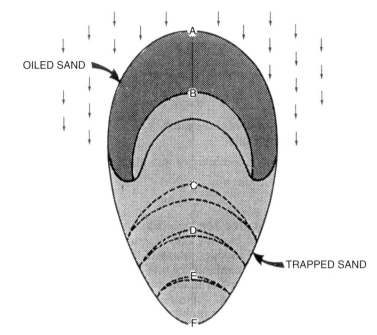

OILED SAND

TRAPPED SAND

PLAN VIEW OF PROGRESSIVE GROWTH STAGES

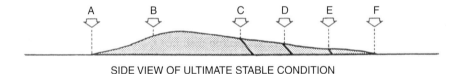

A   B   C   D   E   F

SIDE VIEW OF ULTIMATE STABLE CONDITION

Figure 19.28. Dune stabilization by oiling (after Kerr and Nigra, 1952).

Monitoring and Evaluating Erosion and Desertification (*Red de Estaciones Experimentales de Seguimiento y Evaluación de la Erosión y Desertificación – RESEL*), which covers a large part of the country, is now installed (Rojo and Sánchez, 1996).

At the 1977 World Conference on Desertification (UNWCD) in Nairobi, the United Nations proposed a plan of action for combating desertification. Desertification was defined as the reduction or destruction of the biological potential of land that could finally lead to conditions similar to those of deserts (UNESCO–FAO, 1977). There are many other definitions that emphasize the role of human activity or the climate (Baumer, 1990; Mainguet, 1991; Thomas and Middleton, 1994). The risk of desertification comes from the interaction between natural physical and biological systems and human activity. Such interaction has become more intense because of the explosion of the human population in the 20th century (Goudie, 1986, 1990a).

Physical processes such as water and wind erosion (mentioned above) intervene in the degradation of deserts. But the most important physicochemical processes in arid areas are

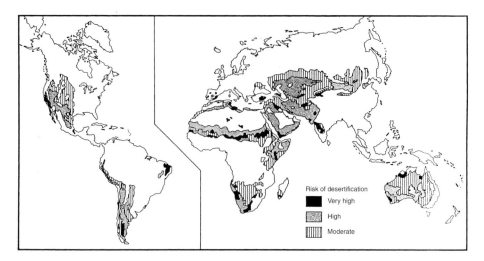

Figure 19.29. The United Nations Conference on Desertification (1977) map of areas at risk of desertification held in Nairobi, Kenya (Goudie, 1990a,b).

those of soil salinization and sodification (sodium enrichment) which result from the accumulation of soluble salts (chlorides, sulphates and carbonates of sodium, magnesium and calcium) (Rhoades, 1990). According to the FAO, a soil is saline when its content in salts exceeds 1–2 per cent in the upper 20 cm. In such soils, accumulated salt rises because of capillary action to give salty coverings and crusts.

Human-induced salinization is a consequence of the irrigation of arid and semi-arid areas (Fig. 19.30) and it affects about one third of the world's 200 million irrigated hectares (Table 19.3) (Goudie, 1990a,b). Its effects include modification of the soil structure, reductions in permeability and porosity, and the reduction or even complete loss of natural vegetation (Fig. 19.31). Salinization can also be linked to the over-exploitation of aquifers in coastal areas, which can allow seawater to gain access to them. The failure of fresh water to reach delta areas because of dams can also be a cause of salinization, as occurs in the Nile Delta (Mainguet, 1991).

The salinization of the Ebro Basin in Spain (Fig. 19.32) is due to the area's dry climate, the salt content of the bare rock, the redistribution of salts, geomorphological conditions, intrinsic features of the soil, and the agricultural practices of the area (Alberto, 1989; Herrero et al., 1993). Among the latter, the levelling and irrigation of alkaline soils are among the most important. Aspersion irrigation or drip-feed systems are the best for preventing salinization. The problem of irrigation-associated salinization is economically very important (Zekri et al., 1990). In the central region of the Ebro Valley there are 5360 km^2 of irrigated land, of which 1000 km^2 are severely affected by salt and another 1500 km^2 require some kind of control (Herrero and Aragüés, 1988; Herrero et al., 1993). Similarly, in the San Joaquín Valley of California, 160,000 ha are affected by salts; by the year 2080, 445,000 ha will be unproductive (Sheridan, 1981).

Rainfall varies enormously in desert areas, and the discharge of rivers can fluctuate widely. This leads to important variations in salinity, as reflected by the River Casamance

Figure 19.30. Channelled water from the San Juan River originally from the Andes. San Emiliano Dyke, San Juan, Argentina. This water is put to several uses in this very dry area.

of Senegal (Fig. 19.33), which affect the availability of drinking and irrigation water. In October 1968, salinity was low, coinciding with the rainy season. But in May 1984, the whole river had a salt content higher than that of sea water (35‰). Close to its mouth, the salinity of the river tripled this value (Farmer and Wigley, Unpub. data; in Mainguet, 1991).

As mentioned above, inadequate agricultural practices and over-irrigation cause salinization, but they can also cause lixiviation. As stated by Mainguet (1991): "It is absolutely vital that irrigation in arid or semi-arid ecosystems be considered a last resort, especially for large projects."

Table 19.3. Approximate percentages of irrigated land affected by salinization in different countries (Goudie, 1990a,b).

Algeria	10–15	India	27
Egypt	30–40	Iran	<30
Senegal	10–15	Iraq	50
Sudan	<20	Israel	13
USA	20–25	Jordan	16
Colombia	20	Pakistan	<40
Peru	12	Sri Lanka	13
China	15	Syria	30–35
Cyprus	25	Australia	15–20

Figure 19.31. Cereals affected by salinization in the area of Los Monegros in the Ebro Basin. El Tormillo, Province of Huesca, Spain.

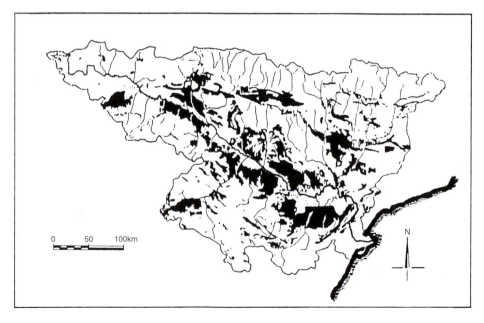

Figure 19.32. Distribution of saline areas in the Ebro Basin, Spain (Alberto, 1989).

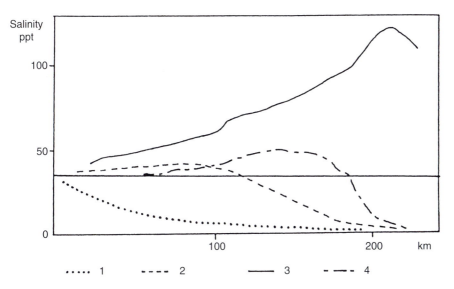

Figure 19.33. Casamance River salinities as a function of distance (in km) upstream from the mouth (Pages, CRODT, Dakar, Senegal). The ordinate is in parts per thousand (ppt). Seawater salinity (35 ppt) is shown as a horizontal line. 1 October 1968; 2 July 1969; 3 May 1984; 4 September 1984. (After Farmer and Wigley, 1985); in Mainguet, 1991, Fig. 44).

Figure 19.34. Experimental plot for the study of water erosion. Four years after its set-up, more vegetation was found inside than outside the plot, which became affected by overgrazing by sheep and goats. Mishor Adumin, Judean Desert, Israel.

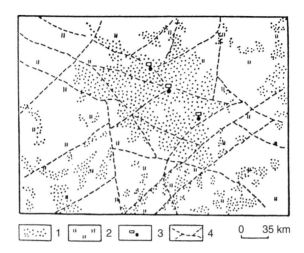

Figure 19.35. Degradation caused by overgrazing. 1. Barren alluvial surface; 2. Grazing land; 3. Livestock sheds and wells; 4. Road and path. (Mainguet, 1991).

Another human-induced cause of the degradation of desert areas is the abusive agriculture sometimes practiced. Sometimes, degradation is the consequence of introducing inappropriate agricultural practices (Thomas and Middleton, 1994), and may lead to the complete loss of productivity. The land is then abandoned and becomes subject to wind and water erosion. Many examples of this can be seen in China, the Sahel, Rajasthan, and elsewhere. *Deforestation* is also a cause of degradation. Deforestation for fuel is severe in some places where electricity or gas is too expensive for the majority of people. Acacias have disappeared from Khartoum in Sudan for a radius of 100 km and wood now has to be brought in from over 500 km away (Goudie, 1990a,b).

The number of cattle in the world increased by 38 per cent between 1955 and 1976, and the number of sheep and goats by 21 per cent (Goudie, 1990a,b). In many areas, the number of livestock animals has exceeded the carrying capacity of the land and *overgrazing* has occurred (Fig. 19.34). Land has also been brought under cultivation which has further reduced grazing areas. The loss of plant cover has brought with it both wind and water erosion. In northern China, steppe vegetation has been degraded by overgrazing and a series of rings can be seen around wells (Fig. 19.35). These are arranged as a central area with a radius of 300–500 m where there is little vegetation (often just bare ground), a second ring with a radius of about 500–1000 m where some resistant grasses and bushes remain, and then the vegetation returns to normal some 1000–3000 m from the well (Mainguet, 1991).

Desertification is a serious problem. Technically it can be stopped and many degraded areas can be restored to production, but the solutions are costly. Preventive measures for saving plant cover are based on introducing new species more resistant to the lack of water and the presence of salts. Water usage must also be improved, and wind and water erosion combated. Ignorance must also be fought, perhaps with social awareness/training programs aimed at all levels of society.

SEVENTH PART
Geomorphology of tropical zones

Chapter 20

The humid tropics: weathering and laterites

1. Introduction

The tropics are the regions located between the Tropic of Cancer (23.5°N) and the Tropic of Capricorn (23.5°S). These limits involve areas where the sun can reach the zenith. As a result, these zones of low latitudes receive a great amount of solar radiation and continuous heat, with important biological consequences such as the development of a dense and widespread vegetal cover, which has a strong influence upon the morphogenesis of these regions. In the intertropical zones there exist both large areas warm deserts and important zones with high rainfall. A separation between different environments is needed, which can be based upon several criteria such as climate, biology or geomorphology, among others. The subdivision is always difficult, especially in continental areas (Reading et al., 1995).

Some authors discriminate the humid tropics by using vegetation criteria, concluding that these regions are characterized by the scarcity or absence of xerophytes and the development of rainforests, deciduous forests and savannas (Fosberg et al., 1961). Other authors use climatic criteria and establish a limit on an average temperature of 18°C for the coldest month (Köppen, 1936). The mean annual precipitation has also been proposed, but great discrepancies exist among researchers. Köppen (1936) used the range of 450–600 mm and Tricart (1974b) between 750 and 800 mm.

About 2400 million inhabitants live in the tropical zones, which means nearly 45 per cent of the world's population. Nearly all these people live in the humid tropics, and about 60% in western and southern Asia. The most important economic sector is agriculture, which obviously depends a great deal on the climatic fluctuations and also on catastrophic events (McGregor and Nieuwolt, 1998).

Most parts of the surface occupied by the non-arid tropics are located on stable cratonic zones, which are basically constituted by Precambrian plutonic and metamorphic rocks, mainly in central Africa and the Brazilian shield. These materials underwent long-lasting denudational processes, which favoured the development of extensive planation surfaces. Due to the low topographic gradient, the erosion velocity is slow and weathering progresses with no difficulty. For this reason, landform evolution over long periods of time can be studied, under conditions of crustal stability (Thomas, 1974; Wirthmann, 2000).

In the opposite situation, regions affected by alpine tectonics and with high relief, mainly in Indonesia, SE Asia, Central America and the Antilles, suffer high denudation rates and weathering cannot deepen. In addition, prevailing rocks are different from the stable regions, like the great calcareous outcrops where singular karst landforms develop.

As previously indicated in preceding chapters, in the arid regions, and especially in the semi-arid zones, typical humid tropical weathering profiles are quite common.

This indicates that in the past these dry zones underwent conditions much more humid than at present. Hence, the recognition of processes and deposits in the humid tropics helps in the interpretation of climatic variations in past times.

2. Climatic characteristics, vegetation and morphoclimatic domains

Climatic parameters play an essential role in the dynamics and intensity of geomorphological processes, as well as in the nature of the resulting landforms (Faniran and Jeje, 1983). For this reason, it is convenient to know the magnitude of these parameters and their variability.

Temperatures in the humid tropics are moderate and do not exceed 34°C, very different from the tropical deserts, with values higher than 50°C. Mean annual temperatures fluctuate between 24 and 30°C, measured at sea level. Seasonal and annual variations are lower than 2°C. This thermal uniformity is broken with increased altitude, which is accompanied by cloud intensification.

In the humid tropics humidity is always high. Relative humidity often exceeds 80%, and normal values throughout the year do not move far from this number. Maximum potential evapotranspiration is recorded in the equatorial zones and varies between 1000 and 1500 mm/yr.

With the exception of very elevated areas, precipitation is in the form of rain. As temperature is quite uniform, seasonality is marked by precipitation variability. Thunderstorms are very common, and contribute some 90% to the total rainfall in savanna zones. Most thunderstorms start in the afternoon and have short duration. The strongest one ever recorded produced an 1870 mm rainfall in a single day, 16 March 1952, in Cilaos on Reunion Island. In the humid tropics 40% of the precipitation has an intensity higher than 25 mm/h and several records are known with intensities of up to 340 mm/h for some minutes (Reading et al., 1995). Precipitation regimes are as follows (Birot, 1973): (1) Tropical areas of equatorial regions influenced by the Intertropical Convergence Zone, rainfall through all the year and one single maximum. Annual average precipitation is about 2000 mm or much higher. (2) Areas within 10° latitude north and south of the Equator, with two dry and two rainy seasons and precipitation ranging between 1000 and 2000 mm. In this humid and dry tropical regime rainy seasons can be shorter and precipitation can reach 650 to 1000 mm. (3) Dry regime, with precipitation between 250 and 650 mm, located on the desert margins.

Vegetation in the humid tropics is a direct response to the existing climatic regime. An equatorial rain forest (*pluvisilva*) can be differentiated, which is constituted by very close trees of up to 40 m high with large perennial leaves and thousands of tree species (Figure 20.1). The high vegetation density generates a continuous shadow on the ground. Coastal vegetation is represented by mangroves. The rainforest mainly develops in the Amazon and Congo basins, Central America and SE Asia and Indonesia. The savanna is a consequence of a humid–dry tropical climatic regime. It is formed by very isolated trees, sometimes grouped in corridors following valley bottoms, that arise upon a grassland of tall graminoids, commonly higher than a person (Figure 20.2). Summer fires in these yellow dry graminoids are very frequent and are responsible for the grassland predominance over the forest. The savanna is mainly located in central and southern Africa,

Figure 20.1. Equatorial rain forest in Amazonia, Manaus, Brazil.

with acacias of horizontal canopy and baobabs, also in South America (*campos cerrados*), India, southeast Asia, and Australia (Tricart, 1974). The vegetation cover protects the soil surface from erosion by the interception of up to 20 to 30% of the raindrops. Moreover, it contributes to physical weathering by root penetration into joints, and to chemical weathering through the activity of organic acids. Vegetation also protects soils from water erosion and helps in preventing mass movements, which mainly occur during periods of intense rainfall.

Figure 20.2. Savanna with graminoid grassland and dispersed trees in Nairobi National Park, Kenya.

Because temperatures are very homogeneous, climatic subdivisions are made as a function of precipitation. For the humid tropics and transitional areas a limiting value is established with annual precipitation higher than 750 mm (Thomas, 1994). It includes rain forest (Af and Am) and savanna (Aw and As) climates (Köppen, 1936).

Morphoclimatic differentiation is quite similar, because precipitation and the resulting vegetation cover play a determinant role in humid tropic regions. The numerous geomorphic inheritances from other climates existing both in equatorial and in humid and dry climates, however, must be taken into account, although more obvious in the latter, near zones of dry climate. Dominant geomorphological processes in the humid tropics are chemical weathering, mass movements, and fluvial erosion, the latter being much more important in regions of humid and dry climate (Wilson, 1968, 1969).

Without any doubt, the intense deep weathering and its profile evolution constitute one of the most characteristic features of this morphoclimatic zone. Some authors, however, doubt the existence of a tropical climatic geomorphology (Stoddart, 1969c), and others the differentiation of the tropical zone within the global morphoclimatic system (Douglas, 1978; Gupta, 1993). This last author indicated that processes and forms in the tropics depend only partially on climate, and that there exist other important factors such as neotectonics, lithology, episodic and catastrophic events (cyclones), and anthropic activities. Although this statement is true, it must be kept in mind that all these non-climatic processes and factors act throughout the world, and they operate at the same time as climatic causes. This vision would correspond to very populated regions undergoing important neotectonics. This is a quite different situation from the one existing in African and Brazilian regions, where old planation surfaces have favoured the development of thick regoliths, while environmental changes have produced lateritic landforms, etch surfaces, and other features typical of humid tropical zones.

Following Büdel (1977, 1982), humid tropical zones are completely different from other morphoclimatic zones. They do not develop fluvial terraces, gravel plains, or flood plains such as in the ectropics (temperate regions), nor loess mantles. The soil is much deeper than in the ectropics, and contains plenty of duricrusts and hardpans. Inselbergs of variable size are common in the tropics, standing out on extensive erosion surfaces, which are weathered at the same time as they form. Mountain zones are characterized by deep incisions. This author assumes that morphodynamics in tropical lowlands are completely different, where landforms are controlled by the generation of etch surfaces.

Tricart and Cailleux (1965) differentiated an *intertropical zone* with warm temperatures and constant fluvial flows. Depending the rainfall distribution, its annual amount, and the vegetation cover density, they divide this zone into a *savanna domain*, with less precipitation and less dense vegetation cover. Sheet wash is considerable and chemical weathering is strong and discontinuous in time. The *rainforest domain* is characterized by dense vegetation cover and higher precipitation. In this domain chemical and biochemical processes reach a maximum intensity.

3. Weathering in the humid tropics

Weathering is the response of materials that were initially in equilibrium inside the lithosphere to conditions at or near their contact with the atmosphere, hydrosphere and, perhaps more important, the biosphere (Reiche, 1950). The importance of the biosphere is perhaps overemphasized. A great part of the rocks have been submitted to high pressures and temperatures, in the absence of air and water. After outcropping from the surface, they change towards an equilibrium with pressures of about 1 atm, much lower temperatures and the presence of air and water, which are the conditions existing in the contact lithosphere–atmosphere (Carroll, 1970).

Climate constitutes one of the most significant factors in weathering development. Water from precipitation is the most important reactive agent in the weathering process and its quantity is a prime factor in the intensity and type of alteration. In humid tropical regions water attacks crystalline structures and evacuates bases (Na^+, K^+, Ca^{2+}, and Mg^{2+}) as well as silica to some extent. It also affects the velocity of biochemical reactions in plants. Temperature has direct and indirect effects on weathering processes. High temperatures accelerate chemical reactions. Following the Van't Hoff Law, every 10°C increase in temperature duplicates or triplicates the rates of chemical reactions (Tricart and Cailleux, 1965). Moreover, heat increases biological activity. As a result of these temperatures, bacteria consume great quantities of organic matter in decomposition, which brings about the scarcity of humus in these humid tropical environments.

The result of the actuation of weathering processes, especially those related to chemical alteration, gives rise to an intensively weathered rock (Taylor and Eggleton, 2001). This rock decomposition is a consequence of intense precipitation and temperature and also of the vegetation response to these conditions. In these areas natural waters reach pH values ranging between 4 and 6, and very possibly these acid conditions may be due to silicate hydrolysis (Curtis, 1976). The important release of biogenic carbon dioxide supplies great quantities of bicarbonates to the waters. Humification processes produce humic and fulvic acids, which export cations as organometallic compounds (Thomas, 1994). The activity of all these processes produces thick alteration profiles that will be analysed later in this chapter.

3.1. Physical and biological weathering

Weathering processes can be divided in three types: physical or mechanical, biological, and chemical, and all of them can act simultaneously. Physical weathering breaks the rock by different mechanical mechanisms. Most of the forces applied to the geological materials is external, although they can also accumulate inside the rock.

Many rocks were originated or remained for a long time at considerable depths in the lithosphere and hence, were submitted to high confining pressures. When uplifted to the surface, these rocks suffered a pressure-release by unloading that generates joint-like fractures parallel to the topography; a process known as *sheeting* (Figure 20.3) (Gilbert, 1904). Joint spacing increases with depth and is better recognizable in massive rocks such as granites, conglomerates, sandstones, and so forth. The difficulty of recognizing this type of jointing in non-massive rocks is probably due to the existence of old bedding planes that would distribute expansion stress (Ollier, 1965). Other authors suggest that sheeting is a consequence of lateral compression (Twidale, 1973, 1982a; Vidal Romaní et al., 1995; Vidal Romaní and Twidale, 1998). As will be discussed later, this jointing has great importance in the landform development of tropical cratonic areas (Birot, 1968a,b).

Figure 20.3. Sheeting developed in granites of the Yosemite Valley, California. Photo: J. López-Martínez.

Insolation weathering appears in the scarce rock outcrops existing in the humid tropics, which can experience temperature rises of up to 50°C (Vischer, 1945), and are therefore subject to thermal expansions and contractions. Wetting and drying cycles can also impose a significant weathering factor on rock outcrops. Frost shattering operates with great

Figure 20.4. Termite-mounds in the Masai Mara savanna of Kenya.

intensity in the high parts of stratovolcanoes, which interrupt the extended plains of western central Africa, like mounts Kenya and Kilimanjaro. It also acts on the high Andean peaks of Peru, Equador and the Nevados of Columbia.

Rock disintegration is controlled by the rupture of materials, due to the pressure exerted by growing roots. The carbon dioxide supply to the solutions is also very important, which results from the gas exchange or respiration of the flora and fauna existing in the soil and from decomposition of organic matter. The fauna plays an important role in transferring and mixing materials, moving the regolith to other areas. In the humid tropics this last work is carried out by termites and worms and can be responsible for the sorting of the upper part of the soil. There exist several thousand socially organized *termite* species. They carry clay size particles to build termitaria, which they construct with segregated saliva. The number of termites reaches values of 9 million/ha in the Ivory Coast and their amount decreases with latitude and altitude (Goudie, 1988). Termitaria appear as small mounds (Figure 20.4) and towers (cathedral-termitaria of Tricart, 1974b) (Figure 20.5),

Figure 20.5. Inactive termitaria tower of about 6 m high in the forested savanna of Eastern Bouaké, Ivory Coast.

and they can reach up to 9 m high and a density of 1000/ha. Large termitaria are built by only a few species, while smaller termite-mounds are associated with numerous species. Growing rates are very variable, from 25 to 500 mm/yr (Goudie, 1995). When termites leave the termitaria, it is then exclusively submitted to erosive processes, leading to its destruction. Although the termitaria material is much stronger than the underlying soil it is finally eroded, normally in less than 10 years (Goudie, 1988). Termites also build small filiform corridors upon the soil and vegetation, made up of cemented soil particles. These accumulations give an idea of the amount of material involved in the removal process. Following calculations made by several authors in Nigeria, Senegal, and Kenya, the accumulation rate ranges between 300 and 1050 kg/ha/yr (Goudie, 1988).

Terrestrial *worms* also build constructions like termites although of lesser size, hardly reaching 30 cm in height. Earthworms excavate burrows of up to 70 cm depth, which indicates the importance of these organisms in the removal and mixing of soil components. The amount of removed material is estimated as 5 to 25 mm/yr, greater than the quantity moved by termites (Goudie, 1988).

All these weathering processes have a certain importance, but chemical weathering plays the most relevant role in the disintegration of minerals and rocks in humid tropical zones.

3.2. Chemical weathering

Chemical weathering results from a change in the chemical environment (Loughnan, 1969), because rocks formed in high pressure and temperature conditions and in the absence of air become unstable when put in contact with the atmosphere. Water, oxygen, and carbon dioxide attack the rock minerals through exothermic reactions. Most soluble products are eliminated by dissolution, leaving a residue rich in less soluble components. This residue is structured to produce neoformed compounds, mainly clay minerals and iron and aluminium oxihydroxides or sesquioxides, which are in a more stable equilibrium with the environmental conditions.

Jenny (1950) proposed a disintegration mechanism for the mineral structures. Following the Pauling Rules, the sum of the positive charges in a crystal must be equal to the sum of negative charges, but on the crystal surface valences are not compensated, resulting in an unstable equilibrium. When crystals are in contact with water, the OH^- ions join to the existing cations and the H^+ ions with oxygen and other negative ions. The H^+ ion has a small ionic radius and can occupy cation places in the crystal surface. Cations are then released as hydroxides. In addition, the loss of H^+ ions modifies the pH of the solution.

3.2.1. Weathering of silicates

Silicates are the most common minerals in rocks. Approximate proportions of mineral species in the Earth's surface are: feldspars (30 per cent), quartz (28 per cent), clay minerals and micas (18 per cent), calcite and dolomite (9 per cent), iron oxides (4 per cent), pyroxenes and amphiboles (1 per cent) and other minerals (10 per cent) (Leopold et al., 1964). Cations released in the weathering reactions can be easily washed away or remain in the residue. This is due to the physico-chemical variations of the solution. In this section a brief discussion is presented on these factors and their incidence on the mobility and precipitation of the most frequent ions in silicates.

Following geochemical studies, fresh water pH fluctuates between 4 and 9 (Mason, 1958). Figure 20.6 presents the solubility of some of the most common compounds as a function of pH (Loughnan, 1969). Sodium and potassium are soluble under any pH value and calcium and magnesium hydroxides are also soluble under the pH of natural waters. Instead, titanium and aluminium oxides and ferric hydroxide are insoluble in the marked intervals. Only ferrous hydroxides, titanium, and calcium carbonates are influenced by the pH of natural waters. The solubility of silica is low, although under a clear alkaline pH it is washed away with a certain facility. Its solubility curve refers to amorphous silica and not to quartz, which has a solubility 20 times lower than the amorphous silica (Carroll, 1970). Nevertheless, several environments exist where acidity and alkalinity reach very high values. For example, aluminium can be leached in the aluminates form in very basic playa–lake environments. In soils with great quantities of organic matter pH can reach values lower than 4 under which aluminium can be removed. Sulphur oxidation also makes waters reach very high acidities.

The oxidation–reduction potential (Eh) has a clear incidence on the solubility of polyvalent cations, such as Fe and Ti. Chemical elements under an oxidation state are in equilibrium with the environment. Because solubility varies as a function of the different valence states, the redox potential affects the mobility of the chemical element. Figure 20.6 shows the solubility of trivalent and divalent iron. Ferric hydroxide is only soluble with

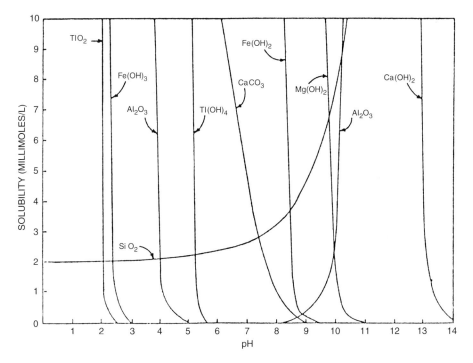

Figure 20.6. Solubility in relation to pH for some components released by chemical weathering (compiled from several authors by Loughnan (1969)).

a pH lower than 3, whereas ferrous hydroxide is soluble with a pH of up to 8. Hence, the value of the redox potential is essential to the mobility of iron. The main factors affecting Eh in weathering environments are the availability of atmospheric oxygen and the presence or absence of organic matter (Loughnan, 1969). Oxidation is an exothermic reaction that takes place above the saturation zone. Below the water table a reducing environment without oxygen prevails. For this reason, in these environments iron appears in its ferrous state and is related to the gley horizon, generated in humid areas without any drainage. The organic matter is a powerful reducing agent due to its rapid oxidation which produces carbon dioxide. Potassium is a very soluble cation, although it is usually retained in the structure of clay minerals of the illite type. Nevertheless, under conditions of high precipitation it is dissolved, without leaving any remains in the system.

Chelation consists in the joining of a cation and a complexing agent (chelate), such as the EDTA. As a result an organometallic compound is generated, bearing a ring-shaped structure with the cation inside (Bland and Rolls, 1998). The cations fixed by these chelates can be divalent or trivalent and the organometallic complexes are quite soluble. In nature they are produced by the generation of humic and fulvic acids associated by the decomposition of plants and root secretion (Keller, 1957).

Ion exchange is the most important process in chemical weathering (Carroll, 1970). It consists of the reaction between the ions of a solution and the ones existing in minerals. Clay minerals are the main agent responsible for the exchange capacity in rock weathering. If the mineral is not electrically neutral, a cation adheres to the clay surface, neutralizing the negative charge. A frequent replacement is Si^{4+} by Al^{3+}. Each clay mineral has a different ion exchange capacity and minerals of the montmorillonite group reach the highest values.

Hydration is the addition of water molecules to a mineral structure. When hydrated, iron oxides transform into iron oxihydroxides. Water is easily introduced into the crystalline structure of expansive clays. The transformation of anhydrite ($CaSO_4$) into gypsum ($CaSO_4 \cdot 2H_2O$) also involves a considerable increase in volume.

The most important factor controlling the disintegration speed of crystalline structures and resulting products is the quantity of *solution water*. This water removes the soluble cations resulting from hydrolysis processes. The quantity of dissolution depends on the water volume in contact with the mineral surface and on its solubility. When precipitation is important and takes place over long periods, the most resistant minerals, such as quartz, can disappear by dissolution. The Weipa bauxites of North Queensland, Australia derive from the leaching of a substratum constituted by kaolinitic sandstones with 90% quartz and 10% kaolinite. Monsoonal precipitation give rise to bauxites with 5% quartz and small quantities of kaolinite (Loughnan and Bayliss, 1961).

Several investigations have been made for determining the order by which the most common cations are lost. Polynov (1937) compared the composition percentage of the solved matter in rivers, for drainage basins on igneous rocks with the composition of the latter, obtaining the following losing order: $Ca^{2+} > Na^{+} > Mg^{2+} > K^{+} > SiO_2 > Fe_2O_3 > Al_2O_3$. Other studies indicate that this order can vary depending on the rock type (Reiche, 1950; Carroll, 1970).

Crystalline structure and atomic bond strengths play an important role in mineral weathering. During magmatic crystallization the minerals that first form are the ones with the highest melting point and the largest number of weaker ionic bonds. The sequence is

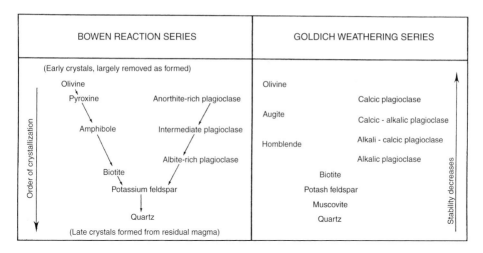

Figure 20.7. The reaction series of Bowen and the weathering sequence of Goldich (adapted by Ollier and Pain (1996), Fig. 4.2).

represented by the crystallization series of Bowen (1928). Goldich (1938) obtained the weathering sequence for the most common silicates from a detailed study on the weathering of several igneous and metamorphic rocks (Figure 20.7). The Goldich sequence is the inverse of the Bowen series, because the most easily weathered minerals are the first that crystallise in a silicate melt. The most stable minerals crystallize later in the cooling sequence and have a greater number of the more stable covalent bonds.

3.2.2. Dissolution of carbonates

Carbonates occupy important outcrops in alpine belts affected by humid tropical climates. Their weathering is due to solution processes, which have been analysed by many authors (Jennings, 1971; Bögli, 1980; Jennings, 1985; Trudgill, 1985; Ford and Williams, 1989, among others).

Calcite is the dominant mineral in carbonates, although it can also be accompanied by dolomite, whose dissolution is five times lower than that of the calcite. This last mineral is slightly soluble in distilled water, although its solubility increases with the dissolution of carbon dioxide in natural waters. Temperature affects dissolution of CO_2 and is maximum at 0°C. This relation is summarized in Henry's Law in which the total amount of limestone that can be dissolved under saturation equilibrium per unit of water volume is a direct function of the CO_2 partial pressure of the air in contact with water, and an inverse function of water temperature (Figure 20.8). The solution at the point located on the saturation curve of 30°C can cool (towards C) and hence more limestone will be dissolved. If it is heated (towards W) some calcium carbonate will precipitate. If it is put in contact with air containing less CO_2 quantity (towards L) it will lose some CO_2 and will precipitate calcium carbonate and, finally, if the solution finds an air with a higher proportion of CO_2 (towards H) it will take it and more limestone will be dissolved.

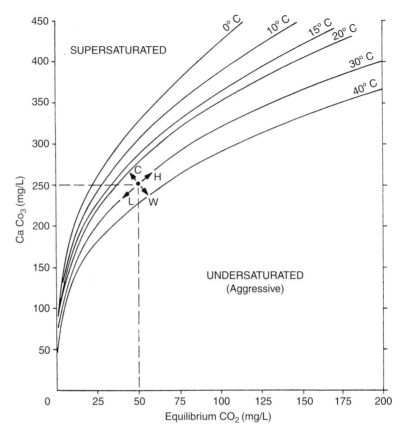

Figure 20.8. Saturation equilibrium curves for solution of calcium carbonate at different temperatures as a function of equilibrium carbon dioxide in solution (after Trombe (1952)).

The effect of temperature on carbon dioxide dissolution is very small, because at about 30°C it dissolves three times less than at 0°C. Air in the soil, however, can reach CO_2 values of up to 30 to 60 times the content of a normal atmosphere (0.03% in volume). In only weakly aerated tropical soils values of 20 to 25% CO_2 can be obtained. This seems to be due to the decomposition of organic matter and to root respiration. Therefore, the biogenic carbon dioxide is the primary factor in the dissolution of carbonates and its content is a function of temperature and precipitation, factors that activate the vegetation development. As a consequence, in humid tropical zones chemical corrosion of limestones reaches very high values.

3.3. Weathering degrees and products

Weathering degrees of the alteration products are variable and several indexes are used to differentiate them (Taylor and Eggleton, 2001). These are based on qualitative scales of friability (Ollier, 1965, 1984), on the mechanical properties of the regolith (Lumb, 1962,

1983), and on the chemistry and mineralogy of the weathering profiles (Grant, 1969; Sueoka, 1988; Geological Society, 1990). As an example, Grant (1969) correlated abrasion pH to the relation between the most mobile elements (Na, K, Ca, and Mg) and the residual clay minerals. The lower the index, the more intense the weathering. Sueoka (1988) proposed the chemical weathering index, which consists of dividing the sum of the molecular weight percentages of Al, Fe, and Ti oxides plus H_2O, by the rest of elements (alkali, alkaline earth cations, and silica). Thomas (1994) included other indexes related to the weathering degree.

As a consequence of the destruction of crystalline structures by weathering processes, several cations are released. Those with high solubility are leached by water, while others remaining in the residue are restructured, giving rise to neoformed minerals (Bland and Rolls, 1998). The most common ones are clay minerals, quartz, and oxihydroxides of iron and aluminium, which form lateritic hardpans.

Clay minerals are hydrated silicates of aluminium, iron, and magnesium ordered into sheets, the reason for which they are named "phyllosilicates." The tetrahedral sheets are formed by Si surrounded by four O ions, while the octahedral layers consist of six O or OH ions located around Al, Mg, or Fe cations that are placed in a central position. These two basic sheets can be grouped in three different ways. First, a tetrahedral sheet joints an octahedral one to form structures of the type 1:1. Clay minerals of this type are those of the kaolinite–halloysite group. Secondly, an octahedral sheet can be sandwiched between two tetrahedral sheets. This type of arrangement gives rise to the 2:1 phyllosilicates, which are represented by the smectite group (montmorillonite, beidellite, and nontronite) and the illites. The third ordering case results from the joining of an octahedral sheet between adjacent 2:1 sheets, and this type is known as 2:1:1 clay minerals, among which chlorite is the most common.

The genesis of the clay minerals is complex due to the wide variety of weathering environments. They can also originate as a result of weathering of non-phyllosilicate minerals. Cations, silicon and alumina are released and they reorganise to from clay minerals. When the altered mineral is a phyllosilicate, weathering is mainly produced in a solid state (Birkeland, 1984).

Studies by Sherman (1952) on basalt weathering in Hawaii indicated an increase in the clay percentage with precipitation, as well as changes in the mineralogy of the rock. Figure 20.9 shows how montmorillonite is a typical mineral of low precipitation regimes that transforms into kaolinite under conditions of higher leaching. When precipitation is high the soil desilicifies and iron and aluminium oxihydroxides are formed. The figure also indicates that if a dry period occurs, the alteration residue enriches in iron and titanium oxides. When precipitation takes place throughout the whole year the main constituents are aluminium oxihydroxides.

Taking into account the dominant weathering products, several weathering zones can be distinguished in the humid tropics (Strakhov, 1967) (Figure 1.8). In this sense, Pedro (1966, 1968) differentiated:

1. *Bisiallitization zone* with the formation of clay minerals of the 2:1 type (montmorillonite, $4SiO_2 \cdot Al_2O_3 \cdot H_2O$), in areas with a precipitation lower than 500 mm.
2. *Monosiallitization zone* with kaolinite ($2SiO_2 \cdot Al_2O_3 \cdot 2H_2O$) as the predominant mineral in the regolith, in zones with a precipitation of 500 to 1200/1500 mm.

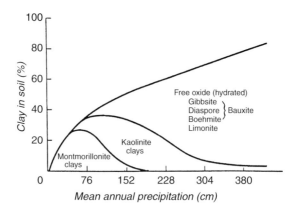

Figure 20.9. Progressive development of clay types in soils of Hawaii, under a continuously wet climate (after Sherman (1952), in Birkeland (1984), Fig. 11.12).

3. *Allitization zone* in which gibbsite ($Al(OH)_3$) appears with kaolinite in regions with a precipitation higher than 1500 mm.

Sanches Furtado (1968) indicated that kaolinite is the most abundant mineral in the humid tropics. He also established a geographical differentiation in several regions as a function of prevailing minerals and precipitation. Jenny (1941) analysed the role of temperature in the generation of clay minerals. He indicated that the clay content in regoliths of basic rocks is a direct function of temperature. At 10°C average annual temperature we obtain 15% of clays in the soil, while at 16°C it rises to 50%.

Duchaufour (1977) differentiated among three main types of alteration in the tropics, which broadly characterize different climatic zones. According to this author, *fersiallitic soils* predominate in zones with precipitation between 500 and more than 1000 mm and temperatures between 13 and 20°C. These soils also have smectites and the beginning of desilicification to generate kaolinite, with the iron appearing as oxihydroxides with nontronite, which is an iron variety of smectite. *Ferruginous soils* generate in regions with higher temperature and precipitation, and most parts of the neoformed clays belong to the kaolinite group. Finally, the *ferrallitic soils* correspond to the terminal phase of the evolution and alteration of soils, when primary minerals, except quartz, have been completely hydrolysed and have released Fe and Al oxihydroxides. Climatically they develop with an average annual temperature higher than 25°C and precipitation above 1500 mm.

In general, the type and amount of clay minerals vary with depth, although some profiles do not present substantial changes. These variations must be related to the intensity of leaching, which removes many ions and silicon down to important depths. This is a quite common modification. In this way, surficial gibbsite passes with depth to kaolinite or, if this clay mineral is already at the surface, the transition is to montmorillonite (Loughnan, 1969). These sequences can be in equilibrium with environmental conditions, although they can change. If the soil profile is eroded, the montmorillonite forms part of the surficial outcrops, but if there is an aggressive climate the montmorillonite transforms into kaolinite at the surface (Figure 20.10). If the surface lowering due to denudation is

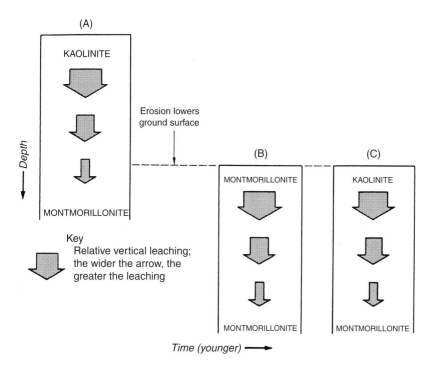

Figure 20.10. Vertical distribution of clay minerals as a function of leaching conditions within the soil. (A) Leaching conditions favour the formation of kaolinite and montmorillonite at different levels in the soil profile. (B) Rapid lowering of the surface by erosion results in the montmorillonite outcropping in a soil environment of high leaching. (C) With time, the montmorillonite in the surface becomes desilicified and alters to kaolinite. In the figure, the wider the arrows, the greater the leaching (after Birkeland (1984)).

slow and is balanced with the speed of kaolinite generation, the profile modifies its mineralogical distribution (Birkeland, 1984).

Formation and transformation of clay minerals in the weathering profile are slow processes. Hence, the study of clay minerals can be useful for making paleoclimatic reconstructions (Birkeland, 1974). As previously indicated, the montmorillonite forms in zones of low precipitation, whereas kaolinite and halloysite generate in regions of high leaching. If we find kaolinite in soils under an arid climate, we can deduce that the area underwent a humid tropical climate in the past. On the other hand, when an arid zone bearing montmorillonite in the soil profile undergoes a climate change towards high leaching conditions, the montmorillonite generated under that climate transforms into kaolinite. In consequence, we can recognize humid to arid climatic changes, but not the opposite (Pedro et al., 1969; Singer, 1979/1980). Because mineral transformations are slow, for these changes to occur, the climatic change must have a long duration. If the climatic change is brief then no detectable modifications occur. In volcanoes located on humid tropical zones, pedogenesis is relatively fast, although it is frequently interrupted by new volcanic emissions that bury and fossilize the soils. Under these circumstances the endogenous exhalative activity is much more rapid than pedogenesis and if some climatic change occurred, it would not be recorded in the soil (Figure 20.11).

Figure 20.11.   Reddish paleosoils buried by volcanic ashes in the Gregory Rift Valley, Kabernet, Kenya.

3.4. The weathering profile

Many landforms exhibit a thin layer of weathered material, whereas others are covered by an important alteration mantle, whose physical and chemical characteristics are of great importance in the landform evolution (Gerrard, 1988). *In situ* altered rock is named *saprolite*, although the more general term *regolith* is commonly used, which also includes weathered rocks that might have suffered thin mass movements (creep) or the activity of animals and plants (Ollier, 1984). French researchers call it *alterite*. The sandy material resulting from granite alteration is known as *grus* or *lem*.

The limit between the unweathered rock and the altered material can be sharp as in some granites, or diffuse as in some limestones and slates. Ruxton and Berry (1957) denote this limit as the *weathering basal surface*, although this term has been criticized due to its static implications. Because alteration deepens, this is a dynamic limit, and therefore it seems preferable to name it the *weathering front* (Mabbutt, 1961a).

3.5. Weathering depth

Regolith development and deepening is a function of factors such as climate (temperature and precipitation), biology (vegetation), geomorphology (surficial stability), hydrology, geology (rock type and joint density) and chronology (climatic and tectonic changes) (Faniran and Jeje, 1983; Thomas, 1994). The weathering depth results from a balance between weathering intensity and surficial erosion, which is closely related to topography.

In the tropics overland flow erosion is low (Fournier, 1960; Douglas, 1969), which favours weathering. Weathered material is easily eroded on slopes, but where these are gentle, or where it develops on plains, erosion is very small. As a consequence, the maximum weathering depth will be reached in plane zones with a climate suitable for rapid weathering, smooth relief, and vegetation that hinder run-off. In these zones erosion is minimum (Ollier, 1984).

Inside the profile, weathering reaches its maximum intensity in the aeration zone. In the saturation zone processes like reduction, hydrolysis and chemical substitution also take place, although with a lesser reaction velocity (Gerrard, 1988). These circumstances led Lelong (1966) to differentiate between an upper alteration zone, where water flows easily and increases hydrolysis processes, and a lower alteration zone in which percolating water slowly flows through joints. All these processes help to explain the great thickness of altered rocks.

Geological research in humid tropical zones focuses on outcrops of unaltered rock. On the contrary, when the regolith is studied with an applied finality (civil engineering, mining, agronomy, and others) it is necessary to know the surficial extent of rock outcrops and the regolith depth. This is obtained by mechanical drilling supported with geophysical methods. Quarry activities (Figure 20.12), open-pit mining works, terrain excavation for road construction, and so forth, are excellent places for studying weathering depth. Ollier (1965) recorded depths of 150 m during some tunnel drillings for a hydroelectric project in Victoria, Australia. Thomas (1965) indicated a depth of 100 m in Nigeria and Ollier (1960) found the same in Uganda. In South America, Nagel (1962) on metasediment ridges in Brazil and Feininger (1971) on quartz diorites in Colombia, determined depths more than 100 m.

Figure 20.12. Weathering profile observable on a granite quarry. Note the ferrallitic zone in the upper part, thinning towards the valley, Aparecida, Sao Paulo, Brazil.

In the arid and semi-arid zones of Africa (Thomas, 1974) and Australia (Mabbutt, 1965b), in temperate regions of the USA (Kaye, 1967), and Europe (Demek, 1965; Gutiérrez and Rodríguez, 1978; Molina and Blanco, 1980) deep weathering profiles are relatively common. It is obvious that the origin of such thickness is not possible in the climatic conditions under which they remain. In the opinion of numerous authors, they must be interpreted as generated under humid tropical climates, with a greater chemical activity. Therefore, the existence of thick weathering profiles located in zones far from the humid tropics is a good indicator of climatic change.

Thomas (1974) has analysed the spatial distribution of regoliths. The influence of rock structure is apparent because it affects weathering depth. In this sense, the density of splitting planes existing on the rock massif controls weathering thickness. The maximum depth is related to the maximum density (Thomas, 1966) (Figure 20.13), which sometimes corresponds to intersection zones between fracturing systems. For this reason the weathering front is very irregular and bears no relation to the topographical profile. On the contrary, the deepest weathering develops on the fluvial divides. This can be due to the fluvial erosion of a previously thick weathering profile (Figure 20.14) or to a faster weathering on the divides (Thomas, 1994).

3.6. Weathering zones

The alteration profile is not an unstructured regolith. Instead, several differentiations can be made within the profile, especially apparent on granitic rocks. The works by Ruxton and Berry (1957, 1961) and Berry and Ruxton (1959) on granite rocks in Hong Kong were the first key studies in establishing a zonation of weathering profiles. Wilhelmy (1958) studied similar topics in eastern Asia, Africa, Central and South America, the Iberian Peninsula, and Corsica. Mabbutt (1961b) analysed them in western Australia. Inside the profile the number of fresh rock blocks decreases towards the higher parts whereas their angularity increases with depth. In some cases joint blocks present a concentric structure formed by shells of residual materials, with a corestone that can remain totally fresh and coherent. This disposition is known as *spheroidal weathering* (Figure 20.15 and Figure 20.16). Its origin is thought to be due to more rapid chemical attack on corners and edges of joint blocks, compared to joint-block faces, as well as the migrations of elements within the rock (Ollier, 1984). The upper parts of the profile are easily eroded down to the zone where blocks predominate. The differentiation into zones is based on the alteration degree of the granite rock and on the presence of unaltered blocks inside the profile (Figure 20.17). All the previously cited authors apply these criteria to establish divisions within the weathering profile. Gerrard (1988) made a wide compilation of classifications of weathering profiles presented by different authors. Ollier (1984) proposed the most widely used zonation (Gerrard, 1988):

1. Soil.
2. Structureless regolith.
3. Regolith retaining rock structure (Figure 20.18).
4. Regolith with rounded corestones.
5. Regolith with angular, locked corestones (Figure 20.19).
6. Fresh, unweathered rock.

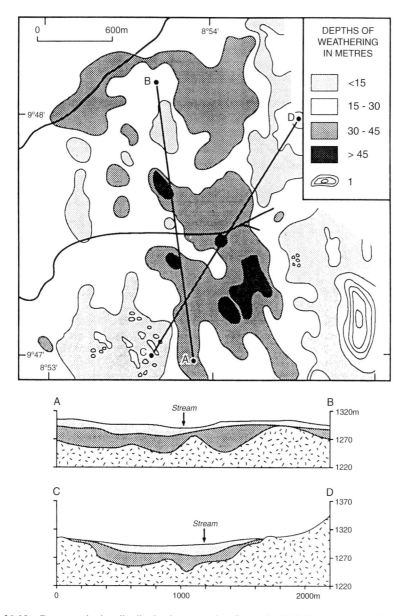

Figure 20.13. Deep weathering distribution in metres, in a fine-grained, biotite granite, near Jos, northern Nigeria. 1 Outcrops of unweathered granite showing contours at 15 m intervals. On the cross sections (A–B, C–D) saprolite is ruled (after Thomas (1966), in Thomas (1994), Figure 3.16).

Obviously, the zonation of weathering profiles on sedimentary, metamorphic and volcanic rocks is very variable, although the Ollier profile type is also applicable. In all of them rock fissuration plays a fundamental role. Other profile divisions can be made based upon chemical, mineralogical, hydrological, or engineering criteria, depending on the objectives (Ollier, 1984).

Figure 20.14. Thick weathering profile incised by the Mathioya River, Kenya.

Figure 20.15. Spheroidal weathering in acid volcanic rocks, Fort de France, Martinique Island, Lesser Antilles.

Figure 20.16. Spheroidal weathering in basic lavas of the Gregory Rift Valley, Kabernet, Kenya.

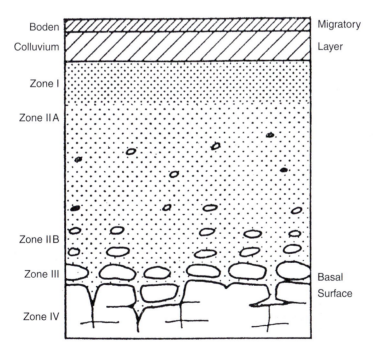

Figure 20.17. Characteristic weathering profile over jointed granite (modified from Ruxton and Berry (1961)).

Figure 20.18. Weathering profile on granite. Primary and *schlieren* structures can be seen together with spheroidal weathering. The photograph was taken on a quarry where altered granite is exploited for construction in La Alberca, Salamanca, Spain.

Figure 20.19. Weathering profile where the transition between the rounded and the angular block zones can be observed, Spanish Central System, Villacastín-Ávila road.

3.7. Weathering rates

Chemical weathering processes exist in other climatic zones, but in the humid tropics alteration velocity is much higher (Muxart and Birot, 1977). In this sense, and in addition to the study of these processes, the quantification of rock weathering is also necessary.

The methods applied for this purpose are varied. Quantitative data can be obtained from the alteration of building and ornamental stones, whose age of installation is known. Numerous authors have carried out these types of works under different climates (Ollier, 1984). Approximate data can also be obtained in the laboratory by accelerated aging. Values recorded in the beginning of the weathering process are the highest. Therefore, recorded weathering rates will be greater than those existing throughout the whole alteration, and they have been estimated to be about 200 to 400 times higher than the ones obtained in the field (Swoboda-Colberg and Drever, 1993). Nevertheless, one of the most utilized methods is based on the quantification of the solid and solved discharge, measured in experimental plots of some tens of metres long. The chemical composition of the weathering residue has also been studied (Nahon, 1991). The experimentation period is variable and normally fluctuates between 1 and 5 years. In humid tropical savanna zones of Cameroon a surficial lowering rate has been obtained of 1 to 2 mm/ka (Birot, 1978), and 1.2 mm/ka in Ivory Coast (Roose and Lelong, 1976). In Burkina Faso, Senegal and Ivory Coast values fluctuate between 1 and 8 mm/ka (compiled by Goudie, 1995). Another method consists in the study of the bedload, suspended, and dissolved load discharges from small basins. In the study of three experimental basins in New England in western Australia, it was found that the dissolved load is ten times greater than the suspended load, and the bedload is virtually absent (Zakaria, 1977; in Ollier (1984), p. 209). Obtained denudation values are 14.4 mm/ka for basalts and 4.5 mm/ka for granites (Ollier, 1984). Data from Rhodesia indicate a rate of surface lowering in granites of 15 mm/ka in rainy zones and 6 mm/ka in drier zones (Owens and Watson, 1979). Erosion values obtained on ultrabasic rocks of New Caledonia are about 28 mm/ka (Trescases, 1973, in Muxart and Birot (1977)). A compilation made by Thomas (1994) on alteration velocities gave an estimation of 2 to 50 mm/ka. Finally, weathering rates over long periods of time can be calculated from absolute dating of geologic materials (pyroclasts, lavas, etc.).

Values obtained from experimental methods indicate that most regoliths formed during the Quaternary because average weathering rate values are about 20 m per million years. Birot (1978) gave values for the alteration of 1 m granite: 50,000 years in a rainforest, and 70,000 years in a savanna. All of them are similar to that proposed by Thomas (1994). When considering these long time periods, it must be kept in mind that the climatic changes can have occurred during the profile development. In dry periods the weathering rate slows, whereas denudation increases.

3.8. Laterites

Several denotations exist in the literature for the term "laterite," such as "duricrust," "plintite," "ferric laterite," "ferricrete," "latosol," and others (Thomas, 1974), and it is difficult to define. In the 19th century the physical properties, strength, and colour of laterites were already known. Buchanan (1807) described a material soft enough to be cut into blocks with an iron instrument, although it hardened when dried. This is the reason why it is used for making bricks, from which the Latin term laterite is derived. In the transition to the 20th century laterite begun to be defined as a function of its chemical composition, by using the content of iron, aluminium and the relationship between SiO_2, Al_2O_3, and Fe_2O_3. Although numerous laterite definitions have been proposed (for example, Alexandre and Cady, 1962; Sivarajasingham et al., 1962; Norton, 1973; Schellmann, 1981), little difference exists among them. A laterite is considered to be formed in environments with intense chemical weathering, lacking alkali and alkaline earth elements, enriched in iron and aluminium, and with important quantities of quartz and kaolinite. It hardens by wetting and drying.

A controversy exists between the pedological and the geomorphological concepts of laterite. Pedologists understand a lateritic soil as a residual one, iron-enriched in its upper part under conditions of wetting and drying, whereas for geomorphologists laterites generate under special geomorphic conditions and the precipitates are produced in the water table fluctuation zone. For this reason, pedologists consider this type of laterite as a podsolic tropical soil. All this has led to a differentiation between pedogenetic laterites and the ones generated by a fluctuating water table (McFarlane, 1976). This controversy transforms into a difficult agreement when trying to explain soil thickness of 60 m or more (Hays, 1967), in which case it is hard to argue for a pedogenetic origin. Following Goudie (1973), this last type requires a thin horizon, although in most cases this does not occur.

3.8.1. Constitution and structure

Chemical analyses of Al_2O_3, Fe_2O_3, SiO_2, and clay minerals from laterites can be represented in triangular diagrams, which permit the making of compositional differentiations. Some authors use Fe_2O_3, Al_2O_3, and kaolinite percentages (Bardossy and Aleva, 1990), while Dury (1969) and Schellmann (1981) substituted SiO_2 in the diagram for kaolinite (Figure 20.20). These studies show a wide compositional variety,

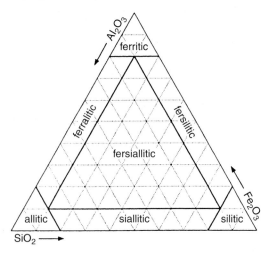

Figure 20.20. Ternary diagram with different types of duricrusts as a function of their chemical composition (after Dury (1969), in Thomas (1994), Figure 4.2).

even within a single level. Thomas (1974) proposed a terminological simplification by using three terms: bauxite for a duricrust rich in aluminium, silcrete where the silica content is high, and the term laterite was restricted to materials of intermediate composition.

Because laterite is generated from a residue resulting from the chemical weathering of a rock, we can expect to find a direct relationship to the parent rock composition. At a regional scale, some variations in the iron content have been recorded, while at a minor scale no clear relations exist between the laterite chemical composition and the parent rock. In this sense, the generation of bauxite is not always favoured by rocks rich in Al_2O_3 (McFarlane, 1983).

The main neoforming minerals in laterites are iron and aluminium oxides and hydroxides, such as hematite (Fe_2O_3), goethite ($FeO \cdot OH$), gibbsite [$Al(OH)_3$], boehmite ($AlO \cdot OH$), and diaspore ($AlO \cdot OH$), and clay minerals such as kaolinite and halloysite [$Al_4Si_4O_{10}(OH)_{18}$], both bearing the same composition. All the neoformed minerals tend towards a higher stability according to the environmental circumstances and in humid tropical zones they modify mainly by leaching. The kaolinite can proceed from feldspar alteration or from a neoformed montmorillonite. The goethite directly derives from biotite and hornblende, and the hematite from goethite by dehydration. Feldspar weathering can directly originate gibbsite, although normally this mineral proceeds from the desilicification of kaolinite.

Laterites present different *structures*, frequently bearing *nodules* without inner structure, and *pisoliths* (diameter > 2 mm) and *ooliths* (< 2 mm) with concentric layers. These can appear as scattered or packed concretions, forming a cemented oolithic or pisolithic laterite. The concentric structure forms by enrichment in the water table fluctuating zone. The pedogenetic pisoliths are composed of loose material and present irregular form and size, with a manganiferous external layer (McFarlane, 1976). Other structural types in laterites are vermicular, vesicular, cellular, tubular, and other types, whose differentiation is based on the morphology of hollows. This terminology is not clearly defined and McFarlane (1976, 1983) only uses the term *vermiform* for characterizing laterites with inner tubes and cavities (Figure 20.21). The diameter of the conduits is about 2 to 3 cm and they are commonly filled with kaolin. The origin of such structures is complex. It is ascribed to termite activity (termite armouring of Erhardt, 1951) and to plant root penetration. McFarlane (1976) considered that they result from precipitation and they only appear in laterites formed by water table fluctuation. This author also indicates that pisolithic laterites are immature laterites, while the vermiform ones are mature laterites and derive from the former. Therefore, pisolithic structures can be found alternating with vermiform ones.

3.8.2. Factors affecting generation

Laterites and bauxites appear in a wide variety of rock types (Mabbutt, 1961b; Goudie, 1973, among others), although rock mineralogy substantially influences the composition of lateritic rocks (Thomas, 1994). Nevertheless, there exist zones where a given lithology favours laterite development, whereas in other sites the same lithology inhibits its generation. This seems to indicate the existence of other more determinant factors in its development (McFarlane, 1976).

Figure 20.21.   Vermiform laterite, Eastern Bouaké, Ivory Coast.

Topography is one of the factors that substantially affect laterite development. In steep slopes runoff is important and erodes weathering products, whereas in gentle slopes water easily penetrates into the soil, dissolving the most soluble constituents and accumulating the less mobile ones. As a consequence, in its generation laterite is associated with zones of low relief. Some laterites, however, can be seen crowning mesas and platforms, which are interpreted as produced by incision of low-relief laterites. Detrital laterites can also develop at the foot of these high-relief areas (Figure 20.22). Some authors think that laterites form with the progressive lowering of the land surface, whose final product is a planated surface capped by a thick layer of residual materials. Therefore, laterite delays the formation of this planation (McFarlane, 1976, 1983).

The *climate* prevailing in the humid tropics is the most adequate for laterite generation, although they also seem to form under temperate climates (Taylor et al., 1992). Soil temperatures, close to 27°C, are the most suitable for the development of lateritization processes (Thomas, 1994) because at these temperatures percolating water disintegrates rocks more easily and silicon dissolution is faster. The most suitable conditions for precipitation are those of humid and dry tropical climates, because only humid climates do not favour the oxidation–reduction conditions needed for iron mobility and precipitation (Nahon, 1986).

The *biological* factors include the vegetation influence and the activity of organisms and organic compounds. Laterites commonly appear associated with grasslands (savanna), although they have also been recognized in rainforest zones (McFarlane, 1976). Thomas (1994) indicated that bauxites are related to humid conditions of rainforest vegetation,

Figure 20.22. Lateritic breccia over weathered gneiss, Machakos, Kenya.

whereas the generation of laterites is favoured under drier conditions. The rainforest maintains soil humidity and the vegetation mass produces an important raindrop interception and reduces sheet wash erosion. Plants absorb silicon, which is generally stored in the leaves. At the same time, different micro-organisms are involved in aluminium, iron, and silicon mobility (McFarlane, 1987). Aluminium and iron can also be leached from chelates, as organometallic compounds.

Most laterites are associated with wide planations and the time needed for the generation of these forms covers millions of years. Ahnert (1970) estimated the time required for their development to be about 10 million years. In addition, several studies have been made about the parent rock thickness necessary for the generation of a given thickness of laterite. By taking into account that granites have an average proportion of 2% of $FeO + Fe_2O_3$, Trendall (1962) calculated that for generating a laterite 9 m thick, weathering would be required of 180 m of the Buganda granite (Uganda), which would take several million years. Humid tropical zones, especially those of the savanna type closer to desert margins, have been subjected to important *climatic changes*. For this reason, it is common to find extensive lateritic mantles in semi-arid zones. Moreover, many lateritic deposits of Late Cretaceous and Paleogene ages have been recognized in extratropical areas. The basal deposits of the Tertiary sedimentation in the Duero Basin of the Zamora–Salamanca region in western Spain are constituted by transported laterites, kaolinitic sandstones, and silcretes (Martín-Serrano, 1988). Because the upper concomitant levels have a Paleocene age, it is deduced that prior to Tertiary times the Hesperian Massif was exposed to a humid tropical climate. The ferruginous Buçaco sandstone in Portugal (Birot, 1949a,b) must constitute a correlative sediment to this tropical climate.

3.8.3. The laterite profile

Laterite forms part of a group of weathered materials that are organized into zones or horizons, forming what is known as the laterite profile. An ideal profile was elaborated by Millot (1964) (Figure 20.23), inspired by differentiations made by Walther (1915) and reported by Thomas (1974) and Faniran and Jeje (1983) in their respective treatises. Above the lateritic horizon there lies a brownish loose material, slightly humic and with ferruginous concretions and fragments. The *laterite*, with a dark ochre colour, presents a thickness ranging between some metres to 60 m with dominant vermiform and pisolithic structures. Below the lateritic horizon there appears the *mottled clay* zone with red ferruginous blots 1 to 9 m thick, although in some profiles it is lacking. The transition to the laterite can be sharp or gradual (McFarlane, 1976). The lowest horizon is the *pallid zone*, also named lithomarge, constituted by kaolinitic clays and quartz grains, and of variable thickness (5 to 60 m). Its white colour is due to iron leaching and, much like the motted clay zone, it is absent in some profiles. In general, if the pallid zone presents an important thickness, the laterite thins and vice versa (Loughnan et al., 1962). Downwards there is a *transition zone* of variable thickness, constituted by weathered material rarely containing fresh rock blocks, which lies upon the parent rock.

This standard or ideal profile serves as a guide for recognizing the different zones that can exist in lateritic profiles although, as mentioned above, some horizons may be missing in a given profile. Laterites or bauxites of great thickness in some cases lie almost directly on fresh rock, whereas thin laterites commonly rest on very thick regoliths (McFarlane, 1983).

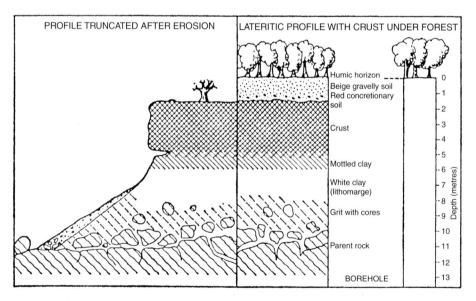

Figure 20.23. Lateritic profile under shade forest, and its equivalent truncated by erosion in Sudan (after Millot (1964)).

3.8.4. Laterite genesis

One of the first hypotheses indicates that laterite originates from a weathering profile where the pallid zone looses iron, which is transferred to the laterite. This iron enrichment is produced by capillarity and by seasonal fluctuations of the water table. The capillary rising seems to be ineffective and overemphasized (Goudie, 1973). Moreover, when the pallid zone presents a thickness of about 60 m the capillarity mechanism results are hardly acceptable. Water table oscillations, especially in climates with humid and dry seasons, favour the movement of dissolved iron. When the groundwater table rises, the iron moves and precipitates in the higher parts of the profile (McFarlane, 1976, 1983).

Another explanation for the iron enrichment results as a consequence of its lateral movement from higher parts in the slopes towards the valleys (Maignien, 1966) (Figure 20.24). By this mechanism iron concentrates in the lower parts of slopes and in the valley floors. If erosion processes indicated in the figure act over a considerable time interval, a relief inversion can be produced by which laterites existing in low areas occupy divide positions (McFarlane, 1976) (Figure 20.25).

In order to explain the strong thickness of the pallid zone, Trendall (1966) proposed topographic surface lowering through time, by which the profile is displaced downwards. These ideas were developed by McFarlane (1983) and by Tardy and Roquin (1992) when interpreting thick laterites lying upon a regolith. Instead of considering the weathering profile as a static column, they estimated that it lowers by losing solutes and fine particles, and then horizons are continuously changing. Hence, the weathered rock that supplies iron locates in progressively lower topographic positions, as happens with the laterite. After studies made in the lowlands of Uganda, McFarlane (1976, 1983) proposed a set of

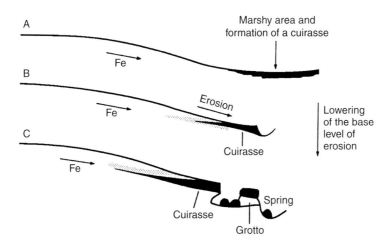

Figure 20.24. Lateral migration of iron oxihydroxides and laterite formation. (A) Migration of Fe in groundwater by organic acids and accumulation in marshy valley floors. (B) and (C) Incision of drainage with a falling water table, which leads to induration and erosion of the duricrust (Maignien (1966), in Thomas (1974), Figure 13).

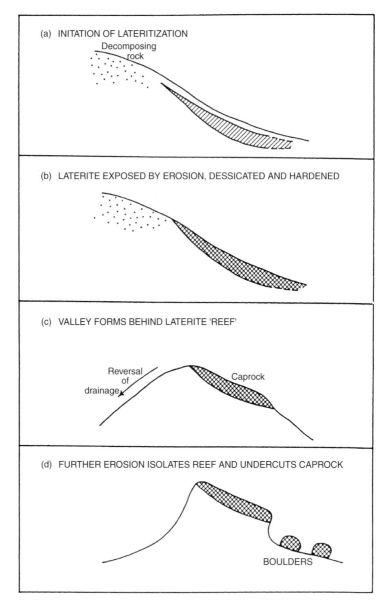

Figure 20.25.   Relief inversion by erosion of slope-valley laterites (Clare (1960), in McFarlane (1976), Figure 20).

several developing stages for laterites generated by water table oscillations (Figure 20.26). In this evolutionary sequence, a topographic surface lowering is required, with continuous segregation developed during the time intervals of water table oscillation that give rise to a pisolithic lateritic layer. When relief smooths, these pisolithic lateritic layers transform into massive vermiform laterite. Therefore, profile and land surface evolutions are interdependent.

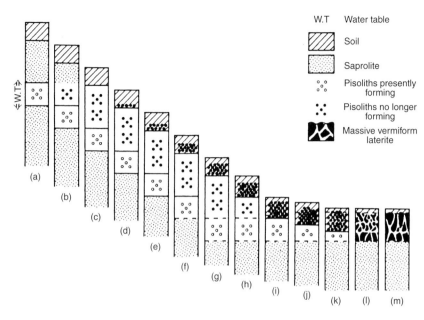

Figure 20.26. Evolution of laterites generated by water table fluctuations. (a) Iron segregation to form pisoliths within the narrow range of oscillation of the groundwater table. (b) and (c) The lowering of the groundwater table lowers the locus of pisolith formation, leaving earlier-formed pisoliths in the vadose zone. (d), (e), (f), and (g) The pisoliths accumulate at the base of the soil. (h), (i), (j), and (k) The water table stabilises and continued leaching reduces the vadose zone to bring the residual sheet of pisoliths into the zone of intermittent saturation. (l) and (m) The pisolithic residuum is altered to goethite-rich massive vermiform laterite (after McFarlane (1983)).

3.9. Bauxites

They are also denoted as aluminic laterites and alcretes, and are considered to be the final product of deep weathering because SiO_2 and Fe^{2+} have been intensely leached (Thomas, 1994). They can be recognized in the geological record of numerous places on the Earth. Conditions for their development are related to humid tropical climates, porous parent rocks, good drainage, and enough time for generation (Ollier and Pain, 1996). Bauxites can be generated by alteration of rocks with aluminic silicates, such as basic igneous (western Africa) (Valeton, 1972; Boulangé and Millot, 1988), metamorphic (Malaysia) (Allen, 1972), and volcanic rocks (Hawaii) (Sherman, 1952). These bauxites can be produced by the direct weathering of alkaline rocks, without any volume loss (Boulangé and Millot, 1988), or by indirect weathering, where kaolinite desilicification gives rise to gibbsite (Schellmann, 1977). Detrital bauxites have also been recognized, originating as breccias or nodular deposits (Allen, 1972). Where tropical weathering affects carbonate rocks, then karstic bauxites can be produced (Bardossy, 1981). Bauxites also develop upon sediments (Tertiary arkoses of Weipa, Queensland, Australia) (Loughnan and Bayliss, 1961) and can be generated by sedimentation of aeolian dust (Brimhall et al., 1988).

Many bauxites can be considered as derived from leaching of old laterites. Tardy and Roquin (1992) indicated that some aluminic laterites present two layers with gibbsite enrichment: one near the saprolite profile base, almost in contact with the parent rock, and the other at the profile top, near the soil surface.

Bauxitization increases with fluvial incision. It has also been observed that bauxites show a better development along fault lines, which facilitate leaching processes. In many cases the bauxite outcropping on an escarpment laterally wedges towards the lateritic mass. This indicates that the bauxite does not form a continuous layer and is genetically related to the escarpment evolution (McFarlane, 1983).

About 10% of the bauxites belong to the karstic type (Ollier and Pain, 1996). In these cases the bauxite fills fissures, depressions, and cavities of the karstified carbonate rock. Initially bauxites were thought to constitute insoluble residues from the dissolution of calcareous massifs. It is difficult, however, to explain the source of important aluminium quantities. Following Bardossy (1981), karstic depressions produced by differential dissolution receive aluminium-rich sediments supplied from surrounding zones. Moreover, bauxitization is favoured by the good drainage conditions of these zones, which are normally protected against erosion. Bardossy also indicated that, while the karstic depression progressively grows in size due to carbonate dissolution, aluminium content increases by leaching of the renewed external supplies.

Bauxites appear in the geological record of a great many sites in the world (Valeton, 1972). In Spain, bauxites have a karstic origin and are mainly located in the Jurassic-Lower Cretaceous limit of the South Pyrenees, in Late Triassic sediments of the central Catalonian Coastal Ranges and in the Lower Cretaceous of the northeastern Maestrazgo region (Combes, 1969).

Formation of karstic bauxites has progressively increased since Paleozoic times (Bardossy, 1981), with a notable increment during the Cenozoic probably related to an increase in carbonate rocks. Neogene karstic bauxites constitute the fourth part of the total existing in the world.

Chapter 21

Tropical landforms

1. Introduction

Tropical zones affected by the Alpine Orogeny are constituted of high relief with great volcanic cones like Nevado del Ruiz in Colombia, Cotopaxi in Ecuador, La Soufrière in Guadalupe (Figure 21.1), Pinatubo in the Philippines, Merapi in Java, and many others. These tropical regions correspond to the mountains of Central America, the Antilles, southeast Asia and Indonesia. In the three last areas dissolution processes affect great extensions of carbonate rocks, giving rise to characteristic tropical karst landforms with towers and conical hills. These tropical zones present a remarkable tectonic control because they still undergo Quaternary crustal instability. Weathering products generated in these areas are easily exported by fluvial action. Hence, these regions are characterized by undergoing vertical tectonic movements that generate relief, important volcanic activity of the explosive type, and an important erosive fluvial activity (Wirthmann, 2000).

Instead, regions on shields or cratons of Africa, Brazil–Guyana, India and Australia are characterized by tectonic stability, at least during recent geologic times. These circumstances have favoured the development of extensive planation surfaces mainly upon Precambrian rocks, typical of these tropical shield regions. As a consequence, relief is commonly smooth and weathering products undergo weak water erosion. The result of these conditions is the development of thick weathering profiles that sometimes can exceed 100 m. Base-level lowering produces deep incisions in the regolith, originating a set of landforms related to differential weathering and erosion processes (Faniran and Jeje, 1983). Finally, the *rift valley* system is located in eastern Africa and is associated with important volcanism. The most outstanding examples of this igneous activity are Mount Kenya (5194 m), and Kilimanjaro (5895 m) (Figure 21.2) in Tanzania.

2. Erosion and sedimentation in fluvial systems

Water courses in the humid tropics transport little coarse bed load. This is due to intense alteration and that the regions were not glaciated in the recent past as were so many of the temperate areas of the Northern Hemisphere. The regolith is constituted of sand, silt, and clay particles and, therefore, rivers mainly evacuate materials of these sizes and dissolved products. These sediments have little abrasive power and small erosive capability. Little research exists on the bed load of tropical rivers and hence very few data are available about this topic. More research work was produced for mountain zones, where sediments are rapidly removed by major flows which are able to transport all the solid discharge to lower parts of the basin by incising into the existing rocks.

Figure 21.1. La Soufrière volcano with a domal summit. South Guadalupe Island (Lesser Antilles). Photograph taken in March, 1994.

Figure 21.2. The great Kilimanjaro strato-volcano (5895 m high) in Tanzania, rising over the plains of the Amboseli National Park of Kenya.

This behaviour sharply contrasts with the morphodynamics of rivers flowing through low relief zones, where sediment evacuation by rivers is mainly done by suspension and dissolution (Thomas, 1994).

In the preceding chapter the main characteristics of precipitation were analysed. In this chapter hydrological, erosional and depositional aspects of fluvial systems are discussed. During rain the soil can reach its infiltration capacity and then overland flow occurs. The run-off coefficient expresses the percentage of precipitation flowing on the surface, and knowledge of it is very important for understanding the dynamics of geomorphological systems. Calculation of the run-off coefficient is usually made on plots, micro-basins, and fluvial catchments. Results vary substantially through time due to the important role that storm duration and intensity play on the water infiltration.

2.1. Soil-loss quantification

By using our data about soil loss and other data compiled from other authors, results indicate they strongly differ depending on the studied zones and are controlled by several factors (Fournier, 1960, Walling, 1984, Douglas and Spencer, 1985b). Nevertheless, in small *plots* (Figure 21.3) on slopes with rainforest or savanna vegetation and surfaces sloping from 5 to 15°, similar values of soil loss are obtained. Data obtained by Kesel (1977a) in Guyana and by Nortcliff et al. (1990) in Ivory Coast fluctuate between 0.2 and 8.5 T/ha/yr. Where vegetation undergoes an important removal by human cutting or by fire, these values strongly increase (Lal, 1986). This type of vegetation totally covers

Figure 21.3. Experimental plots for studying water erosion. Katumani Experimental Station, Machakos, Kenya.

the soil with a continuous and dense layer of fallen stems and leaves that can reach 25 cm thick (Ruxton, 1967), preventing the effects of raindrop impact and splash. Therefore, particles are removed by water flow under the layer of fallen leaves.

Research carried out on fluvial basins reveals that suspended load varies with rain regime, and from one storm to another. Moreover, suspended load rapidly increases with water discharge (Douglas, 1977). Data about suspended load obtained from experimental plots are not easily comparable with values obtained from *small basins*, due to the implication of other types of process. In Nigeria, for basins smaller than 20 km^2 in rainforest zones, values obtained of *suspended load* fluctuate between 0.4 and 29.5 T/km^2/yr (Ogunkoya and Jeje, 1987). In areas of high relief, such as eastern and southern Africa, sediment loss considerably increases to values of 100 to 1000 T/km^2/yr (Walling, 1984). On the other hand, where basins exceed 10,000 km^2, obtained values on transported sediments decrease substantially (Pilgrim et al., 1982). In this way, the Congo River only records 11.3 T/km^2/yr and the Niger River 33.1 T/km^2/yr (Milliman and Meade, 1983).

Data compiled by Douglas and Spencer (1985b) about suspended load, run-off, basin area, and vegetation cover percentage on *tropical rivers* of different locations in the world can be used for making some considerations. The Amazon River begins in the Peruvian Andes and at the basin head it has numerous and important affluents excavating the Andean Cordillera. These tributaries have a great erosive power and denudation velocities are very high, whereas in the medium and low courses erosion is moderate. Recorded values of suspended load (Figure 21.4) fluctuate between 0.3 and 115.9 m^3/km^2/yr

Figure 21.4. The confluence of Negro and Solimoes rivers, affluents of the Amazon River. The Solimoes River transports an important suspended load, while the one of the Negro River is very small. This confluence is called "encounter of the waters" and is located some kilometers East of Manaus, Brazil.

(expressed in volume) (Gibbs, 1967). The greatest erosion values correspond to islands of SE Asia, Indonesia and Papua-New Guinea, where slopes and runoff are very high. The Cilulung River of Indonesia annually transports 8000 m^3/km^2, the highest value ever recorded. In areas of low relief, represented by the cratonic regions of Brazil, Africa, India, and Australia, recorded values are much lower, about 30 m^3/km^2/yr. Finally, in the lower Amazon tributaries, where relief is very low, suspended load is remarkably lower and ranges between 0.3 and 3.8 m^3/km^2/yr.

Studies on plots and small basins carried out in Malaysia indicate that *dissolved load* represents 27.4 per cent of the total sediment discharge (Douglas and Spencer, 1985b). In fluvial basins chemical denudation varies between 143 T/km^2/yr for the Yamuna River, affluent of the Ganges River, and 9 T/km^2/yr for the Congo or Zaire River. The Amazon River presents intermediate values, of 35 T/km^2/yr (Sarin et al., 1989). Nevertheless, average values of dissolved load for the affluents of the low Amazon basin are of 5 to 8 mg/l, very small if compared with the global average concentration in rivers, which is about 120 mg/l (Walling and Webb, 1986). The highest dissolved concentrations appear in rivers crossing evaporitic formations, with values of up to 60,000 mg/l.

By considering the average values from all rivers of the world, the resulting relationship between suspended and dissolved load is 3.5:1 (Meybeck, 1979). The highest values are those of the Ganges–Brahmaputra rivers, which bears a relationship of 11:1 because of the great supply given by the Himalayan range. On the other hand, the Congo River basin has little altitudinal contrast and hence the relationship of this river is of 0.9:1 and dissolved load exceeds that transported in suspension (Walling and Webb, 1986).

2.2. Slopes and stone lines

Despite the great extension occupied by slopes within these environments (Young, 1972), few rigorous studies exist about them. This can be due to the high vegetation density, which hampers their study, and that a great part of the observations made on them are based on visual estimations rather than on accurate measurements of angles and forms in the field (Chorley, 1964).

Relief in shield areas is sometimes interrupted by long and deep *escarpments* linked to different factors and processes. Differential erosion is recognizable on escarpments on sandstones, of which the most spectacular ones are those of Roraima in southeast Venezuela, with plunging walls several hundred metres high. In Africa, some of the great escarpments are associated with the stepping of extensive erosion surfaces of different ages (King, 1962). Escarpments of tectonic origin, generated by endogenetic activity during the Cenozoic period, reach their utmost expression along the *rift valley system* of East Africa (Figure 21.5) with vertical flights greater than 1500 m. In general, they form on basic volcanic rocks and have been mostly generated in recent times because they appear relatively unaltered by erosion (Nyamweru, 1980). Within the prelittoral Tertiary graben system of the region between Sao Paulo and Rio de Janeiro in Brazil important escarpments can be recognized in the Da Mantiqueira and Do Mar ranges, with heights close to 1000 m (Petri and Fúlgaro, 1983).

In the cratonic areas, bared hills on crystalline rocks are plentiful, commonly forming *inselbergs*. In some cases they constitute grouped hemispheric domes, named *half-oranges*

Figure 21.5.   View of the Gregory Rift Valley bottom and its western border, seen from the Menengai Caldera. Nakuru, Kenya.

(in Portuguese, *meias laranjas*), or the sugar loafs (bornhardts), that are very common in several states of eastern Brazil. Their prevailing morphology consists of convex forms in the upper parts of the slopes and concave profiles in the lower parts; rectilinear elements or, in some cases, hanging faces, develop between them (Savigear, 1960; Thomas, 1965). Where the hill is constituted by blocks the slope profile is complex and forms a stepped microrelief (Pugh, 1966).

Slopes developed on flat zones on weathered materials mainly bear a convex profile, although it can change to a convex–concave one by valley incision. In his study made on the Mato Grosso of Brazil, after using the methodology proposed by Savigear (1952) and later extended by Young (1964, 1970) differentiated among five profile types (Figure 21.6). Profile A has a continuous convex element with low slopes, is associated with minor-order valleys and represents 44% of all the measured profiles. Profiles B and C present gentle convex crests interrupted by steps and incised by valleys. Profiles D and E bear laterites on their crests and show more evident concavities. Following Swan (1972), profiles developed on coarse-grained regolith present convex forms due to their high porosity, while in fine-grained laterites slope morphology acquires a concave form.

The main processes that produce slope accumulations are surficial water flow and mass movements. The vegetation and precipitation regimes considerably influence colluvium removal, by which a high intensity rainfall can trigger failures. One of the most important problems, also common in slopes of other morphoclimatic zones, is the absence of dating of slope deposits. Some ages have been obtained from archaeological remains. Relative dating has been established by studying soil sequences, in which different slope deposits stack vertically. At the same time, these colluvial sequences can alternate with alluvial sequences or interdigitate with them. Climatic changes tending to more humid or drier

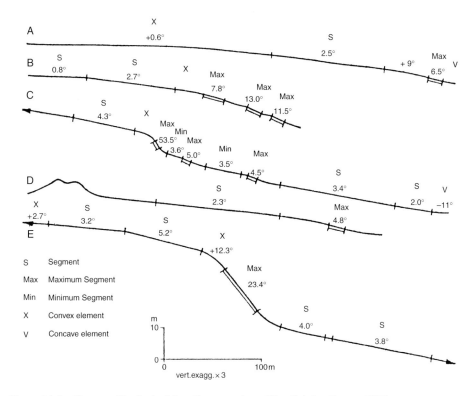

Figure 21.6. Slope profiles in the Mato Grosso region of Brazil (after Young, 1970).

periods are recorded in slope deposits of rainforest zones (Thomas and Thorp, 1992) and more commonly on those of savanna areas (Watson et al., 1983, among others).

Stone lines constitute a continuous or discontinuous layer of little weathered clasts, normally made of quartz, quartzite, and laterite fragments, with a thickness between 0.1 and 1 m. They are located within the slope deposit and occur 0.5 to 2 m below the surface (Figure 21.7), although in SW Nigeria they can occur at 5 m depth. Stone lines usually have a rectilinear outline, despite the fact that in some cases they present many undulations. In general, there only exists a single stone line, although Fölster (1969) recognized three layers in slopes of Nigeria. Stone lines are mainly described for only humid tropical zones although they have been also cited in temperate climates.

Several hypotheses have been proposed about their genesis. One of them considers stone lines as a residual layer resultant of intense *sheetwash erosion* under a dry period during which fine material is exported. Stone accumulation on the slope takes place at the same time weathering deepens. Afterwards, during the transition to a humid period, the stone line is covered by fine material (De Ploey, 1964; Vogt, 1966; De Ploey and Poesen, 1989). It has been also proposed that the evacuation of fine and dissolved material occurs by underground waters under the stone line (Young, 1976; McFarlane and Pollard, 1989).

Figure 21.7. Stone line located a meter below the surface. The weathered granitic substrate can be seen in the foreground. Machakos, Kenya.

Other authors indicate that *creep*, which affects soil and quartz clasts coming from the abundant small dikes, removes and redistributes the stones. These are concentrated at the saprolite-soil contact, forming the stone line (Berry and Ruxton, 1959). Creep is more active in some soil discontinuities, developed at 0.5 m depth, and can be accompanied by small laminar slides and interstitial flow, which helps in evacuating fine material (Moeyersons, 1989). All this produces clast removal towards the discontinuities, giving rise to the stone line.

Another hypothesis, supported by geomorphologists and pedologists, is based on the activity of *termites*, which remove the fine fractions of the soil (<2 mm). These particles are transported from the underground to the surface, leaving behind the coarse fragments, which constitute the stone line (Williams, 1978; De Dapper, 1978, 1989). Several arguments exist in favour of this interpretation: the size of the particles located over the stone line is the same as the one of termitaria; termite galleries are usually observable under the stone line, and the stone lines present synformal outlines under a termite tumulus (Figure 21.8). Nevertheless, there exist regions such as Madagascar, where stone lines are common whereas termites are rare or do no exist (Young, 1972); in other areas, rounded stones, allocthonous fragments or prehistoric artifacts contradict this interpretation. Thomas (1994) compiled data about the speed of generation of the upper soil parts by termite bioturbation. It was estimated that a layer 3 to 5 m thick is formed in 6000 to 10,000 years, although this was estimated without knowing the account of surface erosion. After studying the Northern Territory of Australia, Williams (1968) calculated a generation speed of 0.03 mm/yr, a much lower value after considering the overland flow erosion.

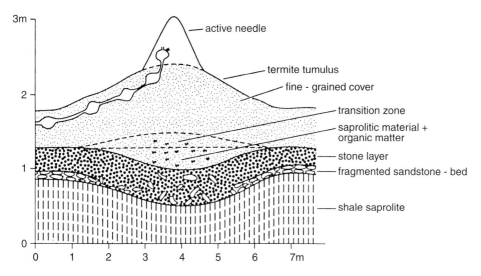

Figure 21.8. Termite activity under a stone layer near Lubok Perong, Perak Terap, Malaysia (after De Dapper, 1989, in Thomas, 1994, Figure 8.33).

Stone lines interpreted as a residual layer lead to the past development of a climatic change towards more arid conditions. Bigarella and Andrade (1965) made these circumstances evident in Curitiba, Brazil, where they interpreted stone lines as palaeopavements. Values about erosion and formation of the upper soil parts by termites vary between thousands and hundreds of thousand years. Within these time intervals important climatic changes occur, which affect the regional geomorphic processes, fauna, and vegetation. It seems that in order to explain the genesis of stone lines, all the above-cited processes do not act separately nor maintain the same intensity during these long periods of time. Nevertheless, an order of actuation can be established, which begins with an aridification that leads to the evacuation of fine material to the valley bottoms. In the meantime, the coarser fragments form the stone line that will be subsequently buried by fine material of mechanical or biological origin (De Dapper, 1989).

2.3. Fluvial erosive landforms

In some fluvial basin heads the vegetation cover may have diminished due to natural or anthropic causes. These circumstances commonly trigger gullying processes on slopes, which progress with great celerity (Morgan, 1986) (Figure 21.9). The *gully* head retreats by *sapping* and if the material under erosion is relatively uniform the gully walls remain vertical. Where resistant rocks constitute the wall top, the base is eroded more rapidly and even the water, while falling in the channel head, exerts a cascade effect that emphasizes wall *undercutting*. This produces falling of material from the top and the subsequent gully head retreat. The *piping* process is relatively common in gully systems and actively collaborates in their development (Jones, 1981). Collapse of subsurface tunnel roofs in

Figure 21.9. Meandering gully developed on non-consolidated sediments. Kathorin Basin, Baringo District, Kenya.

the channel head produces gully retreat (Crouch, 1983) (Figure 21.10). Collapsing can also occur in the gully walls and bottom, leading to its deepening and widening. Active piping can generate gullies as well (Löffler, 1974; Jones, 1990, among others). A sum of factors combine to promote piping, such as soil cracking, hydraulic gradient, exchangeable sodium percentage (ESP), presence of expansive clays, and other factors. Their respective role has been analysed when considering piping in the previous section on arid regions. Gully-head retreat has been studied by comparing aerial photographs of 1956 and 1972 in Nigeria (Jones, 1990; in Thomas, 1994) and values of 20 m/yr were obtained for big gullies. By using microtopographic profilemeter in New South Wales, Australia, Crouch (1990) calculated a 1 m/yr retreat. Oostwoud and Bryan (1994) applied the sediment budget concept to the study of gullies in Lameluk of the Baringo District, Kenya, and obtained retreat values fluctuating between 0.5 and 10 m/yr.

All the gullies mentioned above correspond to cratonic areas, where relief is commonly smooth or is associated with planation surfaces. Areas related to lithospheric plate interaction during the Alpine Orogeny, however, present a very uneven relief. In these areas fluvial activity shows a great erosive potential and an important transport of materials towards lower zones. A net of deep gullies with abrupt walls (Figure 21.11) represents this dynamics.

Rivers crossing the extensive plains of tropical cratonic areas flow upon an alteration mantle of varying thickness. For this reason, fluvial incision makes the rock substratum outcrop in some cases (Figure 21.12). Rivers mainly transport suspended fine particles, which conditions channel morphology. *Straight channels* are commonly developed, belonging to system 1 of Schumm (1981), with low gradient, relatively uniform width and scarce bed load. Stream velocity and energy are low. Channels are usually narrow and

Figure 21.10. Collapses due to piping affecting a gully head, walls and bottom. Njemps Flats, Baringo District, Kenya.

deep and the margins are quite stable due to their clay-silty nature. As a consequence, these are stable channels with little lateral migration. Meandering rivers (system 3a of Schumm (1981)) also develop, where bed load is very small in relation to the total load and channel gradient is somewhat greater. These are *sinuous channels* (Figure 21.13) with suspended load, similar width and stable margins due to their constitution, although meanders can be cut off.

In some cases river longitudinal profiles show clear changes associated to a gradient increase, which are represented by *rapids* on the rocky channel bed. Another type of modification consists in a profile leap represented by vertical waterfalls, cascades, and cataracts of diverse origin. These abrupt changes in the river longitudinal profile constitute *knickpoints*, which tend to migrate upstream until their disappearance (Petts and Foster, 1985). Most rapids are structurally controlled by predominant fracture systems (Howard and Dolan, 1981). Pits and potholes form on rapids when water flow concentrates at a given point during a considerable time span. They are generated by water eddies and

Figure 21.11.   Deep gully network incising on acid volcanic materials on Dominica Island in the Lesser Antilles.

Figure 21.12.   The Mara River, a straight channel cut in the weathering mantle of Loita plains. The amphibolite rock substratum can be seen in the foreground. Masai Mara National Reserve on the Kenya–Tanzania frontier.

Figure 21.13. Flat plain where a meandering channel cuts in the regolith. Several *tor* morphologies can also be observed. Campo de Azálvaro, Southern Duero Basin border, Avila, Spain.

develop considerable depths, abrupt walls, and spiral furrows. Pebbles can be found in their bottom.

Cascades correspond to a sudden interruption of the river longitudinal profile, in which water falls vertically (Figure 21.14), while *cataracts* can be considered as a group of cascades (Figure 21.15). Undoubtedly, they are the most spectacular fluvial forms and are the object of tourist attractions. Besides, they can supply hydroelectric energy. Their origin is related to eustatic oscillations, tectonic uplift and different resistances to erosion of the channel bedrock. The biggest cascade in the world is the Salto del Angel (Angel Falls) in Venezuela, at 980 m high. The Iguazú cataracts, located in the intersecting frontiers of Brazil, Paraguay, and Argentina, have an 84 m fall and cut Upper Jurassic-Lower Cretaceous basalts of the Paraná plateau. In the humid tropics the Victoria cataracts of the Zambezi River, on the frontier between Zambia and Zimbabwe, are also important.

Schwarzbach (1967) elaborated a genetic classification between destructive cascades, in which water flow erodes the valley head, and constructive cascades, in which water precipitates $CaCO_3$ and the cascade advances with the course of time. The cascades and cataracts mentioned above belong to the destructive type and have originated on horizontally arranged materials with different resistances to erosion.

Erosion on cascades of tropical rivers is much less than the ones of higher latitudes (as, for example, the Niagara Falls), due to the small bed load of tropical rivers and, hence, reduced abrasion power. Bed load percentage considerably increases in tropical zones with a marked rough topography relief. Clasts are only transported small distances, however, due to the size reduction by chemical weathering and abrasion (Douglas, 1977). Cascade and cataract evolution depends on *knickpoint* recession, which is a function of the water discharge variations and of rock resistance. Sometimes cascade or cataract erosion progresses more rapidly following weakness lines (Figure 21.15). Studies on the evolution

Figure 21.14. Thompson Cascade, developed on basalts of the Eastern Gregory Rift Valley margin, Nyahururu, Kenya.

of the Niagara cataract show that its head progressively attained a curved morphology, which is believed to be the most stable form (Philbrick, 1970). Escarpment retreating rates measured on different cascades and cataracts of the world fluctuate between 0.25 and 1.64 m/yr (Douglas, 1977). Deep *plunge pools* develop at the cascade foot on the rock bed. They have varying size as a function of waterfall height, solid load size and rock resistance. The depth of these closed depressions is an inverse function of the retreat rate of the cascade head (Philbrick, 1970). When hitting the bottom, water partly flows retrogressively and produces an intense erosion at the wall base that can even cause collapse.

2.4. Fluvial sedimentary landforms

In regions of low relief within the humid tropics many valley heads present flat-bottom elongated depressions, without any evident channel, seasonally flooded and covered by

Figure 21.15. Iguazú Cataracts during low waters, on basalts of the Paraná plateau. Note the curved border form. Garganta del Diablo (Devil's Throat), located on the frontier limits of Brazil, Paraguay, and Argentina.

herbaceous vegetation (Mäckel, 1974). These areas are named *dambos* in central Africa and receive other names in different tropical regions of the world (Thomas and Goudie, 1985; Boast, 1990; Thomas, 1994). They are typical of savanna climates, with a precipitation between 600 and 1500 mm, although they have also been recognized in rainforest zones.

Dambos have concave margins of gentle slope, usually between 2 and 6° (Acres et al., 1985). The substratum weakness lines influence their plan morphology and they align following these prevailing directions. Soils developed on dambos are hydromorphic or gley soils (Young, 1976), with a higher organic matter content than the watershed soils, which increases towards the dambo centre.

No virtual shrub vegetation or trees exist on dambos, due to the close proximity of the water table to the surface. A wide variety of herbaceous vegetation is recognizable on dambos and their percentage increases towards their centre as a consequence of the higher persistence of the saturation zone. Vegetation changes through the year as a function of the existing precipitation regime, which leads to water-table oscillations (Mäckel, 1974). A vegetation zoning is observed, from tree species in the surroundings, sharply limited by herbaceous vegetation and the prevalence of grass and other hydrophile plants on sandy soils outcropping in the central zone with a fluctuating water table (Mäckel, 1985).

Dambos mainly develop in central and southern Africa upon the African and post African erosion surfaces (King, 1962), developed on granitic and metamorphic rocks. Land flatness favours dambo formation and promotes the water table to flow out as springs or remain close to the surface (Boast, 1990). Moreover, a plane relief favours water penetration and the development of thick saprolites, which following Raunet (1985) constitute the location of a great part of dambos. In the Taubaté Basin of southeastern Brazil, Coltrinari and Nogueira (1989) indicated that slopes limiting dambos are stepped and present small rounded depressions, with or without hydrophilic vegetation. Possibly their origin may be related to slides in the Tertiary sediments.

For some authors, the origin of dambos is related to the drainage network dynamics and evolution, while for others it is associated with differential chemical and biochemical weathering. The former argue for an incision phase in the valley head, followed by a filling up with a variety of alluvium and colluvium (Figure 21.16). As fluvial activity diminishes, slope washing supplies fine material to the valley bottom. Particle removal seems to coincide with the rainy season, when vegetation cover is smaller (Mäckel, 1974, 1985). Meadows (1985) ascribed valley incision and infilling phases in Malawi to climatic changes.

Another hypothesis states that dambos formed independently of fluvial networks as some of them are not integrated into the drainage system. It is believed that the main mechanisms are chemical and biochemical weathering (McFarlane, 1989; in Boast, 1990). The existence of topographical irregularities is thought to be originated by differential weathering. These are controlled by a higher fracture density in the rock substrate, which

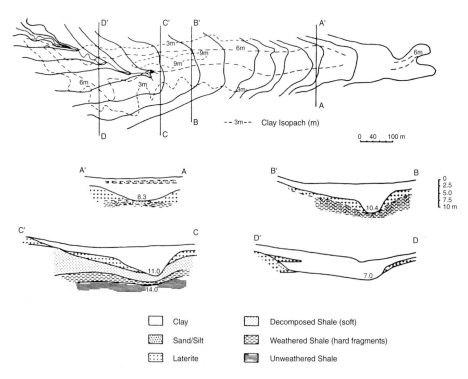

Figure 21.16. Dambo plan morphology and cross-sections in Kankamo (Zambia). Note the buried channel morphology and the sedimentary infill of sand, silt and clay (Clark, 1974, in Thomas, 1994, redrawn by Clarke, 1974, in Fookes and Vaughan, 1986, Figure 11.17).

favours water penetration and, hence, rock alteration. Solutes are leached towards the low parts in the profile and the underground flow runs towards the water courses. All this leads to a slow land-surface lowering, which gives rise to zones with poorly contrasted highs and lows. In this sense, a fluvial network rejuvenation affects dambos because incision produces water-table lowering and increases leaching due to the rising hydraulic gradient (McFarlane and Whitlow, 1990). McFarlane (1989) in Boast (1990), established an evolutionary sequence starting from a small rounded depression that transformed into a dambo due to surface lowering. Drainage network rejuvenation brings about a transformation of big dambos into smaller linear forms. Finally, if the water table keeps on lowering, dambos increase their size individually.

Although *alluvial fans* reach their maximum expression in arid and semiarid zones, they also appear in the humid tropics. Nevertheless, in these environments they usually acquire a lesser development. Their study also becomes difficult due to the vegetation cover. Fan sedimentation originates by changes in the flow hydraulic geometry. This is due to a sudden decrease in the channel slope, which transforms the flow from a confined state to an unconfined one. Fans present a triangular plan shape with convex transverse profiles and concave radial ones. They can coalesce with other nearby fans.

Alluvial fans occur in some African areas, although in general they are of a small size (Faniran and Jeje, 1983; Thomas, 1994). Instead, fans from the Pantanal region (Brazil) occupy a considerable extension and one of them reaches up to 50,000 km^2 (Klammer, 1982a). Fans can develop at the foot of mountains, such as in the high regions of eastern Papua-New Guinea (Blake and Paijmans, 1973), where the coarsest fan deposits become altered very rapidly and with later erosion are transformed into fine particles. Some fans are located at the base of plateaux, such as the ones described by Hill and Rackham (1978) in the Jos Plateau of central Nigeria. These authors also cite fan development within the plateau, between granitic residual forms. Fans have been also observed at the foot of labyrinthic platforms in eastern Zambia and SW of Sierra Leone, where they form iron-cemented breccias (Thomas, 1983). Small fans appear in the limit between residual hills and plains, although they have a little sediment supply due to the scarce development of the regolith upon the hills (Thomas, 1994).

One alluvial fan type cited by Tricart (1974a) is the one located at the foot of gully zones as a consequence of anthropic tree felling, and occur in Madagascar, Brazil, and Hong Kong. These deep gullies are called *lavakas* in Madagascar and *voçorocas* in Brazil. In Madagascar alluvial fans are basically constituted by clay and micaceous sands coming from the erosion of thick saprolites. In the Betsiboka basin, 60,000 km^2 in area, sediment loss is as much as 15,106 m^3/yr.

Areas with high relief, developed during the Alpine Orogeny and many of them located on insular zones, normally present sharp topographic gradients. The large gullies, with a radial pattern in many volcanic islands, rapidly remove great amounts of sediments that are deposited in zones of changing slope, usually on the coastal fringe. This is the case of the alluvial fan produced by the Rivière Sèche on Martinique Island, Lesser Antilles (Figure 21.17). A natural cut allows observation of a sediment thickness of about 12 m, constituted by stratified heterometric deposits. At the base there are several car tires inside a laminated deposit, possibly from the 1960s or 1970s. It can be deduced from all this that erosive activity is very intense and rapid, due to the heavy precipitation and frequent cyclones (700 mm/day in August 1970, D'Ercole, 1994) and to the presence of pyroclastic

Figure 21.17. Alluvial fan deposit including a group of tires near its base and incised afterwards. This disposition indicates the great celerity of geomorphic processes. Rivière Sèche, Martinica Island, Lesser Antilles.

material with an important chemical weathering in the drainage basin. After the fan sedimentation a deep incision took place, which permitted investigation of the deposit. This alternation of intense aggradation and incision has occurred in only a few decades.

Forms resulting from sedimentation in *flood plains* of tropical rivers are similar to those generated on other morphoclimatic zones. As they are covered by vegetation, these areas are difficult to study and hence few data exist about them. Most solid load of tropical rivers is made of sands, although when river slope is important the bed load is mainly formed by gravels (Figure 21.18). Fine sediments deposited by vertical accretion can generate very stable channels due to cohesion of silt–clay particles in the margins. Most of the time the flood plain is not affected by river dynamics, but during big floods, frequent in these areas, intense erosion and sedimentation take place.

Braided fluvial systems develop gravel and/or sand bars separated by low sinuosity channels. As bars have low height, they are completely covered by large floods, which modify their morphology and location. Meandering rivers (Leopold et al., 1964) develop,

Figure 21.18. Gravel bed load on a braided river flowing from the Irazú volcano slope (3432 m). Sucio River, Braulio Carrillo National Park in Costa Rica. Photo: F. Gutiérrez.

by lateral accretion of sand and gravel, *scroll bars* separated by elongated muddy depressions or *swales*, which fill up with well stratified fine material. Both elements constitute *point bar* deposits. Bank overflow by a meandering river produces *natural levees*, which constitute sediment wedges of fine sand gently dipping to the flood plain. During an important flooding episode, water distributes following small channels that cut previous sediments forming sandy tongues or *crevasse splay deposits*. In the lower parts of the flood plain silt and clay sediments accumulate in ponds and shallow lakes, forming *backswamp deposits*. Figure 21.19 shows a map with the distribution of some of these subenvironments in the Rufiji River of Tanganyika. Regarding the extension of backswamp zones, the most important region is the Tonle Sap or Great Lake, in Sri Lanka, which is kept flooded by the Mekong River from May to October (Douglas, 1977).

Sedimentary processes are very complex and vary abruptly, partly due to the important seasonal variations of these rivers. The Auranga River, in northwest India, develops under a monsoon climate with four rainy months (June to September) and storms with an intensity of up to 200 mm/h (Gupta and Dutt, 1989). During the dry season the river becomes braided with low sandy bars. During floods braided bars are destroyed and the river transforms into a meandering one with the generation of gravel and sand point bars.

2.5. Mass movements

These comprise all the material motions induced by gravity on slopes. They can be produced under any climatic environment, though in the humid tropics and also in some

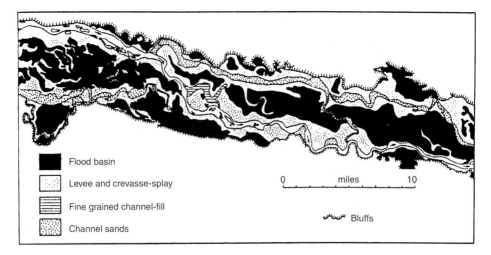

Flood basin

Levee and crevasse-splay

Fine grained channel-fill

Channel sands

0   miles   10

Bluffs

Figure 21.19. Map of part of the Rufiji (Tanganyika), showing the distribution of several fluvial subenvironments (Anderson, 1961, in Allen, 1965).

periglacial zones, they constitute the dominant geomorphic process (Peltier, 1950; Wilson, 1968, 1969). In some regions mass movements are not very common, whereas in others their great abundance makes them an important factor in landform development (Figure 21.20). In the mountainous Taiwan Island, which experiences high intensity tropical precipitation, 7810 slides occurred in a 104 km^2 area between 1965 and 1977 (Lee, 1981; in Thomas, 1994).

The term *slide* refers to a rapid mass movement separated from the underlying stable part of the slope by a slip plane. In the high parts of the slope small and slow *creep* movements can take place, affecting unconsolidated materials. Sliding surfaces serve as indicators of mass movements. In the humid tropics, however, they are rapidly covered by vegetation (Figure 21.21), which hinders their recognition. Many slides have a latent or dormant character and can reactivate under adequate conditions. Human activities like tree felling, incorrect constructions, and so forth, can trigger slides. In this sense, mass movements constitute an important geomorphic hazard when affecting human populations and their activities. Finally, it must be pointed out that, despite the great number of slides produced in humid tropics, there is a lack of deep knowledge about them. The reasons for this are the difficulty of their study and the high cost of monitoring devices.

Many *factors* influence slide generation: mechanical properties of rocks and regolith, precipitation, vegetation, slope gradient, seismicity, anthropic action, and other factors (Varnes, 1958, 1978; Zaruba and Mencl, 1969; Hutchinson, 1968, 1988; Crozier, 1986). It is necessary to understand the role of each factor in order to know the sensitivity of a given area to undergo slides. When several factors increase their intensity at the same time, slide triggering is considerably favoured. This wide variety of factors responsible for the origin of mass movements is an indicator of the high diversity of slide types.

The most important characteristics regarding the *mechanical behaviour of regolith* are inherited discontinuities, texture, and clay mineral types. Obviously, the intensity of weathering and saprolite structure also have an influence, because slopes covered by

Figure 21.20. Numerous slides and debris slide-flows on steep slopes of the Central Andean Cordillera, affecting the railway track between San Lorenzo and Ibarra, Ecuador. Photo: F. Gutiérrez.

weathered boulders can evolve by the means of the rocks moving. *Grus* presents a high porosity with low shear strength and where saturated by water it looses its cohesion. In laterites on granitic rocks the sliding plane at depth is normally close to the contact between fresh rock and regolith (Tricart, 1974b). Mass movement is commonly produced through rotational slides or slumps and block glides (Durgin, 1977). These slide types prevail on the "*meias laranjas*" dome landform of southern Minas Gerais, Brazil (Tricart, 1974b). The existence of relict discontinuities in the regolith, including the basal weathering surface and old sliding planes, favours water penetration which reduces shear strength and can trigger translational slides. Some examples have been cited in the Sierra de la Costa, Venezuela (Garner, 1960; in Tricart, 1974b).

Kaolinites existing in the regolith have a low shrinkage limit, by which they are not cracked during dry periods. The presence of clay minerals of the 2:1 type (montmorillonite), however, gives rise to an important contraction with fissure generation, which allows rapid infiltration and the change to a liquid state, favouring slide generation.

Figure 21.21. The crown of a slide covered by vegetation. Saint Denis, Martinique Island, Lesser Antilles.

On the other hand, kaolinitic saprolites present high plastic and liquid limits, whereas in montmorillonitic saprolites these are lower. As a consequence, when the liquid limit is attained during heavy rainfall, numerous failures of different types are triggered, as it happened in the Andean piedmont of Venezuela, where 899 mm was recorded at La Fria during February and March of 1972 (Tricart, 1974b). Where slope gradient is significant, dissection prevails and clay minerals are of the 2:1 type because a long time period is needed for kaolinite generation. The presence of significant quantities of swelling clays exacerbates heaving processes and reduces shear strength in the slope material. Besides, when the dip angle of bedding in sedimentary rocks is similar to the slope angle, translational slide triggering is favoured (Figure 21.22).

In the humid tropics a major annual *rainfall* is produced, sometimes through rainstorms of variable intensity. Water penetrating into the regolith moves towards a zone of lower potential gradient, by the means of saturated or non-saturated flows. The former takes place when all the regolith pores are filled by water, which is common in the lower parts of the saprolite and also in the upper parts after an intense precipitation. The latter prevails during most times and is characterized by the presence of water in the small pores with a very slow flow (Selby, 1993). An increase of pressure in the pore water produces a change in strength by diminishing cohesion and internal friction, which favours mass movements. A rapid increase in the water pore pressure can produce a sudden liquefaction, especially in fine and silty sands (Zaruba and Mencl, 1969).

High intensity precipitation is not common in the humid tropical climates (Lal et al., 1981), but it is evident that mass movements are triggered by low frequency and intense rainfall. Moreover, deep failures (Figure 21.23) are clearly influenced by preceeding rain

Figure 21.22. Translational slide that affected the road from Cochabamba to Santa Cruz (central Bolivia). The slide hazard at this point was solved by digging a tunnel. Photo: M.C. Maldonado.

Figure 21.23. Deep slide that cut a road for a long time in Saint Denis, Martinica Island, Lesser Antilles.

or by long-lasting precipitation (Thomas, 1994). In March 1974 it rained 742 mm in 16 days in Tubarau in southeast Brazil, of which 240 mm were recorded on the last day. Catastrophic consequences occurred, with numerous landslides and flooding, with 25 people died in one single slide (Bigarella and Becker, 1975).

Brand et al. (1985), in their studies on the relationship between precipitation and slope failure in Hong Kong, indicated that if rain intensity exceeds 70 mm/h important failures are triggered, whereas if it does not reach 100 mm/day failures are very scarce. These authors also emphasised the role played by preceeding rains, because they condition the development of surficial or deep slope failures.

Research on the influence of *vegetation* on slope stability was begun in the decade of the 1960s. Until then many authors had considered it as a factor with a low incidence, but at present it has been demonstrated under several circumstances that vegetation has an important influence (Greenway, 1987). Nevertheless, tree and shrub afforestation has been used for more than a century for controlling erosion and trying to stabilise slopes. So (1971) studied 700 failures triggered by an intense rainstorm in Hong Kong and suggested that such mass movements are more commonly located on the more vegetated slopes.

When studying the role of vegetation on slope stability, Greenway (1987) established a distinction between hydrological and mechanical mechanisms. Among the former, the beneficial role of raindrop interception can be mentioned, which produces water loss by absorption and evaporation and hence reduces infiltration. On the other hand, roots and stems increase infiltration capacity and subsurface flow. Roots also take water from the soil and diminish pore water pressure. An adverse factor is the loss of soil moisture, which leads to soil cracking and to a higher infiltration capacity. Physical mechanisms include reinforcement by roots, which increase soil shear strength. Factors that also must be taken into account include root-type, penetration depth, lateral extension, and root strength. Tsukamoto and Kusakabe (1984, in Greenway, 1987), in their study of the influence of vegetation on slopes in Japan, developed a simple slope classification in four types (Figure 21.24). Type A slopes have a thin soil thickness where roots do not penetrate into the bedrock; the soil-rock interface can act as sliding plane. Type B slopes are similar to type A except that roots penetrate into the bedrock. Type C slopes have a thicker regolith and contain a transition layer in which soil density and shear strength increase with depth. Roots penetrating this layer provide a stabilizing force to the slope. Finally, type D slopes have a very thick regolith and roots are "floating" within the soil, with little mechanical influence of stability. Below the root zone deep slip planes can be produced.

Roots also reduce susceptibility to erosion. Tree weight depends on the species, diameter, height, and separation. Overload on a slope increases shear stress and pore water pressure. In a broad sense, if the slope angle is small then overload increases stability. Finally, forces of instability derived from wind action can be transmitted from the vegetation to the soil. Strong winds can pull out trees, especially those with superficial roots.

In slopes with low *inclination* very few failures occur. In the African humid tropics several authors (Thomas, 1994) have indicated that slope instability appears with gradients higher than 26 to 28°, although these values can substantially differ in other regions of the world. According to different authors, most slides develop between 28 and 45°. When slopes are very steep failures are very numerous, especially in relation to heavy rainfall (Figure 21.20 and Figure 21.25). Slopes can modify their inclination by fluvial or marine

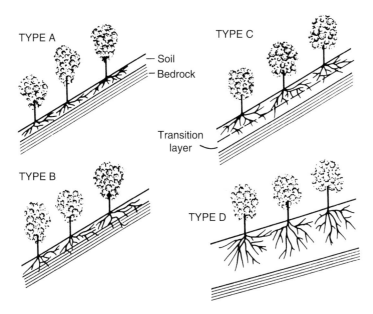

Figure 21.24. Slope classification scheme based on root reinforcement and anchoring (Tsukamoto and Kusakabe, 1984; in Greenway, 1987, Figure 6.14).

sapping and exceptionally by tectonic activity; the increase of topographic gradient induces a change in the stress state, which can trigger failures.

Humid tropical zones located in alpine orogenic belts are exposed to sporadic *earthquakes*, which affect slope stability. Earthquakes produce rock oscillations with different frequencies and, hence, the sudden stress modification can disturb slope equilibrium. Liquefaction can take place in silts and loose sands. In Papua-New Guinea, Simonett (1967) studied slide frequency and pointed out that the number of slides decreases logarithmically with distance to the epicentre. When earthquake magnitude is important, the number and size of slides increase.

Investigations on plots and experimental basins indicate that zones bared of vegetation are exposed to higher erosion than vegetated areas. Deforestation resulting from abusive tree felling or man-made fires, the latter very frequent in the Amazon rainforest, influence the rate of mass movement. This *anthropic action* basically affects saprolites, because in dry periods desiccation cracks favour water infiltration during rainfall. These processes have been interpreted as partly responsible for the triggering of numerous slides in the widely deforested area of Serra do Mar, between Sao Paulo and Rio de Janeiro (Tricart, 1974b). Similarly, the absence of forest mass brings about the lack of root reinforcement in the regolith, the deprivation of bedrock anchoring and the loss of water absorption. As a consequence, slopes undergo a shear strength decrease. These circumstances favour slide development during intense rainfall. On 23 February 1970 a 3-hour rainstorm of 100.7 mm in the Uluguru mountains (Tanzania) led to the generation of 1000 new superficial failures and the reactivation of other previous ones (Temple and Rapp, 1972). The work done by So (1971), however, posed doubt on the role of vegetation as an essential factor in failure

Figure 21.25. Slide on a steep slope with rainforest (*pluvisilva*) vegetation. Pitones del Corbert, Martinica Island, Lesser Antilles.

development. This author indicated that 35% of the 700 slides produced in June 1966 in Hong Kong developed on forested areas, which occupied 8.4% of the region affected by mass movements.

3. Landforms developed on laterites

Landforms related to laterites appear in the humid tropics and also in semiarid zones adjacent to savanna regions. Landforms mostly result from fluvial incision of lateritic deposits, which produce tabular forms capped by laterites. These can last for long periods of time due to the high resistance of laterites to erosion.

Tabular forms are *platforms* or *mesas* of limited dimensions, standing out in the landscape and with angles lower than 0.5°. Their borders are formed by steep scarps that retreat basically by weathering processes, leading to a reduction of the platform extension. Many examples exist in the literature about these tabular landscapes (Figure 21.26) (Maignien, 1966; Thomas, 1974, 1994; McFarlane, 1976). Fluvial incision in a weathered zone can produce a new valley. Fe^{2+} flowing to the valley gives rise to an iron enrichment that leads to the generation of a new laterite level. Successive fluvial incisions generate a terrace or *bench* morphology (Figure 21.27). Destruction of the laterite scarp supplies fragments to the lower parts, forming *slopes* with loose laterite rubble (Figure 21.28) and newly cemented colluvial deposits that can connect with pediments (Figure 20.22). In general, laterites harden as a consequence of fluvial incision. When dried they become very resistant to external agents and after long erosional periods they occupy dominant

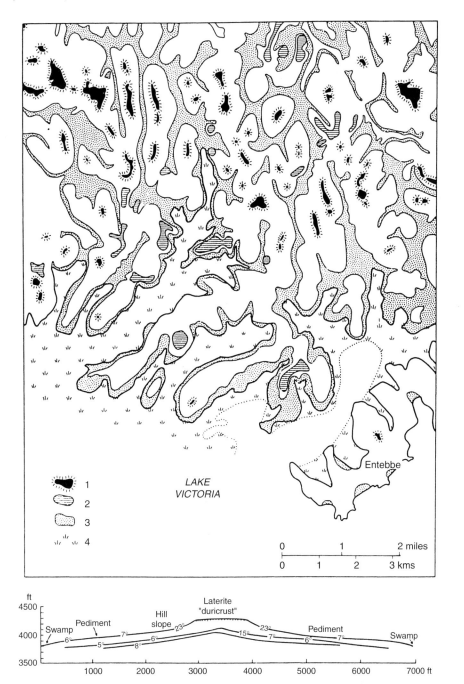

Figure 21.26. Landforms resulting from ferricrete duricrust dissection in Uganda. 1, flat summits capped by laterites (possible remnants of the African Surface); 2, intermediate benches; 3, alluvial silts and lacustrine beds; 4, papyrus swamp (after Pallister, 1956).

Figure 21.27.   Bench morphology on laterites in Nigeria. Photo: J. López Martínez.

Figure 21.28.   Slope with loose laterite rubble in a forested savanna. Eastern Bouaké, Ivory Coast.

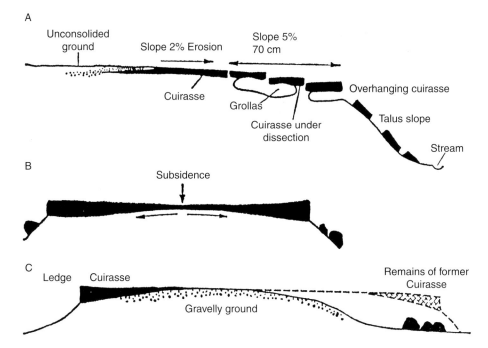

Figure 21.29. Some characteristics of the morphology of lateritic hillslopes (from Maignien, 1966, in Thomas, 1974, Figure 15).

positions, whereas initially they were restricted to valleys. This variation in their relative topographical situation is a case of *relief inversion* (Figure 20.25).

The process of laterite fragmentation (Figure 21.29) is mostly due to undermining by mechanical and chemical removal produced by the underground water that flows below the laterites. This leads to the development of a slight bending in the border of the platform or mesa, called *cambering* (Moss, 1965). Subsurface water flow, favoured by laterite fracturing and regolith porosity, can generate passages and caves of variable size (Ollier, 1965). This subsurface network can evolve producing ductile or brittle subsidence and giving rise to *closed depressions* of metric to hectometric size. These pseudokarstic forms are scattered upon the horizontal surfaces of the tabular hills and can develop diffuse or sharp borders (De Chetelat, 1938; Goudie, 1973).

4. Tropical planation: etched plains

A great part of the tropical zone is covered by extended plains (Figure 20.2) upon which detached hills (*inselbergs*) or mountain ridges stand out. There exist two main types of plains: erosive and depositional; the former develop on very variable lithologies and structures, whereas the latter constitute accumulation zones of alluvial and lacustrine sediments.

Throughout the history of geomorphology different theories have been proposed, giving rise to terms such as *peneplain* (Davis, 1899), for humid zones, and *pediplains*, by extension of pediments, for arid zones (King, 1942; 1953). Such plains form part of the present landscape. These terms and their associated genetic implications have been analysed in previous chapters.

In general, peneplains do not form perfect flat surfaces, but rounded and undulating hills of low height (Vidal Romaní and Twidale, 1998). They develop on unaltered rocks and some think that they can be recognized in South Africa, Transvaal, Namibia, and Brazil. As will be discussed later, they are not exclusively limited to temperate zones. Twidale (1983) indicated that peneplains are being progressively degraded and should be denoted as *ultiplains*. Changes are so slow that the survival of antique forms occurs within the present landform (Twidale, 1976a,b). Some authors consider that "pediments, and by extension pediplains, are not genetically nor temporaly distinct from peneplains. They usually coexist … (and) are simply parts of a great group of landforms called planation surfaces" (Vidal Romaní and Twidale, 1998, p. 101).

In the humid tropics there also exist *exhumed plains* or *surfaces*, generated by the stripping of geological units that fossilize erosion surfaces. A classical example occurs in the Valley of the Thousand Hills, Natal, South Africa, where the erosion of the horizontally-bedded Table Mountain sandstone uncovers the underlying surface and some dome *inselbergs* (Thomas, 1978).

Where the alteration mantle is eroded by fluvial action, the weathering front outcrops as an *etchplain surface* (Wayland, 1933), referred to by other researchers as a chemical corrosion surface (Vidal Romaní and Twidale, 1998). Numerous denominations have been proposed for these forms, which were compiled by Thomas (1989a). Wayland estimated that the regolith thickness fluctuated by approximately 3 m and, hence, during the next erosional cycle triggered by a rapid uplift, the regolith was completely stripped giving rise to a new surface more or less parallel to the former. These alteration and subsequent erosion processes can take place many times. Wayland's ideas mainly differ in the weathering mantle thickness, which at present is estimated in tens or hundreds of metres.

Büdel's (1957) concept of *double planation surfaces* (run-off surface and basal weathering surface) is related to the etching idea. He applied his theories to zones with a tropical seasonal climate and tectonic stability (Figure 21.30). His model starts from a plain whose weathering depth fluctuates between 30 and 60 m that is eroded giving rise to lower plains with a scatter of small rock domes (*ruwares = whalebacks = knobs*) 1 to 3 m high or *shield inselbergs*. Rounded or parallelepipedal blocks from the lower parts of the weathering profile can remain detached during the surface lowering. Valleys with gentle slopes or dambos drain these plains. In summary, it is deduced that soil erosion operates on the run-off surface, whereas the development and deepening of the saprolite takes place in the basal weathering surface. Under seasonal climates, where Büdel (1957, 1965, 1977, 1982) focused his research, these plains have a nearly null gradient (0.2° in the Tamil Nadu plains, southern India). Rainwater hardly finds routes for hierarchical drainage organisation and hence a high water amount infiltrates into the regolith. During dry seasons clay soils get cracked; at the same time, roots and other bioturbation conduits constitute excellent ways for water penetration with the arrival of the rainy season. This "cycle" favours weathering development along the whole saprolite and its basal deepening;

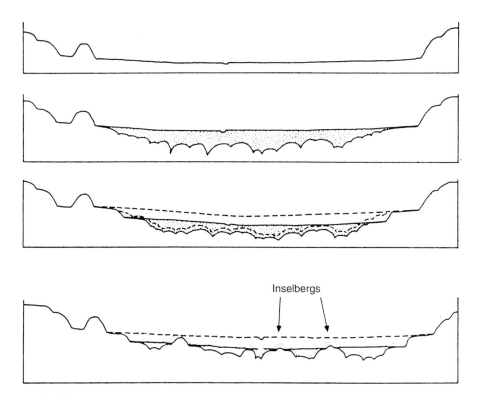

Figure 21.30. Deep weathering and denudation in humid tropics (simplified from Büdel (1957)).

similarly, surface run-off associated with rainfall can export a sheet of sediments, which lowers the run-off surface. Büdel's ideas were soon assimilated by scientists and applied to all the cratonic humid tropics of the world (see Thomas, 1974; 1989a,b; 1994), and even to extra-tropical zones with thick alteration profiles in temperate and cold regions (Bakker and Levelt, 1964; Bakker, 1967; Gellert, 1970; Dury, 1971; Büdel, 1977, 1982; Söderman, 1985; Hall, 1986, 1988). The latter indicate the survival of etchplains whose age for Europe is considered as Upper Mesozoic–Paleocene (Thomas, 1994). Nevertheless, in post-Miocene times a sandy alteration profile was generated under temperate climates (Bakker and Levelt, 1964; Hall et al., 1989), which can induce significant confusions to the interpretations.

Ollier (1965, 1984) has pointed out that saprolite has very variable depths and non-parallel mantles as suggested by Wayland and Büdel. The regolith forms deep depressions separated by bulges constituted by unaltered rocks. The erosion of the saprolite may produce low zones surrounded by fresh rock highs. Saprolite plays an important role in relief configuration due to the slow and continuous lowering of the etchplain, while surrounding hills become relatively higher (Bremer, 1971, 1985), as a consequence of the progressive removal of weathered rock. This process of depression genesis is related to densely fractured rocks, which facilitate water penetration through discontinuity planes. Büdel (1977, 1982) referred to this process as "divergent weathering."

Thomas (1965, 1974, 1994) proposed a classification of plains and etchplains as a function of denudation rate of the lateritic profiles. In his last version he introduced some modifications, although this differentiation has not received great acceptance. Thomas (1989, 1994) and Pavich (1985) compared the basal surface lowering and the differential erosion of the regolith with the dynamic equilibrium concept of Hack (1960, 1975), by which there is presumed to exist a balance between regolith erosion and generation. On the other hand, other authors (e.g., Ollier, 1991) estimated the landform genesis as a group of events developed through time. Ollier (1981) considered the Uganda territory as formed by a single etchplain (Figure 21.31). On the Buganda surface, which is equivalent to the Gondwana surface, thick alteration profiles have been preserved. In the north, the regolith

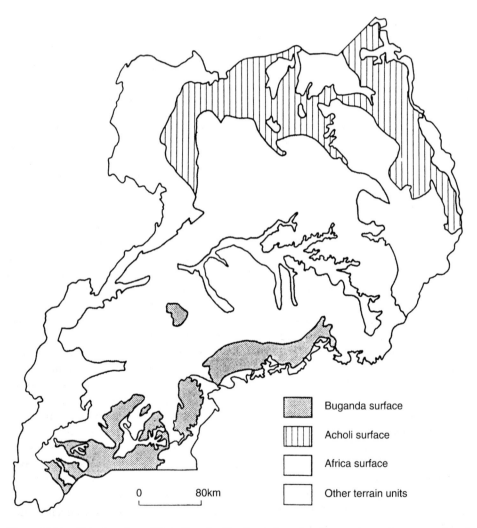

Figure 21.31. Stripping of regolith in Uganda. The Buganda surface has erosion-surface remnants and deep weathering. The African surface is partially stripped. The Acholi surface is stripped of old regolith and modern soils are forming on fresh rock rather than saprolite (after Ollier, 1981).

has been completely stripped and the Archoli surface developed upon fresh rock, with numerous inselbergs along its rim. On the African surface, the most extensive one, a deep saprolite is eroded and several inselbergs outcrop, indicating a very irregular basal weathering surface. Several Miocene volcanoes are located upon some points of this surface. In the opinion of Ollier, this last surface was formed, at least, in the Lower Tertiary period while the Buganda Surface must be of a Mesozoic age or older.

5. Inselbergs

This is a German word that means island mountain and was first used by Bornhardt (1900) to describe abrupt rocky hills rearing from the tropical plains of Tanganyika (today Tanzania). They basically develop in cratonic tropical and subtropical regions, especially on granites and metamorphic rocks. Nevertheless, they can also occur on sedimentary rocks like the arkoses of the famous Ayers Rock (Uluru) monolith or the Olgas conglomerates, both in the central desert of Australia (Twidale, 1978, 1982a; Vidal Romaní and Twidale, 1998). The study of these forms is complex because it comprises of factors like climate, lithology and rock structure, which at the same time determine the degree of rock weathering. Similarly, they raise basic genetic questions, such as those inherent to the exhumation of the basal weathering surface or to models of slope retreat. Research on inselbergs must also include the effects of climatic changes and the persistence or duration of these forms. As a consequence, many authors have focused their studies on diverse aspects of inselbergs.

Inselbergs have raised numerous problems related to different definitions and terminology. Twidale (1968a,b) defined inselbergs as isolated residual hills that stand out in the tropical plains. Later, Twidale (1982a) and Vidal Romaní and Twidale (1998) considered that inselbergs can occupy any topographic position. When studying the *morros* and sugar loafs of the Rio de Janeiro zone, Birot (1958) already indicated that crystalline domes are located in variable positions on the landscape. Thomas (1974, 1978) agreed with the idea of not necessarily relating inselberg and plain. If this opinion is agreed with, the initial definition of inselberg becomes rather distorted. Twidale (1968) also indicated that inselbergs are located on tropical plains, although the existence of inselbergs on the European temperate climates was known long ago (Penck, 1924). Therefore, either inselbergs form under different climates, or they are paleoclimatic indicators. In order to accurately define the situation of inselbergs in the world, Kesel (1973) analysed existing publications and calculated the percentages of inselberg distribution under different climates: 40% in savanna climates; 32% in arid and semiarid zones; 12% in humid continental and subarctic climates and 6% for each of the humid tropics, subtropics and Mediterranean zones. This author concluded, as did King (1957a), that inselbergs can occur under nearly all climates. He recognized, however, that the humid and dry tropics are favourable regions for inselberg formation, as previously suggested by Cotton (1961, 1962). The application of numerical parameters to the identification of inselbergs, as compiled by Thomas (1978) from several studies, does not provide determinant values and, following this author, can be applied to numerous hills.

The definition proposed by Young (1972) has a more general character and the inselberg denotation is applied to isolated hills with steep slopes that arise abruptly in zones with moderate relief. This author distinguishes several types: mesas on horizontal strata or laterites, hills with rectilinear slopes (typical of arid zones), and convex–concave slopes covered by regolith, rock domes, and boulders (*tors*) or castellated forms (*castle koppies*). Thomas (1976, 1978, 1994) made similar differentiations. The German authors distinguished between inselbergs of position (*fernlinge*) and inselbergs of resistance (*hartlinge*). The former have been preserved because of their situation on water divide zones, where there is little erosion, whereas the latter originate because of the great resistance to erosion of the rocks. Nevertheless, two main types of inselbergs can be differentiated: *boulder inselbergs* or *tors* and *domed inselbergs* or *bornhardts*.

5.1. Boulder inselbergs (tors)

The morphology of boulder inselbergs is varied and controlled by fracture systems. They develop on gneisses, schists, quartzites, diabases, sandstones, and arkoses, although they are more abundant on granites (Gerrard, 1988). The boulders (spheroidally weathered corestones) occur in diverse topographical positions and their size and form depend on the fracture type and spacing. In Australia they vary in size from 0.25 to 33 m, although the most frequent sizes are 1 to 2 m. Joint blocks commonly develop a spherical or ellipsoidal corestone form from preferential decomposition of the corners and edges over the faces, and in some places, present parallepipedic trends (Figure 21.32). Some blocks remain in an unstable position and are named "perched" or "balanced blocks" (Vilaplana, 1987; Pedraza et al., 1989; Vidal Romaní and Twidale, 1998), which frequently end up by falling down the slope. Groups of angular blocks sometimes appear on summit zones, simulating

Figure 21.32. Tors developed on granites. Northern side of the Serra da Estrêla, Portugal.

an old fortress, and are named castellated forms (*castle kopjes = koppies = rock towers*). Some castle koppies are dominated by spheroidally-weathered corestones. On the other hand, rounded boulders can occur in any climatic environment.

The *origin* of the boulders supposes a deep differential weathering, which deepens with higher intensity in zones of intense fracturing. Afterwards, an environmental change brings about the saprolite erosion. As a result, the weathering profile base outcrops, characterized by the presence of boulders that originate the above-mentioned forms (Figure 21.33) (Linton, 1955). This author referred to this sequence of processes as a two-stage hypothesis. Boulder generation can be observed on roadcuts and lineal works. The balance between weathering and denudation must be kept in mind. If weathering is more intense than erosion, then neither the weathering front nor the associated blocks outcrop. On the other hand, when erosion is more intense than weathering, then the profile is dismantled and basal blocks appear (Vidal Romaní and Twidale, 1998). Outcropping boulders are exposed to subaerial weathering. The regolith can be removed by fluvial action and give rise to block channels more than 10 km long, such as the Homen River of north Portugal. Saprolite exportation can also take place by marine, glacial and eolian erosion (Vidal Romaní and Twidale, 1998).

In general, the development of many boulder inselbergs is related to the decay of blocks located within the inselberg. The main processes are block-weathering, grus erosion and land surface lowering, which can lead to the disappearance of the boulder inselberg (Figure 21.34) (Thomas, 1978).

As indicated above, boulders have a wide distribution all over the world and, because deep weathering is required, in many zones they are assumed to be a very old alteration

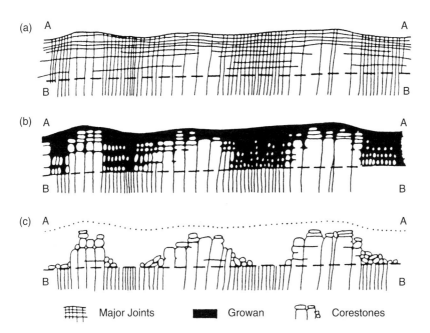

Figure 21.33. The evolution of the tors and granite landscapes based on the drawings of Linton (1955).

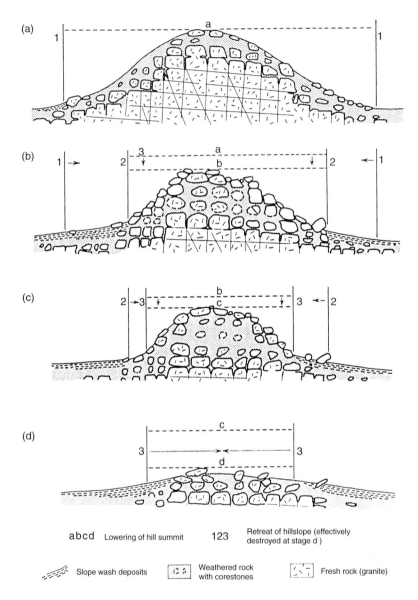

Figure 21.34. A sequence of development for boulder inselbergs, involving decay of interior joint blocks within the zone of circulating vadose water, together with lateral weathering penetration leading to slope retreat. The first stage (A) may be reached without recourse to deep preweathering, but this is not precluded. The stripping involved in the transition A–B can be related to a climatic change (after Thomas, 1978).

of the bedrock. Linton (1955) suggested an intense chemical weathering during Tertiary times for the Dartmoor granite of southern England. In this case regolith removal took place by periglacial processes, in the same sense as proposed by Demek (1964) in the Bohemian Massif (Czechoslovakia). Later on, Czudek and Demek (1971) interpreted tors as originated by weathering and removal under periglacial envronments. Mineralogical analyses of the Dartmoor grus make evident the absence of alteration of feldspars and a low clay content (Doornkamp, 1974). Consequently, in European latitudes a "sandy weathering product" can form (Bakker, 1967), which can involve producing blocks.

King (1948) proposed that tors, the same as dome inselbergs, originate by slope retreat in arid and semiarid climates, giving rise to a pediment at their feet. Similar ideas, although developed for periglacial environments, were elaborated by Czudek (1964) (Figure 8.11).

From the above discussion, there is no doubt that there exist several hypothesis for explaining the origin of such boulders. Two main tendencies can be deduced: one related to deep weathering and later erosion (the two-stage hypothesis of Linton) and the other related to slope retreat in warm and cold deserts.

5.2. Domed inselbergs (bornhardts)

As previously noted, the term *bornhardt* is due to Willis (1936), for defining crystalline domes in the honour of the German explorer Bornhardt (1900), who first described inselbergs. They have a dome form and develop on resistant massive rocks, mainly granites, although also on sandstones and conglomerates in central Australia. Twidale (1982a) considered the Mallos de Riglos conglomeratic monoliths in the Spanish Southern External Pyrenean ranges (Figure 21.35), to be bornhardts. This assumption can be extrapolated to the conglomeratic mountains of Montserrat, Montsant,

Figure 21.35. Monoliths developed on massive horizontal conglomerates in the southern limit of the external Pyrenean ranges, in contact with the Ebro Basin. Mallos de Riglos, Huesca, Spain.

Iregua River, and elsewhere, which form part of the marginal Tertiary units of the Ebro Depression in Spain.

Bornhardt slopes are commonly steep and with a very variable size, from some metres to several hundreds of metres (Figure 21.36). Their plan form presents circular or elliptical morphologies, commonly affected by fracturing (King, 1948; Twidale and Bourne, 1978). Where elongated and with low height, they are called "whalebacks" or "ruwares." Some bornhardts appear isolated, while others cluster to form hemispheric groups, known in Brazil as half-oranges (Birot, 1958). Contacts between the rock monolith and the bornhardt foot can be sharp, with a clear *knickpoint*, or transition with no evident slope modification. Several causes exist for explaining a sudden slope change: contact between rocks of different mechanical properties, contrast between fresh rock and saprolite (Ollier, 1960; Thomas, 1965) and intense weathering at the slope base (Ruxton, 1958; Twidale, 1962; Mabbut, 1966; Twidale and Bourne, 1978).

Bornhardts can occur in different climatic environments (Kesel, 1973), from equatorial rainforest zones, or intermediate latitudes (Figure 21.37 and Figure 21.38), to the north of Norway (Büdel, 1978). For the Hercynian Massif of the Iberian Peninsula, numerous works exist about granite bornhardts (Birot, 1949b; Birot and Solé, 1954a,b; Brum Ferreira, 1978; Coudé-Gaussen, 1981; Sanz, 1988; Pedraza et al., 1989; Vidal Romaní and Twidale, 1998, among others). They have also been recognized in granites of the Catalan coastal ranges (Roqué and Pallí, 1998).

A set of factors exists that influences the location of domed inselbergs. Most studies focused on checking bornhardt location in relation to rock type have concluded the absence of lithological variations in their piedmont. Some studies, however, highlight a

Figure 21.36.   Domed inselberg several hundreds of metres high. Abuja, central Nigeria. Photo: J. López Martínez.

Figure 21.37. Dome forms and other granite morphologies capped by the bell-like monolith of El Yelmo. La Pedriza de Manzanares, Sierra Guadarrama, Madrid, Spain.

Figure 21.38. Bornhardt with summit exfoliation joints. Yosemite National Park, California. Photo: C. Sancho.

relation between inselberg situation and rock type (Shroder, 1973). Jeje (1973) indicated that in zones of southeastern Nigeria, inselbergs are located on biotitic porphyritic granites (Figure 21.39). In his research near Johannesburg, Brooks (1978) pointed out that tors and castle koppies appear in a transitional zone between gneisses and migmatites and homogeneous porphyritic granodiorites, whereas in other zones they prevail on biotitic porphyritic adamellites. In summary, this author indicated that inselbergs are more common on rocks that have suffered potassic metasomatism or on potassium-rich plutons, or on those containing primary low-temperature quartz. Such relations, however, do not occur between inselberg location and lithology.

Most inselbergs are mainly located in areas of crustal stability (Kesel, 1973), although they can also appear on tectonically active ranges. One of the most classic examples is the Pic Para, in southeastern Brazil, which forms part of a tectonic *horst* (Birot, 1958). Twidale (1973, 1982b) and Vidal Romaní and Twidale (1998) indicated that sheeting produces lateral compressive stresses that exceed those generated by the loss of loading. The acceptance of this hypothesis implies that domes controlled by sheeting must be generated by this mechanism. Numerous data on stress measurement in tunnels and wells confirm this idea (Twidale, 1982b).

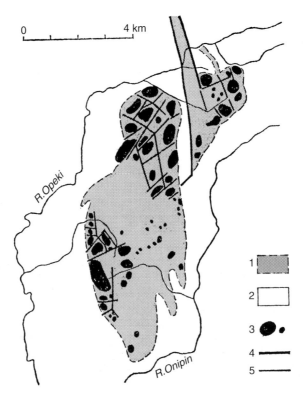

Figure 21.39. Residual hills in southwestern Nigeria. 1: porphyritic granite with biotite; 2: gneisses and schists; 3: inselberg, small domes and *koppies*; 4: fault; 5: fracture (after Jeje, 1973).

At the end of the 19th century inselbergs were interpreted as generated by glacial, marine and aeolian processes (Twidale, 1982; Vidal Romaní and Twidale, 1998). Later on, fluvial erosion was considered as the main cause. At present, two main theories explain bornhardt generation. Several authors (Cotton, 1942; Howard, 1942; Pugh, 1956) have proposed the *scarp retreat hypothesis*, although it was King (1942, 1948, 1949, 1953, 1962) who led this idea with the most emphasis. More recently, King's points of view have been supported by Kesel (1973, 1977b) after studying deserts of the southwestern USA, and by Selby (1977) in the Namib Desert. Criticism against King argues that none of his work was done in humid tropics, zones with a great profusion of bornhardts.

The hypothesis deals with inselbergs of position (*fernlinge*) that for extended periods of time have experienced parallel scarp retreat. This process has led to a pediment enlargement that can finally give rise to pediplains (Figure 21.40). Inselberg development by pediplanation involves a progressive decrease in size, until its complete disappearance occurs. Most researchers restrict these processes to arid and semiarid environments, where their effects are most profound (King, 1957a). For this author, weathering at the scarp foot leads to an increase in its steepness and to slope retreat. In warm deserts like the Namib, however, two opposite interpretations can be postulated. Selby (1977) suggested that inselbergs originated by pediplanation, while Ollier (1978) considered they have a great antiquity and form part of an exhumed topography. Consequently, there is a need for a deep knowledge of past geological times, especially about climates, in order to avoid such controversial interpretations.

The other theory about the origin of domed inselbergs is based on the *exhumation* of a deeply altered rock massif. When the basal weathering surface outcrops, domical forms appear indicating an origin under the alteration mantle. The hypothesis is quite similar to

Figure 21.40. Inselberg standing out over extensive plains. Narok Plateau (1825 m), southwestern Kenya.

the "two-stage" one proposed by Linton for boulder inselbergs and tors. This theory has been checked in quartz diorite pit fronts in southern Cameroon, where a 50-m dome is surrounded by weathered material (Boyé and Frisch, 1973). Some roadcuts also permit the observation of these incipient domical forms (Thomas, 1978). Domes are located on the less fractured parts of the weathering front. Stripping is progressively carried out by fluvial action and domical forms slowly appear on the land surface as small cupolas, which announce an embryonic or incipient domed inselberg (Thomas, 1978; Twidale, 1982) (Figure 21.41).

In general, small bornhardts are related to a single episode of deep regolith excavation that can reach 100 m. When trying to explain great dome inselbergs, however, difficulties arise (Thomas, 1965; 1974). Nevertheless, if a continuous weathering at the slope foot is admitted, then stripping produces a surface lowering of the surrounding plains and bornhardts several hundred metres high can be satisfactory explained (Figure 21.42). Consequently, it is a matter of repeated alternations between weathering and denudation phases throughout a long period of geological time (Willis, 1936; Büdel, 1957; Twidale, 1964, 1982; Thomas, 1978, 1994). Thus, many great domed inselbergs are geologically very old.

The scarp retreat theory does not explain some forms related to the contact rock-plain and rock-saprolite. Instead, the exhumation hypothesis justifies the existence of embryonic domes and microforms generated at the weathering front (polygonal weathering, tafoni, gnammas, pseudokarren, caves, and so forth) (Twidale, 1982a; Thomas, 1994; Vidal Romaní and Twidale, 1998). It also explains the genesis of stepped slopes (Twidale and Bourne, 1975, 1978) and mushroom profiles (Twidale, 1962), which together with tafoni lines at different heights in Ayers Rock (Uluru) (Twidale, 1978), indicate an older

Figure 21.41.   Embryonic dome forms on a savanna region. Masai Mara, Kenya.

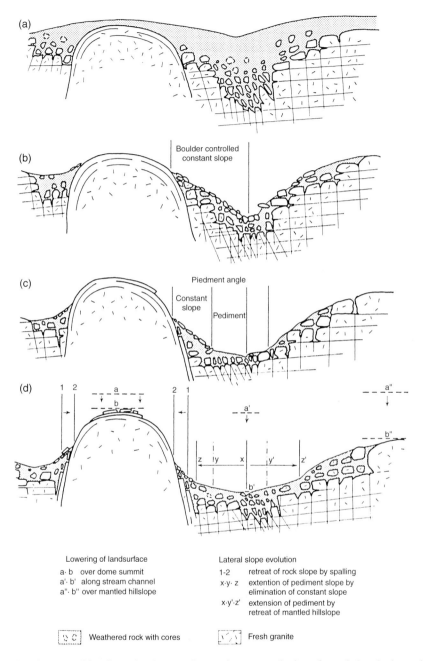

(a)

(b)

Boulder controlled
constant slope

(c)

Piedment angle

Constant
slope

Pediment

(d)

Lowering of landsurface		Lateral slope evolution	
a· b	over dome summit	1-2	retreat of rock slope by spalling
a'· b'	along stream channel	x·y· z	extention of pediment slope by elimination of constant slope
a"· b"	over mantled hillslope	x·y'·z'	extension of pediment by retreat of mantled hillslope

Weathered rock with cores   Fresh granite

Figure 21.42. Domed inselberg development by continuous weathering of mantled rock slopes during dissection (after Thomas, 1978).

subsurface weathering. These forms also reveal the amplitude of lateral corrosion on slopes of massive bornhardts during different stages of base-level standstill.

Regardless of whether bornhardts may have originated by scarp retreat or by exhumation, they require a long time to be formed. Their age, however, is a function of the velocity of geomorphic processes involved in their genesis and of the size of the form. Hence, we can find dome inselbergs recently exposed as well as bornhardts tens of millions of years old. Another problem arises when trying to explain the survival of great bornhardts for long periods of geological time. It has been interpreted that, since the inselberg domical form is controlled by sheeting, precipitation water flows as if on an umbrella, producing very weak alteration and low erosion (Thomas, 1974). Mesozoic or Early Tertiary ages have been proposed for many dome inselbergs. Savigear (1960) considered that inselbergs in Nigeria were exhumed from a Cretaceous sedimentary cover. Barbier (1967) indicated that domes (sugarloafs) in the Tassili Massif, South Algeria, have also been exhumed from below an Upper Paleozoic sandstone. Chaput (1971) pointed out that the pre-Triassic surface in eastern Sierra Morena in Spain is partially exhumed, with the outcrop of quartzitic inselbergs. Twidale and Bourne (1975) and Twidale (1978) used stratigraphical and geomorphological arguments to suggest that the famous Ayers Rock (Uluru) monolith dates back to the Mesozoic age.

As the age of inselbergs is sometimes very old, these forms have been exposed to endogenic and exogenic activity during long periods of time. They may have been affected by the motion of lithospheric plates, with the modification of their latitude and even climate, as for example, in India. As far as the origin of inselbergs is concerned, climatic changes affect the balance between weathering and erosion velocities. Periods during which weathering prevails correspond to geomorphological and ecological stability (biostatic stages) and when this equilibrium is interrupted then erosion sets in (rhexistatic stages) (Erhart, 1967). In the European middle latitudes a wide and deep ferralitic weathering has been recognized, especially during Eocene times (Millot, 1964). This alteration is confirmed in the sedimentary record of Tertiary basins and gives convincing arguments for an old deep weathering of zones where inselbergs have been formed.

6. Tropical karst

Landforms developed on limestones present clear analogies with forms developed on granites in humid tropical climates (Twidale, 1990). The activity of the etching processes can be also recognized in carbonate rocks and is extensive in karstic corrosion plains and their tower forms. There is even a clear similarity between the half-orange morphologies on granite landforms and the domical forms of the tropical cone karst. These circumstances induce one to consider a matter of form convergence or equifinality (Thomas, 1994).

Lehmann (1936) was probably the pioneer in studying the climatic differentiation of karstic landforms. In the Karst Symposium held in Frankfurt, Germany, in 1953, Lehmann indicated that every climate has its own karstic development, although factors like composition, thickness and carbonate rock fracturing must be also taken into account.

During the symposium a karst climatic zonality was recognized, including the differentiation of karst landforms of humid tropical zones, among others. These forms mainly develop on tropical zones with carbonate outcrops associated with alpine ranges; the Greater Antilles, some regions surrounding the Caribbean Sea, southeast Asia, Indonesia, New Guinea and several African zones.

Some significant differences exist between tropical karst and other karst types developed in other climates. Firstly, carbonate dissolution is more rapid due to the higher content of biogenic CO_2 (Birot, 1954, 1966; Ford and Williams, 1989). Secondly, strong precipitation produces intense slope gullying, giving rise to irregular closed depressions. Finally, high evaporation rates induce calcium carbonate precipitation forming very hard caprocks on the steep slopes, which often is considered as the main cause of their great inclination (Sweeting, 1972).

Tropical karst landforms are usually divided into two types: *kegelkarst* (*cone karst*) and *turmkarst* (*tower karst*). Differentiation is made as a function of hill morphology. In the future it is possible that the classification be based, more logically, on the morphology of closed depressions (Sweeting, 1972). Balazs (1973) proposed a tropical karst differentiation mainly based on morphometric considerations and distinguished three types: doline karst, cone karst and tower karst. Cone karst hills are 30 to 120 m high and have a density of 15 to 30 hills/km^2, whereas in tower karst heights fluctuate between 100 and 300 m and their density is about 5 to 10 hills/km^2.

Many factors influence the development of these forms. Limestone types seem to play an important role. Jamaican cone karst hills are located on very fissured limestones (Sweeting, 1958; 1972), whereas karst towers in Puerto Rico (Monroe, 1964) and those of Sarawak (Malaysia) (Wilford and Wall, 1965) develop on crystalline limestones with soft interbedded levels. Bedding strike and dip affect form development in such a way that hills and depressions opened at their toes (dolines, uvalas, poljes, and valleys) elongate following those directions. In the same sense, splitting planes affect landform development.

No clear relationship exists between total precipitation and type of tropical karst. In his studies on New Guinea, Verstappen (1964) indicated that a precipitation of at least 1500 mm is needed for the cone karst to develop. In China, the cone karst region of Schuicheng has an average annual precipitation of 1230 mm and the tower karst zone of Guilin, the most spectacular one in China, has an annual rainfall of 1870 mm (Sweeting, 1995). In order to make comparisons, seasonality, intensity of precipitation, and temperature also need to be known, because altitudinal variations clearly influence cone karst development.

Water infiltration is favored by the open-joint network, which controls surface drainage. Despite allogenic rivers, which usually have large discharges, the drainage pattern is constituted by dry valleys, blind valleys ending in a ponor, and steephead valleys that begin in a spring (*cul-de-sac*). Superficial associated forms are bare and covered karren, dolines of different types, uvalas, and poljes. The rapid and intense infiltration of water into the karst massif gives rise to an important cave system, mainly horizontal, embracing a long and complex history of tropical karst development (Barbary et al., 1991; Sweeting, 1995). Caves generate more easily in the zone of fluctuating phreatic level, characterised by a greater dissolution (Jennings, 1985).

6.1. Cone karst

This is constituted by numerous conical or hemispheric hills (Figure 21.43), separated by closed depressions called *cockpits* in Jamaica (Sweeting, 1958). In Java, a hill density is of about 30 cones/km^2 (Lehmann, 1936). Cone height fluctuates depending on the sites and the highest ones at 100 to 130 m occur in Jamaica. Slopes present inclinations between 40 and 60°. Cones can be symmetrical or asymmetrical, due to control by dipping strata dipping or to fracturing. Commonly slopes are covered by a quite resistant duricrust with thickness of up to 5 to 10 m in Puerto Rico (Monroe, 1966). This deposit originates after brief storms during which water enriched in CO_2 dissolves calcite, which then evaporates and $CaCO_3$ precipitates, especially on the windward slopes. Cone summits can be affected by an elongated depression of up to 100 m, named *zanjón* in Puerto Rico. It results from

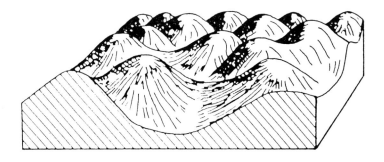

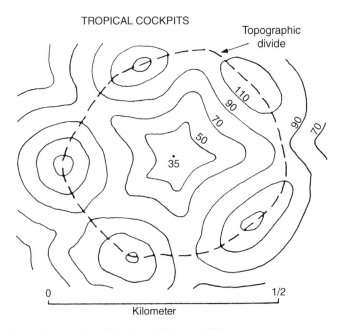

Figure 21.43. Conical karst and cockpit (from Williams, 1969).

the dissolution and enlargement of a splitting plane, for which it can be considered as a structural karren.

Cockpits are closed depressions with a star plan-form that develops at the cone foot and are separated from each other by hill passes. This morphology is due to the intense tropical heavy rains that cut gullies on slopes. Morphometric studies on the cone karst of New Guinea seem to confirm this interpretation (Williams, 1972). In some cases, cockpits are oriented following fracture lines. Where several cockpits coalesce, a *glade* is generated (Sweeting, 1958; 1972), a Jamaican name equivalent to the uvala of the temperate climate karst. Glades can present an irregular bottom, with small depressions inside.

The origin of cone karst is problematic and the ideas of Lehmann (1936), developed by the Java karst, are partially admitted (Figure 21.44). This process is thought to start with a plane surface that suffers upwarping by endogenous activity. This relief variation produces drainage rejuvenation and strong gully incision on limestones. A later stage consists of drainage disintegration, with the development of closed depressions and cone generation. This hypothesis is mainly applied the last stages of evolution. Following Sweeting (1972), it is possible that a better knowledge of these depressions might supply new data for solving the genesis of cone karst.

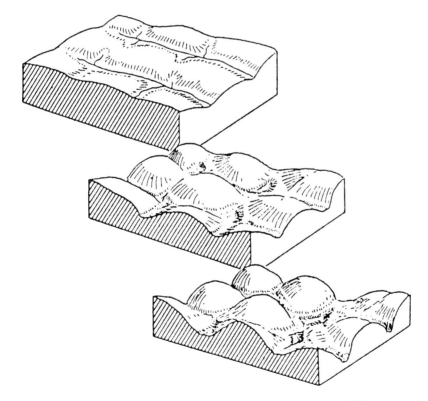

Figure 21.44. Evolution of conical karst in Goenoeng Sewoe, Java (Lehmann, 1936).

6.2. Tower karst

This is also called "pinnacle karst" and in the Spanish speaking Greater Antilles, mogote or pepino karst (Figure 21.45). This karst type develops in the Caribbean zone, Indonesia, New Guinea, and especially in southeast Asia, where it reaches an extent of 600,000 km² in southern China and northern Vietnam (Sweeting, 1972). Tower karst is constituted by steep hills, isolated or grouped, that can exceed a height of 300 m. Towers are usually surrounded by a river, lake or alluvial plain.

Two main tower-karst types are recognized (Sweeting, 1972). The first one is formed by alluvial plains with underlying limestone over which carbonate towers stand out (Figure 21.46). In the second one, towers arise on a plain over non-carbonate rocks, like granites in central Jamaica (Sweeting, 1958; 1972) and Malaysia (Jennings, 1985). The Guilin tower karst in China is the most spectacular one and develops on a 2600-m thick limestone unit of middle and upper Devonian and upper Carboniferous age (Sweeting, 1995). Two varieties have been distinguished (Yuan, 1987): the *fengling* type is the most extensive one and is represented by isolated towers with abrupt walls arising from plain limestone surfaces covered by Quaternary alluvium. The *fengcong* type is formed by a scatter of tower groups, which stand out upon a rocky basement. Slope form and inclination depend on limestone type. Shear-isolated peaks prevail on crystalline limestone, whereas domical forms are associated with dolomites. The statistical study of towers in the Guilin area results in an average height of 74 m and a density of 1.23 to 1.59 towers/km². In the *fengcong* type, towers are located on high zones and slope inclination fluctuates between 30 and 70°, depending on the rock type on which they form. Depressions of varying size

Figure 21.45. Mogotes in the karst of Sierra de los Organos, the westernmost ranges in Cuba. The plain forms part of the Viñales polje. Photo: J.L. Peña.

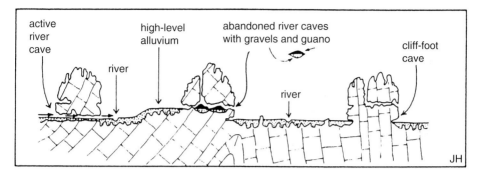

Figure 21.46. Schematic section through tower karst of Kinta valley, Malaysia (Jennings, 1985).

appear between peaks, and some of them are several kilometres long. They present steep slopes and are interpreted as old fluvial valleys (Sweeting, 1995).

In a broad sense, the different tower karst types are controlled by the fracturing network, which affects these monoliths through the development of very deep corridors and *bogaz* (Figure 21.47). Dolines of abrupt walls are also recognized, possibly related to cave collapse. Precipitation of $CaCO_3$ takes place in the tower surface and generates hard calcretes, resistant to erosion. This deposit protects towers against karstic corrosion and hence slope evolution is very slow, favouring the endurance of form.

Lakes, swamp zones and fluvial plains develop at the foot of *fengling* types of towers. The contact with the tower base is characterized by a great geomorphological activity. Horizontal grooves can occur close to the scarp foot, hundreds of metres long, which represent old lacustrine or fluvial levels. Overhanging slopes sometimes appear in relation

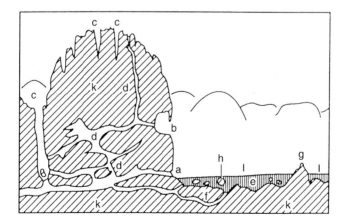

Figure 21.47. Characteristics of tropical karst hills (towers) of Sierra de los Organos, Cuba. a: active hillfoot cave; b: abandoned caves with speleothems; c: deep karstified fissures; d: vadose cave systems, related to previous levels of plains; e: alluviated valley floor; f: suballuvial phreatic cave systems; g: limestone knobs; h: detached limestone blocks; i: flat surface of intervening plain, frequently flooded; k: limestone tower (Lehmann, 1955).

to grooves, which usually collapse, but the most common feature is the presence of caves in the contact tower-plain, with increases of phreatic or vadose water that accelerates limestone dissolution in this zone and as a consequence slopes become steeper. This zone of basal undermining can be emphasized by fluvial erosion. If this activity decreases then the angle of the basal slope segment declines (Gerstenhauer, 1960) and the tower degrades.

In the Guilin karst of southern China, caves develop with greater profusion in crystalline limestones. They normally form nearly horizontal cavities with large collapse chambers that distribute with varying numbers at diverse heights. The first three cave levels, located at ± 0; 7 to 15 and 25 to 35 m over the water level, occur all along the Guilin region, although other cavity systems develop at 50, 90, and even 300 m. Lower caves are recent, while the ones located at higher altitudes are older. This indicates that towers have been growing at the same time the plain lowered, mainly by karstic corrosion. It is believed that their evolution started in the upper Pliocene period. They present middle Pleistocene hominoid remains in the 90 m cave (*Gigantopithecus*). Many data exist on absolute dating, basically made by U/Th and C-14 techniques. Palaeomagnetic studies indicate that sediments from one cave are at least 900,000 years of age (Sweeting, 1995).

Plains originate by limestone dissolution planation under alluvial sediments, ponds, or lacustrine zones. Planation surfaces are controlled by phreatic level oscillations (Ford and Williams, 1989). They are equivalent to the marginal or border plains of polje bottoms in temperate regions. Following Thomas (1994), their evolution is quite similar to the one of etchplains, where surface plain lowering is accompanied by tower growth.

Chapter 22

Applied geomorphology in tropical regions

1. Introduction

Tropical zones embrace a great variety of landforms and ecosystems that include important resources, although they also bear numerous hazards (Reading et al., 1995). Tropical climates control the life and economic activity of people at a higher level than in middle latitude climates. In the last 50 years a great demographic explosion has taken place in tropical regions. At present some 2400 million inhabitants live in these areas, representing some 45% of the world's population. About 60% live in southern and eastern Asia. Many of these nations are developing countries with a mainly agricultural subsistence economy (McGregor and Nieuwolt, 1998). The need for food has led to agricultural extension and intensification, together with an increase in deforestation practices. All this has resulted in an increase of soil erosion (Figure 22.1), flooding and mass movement hazards. These circumstances make it necessary to apply soil conservation practices (Douglas and Spencer, 1982).

The thick regoliths developed in the humid tropics have been studied in detail due to their implications for engineering activities (Lovegrove and Fookes, 1972). They have a varying depth, structure, and mineralogical composition, and abrupt changes are common within the regolith (Douglas, 1986). All these variations considerably complicate engineering works.

Humid tropical areas include numerous mineral, hydrocarbon, forest, and agricultural resources whose exploitations affect different geomorphic processes, usually by triggering or accelerating their activity. Furthermore, many areas are located on tectonically active zones and hence can be affected by earthquakes and tsunamis. Volcanic hazards are also very important, mainly the explosive type, with the emission of burning clouds, ash falls, and generation of lahars. Finally, hazards with possibly the highest occurrence are those associated with tropical cyclones, with strong destructive winds, floods, and slides related to the cyclonic rains, as well as onshore storm surges caused by wind and atmospheric pressure changes. An example of many of these hazards is represented by the great human and material losses produced by the Mitch hurricane in Nicaragua and Honduras at the end of October and beginning of November 1998.

2. Regolith characteristics. Some implications for engineering geology

A wide variety of rocks outcrop in these areas, although granites and other chemically resistant quartz-rich rocks are among the more abundant in outcrop. For this reason, most researchers have focused their studies on rocks of granitic compositions. After rock

Figure 22.1. Gully head developed on a partially deforested area, weakly affecting the settlement and crops, Muranga, Kenya.

weathering, the resultant regolith includes two main particle sizes: one is constituted by non-altered quartz grains and the other is represented by the clay fraction resulting from feldspar and mica weathering. The regolith presents a porosity of up to 60%, high permeability and angles of internal friction between 20 and 30° (Dearman and Shibakova, 1989). During heavy rains saturation can be attained, giving rise to catastrophic rock falls. Towards their base, alteration profiles include blocks of non-altered rock of varying size. These blocks pose problems in road cuts and also in slope stabilisation (Douglas, 1986).

Weathering depth is quite variable. Prevalent clay minerals in the saprolite are kaolinite and halloysite under high precipitation conditions. Montmorillonite appears in the transition zones to semiarid regions or in poorly drained areas. High clay contents contribute to steep slope instability and landslide generation during heavy rain periods (Thomas, 1986).

Chapter 20 included the analysis of the weathering profile and its differentiation into zones (Ollier, 1984), based on grus structure and composition, and on fresh rock morphology. The same criteria have been adapted to design a classification of alteration profiles with an engineering purpose (Dearman et al., 1978). The Tropical Soil Committee of the International Society of Soil Mechanics and Foundation Engineering (ISSMFE, 1985) developed a classification based on these principles, suitable enough for engineering works and especially useful when estimating loading capacity for foundation purposes (Dearman and Shibakova, 1989). This classification appears in Table 22.1. The six different grades greatly resemble the zones proposed by Ollier.

Three-dimensional knowledge of regolith is of prime importance for making engineering works. Drilling and geophysical exploration are carried out with this

Table 22.1. Classification of residual tropical soils by degrees of weathering (ISSMFE, 1985).

Grade	Degree of decomposition	Field recognition	Engineering properties
VI	Soil	No recognisable rock texture; surface layer contains humus and plant roots	Unsuitable for important foundations. Unstable on slopes when cover is destroyed
V	Completely weathered	Rock completely decomposed by weathering in place but texture still recognisable. In types of granitic origin, original feldspars completely decomposed to clay minerals. Cannot be recovered as cores by ordinary rotary drilling methods	Can be excavated by hand or ripping without use of explosives. Unsuitable for foundations of concrete dams or large structures. May be suitable for foundations of Earth dams and for fill. Unstable in high cuttings at steep angles. Requires erosion protection
IV	Highly weathered	Rock so weakened by weathering that fairly large pieces can be broken and crumble in the hands. Sometimes recovered as core by careful rotary drilling. Stained by limonite. Less than 50% rock	Similar to grade V. Unlikely to be suitable for foundations of concrete dams. Erratic presence of boulders makes it an unreliable foundation stratum for large structures

(Continued)

Table 22.1. (Continued)

Grade	Degree of decomposition	Field recognition	Engineering properties
III	Moderately weathered	Considerably weathered throughout. Possessing some strength – large pieces (e.g. NX drill core) cannot be broken by hand. Often limonite-stained. 50 to 90% rock	Excavated with difficulty without use of explosives. Mostly crushed under bulldozer tracks. Suitable for foundations of small concrete structures and rockfill dams. May be suitable for semipervious fill. Stability in cuttings depends on structural features, especially joint attitudes
II	Slightly weathered	Distinctly weathered through much of the rock fabric with slight limonite staining. Some decomposed feldspar in granites. Strength approaching that of fresh rock. More than 90% rock	Requires explosives for excavation. Suitable for concrete dam foundations. Highly permeable through open joints. Often more permeable than the zones above or below. Questionable as concrete aggregate
I	Fresh rock	Fresh rock may have some limonite-stained joints immediately beneath weathered rock	Staining indicates water percolation along joints; individual pieces may be loosened by blasting or stress relief and support may be required in tunnels and shafts

objective (Ollier and Pain, 1996). Seismic refraction profiling gives good results for obtaining regolith depth but only barely defines some of these zones (Kesel, 1976). The hammer seismic technique is the most adequate for calculating regolith thicknesses less than 30 m and has been used to distinguish regolith from colluvium in slopes, in order to know their evolution (Mills, 1990). The electric method allows the differentiation of layers with varying electrical resistivity and their thickness within the regolith. Both land and airborne electromagnetic devices permit the obtaining of three-dimensional maps of the regolith by the means of measuring the vertical conductivity distribution (Street and Anderson, 1993). Ground-penetrating radar techniques (GPR) also allow three-dimensional mapping up to depths of about 40 m (Mellet, 1990).

3. Laterites for construction

Laterite is a rock of very diverse uses and is resistant to erosion. It has been used in the construction of temples in Cambodia, in the building of the San Marcos Spanish Castle in Saint Augustine, Florida, and in the construction of many routes in tropical zones, such as the resistant roads of Gabon (Persons, 1970).

As far as engineering is concerned, it is worth differentiating between laterite, which is an irreversibly hardened rock, and lateritic soil (latosol), which softens when moistened and hardens when dried (Persons, 1984). Lateritic soils are easily extracted with axes and pickaxes in quarries. Bricks obtained are dried in the open air and need at least 2 years for hardening (McFarlane, 1976). Hardened laterite bricks have been employed for wall construction, revetments, small dams, sewers, channels, and pavements. Its nice texture and resistance make this rock an attractive material in architecture and engineering (Persons, 1970, 1984). Its use as a construction material has declined in most tropical countries, however, and only persists in those where laterite is the only existing resistant rock. Numerous remains of ancient civilisations have been preserved in which springs, breakwaters, moats, sewers, and so forth, which were constructed of such materials, have functioned for hundreds of years (Persons, 1970).

Laterite is also used in the construction of road and airport subgrades. In the streets and radial roads of Lagos, Nigeria laterite is commonly used due to its resistance and proximity to quarries. In Brazil and different countries of the Guinea Gulf the construction technique is as follows (Figure 22.2): oblong blocks of laterite are placed on an excavated base and the voids are filled with crushed laterite, whose sizes vary between 0.1 and 10 mm (Persons, 1970, 1984). Afterwards, it is compacted and a 5-cm layer of crushed laterite is added. In the humid tropics it is necessary for the road subgrade to have good drainage in order to prevent deformation in the bearing surface. Some lateritic soils are sensitive to moisture and lose strength with an increase in humidity. For this reason, it is necessary to ventilate the lateritic soil for enough time before its completion.

Dams made with rock fills and constructed with ground and compacted laterites present great resistance. This is in part because laterite particles are able to cement together and so produce an impermeable surface. On the other hand, where no laterite is available, dams are filled with diverse rocks that are still permeable after compaction and need a later impermeabilisation (Persons, 1970, 1984).

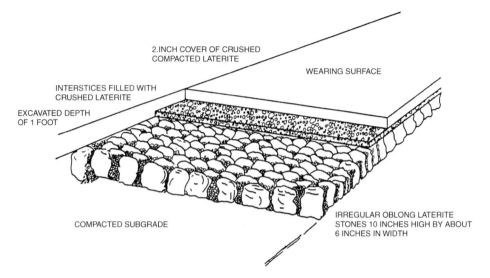

2.INCH COVER OF CRUSHED
COMPACTED LATERITE

WEARING SURFACE

INTERSTICES FILLED WITH
CRUSHED LATERITE

EXCAVATED DEPTH
OF 1 FOOT

COMPACTED SUBGRADE

IRREGULAR OBLONG LATERITE
STONES 10 INCHES HIGH BY ABOUT
6 INCHES IN WIDTH

Figure 22.2.   Scheme of laterite use as road subgrade (after Persons, 1970).

4. Supergene mineral ores and placers

Many mineral deposits are a consequence of weathering processes. Metallic elements are easily removed during alteration and can be transported to the sea or precipitate within the weathering profile. If they concentrate they can generate a mineral ore with economic interest. This process is known as *supergene enrichment*and is of prime importance with sulphide ores (Ollier, 1977, 1984; Ollier and Pain, 1996).

In a supergene ore several zones can be differentiated (Figure 22.3). The upper part of the weathering profile constitutes the oxidized zone. It can be subdivided into a lower zone with oxides that enriches towards the base, and a leached zone. In the upper part of this leached zone residual iron minerals can concentrate to form a caprock (*gossan*), which constitutes an excellent prospecting sign. Beneath the oxidized zone there appears the sulphide enrichment zone that forms by alteration of the upper material, which precipitates below the water table under reducing conditions. The limit between oxidation and sulphide zones corresponds to the water level location at the time of their formation. The slow lowering of the water table favours the development of an important thickness in the supergene sulphide zone. Water-table lowering can be due to climatic and tectonic changes or to drainage modifications by volcanic eruptions. Due to all these lowering processes, the supergene sulphide zone, of variable thickness, can be located at some metres to hundreds of metres depth, and can even reach more than 600 m depth in several places (Ollier, 1984; Ollier and Pain, 1996).

Laterites can form exploitable ores of iron, aluminium, nickel, manganese, gold, etc. (Figure 22.4). Within the alteration profile the metallic concentrations increase with depth and in the upper parts aluminium oxihydroxides prevail. These metals with high ion charge are transported within the regolith as small amorphous particles that afterwards

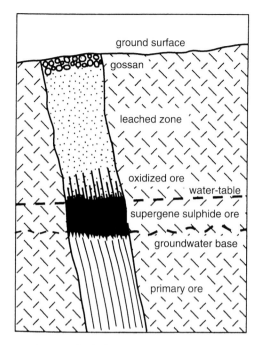

Figure 22.3. Different zones in an ore body formed by supergene enrichment (Ollier and Pain, 1996).

dehydrate and crystallise (Kühnel, 1987). Particle removal towards the lower profile parts and their ulterior crystallisation were confirmed by Aleva (1991) for nickeliferous laterites and by Nahon et al. (1985) for manganesiferous laterites.

Gold from auriferous laterites is very pure (>95% in Au) and precipitates in quartz grain fissures and also around iron oxide nodules (Thomas, 1994). It is believed that gold is released from the sulphide crystalline structure under very low pH conditions. It is afterwards solubilised under the sulphate form and when it deepens into the profile pH and Eh conditions change, giving rise to its precipitation near the phreatic level (Wilson, 1984). The activity of continental brines has also been considered, because when entering into the profile, chlorine waters reach a high acidity. Gold solubilises in the form of chlorides and, under reducing conditions, the increase of Fe^{2+} content makes gold precipitate in the form of small crystals of great purity (Mann, 1984). Similarly, the role of humid acids has been also considered in the removal of colloidal gold.

Many iron ores exist in relation to *karstification*. They are mainly constituted by limonite, hematite, goethite, and siderite, which are an alteration residue of limestones and dolomites. They occur forming discrete accumulations, filling fractures or karstic depression of varying size. Contacts between carbonate rock and iron are very irregular. The carbonate massif usually develops underground caves linked to phreatic flow. In Spain several iron ores exist related to carbonate karst processes; Alquife in Grenada, Sierra Menera in Teruel, Cabárceno in Santander, and Cerro del Hierro (*Iron Hill*) in Seville, all of them developed on carbonate rock of different ages. The iron ore of Sierra Menera is of the interstratal type and karstification of Ashgillian mixed iron/magnesium

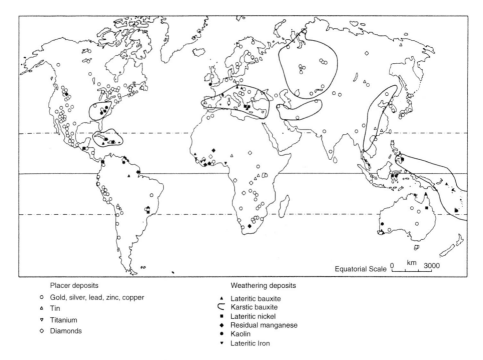

Placer deposits

o Gold, silver, lead, zinc, copper
▲ Tin
▿ Titanium
◇ Diamonds

Weathering deposits

▲ Lateritic bauxite
Ϲ Karstic bauxite
■ Lateritic nickel
◆ Residual manganese
● Kaolin
▼ Lateritic Iron

Figure 22.4. Major mineral weathering deposits and placers formed under humid tropical conditions (Reading et al., 1995).

carbonates gave rise to minerals of secondary iron (Figure 22.5) (Fernández-Nieto et al., 1981).

Nearly 10% of the world's aluminium production proceeds from karstic bauxites (Caribbean region and southeast Asia). They mainly appear filling karst depressions that have been slowly lowering at the same time that the bauxite deposit increased. They are recognised from the Cambrian through the Quaternary, always related to warm climates (Bardossy, 1981).

In the Malayan tower karst, staniferous alluvium penetrates the endokarst and fill voids and fissures. It constitutes an important tin ore where processes of previous karstification and later placer infillings are involved (Sweeting, 1972).

Placers originate by mechanical concentration due to the gravity separation of heavy minerals (denser than quartz), by water, or by wind (Bateman, 1950). For this process to occur, it is required that minerals be released from the altered rock and become concentrated. Most placers are generated in fluvial and littoral environments. The most important exploitable minerals are gold, diamond, ilmenite, rutile, cassiterite, zircon, monazite, gemstones, and others. Although placers have a wide distribution all over the world (Figure 22.4), humid and subhumid tropical environments present optimum conditions for their formation (Thomas, 1994). Most placers result from the activity of present (Figure 22.6) or recent geomorphological processes, such as diamonds in the Central African Republic or cassiterite in Indonesia. Other deposits were generated in former geological times, such as the Tertiary auriferous placers of Las Médulas in León

Figure 22.5. Exploitation of iron oxihydroxides, which result from carbonate karstification, Cantera Corral, Sierra Menera. Ojos Negros, Teruel, Spain.

province, Spain and the Proterozoic deposits of Witwatersrand, in South Africa, or the Triassic diamond-bearing placers of Swaziland (Sutherland, 1984, 1985).

In the fluvial systems, Quaternary climate alternations gave rise to fluvial terraces, which can be exploited because they contain heavy minerals of economic interest and, at

Figure 22.6. Prospecting a gold placer by the use of trays and a suction dredge. Nzi River affluent, Bouaké, Ivory Coast.

the same time, they can be transported to lower terrace levels. Fluvial channel beds are affected by a continuous reworking and channel migration in such a way that coarse gravel sedimentation favours trapping of small diamonds, such as those in Sierra Leone and Ghana (Thomas and Thorp, 1993). In Indonesia and Malaysia cassiterite placers derive from the stripping of weathering profiles on granites and andesites, and appear in colluvium, alluvial fans and fillings of endokarstic cavities (Aleva, 1985; Gupta et al., 1987). A wide variety of economic minerals can concentrate on beach placers, such as ilmenite, zircon, and monazite. Their concentration is made by waves and littoral drift currents. On the other hand, removal of beach sands by wind can give rise to aeolian placers on littoral dunes.

The commercial extraction of minerals can produce important environmental modifications, usually tolerated by the governments of developing countries. A great part of the cassiterite exploitations of alluvial placers in SE Asia are near the coast. Extraction is carried out by dredging and pumping fluvial channel beds. This operation produces intense water turbulence and an important part of the suspended load is spilled into the sea, affecting the life of coral reefs. The Madeira Basin of the SW Amazon River represents another example of impact derived from some metallurgical treatment practices. Dredged gold is amalgamated with mercury and some 5 to 30% mercury is lost in the river. Water pollution has serious effects on the fauna existing in this fluvial basin (Reading et al., 1995).

5. Slope failures and hazards

Mass movements constitute a hazard when threatening people's life and properties. During the last century this hazard has increased due to population growth and to the expansion of areas needed for obtaining essential resources, which on occasion makes people exploit hazardous zones (Crozier, 1986). In addition, other factors have converged, such as population migration to the great cities and the construction of numerous engineering works that modify the landscape and introduce slope instability (Figure 22.7 and Figure 22.8). Slope failures can produce deaths and injury, material losses and environmental damage. In humid tropical zones failures present a very important danger due to the existence of more or less thick saprolites and to the high precipitation. These circumstances favour the removal of surface or deep altered layers, which takes place especially in relation to rainstorms. For this reason, it is necessary to draw detailed geomorphological maps and make boreholes for terrain recognition (Tricart, 1974a).

Heavy rainfall commonly gives rise to slope failures and floods. From the first to the fourth of October 1968, several states of northwest India underwent major monsoon rains that produced slides and flooding, burying or partially submerging the cities of Darjeeling and Jalpaiguri. About 1000 people died and many material losses were recorded, especially oil pipes, roads, railways, and houses (Davis, 1992). This combined action of flooding and slope failure also took place in Honduras and Nicaragua, as a consequence of the rain associated with the Mitch hurricane, between the end of October and beginning of November 1998. Some 85% of the land was flooded, more than 10,000 people died, 15,000 disappeared, and enormous damage was made to the infrastructure (Figure 22.9), houses and crops. The slide produced on the Casitas volcano was channelled by the hill slopes and

Figure 22.7. Rotational slide developed on Pliocene cinders, affecting a road, Sua, Esmeralda Province, Ecuador. Photo: F. Gutiérrez.

Figure 22.8. Bellefontaine slide affecting pyroclastic materials, which have fallen from the roadcut slope, Western Martinique Island, Lesser Antilles.

Figure 22.9. Bridge fallen into the Cocos River as a consequence of floods associated with the Mitch hurricane, during the end of October and beginning of November of 1998, North of Nicaragua, near the Honduras frontier. Photo: A. Gómez Sal.

water addition transformed the pyroclastic deposits into a *debris flow*, which spread out through the croplands existing at the volcano foot. Numerous settlements were buried, causing around 1500 deaths. The debris flow thickness was estimated to be about 2 m (personal communication from Prof. A. Cendrero) (Figure 22.10).

Slope failures constitute periodic and catastrophic erosive processes in some zones. The region between Sao Paulo and Rio de Janeiro in Brazil is crossed by the abrupt ridges of Serra do Mar and Mantiqueira, separated by the Val Paraiba valley. Catastrophic failures have been recorded in 1942, 1956, 1966, 1967, and 1988. As a consequence of the floods and slides of 1942, governmental measures were established in relation to construction practices, forest conservation, and soil use. In March 1956, the city of Santos was affected by heavy rains, with a maximum intensity of 250 mm in 10 h, that triggered numerous rotational and translational slides (Pichler, 1958). In January 1966, Rio de Janeiro experienced rainfall of 472 mm in 72 h and produced mass movements such as rock falls, mud flows and rotational slides (Tricart, 1974a). The disaster produced 1000 deaths. The most spectacular slope failure development in this region took place in the steep Serra das Araras, which forms part of the Serra do Mar and is located 50 km West of Rio de Janeiro. On the 22nd and 23rd of January 1967, 275 mm of precipitation occurred with an intensity of 114 mm/h, accompanied by strong thundering, which triggered more than 10,000 failures in a 180 km^2 area. There were debris slides, debris avalanches, and mud flows that affected water and power supplies and road traffic, cutting the Sao Paulo–Rio de Janeiro highway. About 1700 people died and material losses were estimated at 1600 million dollars (Jones, 1973). Finally, in February 1988, debris avalanches in Rio de Janeiro killed 200 people, and 20,000 lost their homes (Smith and de Sánchez, 1992).

Figure 22.10. Slope failure on the Casitas volcano and subsequent debris flow, triggered by rains associated with Hurricane Mitch, during the end of October and beginning of November of 1998. Debris flow materials buried about 1500 people. Northeastern León city, Maribios volcanic ridge, Nicaragua. Photo: A. Gómez Sal.

In the geomorphological literature many examples exist of catastrophic events of different magnitude produced by slides. One of the most important took place on the night of 29th October, 1959, in the Minatitlán city of southeast Mexico, when massive mudflows buried numerous houses and caused the death of 800 people while sleeping. On the 13th and 14th of November 1963, significant mud and rock movements originated in the Grand Rivière du Nord in Haiti in relation to intense tropical rainstorms. These slope failures affected many villages and killed some 500 villagers and tourists (Davis, 1992).

6. Catastrophic flooding

Floods constitute the most common geomorphic hazard in humid tropics, due to the intense and prolonged rainfall that happens in these zones. The use of floodplains for obtaining agricultural resources leads to a considerable hazard increase. As a consequence of flooding, in Asia about 4 million hectares of land annually receive significant damage, affecting 17 million people (Reading et al., 1995).

Thunderstorms are the major rain producers in low-latitude tropical regions. Figure 22.11 represents the global distribution of thunderstorms with different frequencies: more than 100, between 50 and 100, between 20 and 50, and less than 20 storms per year (Lamb, 1972). A band of higher thunderstorm frequency occurs in areas near to the equator.

Precipitation in the tropics is associated with the convergence of air masses between both hemispheres (Intertropical Convergence Zone), with tropical cyclones, and with

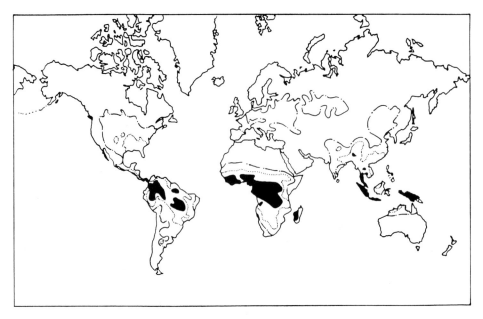

Figure 22.11. Annual thunderstorm frequency. Regions with 100 or more thunderstorms per year are black. The dotted line encompasses regions with more than 50 thunderstorms per year and the dashed-dotted line encompasses regions with at least 20 thunderstorms per year (after Lamb, 1972).

orographic effects (Hayden, 1988). In the first case, air convergence and elevation produce a cloudy strip around the equator. The seasonal variation of this cloud belt gives rise to monsoon rains, which in Asia develop between June and September, with an average of seven monsoon depressions per year (Gupta, 1988). Sometimes these rains can last up to 9 days. Cyclones move from east to west and maximum precipitation takes place on islands or on continental margins, where the land surface is warmed up and topography enhances air uplift and increases convectivity (Hayden, 1988). In Cilaos, Reunion Island, 1870 mm rain fell during a single day (Gupta, 1988). Tropical cyclones with winds greater than 116 km/h are called "hurricanes" and in eastern Asia are known as "typhoons." The orographic effect consists in the rise and condensation of air masses, which generate the maximum diurnal precipitation during the period of higher sun warming.

Gauging stations in humid tropics are scarce and in general they are located on big rivers. Consequently, records of great floods are very limited. Recurrence periods of intense precipitation are much better known, however, and these data can be related to great floods, by which some approximate values can be available (Gupta, 1988).

Drainage basin hydrology experiences deep modifications in urban zones, mainly due to the impermeable surfaces of streets and roofs, which increase flood frequency (Leopold, 1968). This urban-related hydrological change affects urban areas when channel and sewer networks are insufficient for water evacuation. Then flooding of cities occurs, as it sometimes happens in Kuala Lumpur (Malaysia) and Singapore (Gupta, 1984; 1993).

Bangladesh, former East Pakistan, is the country that has suffered the worst catastrophic floods in the humid tropics. It is a small and poor country, with an area of

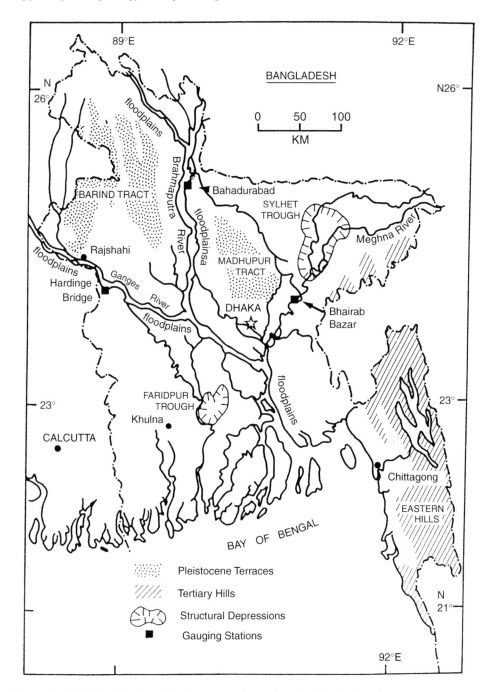

Figure 22.12. Major physiographic features and rivers of Bangladesh (Rasid and Pramanik, 1993).

144,836 km^2 and more than 110 million inhabitants, of which a greater part live upon floodplains and have a *per capita* income of about 140 dollars (Brammer, 1990; Khalil, 1990). The Ganges–Brahmaputra delta is located in the southern region, and it is also fed by the Meghna River (Figure 22.12). The delta contains numerous islands where population density is up to 1000 people/km^2, and agricultural land exploitation is very intense (Carter, 1988). About 60% of the country presents a flooding hazard. Two great closed and shallow structural depressions develop on those wide deltaic plains and floodplains. Elevated zones correspond to Pleistocene fluvial terraces and Tertiary hills (Rasid and Pramanik, 1993). Maximum annual precipitation is 11,615 mm, recorded at Cherrapunji, to the NE of the country. Rainfall decreases from this region towards the west, where it reaches an average value of 1250 mm (Brammer, 1990). In their low courses, the Ganges, Brahmaputra and Megna rivers have valley widths of about 10 km and during floods maximum discharge is 160,000 m^3/s, three times the average value of the Mississippi River. These rivers transport a sediment load of 2.5 × 10^9 t/year (Coleman, 1969) and have a marked seasonality. They record high levels from the beginning of spring to the end of summer, due to the Himalayan snow melting and to the June–July monsoon rains, which can give rise to important flooding.

The Bay of Bengal has a tropical cyclone frequency of 20% (Barry and Chorley, 1987). When cyclones approach the coast they produce a sudden sea level rise due to strong winds and low atmospheric pressure (*storm surge*), and high storm waves. These waves can reach 6 to 7 m in height and penetrate deeply into the coastal fringe, which is constituted by deltaic zones of very low height, and hence houses and crops get flooded. Sea penetration effects are emphasised when the cyclone passing coincides with high tide conditions, as happened in November, 1970 (Smith and Ward, 1998). Strong destructive winds that can exceed 200 km/h and strong cyclonic precipitation adds to the marine flooding (Reading et al., 1995).

The Bangladesh territory has suffered nearly 20 important floods since the middle-20th century (Brammer, 1990). Some have only produced material damage, but in most of them the number of deaths has been very high. The biggest catastrophe happened on the 12th of November, 1970, and it is considered as the worst natural disaster of the 20th century. It was generated by a cyclone that arrived at the coast at 11 p.m. and produced 300,000 deaths (Burton et al., 1978; Carter, 1988). In successive years new catastrophes also took place. In 1987 and 1988 river floods were caused by strong monsoonal precipitation and, according to official data, they produced 1657 and 2379 deaths, respectively (Brammer, 1990). On 30 April 1991, as a consequence of a new catastrophic cyclone, 70,000 people lost their lives in SE Bangladesh and millions lost their homes (Davis, 1992; Reading et al., 1995).

Protection measures against these natural catastrophes have been quite scarce due to the great poverty of the country. Earth walls and tree fences have been constructed, although the former are designed for fighting against salinization. During the 1970 cyclones, some 5% of people were saved in official refuges and about 38% by climbing into the trees. This cyclone was identified 3 days before and monitored by satellite, although no intensity was calculated (Burton et al., 1978). Despite alarms, very few people were evacuated and most of them remained on their lands, perhaps due to religious reasons (Carter, 1988). In the 1991 cyclone, however, meteorological predictions and alarms led to a major evacuation.

EIGHTH PART

Geomorphology and climate change

Chapter 23
Environmental change

1. Introduction

In the course of Earth's history numerous changes in the climate, sea level, vegetation cover, animal population, soils, and landforms have taken place (Goudie, 1992), but it is Quaternary time during which those constant environmental changes, which occasionally take place in short spans, can be detected with more accuracy. In recent times, from the geomorphological perspective these environmental changes are due to the three basic causes that are geological, climatic and anthropic. The difficulty in knowing in detail the factors that trigger these changes derives from the interactions among causes, as often they act jointly, and it is therefore, difficult to interpret the contribution of each separate one to the environmental change.

Geological environmental change is mainly due to neotectonic activity (Figure 23.1), which is significant in areas located on active plate margins, where isostatic compensation and relative displacement between plates of the lithosphere produce significant relief modifications. For example, the deserts of mountains and depressions of the western USA and central Asia (Mabbutt, 1977) that show large crustal instability are affected by such activity. On the other hand, shield and platform deserts, such as the Sahara, India, and Australia, are not significantly influenced by endogenic causes. On the latter continent climate change can be more easily studied in the absence of tectonics.

Even though humans have dwelt on Earth for about 3 million years, their interaction with the environment began to be significant only about 10,000 years ago with the agricultural revolution, and especially with the more recent industrial and medical revolutions (Goudie, 1981b). Humans can significantly affect the vegetation, fauna, soil, water, and climate, which is why is they are considered a geomorphological agent (Dov Nir, 1983), as well as the fact that human activity modifies the Earth's surface (Figure 23.2) and interferes with non-human geomorphological processes.

Climate change seems familiar, because we are used to temporal fluctuations in precipitation and temperature in our lifetime. In addition to these short-term variations, longer and deeper fluctuations occur. These features are illustrated in the different climate change scales shown by Goudie (1992) and Boulton (1993). In those scales climate fluctuations are reflected from intervals of tens of years to the scale of geological time (Figure 23.3).

Long-term climate changes (Goodess et al., 1992) become apparent in temporal scales above 20,000 years, as with Quaternary glaciations. These changes are produced by modifications outside of the climate system, such as changes in the geometry of the orbit of the Earth around the sun (Milankovitch cycles). Short-term climate changes range from 100 to 20,000 years. They are related in part to external causes, but some mechanisms can

Figure 23.1. Nigüelas Fault, located in the northwestern border of the Lecrín valley in the Betic Range, Spain.

Figure 23.2. Open-pit mine works exploitation of Cretaceous lignite in the Val de Ariño, northern Teruel Province, Iberian Range, Spain. Photo: B. Leránoz.

be relevant, namely changes in the oceanic circulation, volcanic activity, and in the gas content of the atmosphere. On the other hand, the term "climate variability" is used for those changes of less than 100 years, which are due basically to internal mechanisms (Goodess et al., 1992).

In most cases, in the landscape there is a collection of landforms resulting from the past activity of the geomorphic processes, which acted under a wide spectrum of environmental conditions that were different from the present ones (Figure 23.4). As indicated above, these processes, coming from tectonic, anthropic, and climatic activity can interact. Therefore, for the study of the climate change, it is necessary to discern the contribution of each one and try to eliminate the "noise" attached to the activity of the first two. Undoubtedly, human action during the last millennium has been very significant in certain areas, as in the Mediterranean countries (Goudie, 1981b, 1990a; Mannion, 1997), and presents very complex problems for determining the real influence of climate change on the genesis of the current landscape. Therefore, the best conditions for researching climate change are those with a history of long crustal stability and sparse habitation. In this way, those areas are a reflection of landform generation due exclusively to geomorphological processes triggered by climate changes.

It is necessary to know the age of the forms related to climate change. Early researchers used relative dating in relation to stratigraphic series, morph genetic sequences, degree of weathering, and soil development (Birkeland, 1984). Recently, above all during the last 20 years, many dating methods have been developed, providing direct or indirect absolute ages for buried or exposed particles. Examples of the first group include radiocarbon, thermoluminescence, uranium series, cosmogenic isotopes, amino acid racemization, IR luminescence, and electron spin resonance. The indirect techniques are mainly based on paleomagnetism, and the isotopes of O and C. They are used depending on the material available, and their precision varies according to the technique utilized. Some of these methods are under experimental development, but in the future they promise to revolutionize our knowledge and comprehension of environmental changes (Stokes, 1997).

2. Some considerations on climate change

As indicated by Chorley et al. (1984), climate change is one of the subjects about which many geomorphologists feel rather ignorant. According to these authors, this is due to the following; in the course of Earth's history continuous climate changes have occurred, change mechanisms act in complex ways, climate changes are apparent under different patterns, and, finally, the lack of knowledge of the most significant aspects of the climate producing erosive and sedimentary processes hinders understanding.

Another important fact comes from the knowledge of the required time for the adaptation of landscape forms to new climate conditions. Some authors point out that some landforms can remain during long periods of the history of the Earth. Twidale (1978) indicated that the well-known Ayers Rock (Uluru) inselberg in the Australian desert has remained from upper Mesozoic–lower Cenozoic to present. On the other hand, areas with rills and active dune fields experience constant and rapid evolution. This persistence or modification of landforms is a function, among other factors, of their spatial location within the desert morphoclimatic region. Thus, landforms located in hyperarid

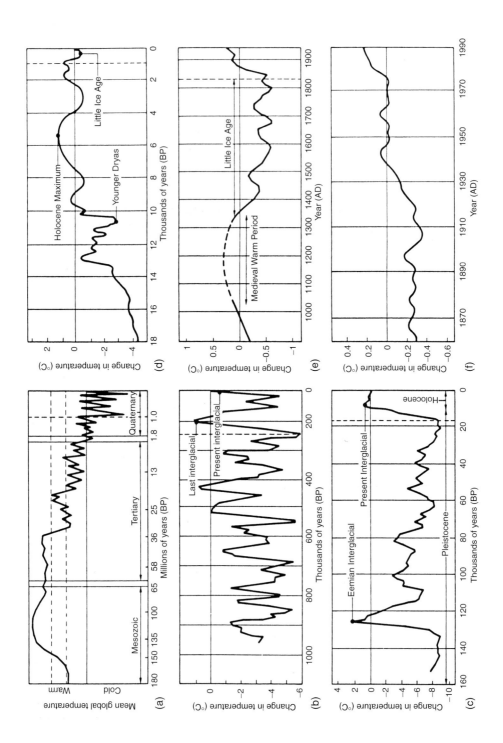

Figure 23.4. Foreground: terraces of the Jachal river, which present a braided morphology. Background: structural landscapes corresponding to the Precordillera of Andes, from where a pediment system develops. The area is located between Rodeos and San José de Jachal, San Juan Province, Argentina.

environments have undergone little or no change through Quaternary climate fluctuations (erosive paralysis of Oberlander, 1997b). In comparison to this, landscapes existing in transition regions will be clearly affected by temporal variations in the global atmospheric circulation, which cause the migration of the boundaries of the climatic zones. The latter are the most suitable for the reconstruction of the climate changes that have occurred in recent periods of the history of the Earth.

Many landforms are in a dynamic equilibrium (Hack, 1960), in which processes act but the active elements of the system are balanced. If any modification takes place, for example, a climate change, the *geomorphic threshold* can be exceeded (Schumm, 1973, 1979), causing instability in the environment. These thresholds can be sharp or gradual (Begin and Schumm, 1984) and can act in any time interval, from minutes to millions of years. Geomorphic thresholds can be triggered by extrinsic or intrinsic causes. An example of the first, or extrinsic type is an abrupt change that eliminates vegetation that leads to

---

Figure 23.3. Climate change during the last 180 ma. (a) Climate change on a 10^7 to 10^8 year timescale, reflecting long-term climate evolution as a result of plate tectonic changes. (b) Change with 10^5 to 10^6 year timescale of glacial/interglacial periods, showing the 100,000 year cycle. (c) Change on a 10^4 to 10^5 year timescale showing the structure of change within a glacial/interglacial cycle, dominated by 20,000 and 40,000 year Earth orbital changes. (d) Change over the last 10^4 years and the transition from a glacial to interglacial period. (e) Change over the last 10^3 years showing the Medieval Warm Period and the Little Ice Age. (f) Change during the last century, with a possible anthropic influence on climate (Boulton, 1993).

intense erosion. Intrinsic thresholds operate inside the system, as when there is a continuous supply of particles on a slope, which then reaches a critical threshold of instability. In some cases, these geomorphologic thresholds can be quantitatively defined and used as the basis for the recognition of potentially unstable landscapes (Schumm, 1979). On the other hand, because connections inside a geomorphic system are so complex, when an external change takes place, it spreads over the system in a complex way (*complex response* of Schumm, 1973). Bull (1991) analysed a collection of important concepts in relation to the response of the geomorphic systems to the climate change (Figure 23.5). This author showed a reaction time of the system when faced with stimulus, followed by a relaxation time, which is necessary to reach the new conditions of equilibrium. The sum of reaction and relaxation times constitute the response time. The time during which the geomorphologic system remains in equilibrium is the persistence time. Therefore, systems can have different thresholds, short to long times of persistence, and very variable times of response. The latter indicate the system sensitivity to change (Brunsden, 1990).

The resistance of the natural systems to climate changes is not well known, so landscape sensitivity to climate changes is a field of research of special interest. "The *sensitivity* of a landscape to change is expressed as the likelihood that a given change in the controls of a system will produce a sensible, recognizable and persistent response" (Brunsden and Thornes, 1979). Sensitivity may vary in space and time and also depends upon the scale (Thornes and Brunsden, 1977). Geomorphologic systems of small areas

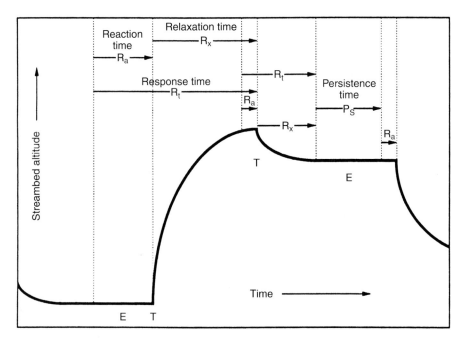

Figure 23.5.   Changes in the height of a fluvial channel produced by aggradation and degradation are used to show the components of response time. R_t is the response time, which is the sum of reaction time (R_a) and relaxation time (R_x). P_s is the persistence time in the new equilibrium conditions. T and E represent threshold and equilibrium, respectively (Bull, 1991).

reach a balance with the new conditions faster and permit carrying out the analysis of short time spans. Temporal changes can be slow and gradual or sudden and catastrophic. In general landforms present a slow response to climate change, so they commonly can be analysed only for long time periods. To understand the more or less rapid changes in the development of landscapes, it is necessary to know as well as possible the magnitude of the processes and their temporal record. All this can lead to the understanding of the sensitivity of geomophic systems (Allison and Thomas, 1993). It is possible to distinguish those landscapes of high sensitivity to external impulse which respond rapidly to climate change, such as fluvial channels. Insensitive systems are those whose response is slow and present a strong resistance to change such as interfluve areas or the plains of Africa and Australia. Insensitive areas commonly correspond to landforms originated on resistant lithologies, whereas in sensitive zones small changes take place more rapidly, as is shown in rilling areas (Brunsden and Thornes, 1979).

In desert areas, activation of the systems, which is necessary to produce a response on the landscape, is generally much smaller than in other climatic zones. This indicates that climatic sensitivity for arid zones landform generation is high because the relaxation time necessary to reflect changes in the forms is low (Shaw and Thomas, 1993). These models of short relaxation time are termed *labile landforms* by Trudgill (1976) (Figure 23.6).

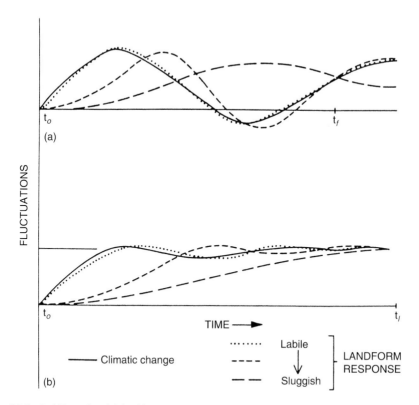

Figure 23.6. Labile to sluggish landform responses over time. (a) Corresponds to a cyclic climate change and (b) a pseudo step-functional change of climate (modified after Trudgill, 1976).

Labile landforms are rapidly modified under extreme events or they have limited resistance to change. Their response is almost parallel to climate change. On the contrary, *sluggish landforms* are subordinated to weak processes or present a large resistance to change.

Gradual climate change is assumed and the hydrologic response is of the same type. Nevertheless, Knox (1972), in a work on southeastern Wisconsin, USA, indicated that

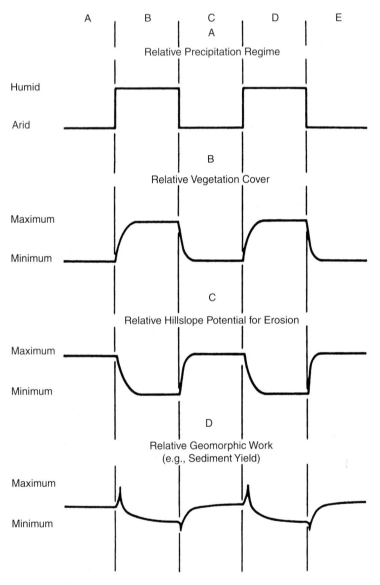

Figure 23.7. Vegetational and geomorphic responses to abrupt changes in climatic regimes. The curves would seem to be most applicable to climatic regions that have annual precipitation between 250 and 1520 mm (Knox, 1972; in Schumm (1977), Figure 3.3).

climate change can be abrupt and, as a result, a response is triggered. This response, although short, can be opposite to the curve of soil loss of Langbein and Schumm (1958). This author (Knox) pointed out that if there is an intense and maintained increase in precipitation, this causes erosion of the channel and a rise in slope gradient (Figure 23.7C). Therefore, in this short period of time an important increase in the loss of soil is also produced, followed by a fall (Figure 23.7D). Nevertheless, these fluctuations are difficult to distinguish from the effect of large flooding with long return period (Chorley et al., 1984).

3. Future climate and the greenhouse effect

The study of the climate trend is nowadays the purpose of many researches, as a result of the increased warming predicted by scientists due to the effect of greenhouse gases. These gases are water vapour, carbon dioxide, nitrous oxide, methane, and chlorofluorocarbonates. They have increased during the last decades, affecting the global climate. The most significant effect is due to the rise of carbon dioxide produced by the combustion of an ever-increasing quantity of coal, petroleum, and other fossil combustibles, as well as woodcutting and forest fires. The CO_2 is almost transparent to short-wave solar radiation and absorbs a major part of the long wave. Therefore, it is one of the factors that produce a greenhouse effect, raising the temperature of the Earth's surface and lower atmosphere. Thus an increase in the quantity of CO_2 leads to global warming (Budyko, 1982).

In the meeting that took place in 1990, sponsored by the United Nations, the Intergovernmental Panel on Climate Change (IPCC) produced a report (Houghton et al., 1990). It was pointed out that according to different models (Figure 23.8) temperatures can raise 0.3°C per decade in the higher assumption, as an effect of the greenhouse gases increase.

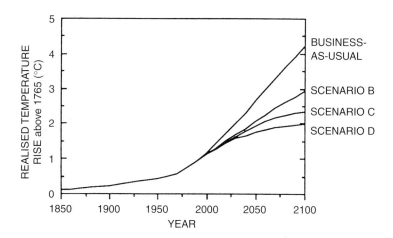

Figure 23.8. Simulations of the increase in global mean temperature from 1850 to 1990 due to observed increases in greenhouse gasses, and predictions of the rise between 1990 and 2100 resulting from IPCC Scenario B, C, and D emissions, with the business-as-usual case for comparison. The rise in temperature is 0.3°C per decade for scenario A; 0.2°C (scenario B); slightly more than 0.1 °C (scenario C) and 0.1°C (scenario D) (after Houghton et al., 1990).

As a result of it, a partial ice cap melting could happen and see level could raise 5 to 6 cm on average in one decade.

These variations will generate intense modifications of natural systems producing important socioeconomic implications that would seriously affect human life. For this reason, during the summit in Rio de Janeiro in 1992 a Covenant on Climate Change was signed (CCC). In the agreement a group of programs was proposed to mitigate future climate change, as well as to promote new technologies to reduce greenhouse gases.

From IPCC in 1990 to the one in 1995 (Houghton et al., 1996), relevant progress has been made. This lies in the application of more powerful simulation models, as well as the consideration of the effects of sulphate aerosols and the reduction of the stratospheric and anthropic ozone, which highly affect climate change (Santer et al., 1996). All that research led IPCC in 1995 to the development of new scenarios for the period 1990 to 2100, on the basis of different hypothesis. For the average scenario (IS92a) an increase of 2°C in the temperature of the air in 2100 is estimated, and 1 to 2.5°C in the case of extreme suppositions (IS92c and IS92e). The sea level rise due to ice masses melting and the thermal expansion of the oceans are calculated for the year 2100 as 50 cm (scenario IS92a) and 15 to 95 cm in the case of other extreme scenarios (Houghton et al., 1990). These new values reduce quite a lot the most catastrophic forecasts of the IPCC in 1990. This important decrease is basically due to the consideration of sulphate aerosols in predictions and to the application of new atmosphere–ocean models.

Nevertheless, Houghton et al. (1996, p. 5) indicated that in the predictions of climate for the future:

> important uncertainties remain which come from vague estimates of future gas emissions, from the representation in climatic processes models (as those associated to clouds, ocean, marine ice, and vegetation) and from the systematic capture of climatic variables. Furthermore, short- and long-term climate changes are difficult to predict, because the nonlinearity of the climatic system, so that, numerous "surprises" can occur.

The most important finding of the IPCC third assessment report, was the statement: "There is now new and stronger evidence that most of the warming observed over the last 50 years is attributable to human activities." This represents a significant strengthening of the analogous statement issued by the IPCC (1996): "The balance of evidences suggest a discernible human influence on global climate." The globally-averaged surface air temperature has increased by about $0.6 +/- 0.2°C$ over the 20th century. Most of this warming has occurred during two periods: 1910 to 1945 and 1976 to 2000, with gradual cooling during the intervening period. The range, reported in the IPCC Third Assessment Report (IPCC, 2001), is higher than the 1.0 to 3.5°C range reported in the Second Assessment Report (IPCC, 1996) probably because a greater range of scenarios are now being used. In the 1966 report only six scenarios were used, whereas now 40 scenarios are used.

Although there is a general consensus on the fact that global warming will occur in the near future, some opinions disagree or express doubts about discrepancies. Dickinson (1986) showed that general models of circulation (GCMs) have many limitations and should be carefully used. Sundquist (1993), in a study about the emissions from ice during the last interglacial period and the absorption balance in the oceans, indicated that there are significant uncertainties in relation to anthropogenic CO_2, considering its later effects.

Paté-Cornell (1996) pointed out that the models used for the study of the effects of global climate change are based on a limited knowledge of the main phenomena (for instance, the role of clouds on oceans). Demangeot (1996) considered that proof of the actual quantity of warming of the planet is not convincing and asked specialists for more precise data on icecap melting, the role of deforestation, and so forth. "Since 1940, the Greenland coastal stations data have undergone predominantly a cooling trend. At the summit of the Greenland ice sheet the summer average temperature has decreased at the rate of 2.2°C per decade. This suggests that the Greenland Ice Sheet and coastal regions are not following the current global warming trend" (Chylek, 2004).

Embleton (1989) indicated that climate changes forecast for the next decades or centuries is extremely difficult. Obviously, CO_2 has significantly increased from 1750, according to data provided by the analysis of ice cores in Camp Siple in Antarctica and from air samples taken since 1958 at the observatory of Mauna Loa in Hawaii. There is no doubt that CO_2 produces an increase in temperature. As indicated above, predictions show an increase of up to 3°C per century (Houghton et al., 1990). If these predictions came true a "super-interglacial" would occur (Figure 23.9), which would then be the event of greatest magnitude in the Quaternary. According to Embleton (1989) and Goudie (1990b),

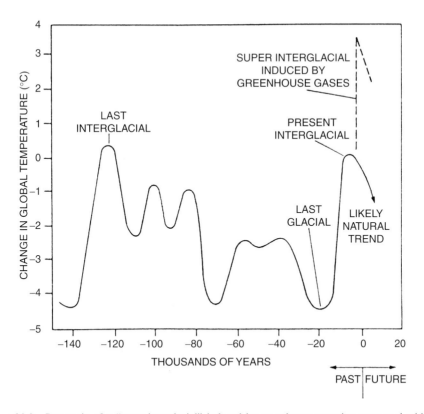

Figure 23.9. Prognosis of a "super-interglacial" induced by greenhouse warming compared with the natural temperature trend of the last 140,000 years (Lamont Newsletter 6, 1984; in Embleton (1989)).

oceanic water, which constitutes an enormous reservoir of CO_2, and marine biota play an important role on the regulation dynamics of atmospheric CO_2. Carbon dioxide flux in the oceans depends on CO_2 pressure, atmospheric concentration of CO_2, and on surface winds. Carbon dioxide pressure is controlled by oceanic transport processes, geochemical processes, and biological processes in the top part of the oceans. Furthermore, solar flux and latent calorific flow have an influence, as do the above-mentioned factors, in climate modelling. Therefore, the system is very complex and the models of the flows in the ocean–atmosphere inter-phase can be subject to large errors (Dickinson et al., 1996). Moreover, there is limited knowledge on the temperature and salinity of deep waters, close boundaries between currents, ocean eddies, and on thermohaline circulation (McBean et al., 1996).

The statistical analyses of the daily difference of temperature between maxima and minima from 2000 stations all over the world for a period of 40 years show that nights are warmer than days, $0.84°C$ and $0.28°C$, respectively. This means that the principal warming is due to the increase of night temperatures. Statistics reveal that *cloud cover* seems to be the cause of this circumstance (Henderson-Sellers, 1992), because clouds regulate night temperatures by controlling long-wave emissions. During the 20th century, clouds have increased 3.4 to 9.4%, possibly because of the increment of condensation nucleus due to air pollution. This could explain the smaller percentage of clouds above oceans with respect to continents. Without doubt, it is necessary to know the variations of cloud cover in order to carry out further analysis of climate change in the future.

Volcanic activity, mainly in the case of acid composition eruptions, which are more explosive, can occur sporadically and has a marked influence on climate. If the eruption is large, volcanic ash is driven into the stratosphere, affecting general atmospheric circulation. In June 1991, the Pinatubo volcano, located in the Philippine islands, erupted. Volcanic aerosols extended all over the planet and the models developed indicated that the average global temperature decreased on the surface up to $0.5°C$ for 2 years and $0.6°C$ in the troposphere for several months. The effects of the eruption were less evident in 1994. The consequences of the volcanic activity, which produced opacity in the stratosphere, had a significant influence on climate, but only sporadically.

Man, by means of industrial activity and agriculture, increases *atmospheric dust* significantly. Also, large quantities of aeolian dust can be deposited on to continents and into oceans from long distances. Dust reaches the upper part of the troposphere and restricts solar radiation. As a result, there is a global cooling (Bryant, 1997).

The importance of anthropic *sulphate aerosols* on climate change is becoming accepted. These aerosols come mainly from coal and petroleum combustion. Records indicate that during the last century the exponential increase in sulphate aerosols is parallel to the rise of CO_2. The high reflectivity of the sulphate aerosols produces a cooling in the troposphere with both clear and cloudy skies (Pearce, 1994; Santer et al., 1996).

Houghton et al. (1996) point out the difficulty for the prediction of *long- and short-term climate changes*. This is what some authors call "climate surprises" (Jouzel and Lorius, 1999). Rapid climate variations have been found for the last glacial age. In cores obtained from the Greenland icecap, several warming phases of about $10°C$ have been found. Those phases lasted several tens of years and gradually turned into colder conditions. Such sequences, the duration of which range from 500 to 2000 years, recurred about 20 times in the last glacial epoch (Bender et al., 1994), and not only in Greenland. On the other hand, during the last 70,000 years, several glacial sedimentary layers coming from icebergs have been identified from cores in the northern Atlantic Ocean. These six synchronic stages are called Heinrich events (Heinrich, 1988; Bond et al., 1993, 1997). The increment of freshwater and the temperature decrease in the upper part of the oceans produced significant changes in some of the climatic parameters. Therefore, the occurrences of rapid climate variations in the past make feasible the triggering of these types of climate modifications in the future.

Without doubt, climate change modelling is uncertain and complex, due to the quantity of parameters needed and the difficulty in comprehension of their relationships. This debate is not about if climate change is due to human activities, but rather how quickly, to what magnitude, and with what regional implications (Weaver, 2003). An alternative methodology for the study of climate changes is the study of the geological records (Petit-Maire, 1999), wherein valuable data about climate change can be obtained.

In order to study the consequences of warming on landscape, researchers can go into the past, mainly to the Climate Optimum or Altithermal, which occurred about 7000 BP, and compare it with the last maximum interglacial at about 120,000 BP, so that the response of the landscape in warmer phases can be analysed. Thus, landscape changes can show the geomorphologic setting for the future and the attached consequences. This prognostic approach is an important contribution from geomorphology.

If the change predicted by IPCC becomes evident, it would affect areas very differently depending on the climatic environment. For example, in arid regions an increase of the existing problems is expected (Boer and De Groot, 1990), such as soil degradation, hydrologic and aeolian erosion, decrease of soil fertility, salinization, sedimentation, and flooding. Moreover, an expansion of semiarid areas, an increase in fire risk, and a decrease in the quality and quantity of subterranean water would be expected.

Chapter 24

Climate change in glacial and periglacial regions

1. Introduction

Glacial and periglacial regions constitute morphoclimatic zones related to cold. As Tricart (1967) showed; "cool countries are those in which the geomorphological action of the water is controlled by its presence as solid state, permanently or periodically." Glaciers are developed in areas in which the snow does not melt completely from year to year, so that snow accumulates and in the end it is converted into ice. Limits of the glaciers are clear and correspond to ice accumulation. Periglacial areas are characterized by frost and thaw, and snow does not remain from year to year. Frozen soils can develop under the Earth's surface.

During the last glacial epoch some 30% of the Earth's surface was overlain by ice, covering vast areas of North America and Eurasia. Also the Antarctic expanded. In mountainous areas, small icecaps were formed and valley glaciers advanced. Nowadays, ice occupies 10% of the Earth's surface. The big advances and recessions of the ice masses have sculpted wide areas, eroding rock massifs and depositing thick accumulations of glacial materials. Movements of the glacier fronts respond to climate changes, so they constitute a source of information for the study of those changes. On the other hand, ice sheets and glaciers have an important influence on the general system of atmospheric circulation.

As a result of the important advances and recessions of the masses of ice during Quaternary, the periglacial domain underwent major fluctuations. During glaciations the extent and location of the periglacial zones were very different from the present. Thus, vast areas that were covered by ice today were periglacial then. Also, the areas in the surroundings of the ice that were periglacial, nowadays are temperate zones. Therefore, changes from glacial to periglacial conditions and from the latter to temperate can be recognized.

On the other hand, there are regions in which the periglacial domain has remained during the Quaternary. They correspond to areas in which the precipitation is so small that they cannot generate an icecap, as is the case of wide areas in Siberia (King, 1976a). These important climatic fluctuations affecting cold areas have generated relict forms, which remain under new climatic characteristics, and other landforms that correspond to current conditions. Within both situations there are many inherited and partially eroded landforms that are difficult to recognize. This aspect adds to doubts on the differentiation of relict and current landforms.

2. The paleoclimate information provided by drill cores from ice sheets and deep sea sediments

The studies of ice and sediment samples from boreholes makes possible the knowledge of past climate changes that have affected the whole planet. Many of these changes have been produced very rapidly. These important knowledge advances on climate change have been developed mainly over the last three decades.

From drill cores made in ice sheets, it is possible to obtain paleoclimate records that contain annual snow precipitation data. Records from the upper part of the core are alternate layers coming from annual precipitation, whereas in lower layers the identification of the alternation is not so simple and the records have to be dated by indirect methods.

Valley glaciers and mountain icecaps have also been cored in Tibet, the Peruvian Andes, and elsewhere, but the data only provide information on recent climate changes, which is different from the results obtained from deep drilling in polar ice sheets. The first deep core was drilled in 1963 to 1966, in Camp Century, Greenland (Dansgaard et al., 1969). In Antarctica, the Vostok drill in 1983 (Jouzel et al., 1987) provided a climatic record of the last 160,000 years, exceeding the ice of the last interglacial. The final section of this core has reached back to 400,000 yr BP Recently the ice has been drilled at Dome C, located 500 km away from Russian station Vostok, in the East Antarctic ice sheet, by a consortium called the European Project for Ice Coring in Antarctica (EPICA). A 740,000 years record is being researched in this project (EPICA, 2004). It is thought that the final section could reach back to 900,000 years or even further (Walker, 2004). Between 740,000 and 430,000 years ago, the climate was characterized by less pronounced warmth in interglacial periods in Antarctica, but a higher proportion of each cycle was spent in the warm mode (EPICA, 2004, McManus, 2004). The result of this research may imply that without human intervention a climate similar to the present one would extend well into the future (EPICA, 2004).

The drill cores, GRIP and GISP2, were carried out in the centre of Greenland, 28 km apart. In the first one no horizontal movement in the ice was recognized, while in GISP2, from 2750 m depth on, disagreements in the interpretation exist, due probably to ice flow. This increases toward the boundaries of the ice sheet.

Oxygen atoms, which are constituents of the glacial ice, have three types of isotopes, ^{16}O, ^{17}O, and ^{18}O. The relative abundance in nature is 0.2, 0.04, and 99.76%, respectively. During water evaporation on the surface of the sea, the heaviest isotope enriches mainly in the liquid phase and decreases in the vapor phase. This process is called fractionating. The resulting vapor has become impoverished in ^{18}O, compared to the former water. Condensation depends on the temperature and the isotopic composition of the precipitation reflects the temperature of the condensation. Therefore, the isotopic composition of the glacial ice can be used to reconstruct the temperature of the original precipitation. The isotopes existing in the ice are determined by melting the ice in the laboratory and then analysing it utilizing a mass spectrometer. The difference between $^{18}O/^{16}O$ and a sample of standard water (Standard Mean Ocean Water, SMOW) is expressed as $\delta^{18}O$. If values are small, the paleotemperatures were low, and vice versa. Because of this, detecting global climate fluctuations is possible (Dawson, 1992; Ehlers, 1996; Benn and Evans, 1998). Figure 24.1 shows data obtained for GRIP, almost 3000 m depth. The table represents the interglacial conditions at 130,000 yr BP (Eemian or last interglacial), several interstades (short warming phases) and the rapid termination of the glacial conditions around 11,000 yr BP. Holocene climate changes have a poor representation, according to Dansgaard et al. (1993), whereas in the later work of Bond et al. (1997) a detailed subdivision has been carried out.

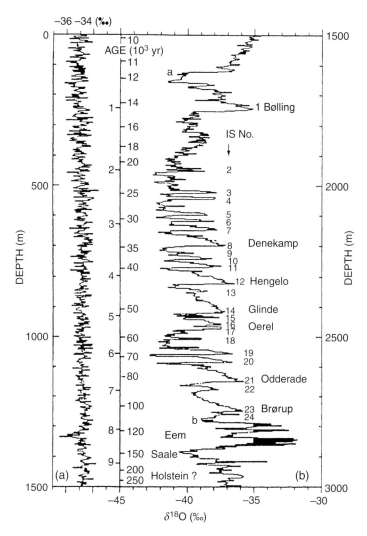

Figure 24.1. Continuous $\delta^{18}O$ record, corresponding to the GRIP Summit, plotted in two sections on a linear depth scale. (a) From the surface to 1500 m deep; (b) from 1500 to 3000 m deep. Measures are taken at each 2.2 m. Glacial interstades are indicated to the right of curve b (Dansgaard et al. 1993).

The *interstadials* are rapid and short-term events that occurred several times between 110,000 and 10,000 years ago. They are called interstadials to be distinguished from cool phases or stadials. In Figure 24.1, 24 interstadials can be recognized. They are called "Dansgaard–Oeschger events" (Bond et al., 1993). Each warm interstadial is attached to a cool one, and the pair, which last about 1500 years, is called a "Dansgaard–Oeschger cycle."

The analysis of the Vostok drill core in eastern Antarctica has revealed important information regarding CO_2 variations in upper Pleistocene and Holocene (Figure 24.2) (Lorius et al., 1988). The temporal variability in CO_2 existing in air bubbles within the ice

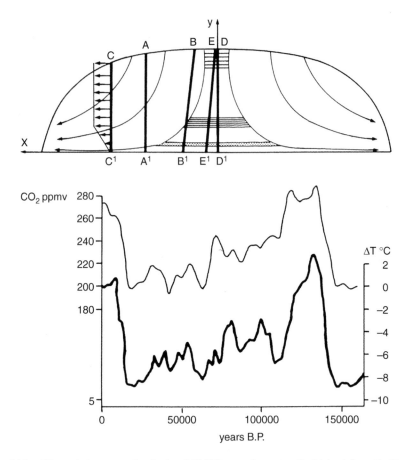

Figure 24.2. CO_2 variation curve for the last 160,000 years (upper part) obtained from the Vostok ice core in Antarctica, and a smoothed curve of the temperature record from oxygen isotopes for the same core (bottom curve) (after Lorius et al. 1988; in: Dawson (1992), Figure 2.5).

presents a good correlation with the fluctuation of the oxygen isotope, which matches the changes in temperature.

The study of the deep sea sediments has provided very relevant data in relation to climate fluctuations in the upper Tertiary and Quaternary eras. In these sediments, there are planktonic foraminifera, which float on the sea when alive, and benthonic foraminifera that live at the bottom of the sea. Their shells are made up of $CaCO_3$ and the oxygen is made up of ^{18}O and ^{16}O. These calcareous shells incorporate oxygen isotopes depending on their concentration in the seawater. When water evaporates, the vapor enriches in ^{16}O and then precipitates on icecaps, the ^{16}O content is higher than in the oceans, which are enriched in ^{18}O. Therefore, analysing the isotopes of oxygen, it is possible to reconstruct the fluctuations in the volume of the ice sheets and thus, the glacioeustatic changes of the sea level. Emiliani (1955) was the first to use this technique. He interpreted that the observed fluctuations mainly reflect the changes in temperature corresponding to glacial and interglacial periods. The analyses of the deep drill cores in the Pacific

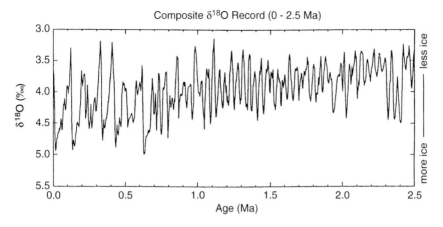

Figure 24.3. Variations in the oxygen isotope compositions in planktonic foraminifera from the equatorial Indian Ocean for the last 2.5 million years (Ma). It can be observed that the global ice volume has increased mainly during the last 750,000 years and that the mean duration of the cool periods has also increased (Crowley and North, 1991; based on data in Raymo et al., 1990).

(Schackelton and Opdyke, 1973, 1976) clearly denote that the record of isotopes of oxygen reflects the whole climate history of the Quaternary (Figure 24.3). There exist several difficulties in obtaining suitable records of oxygen isotopes in those oceanic areas in which the velocity of sedimentation has been relatively small. Also the effect of bioturbation, submarine landslides and turbidity currents hide and make the interpretations complicated.

Other significant changes that are recorded in deep cores from the northern Atlantic to Lisbon latitude during the last 70,000 years are the so-called *Heinrich events* (Heinrich, 1988; Bond et al., 1993, 1997). In contrast to the interstades these seem to be cool events, but also for short periods of time. These events are recorded in the cores by clasts that were transported by the ice, which implies a great presence of icebergs during those events. When the ice melts, icebergs deposit solids on the marine bottom (Table 24.1).

Values for HO–H3 have been obtained from Bond et al. (1997), and those corresponding to H4–H6 from Bond et al. (1993). It is not well known whether or not these processes have been originated by the intern dynamic of the icecaps (MacAyeal, 1993) or by periodic external climate changes (Broecker, 1994). The Heinrich events have been correlated with records of pollen from Europe and North America (Grimm et al., 1993), with loess

Table 24.1. Distribution of the principal Heinrich events (H) and Younger Dryas (YD) for the last 70,000 years (ages in thousand years from present) (Adams et al., 1999).

Event	Age	Event	Age
YD or HO	12.3	H1	16.8
H2	24.1	H3	30.1
H4	35.9	H5	50.0
H6	66.0		

columns from the centre of China (Porter and Zhisheng, 1995), glacial deposits in western USA (Clark and Bartlein, 1995), and with sequences of talus flatirons in northeastern Spain (Gutiérrez et al., 1998b). All this seems to indicate that there is a global climatic influence of the effects of the Heinrich events, at least in the Northern Hemisphere.

3. Fluctuations of quaternary ice sheets and resulting landforms

The advances and recessions of the glaciers have a bearing on morphogenesis, so it is fundamental to know accurately the oscillations that occurred in order to understand the resulting glacial landscape. These glacial landforms are produced by erosion and sedimentation processes and crop out when ice masses retreat. On the basis of existing landforms, areas and flow directions of the ice in ice sheets have been calculated.

The maximum expansions of the last ice sheets have a similar age. Dawson (1992), compiled global data that indicate an age of 15,000 to 20,000 years for the Euroasiatic ice sheet (late Valdai), 18,000 to 20,000 years for the Fennoscandian (late Weichselian), for the Laurentide ice sheet 20,000 to 25,000 years (late Wisconsin), and for the Antarctica ice sheet 17,000 to 21,000 years. The distribution of glaciers and icecaps for the last glacial maximum, with variations of the equilibrium line depending on the latitude and altitude, are shown in Figure 24.4 (Broecker and Denton, 1990).

The warming subsequent to the glacial maximum caused the disappearance of the Eurasian, Fennoscandian, and Laurentide ice sheets, and many valley glaciers and small glacier caps. The response can be observed in the data that show the modification of the general circulation of the atmosphere, in the rise of the sea due to the incorporation of fresh water, and thus, in the modification of the oceanic circulation. Melting events are not synchronous; the Cordilleran ice sheet disappeared 10,000 years ago, the Laurentide ended 6500 years ago, the Fennoscandian melted about 8000 yr BP, and the Eurasian had a shorter life, 9000 yr BP (according to some authors, in Dawson, 1992).

It is difficult to establish a chronology for the *Fennoscandian ice sheet* due to the fact that moraines are very broken and cannot be properly correlated in fiord areas. After 13,000 yr BP, a rapid slimming and retreat occurred, which corresponds to a warm interstadial that lasted until 11,000 yr BP. In southern Scandinavia a big lake was formed between northern Europe and the southern part of the ice. This occurred between 12,800 and 10,300 yr BP, and it is called the Baltic frozen lake (Figure 24.5) (Benn and Evans, 1998). About 10,000 yr BP, the retreat of the ice made possible for seawater to come into the Baltic Sea (Yoldia Stage). After that, there was a rapid recession of the ice, between 10,000 and 9000 yr BP, so that the ice sheet was reduced to a small extension in northern Sweden and a big fresh water lake was originated (Ancylus Lake). Finally, and probably due to the glacioisostatic rebound and to the Flandrian transgression, marine waters went into the Baltic Sea and Botnia Gulf (Litorina sea).

The *Laurentide ice sheet* extended from the Canadian archipelago in the Artic Ocean to the northern states of USA, in the south. It bordered on the Atlantic Ocean to the east and the Rocky Mountains to the west. From this boundary to the Pacific Ocean, the Cordilleran ice sheet developed. As a result of the glacioisostatic depression, large proglacial lakes were originated in the southern part. Ice recession was characterized by sudden retreats that produced catastrophic floods and spectacular modification of the landscape.

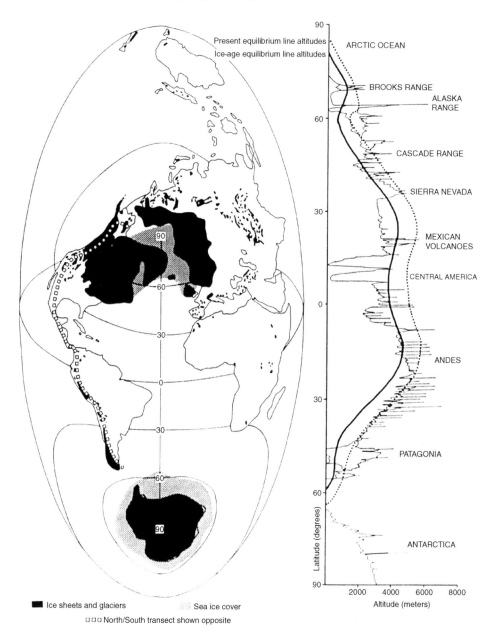

Figure 24.4. On the left: the distribution of the principal ice sheets and mountain glacier complexes during the last glacial maximum. On the right: the worldwide lowering of equilibrium line altitudes during the last glacial maximum is depicted along a north/south transect (open boxed line) (after Broecker and Denton, 1990).

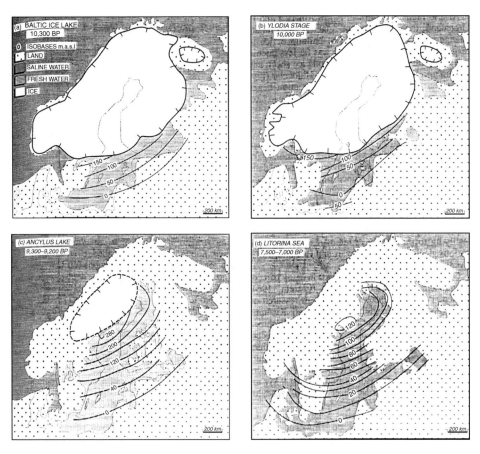

Figure 24.5. Stages in the recession of the Fennoscandian ice sheet and associated marine and lake events. (a) Baltic Ice Lake; (b) Yoldia Stage; (c) Ancylus Lake; (d) Litorina Sea. (According to Eronen, 1983; in: Benn and Evans, 1998).

The history of the melting these two ice sheets is not well known. The southern limit of the Laurentide ice sheet presented several ice lobes (Figure 24.6a) whose advances and recessions seem to be nonsynchronous. The first recession started in the eastern part 17,000 years ago, whereas the east remained active until 14,000 yr BP. The Cordilleran ice sheet started to move back after the many big *jökulhlaup* that broke the ice barrier of Lake Missoula 13,000 years ago (Figure 5.9 and Figure 24.6a). The thinning of the ice continued until its disappearance, 10,000 years ago. In the Laurentide ice sheet, the period 13,000 to 11,000 yr BP marked the disappearance of the ice in the Great Lakes. The most important periglacial lake was Lake Agassiz (Figure 24.6b), which overflowed even into the Gulf of Mexico. Later, the ice sheet recession caused a change in drainage into the western Atlantic Ocean, between 11,000 and 10,000 yr BP. The maximum extension took place between 9900 and 9500 yr BP, covering an area of 350,000 km². Later than this period and until approximately 8000 yr BP, many *jökulhlaup* occurred, some of which were larger volumes than the multiple ones from Lake Missoula. The Laurentide ice sheet

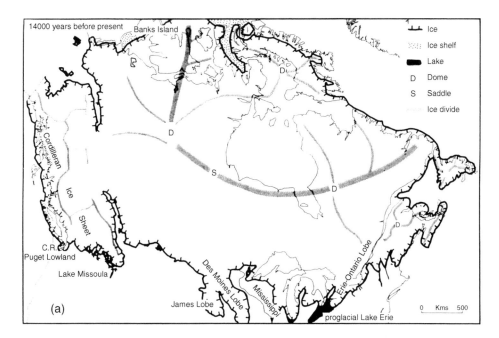

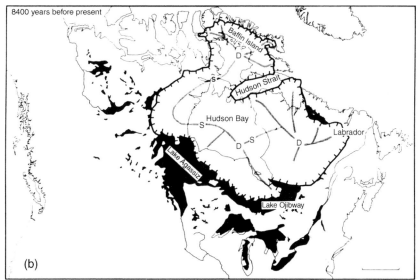

Figure 24.6. Reconstruction of the Laurentide ice sheet. (a) Situation for 14,000 BP, showing positions of major ice divides. C.R. = Columbia River. (b) position for 8400 years ago (after Dyke and Prest, 1987; in: Dawson (1992), Figure 5.6A and Figure 5.9A).

was receding gradually to the northeast and about 7000 years ago only small remnants remained (Dawson, 1992).

The study of the principal groups of glacial erosive and depositional landforms is very relevant and complex, and can indicate the fluctuations and processes produced by the activity of the large icecaps during the Quaternary. These interpretations must be based on precise analogies with the actual environment. Despite the difficulties for interpretation, nowadays there is a detailed knowledge of the principal groups of landforms, the processes involved and the geomorphologic history of the Laurentide and Fennoscandian ice sheets. In the last three decades, important progress on the knowledge of climate change has been made on the basis of core records, deep sea sediments, and ice sheets. This implies a substantial scientific advance concerning the interpretation of the landforms produced by ice sheets (Hughes, 1992).

Erosive morphologies resulting from the activity of ice sheets are a function of a group of basic variables. The factor of greatest effect is the temperature at the base of the ice, because when it reaches its melting point it slides and erodes. Geology plays a variable role; on the one hand, the erosion develops more easily in fractures, whereas on the other hand, the presence of permeable rocks facilitates the loss of water and the decrease of erosion. Topography controls the ice thickness, and also favours the convergence and divergence of the ice flow (Chorley et al., 1984). Nevertheless, it is necessary to take into account that some zones, as in Buchan, Scotland, there was not any glacial erosion, because thick weathered profiles remain intact after having been covered by ice (Hall, 1986). This is one of the arguments used by the followers of the hypothesis of glacial protectionism.

Some authors have calculated the volume eroded by ice in ice sheets of intermediate latitudes. White (1972) considered that its erosive capacity has been very high, because it has eroded and mobilized a rock thickness of more than 1000 m in the Hudson Bay region. The erosive action of the ice sheets is helped by the existence of thick regoliths, as in northern Scotland where there is a regolith thicker than 50 m (Fitzpatrick, 1963). When the front of the glacier advances, it runs into loose material, which is easily evacuated, pushed as does a bulldozer (Bakker, 1965; Thomas, 1994). These regoliths, resulting from preglacial chemical weathering, constitute a good reason for explaining the fine-grained tills, in comparison to those made up of fresh rocks (Feininger, 1971).

Regional erosional landforms are marks of subglacial erosion affecting bedrock. They consist of elongated hills, roches moutonnée and bedrock basins. The erosion is more intense along joints, faults and dikes. This landscape is known as a topography of hills and lakes, so termed in Scotland by Linton (1963). Also large areas exist with this morphology in the Canadian Shield and western Greenland. This topography is characterized by a landscape in a variety of forms, controlled by weak substrate structures, and with widths of no more than 100 m.

The deposits generated by the activity of glacier masses occupy 10% of the Earth's surface. *Depositional landforms*, which are related to icecaps, often are difficult to interpret, because they are the result of the superposition of materials corresponding to several glaciations, in which dates are commonly difficult to obtain.

Moraines formed by icecaps present a simpler pattern than the ones generated in valley glaciers. At first glance, the landforms look straight but their main form is arched, as from the Laurentide ice sheet, where this form is the result of numerous lobes on the frontal side of the icecap (Figure 24.7). In many places, these large moraines are thrust moraines, with hundreds kilometres of length and thicknesses of up to 400 m. Locally, they can exceed 200 m height. The interpretation is complex when deposits of different ages but similar lithology are superimposed. The C-14 dates indicate the different ice-sheet lobes have

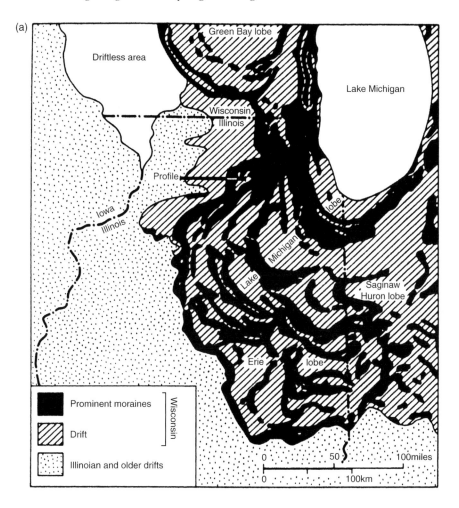

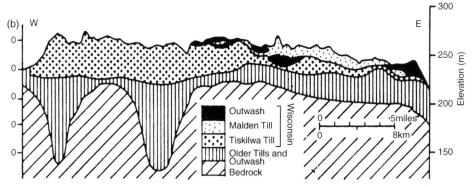

Figure 24.7. Glacial deposits at the southern end of Lake Michigan. (a) At least four lobes in the front of the Laurentide ice sheet have formed this configuration (Frye and Willman, 1973). (b) Profile across some till sheets in northern Illinois (Frye and Willman, 1973; Wickham and Johnson, 1981, in Chorley et al. (1984), Figure 19.15).

experienced numerous cyclic advances and recessions (Clark, 1994). The retreat of the ice sheets produces abundant landforms, such as kames and kettles, fluvioglacial landforms, (eskers and pitted sandar) and glaciolacustrine, drumlins. All these landforms occupy their largest extension within the successive boundaries of the icecap. Excellent examples exist in central Canada (Figure 24.8), where the dividing line of the ice was located in the eastern part and was covered by ribbed moraines, drumlins and eskers. The western periphery is characterized by outcrops of the substrate and a thin layer of glacial materials (Aylsworth and Shilts, 1989). These spectacular examples can be recognized in Labrador in eastern Canada, Scandinavia, Finland and in the Kola Peninsula of Russia (Figure 5.13).

Benn and Evans (1998) tried to make a generalization, at least for the Laurentide ice sheet, of the evolution of these depositional landforms for the last glacial retreat. The first times were characterized by successive advances and retreats of the ice, producing the deposition of a great quantity of sediments and their restructuring, as well as lakes dammed by ice. After that, ice masses suddenly receded and moraines and discontinuous eskers were generated on bedrock. Meanwhile, new dams produced marginal lakes. During the last retreat stages, moraines were more extensive and an important network of eskers, which indicate the flow direction, were produced.

4. Valley and cirque glaciers retreat

Alpine glacier valleys are characterized by very deep vertical slopes, where cryoclastsy acts intensely, providing clasts to the margins and bottom of the valley. Generally, these troughs were developed from periglacial valleys.

In historic times, valley inhabitants worried about the fluctuations of the glaciers, because they produced many disasters affecting human life and properties. The catastrophic events were caused by advances of a glacier tongue and mainly by bursts of lakes dammed by ice or landslides (Tufnell, 1984). This information about the historical events in alpine valleys provides relevant data on the evolution of the valleys (Grove, 1979, 1988). Comparison of paintings and engravings of the 18th and 19th centuries with recent and current photographs supplies relevant data on the fluctuation of the glaciers (Lamb, 1977; Stroeven et al., 1989). On the other hand, during the past few decades precise works on mass balances have been carried out. In those works the variations of glaciers have been related to climate change (Oerlemans, 1989a,b).

Alpine regions are characterized by a landscape with pyramid peaks (horns), edges between glaciers, cirques, and longitudinal profiles with frequent over-excavations. In areas that have undergone large retreats, or that also have limited supply, the predominant landscape is the one with glacial cirques (Figure 24.9). These landforms were developed in all aspects, although shady places are the most prone, because they are sheltered from solar radiation and, therefore, from the thaw of the ice. Often after ice thaw, rock glaciers arise in the cirques.

Valley glaciers are the most spectacular landforms, which reflect the great capacity for erosion of these masses of ice. One of the most significant features is the lateral moraines, which join with terminal ones and generate moraine ridges. These ridges reach a greater thickness if the glacier is stable, although generally, melt water erodes the ridges little by little. In some glaciers, as in the Nigardsbreen glacial valley of Norway, the terminal parts

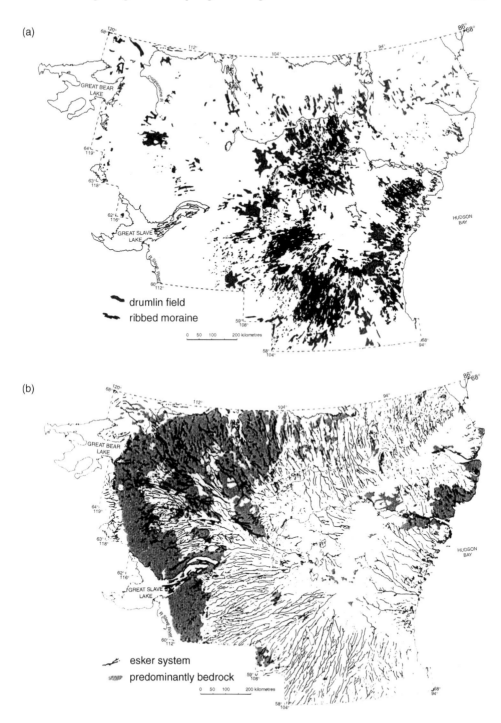

Figure 24.8. Distribution of the glacial landforms in Keewatin, central Canada. (a) Location of the drumlins and ribbed moraines (Rogen moraines). (b) Distribution of eskers and rock substrates (after Aylsworth and Shilts, 1989).

Figure 24.9.   Cirque glaciers, Vilcanota Range, southern Peru. Photo: J.L. Peña.

that correspond to moraine ridges present a good state of preservation and there are numerous dates by means of lichenometry (Figure 24.10).

Some of the most interesting effects of mountainous environments are the lakes dammed by the ice, which have a very ephemeral life. In such cases, the mass of ice of the principal glacier blocks the outlet of a secondary valley. Glacier snout recession triggers the water evacuation from the lake and a new advance can produce another lake. Thus, it is possible to study the fluctuations of the ice, which are related to short-term climate changes. Many lakes dammed by ice are known in high-altitude and high-latitude regions. In Twidale (1976a) many examples are presented.

A particular case of glacier sedimentation occurs when a glacier snout reaches a lake (Figure 2.16 and Figure 2.26) or the sea, a circumstance commonly associated with the effects of climate changes (Figure 24.11). Figure 24.12 shows two of the facies associations proposed by Powell (1981) for the glacier-fiords of Alaska. The depositional model represented in the top part of the figure is characteristic of the margins of a stable glacier slowly receding. In such conditions large fans or moraines are deposited. The bottom part represents the margins of a fiord that receives sediments from a terrestrial glacier. A delta is generated where melt water goes into a fiord, producing a system of top, frontal and bottom layers (Figure 24.13).

Figure 24.10.   The moraine ridge sequence documenting the retreat of Nigardsbreen, Norway, from its Little Ice Age maximum position. Moraines are labelled A–X and numbered 1–19 on the north–east and 1–18 on the south–west sides of the foreland. The historical dates are based upon lichenometry, which was undertaken at sites marked (+). (From Bickerton and Matthews (1992)).

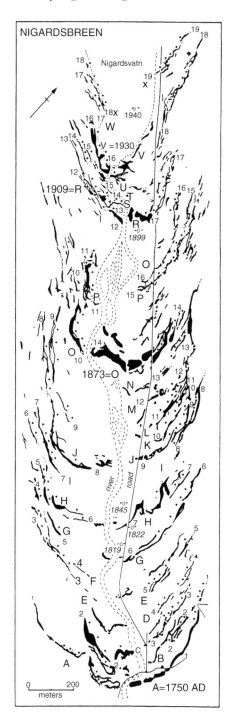

Figure 24.11.   Glacier tongue in the Antarctica ice sheet, flowing into the sea. Whisky Glacier, James Ross Island, Antarctic. Photo: A. Martín-Serrano.

The Glaciar Coronas are the most extensive ice mass on the southern slopes of the Maladeta Range, located in the Central Pyrenees of Spain. Using early and modern photographs and aerial photographs a decrease in length of 325 m, from 1928 to 1998 has been calculated (Chueca et al., 2003). Similar researches focused on glacier retreat have been carried out in the Peruvian and Bolivian Andes as well as in the Italian Central Apennines (Georges, 2004; Ramírez et al., 2001; D'Orefice et al., 2000).

5. Glacioisostasy and glacioeustacy

Glacioisostasy is a consequence of the growth and disappearance of icecaps. When an icecap increases, the lithosphere becomes deformed because of the weight of the ice mass. The magnitude of the depression depends on the thickness of the icecap and the ratio between ice and rock densities, which is about 0.3, so the depression generated is about one-third of the icecap thickness. This depression reaches its highest values under the maximum thickness of the ice and decreases toward the margins. The ice sheet is surrounded by a peripheral depression, whose height results from the ratio $H/11.5$, where H is the height of the icecap (for 1800 m, the depression is 155 m). Farther away from the depression a slight convexity is generated, whose height is $H/100$; that is, 18 m (Figure 24.14) (Walcott, 1970). Its origin is due to a lateral transference of subcrustal material under the icecap. The glacioisostatic depression disappears with thaw, by an elastic rebound, weak at the beginning (restricted rebound), followed by an accelerated uplifting period (postglacial rebound) and, finally by a residual rebound of minimal magnitude (Andrews, 1970).

   The information on glacioisostatic rebound is obtained from studies on beaches and uplifted deltas, that contain seashells and silt and clay deposits with fossils that can be

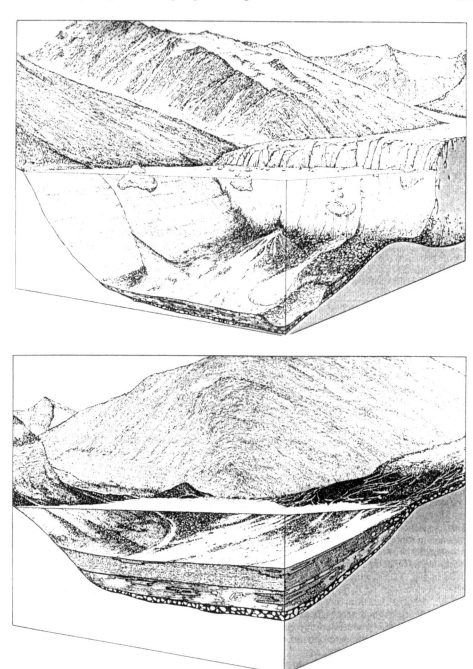

Figure 24.12. Depositional models for retreat based on Alaskan examples. Above: slowly retreating glacier in shallow water; below: terrestrial glacier margin supplying sediment to a delta. (Drawings by R. W. Tope, from Powell, 1981).

Figure 24.13. Bertrand Glacier which supplies sediments to the Argentine Lake, Patagonia, Argentina. Photo: C. Sancho.

dated by means of C-14. It is possible to develop *isobases*, which are isolines that represent the geometry of the surfaces generated at sea level, resulting from the effect of the ice load in a particular period. These isobases make it possible to verify that the Laurentide ice sheet, during the glacial maximum, had several domes associated with a central depression (Andrews, 1970, 1975), whereas with the Fennoscandian ice sheet there was only a central dome (Mörner, 1980) (Figure 24.5 and Figure 24.15), with the highest uplifting in the north of the Gulf of Bohnia. The isobases shape an elliptic morphology revealing that it developed separately from the dome on the British Isles.

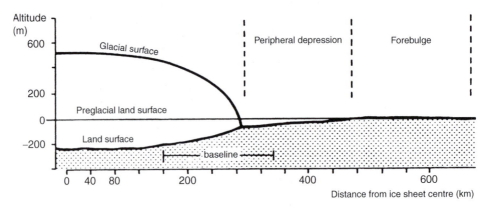

Figure 24.14. The principle of glacioisostasy, showing the depression of the lithosphere below an ice sheet. (Modified from Benn and Evans (1998); reproduced by permission).

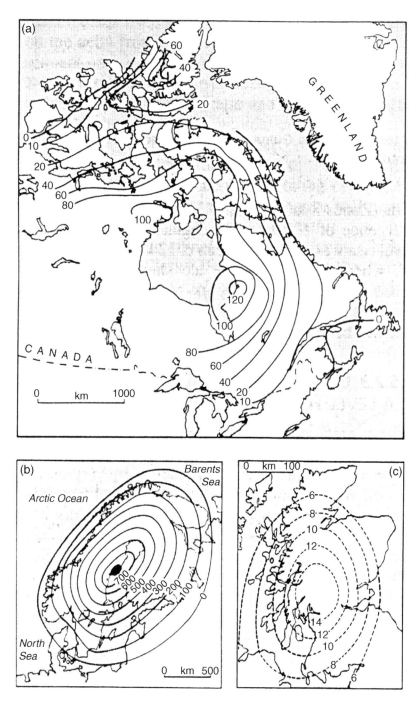

Figure 24.15. Regional isobase maps. (a) Isobase map showing shoreline emergence in eastern Canada since c. 6000 yr BP. Note the evidence for multiple domes. (b) Patterns of absolute uplift in Scandinavia during the Holocene. (c) Generalized isobases for the Main Postglacial Shoreline (7000 to 6000 yr BP) in Scotland. (Andrews, 1970; Mörner, 1980, in: Benn and Evans (1998), Figure 1.40).

Climatic Geomorphology

These data obtained from the studies of glacioisostasy have been used to obtain information on the viscous nature of the mantle. Two models have been proposed; the first indicates that the viscosity of the mantle is relatively constant with depth, and the second suggests an increment of the viscosity on the top of the upper mantle. Values of the second model fit well with the loss of load of the Laurentide ice sheet (Peltier and Andrews, 1983) and for Fennoscandia, the first model fits better (Mörner, 1980). This implies that the processes of the icecap construction and thaw are associated with plastic deformation within the asthenosphere. As a consequence of the increments and decreases of the ice masses existing on the continents, important variations of sea level are produced, called glacioeustatic changes. These fluctuations are recorded by means of the study of isotopes of oxygen in foraminifera from deep sea cores (Figure 24.3).

Other mechanisms coexist with glacioeustasy: the tectonic factors related to the capacity of the oceans; isostasy; hydrostasy, which represents the response of the lithosphere to the load of water in the oceans, and the geoidal effect. The most important are the glacial and tectonic factors, although during upper Quaternary, the glacioeustatic variations were more significant. Without doubt the reconstruction of the ancient sea levels for some areas is quite complex, because several factors are involved.

Fairbanks (1989) carried out a study on the corals of the Barbados Islands. Corals die when they are exposed above sea level. If the sea level rises, new corals grow on older ones. Thus, by means of radiometric dating, it is possible to know the date of the corals, and so, a record of the variations of the sea levels in the postglacial period. This record, together with that carried out for the Indian Ocean (Figure 24.3), can represent the fluctuations of the sea level for a great part of the Earth's surface, because both of them are similar, but cannot be applied in areas close to ice sheets (Boulton, 1993).

The importance of eustasy and isostasy depends on the location in relation to the icecaps. At the boundaries of the caps of ice, isostasy is the principal effect. According to the characteristics of the postglacial contour lines, the surface of the Earth has been divided into six *sea level zones* (Clark et al., 1978, 1980) (Figure 24.16). Zone 1 is located on the ice sheet margins, experiencing constant uplift due to isostatic recovery. Zone 1 to 2 constitutes a transition area, with an initial rise and a later subsidence due to the marginal rebound of the ice sheet. Zone 2 is far from ice sheets and sea level continuously rises, as in Zone 3, but an emerging beach 1000 years ago formed the latter. Zone 4 corresponds to tropical areas where sea has been rising since the glaciation, while in Zone 5 a rise occurred when the large ice sheets melted and then surfaced due to a hydrostatic effect. Zone 6 includes all the continental platforms, except those of Zone 2, and is characterized by a slight rise after the thaw, as a result of the isostatic uplift of the coastal margins. These sea level lines can be affected by tectonic movements.

6. Reconstruction of periglacial environments

Relict periglacial landforms play an important role in the reconstruction of the climates of the upper Quaternary. The degree of modification of the landscape depends on the location and persistence of the periglacial areas, which are more pronounced in the proximity of the ice sheets. Lithology has a considerable influence on the preservation

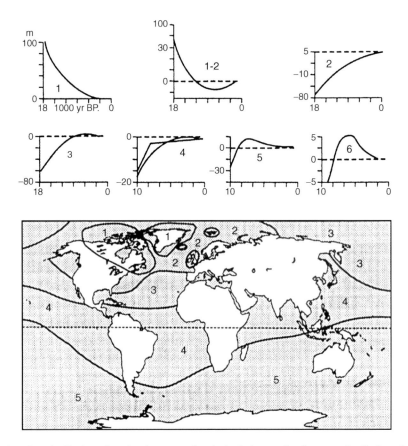

Figure 24.16. Distribution of sea-level zones and typical relative sea-level curves. (In Clark et al. 1978).

of periglacial landforms. There are rocks very susceptible to gelifraction, whereas the periglacial processes easily produce loose materials. These processes vary according to the climate that can be continental, as the one in European Russia, or marine, such as the Iceland climate. The paleoclimatic reconstructions should depend on the actual formational characteristics, but this is not always valid, because there are many differences between the periglacial environments in the Pleistocene of the intermediate latitudes and the ones existing nowadays at high latitudes (French, 1996). The principal reason is due to the insolation. Both latitudes present summers and winters, but a great contrast and speed exists in the change between both seasons. The day–night rhythm is the main difference (Arctic day or night). On the other hand, in the Pleistocene the daytime insolation was higher in intermediate latitudes than nowadays in high latitudes, so gelifraction, frost creep, and ice wedge action probably were several times more intense than in the actual latitudes. Also, in the intermediate latitudes, the effect of the orientation with regard to the sun is very important, so on slopes facing the sun many more ice–thaw cycles occur than on the shaded side. On the other hand, during the

Pleistocene the wind action was more significant at intermediate latitudes than at high ones, because large gradients existed at the boundaries of the ice sheets and vast plains of till and sandar, which provide fine sediments, were developed. Finally, the fluvial activity was more intense in intermediate latitudes due to a higher number of melting days. In short, periglacial environments of intermediate latitudes during the Pleistocene period were fundamentally different from the continental high latitudes, as pointed out above.

Therefore, the principal problem is to find precise indicators of past periglacial conditions. Perhaps the most realistic and best trace is the presence of wedges of ice and sand, which indicate permafrost. Landforms inherited from the thermokarst of the pingos can be also considered as indicators of permafrost, as well as fossil rock glaciers. The rest of the periglacial paleoforms are uncertain indicators of periglacial conditions (French, 1996).

7. Relict periglacial landforms

French (1996) differentiated between the features related to the development of ancient frozen soils in intermediate latitudes and those related to the existence of landforms originated by the action of the frost. The identification of the latter depends on the interpretation, whereas the features denoting ancient permafrost cannot be doubted.

Fossil wedges are relatively common and constitute the most important features for recognizing ancient permafrost. They can constitute ice wedge casts filled by sediments (Figure 7.4), which commonly form polygonal systems and sand wedges (Figure 7.12). It is estimated that winter temperatures on the surface of active ice wedges vary between -15 and $-20°C$ (Péwé, 1983). For this author, the mean annual temperature of the air to produce ice wedge ranges from -6 to $-8°C$. In central and western Europe they are abundant in a wide band and the width and depth increase toward the south of the ice sheet margin. This is understood as a intensification of the periglacial conditions, which seem to be more severe (Poser, 1948). In North America wedge casts are located in the proximity of the Laurentide ice sheet in a relatively narrow band, ranging in their age from 15,000 to 20,000 yr BP, according to several authors.

Ancient *pingos* can be recognized by a ramp, generally ring-shaped, that includes an interior depression. This ramp is the difference from other depressions generated by the thermokarst, as with the palsas. The height of the ramps depends on the size of the initial pingo and varies between 0.5 and 5 m. The diameter of the depression ranges from 200 to 300 m. Remains of pingos occur mainly in western Europe and North America. A classic example is the one studied by Pissart (1965) in the Ardennes of Belgium (Figure 24.17). The irregular form and the interference among pingos show a complex distribution, which is interpreted as pingos originating within an open system. The date of most pingos in Europe and North America is very late glacial, approximately 10,000 to 12,000 yr BP.

Fossil *rock glaciers*, which can be located below the limit of present-day arboreous vegetation, indicate a periglacial climate with permafrost in their origin. This reveals the importance they have as paleoclimate indicators (Barsch, 1988).

Relict morphologies generated by frost are numerous, but only a few provide unequivocal evidences of periglacial conditions. At intermediate latitudes commonly

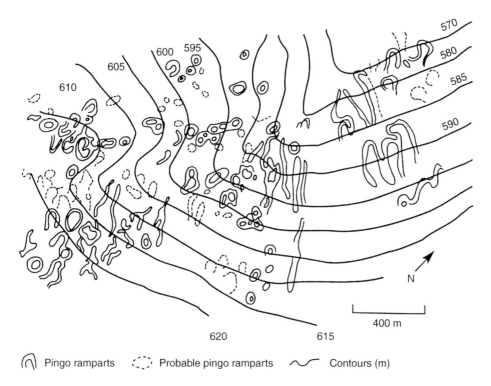

Figure 24.17. Map of the remaining pingos in Hautes Fagnes Moor, Ardennes, close to Malmedy, eastern Belgium (Pissart, 1965).

there are nonharmonically deformed soft sediments, termed *cryoturbations* or *involutions*. They have been considered indicators of permafrost by some authors, but these deformations are azonal and can be generated by different processes (see the section of patterned ground in Chapter 8). Periglacial involutions are formed by thrust of the frost or by thermokarst.

The *fields*, *slopes*, and *streams of rocks (blockstreams)*, as well as the *screes* or *talus*, constitute landforms generated in high areas by the action of the gelifraction. The larger the number of freeze–thaw cycles, the more important the development of these landforms is. In rock accumulations, the activity or inactivity of these landforms must be considered. Lichen colonisation, as in the slopes and river of blocks of the Tremedal massif (Gutiérrez and Peña, 1977), indicate that a deposit is inactive.

Cryoplanation terraces, generated by snowy processes, are common on high European mountains and, above all, in the central region of Alaska (Péwé, 1983). Periglacial *tors* and the related cryoplanation terraces do not clearly indicate permafrost conditions (French, 1996), although some authors think the contrary (Dawson, 1992).

The *grèzes litées* and the *protalus ramparts* are related to seasonal accumulations of snow, according to most of the authors. Finally, *solifluction deposits* are poorly classified, and are very difficult to differentiate from those originating in warmer climates.

8. Climate fluctuations in periglacial regions during the late quaternary

Most of the periglacial regions are over permafrost. It disappears when summer warming exceeds cooling in winter, and vice versa. At present, the periglacial environments occupy 25% of the Earth's surface and during the last glaciation could reach 20 to 25 per cent or more (French, 1996).

Possibly, the most relevant criteria to estimate the periglacial area during the late Quaternary are the distribution of plants and the extension of ecozones. All this implies that there were tundra, steppe and forest fluctuations at this time. Therefore, palynological research is crucial to these effects and has made possible to establish pollen regions and climate periods (French, 1996). Also, faunal remains, such as molluscs and insects, help with the interpretation of the environmental conditions. Nevertheless, the studies leading to this are scarce.

Permafrost evolves very slowly with respect to warming and cooling stages. In the late Pleistocene and Holocene the thaw of the ice sheets and the subsequent sea level rise are processes much more rapid than the loss of thickness of the permafrost (Dawson, 1992). For example, the permafrost in northeastern Siberia has always been present since the late Pleistocene (Washburn, 1979).

The most detailed knowledge that exists on the climate and vegetation fluctuations that have occurred during the late Quaternary exists in *Europe*. Büdel (1951) differentiated three vegetation zones in Europe (tundra, steppe, and forest) and indicated that the majority of the European surface to the south of the ice sheets was affected by periglacial conditions, except for the southern Iberian Peninsula and the Mediterranean coasts. Nevertheless, nowadays it is calculated that the extension of the periglacial band in central and western Europe was about 50 km during the maximum of the last glaciation (Weichselian). This width has been obtained mainly by the analysis of ice and sand wedges and annular ramps of pingos. The most important climate record corresponds to the pollen diagram of Grand Pile (Mook and Woillard, 1982), which shows the first stage of the development of the permafrost was about 90,000 years ago. Later, other permafrost periods, separated by warm interstadials, follow each other.

In *European and Asiatic Russia* an important increase of permafrost occurred during the late Quaternary, reaching up to 700 m thick. The continuous permafrost was located up to 50°N, having in total a dimension of 400 to 600 km (Figure 24.18) (Baulin and Danilova, 1984). The evolution of the permafrost is similar to that in Europe, with growing periods that coincide with cool stages, alternating with warmer periods.

Several data sets reveal that in *North America* there was a relict permafrost corresponding to the last glacial maximum (upper Wisconsinan), because for previous periods the information is very scarce. The permafrost band to the south of the Laurentide ice sheet is narrower than around the Euroasiatic ice sheet. This can be explained by a faster advance of the Laurentide ice sheet toward more southern latitudes, and therefore, areas with intense periglacial conditions are more limited than in Europe. In Figure 24.19 the areas of permafrost are indicated, together with the landforms related to fossil permafrost and other inactive periglacial morphologies that do not require permafrost conditions. On the other hand, at present the permafrost occurs under the 82% of the surface of Alaska and northwestern Canada (Péwé, 1983). On the Alaskan coast, many pingos corresponding to a closed system occur, whereas in central Alaska pingos of

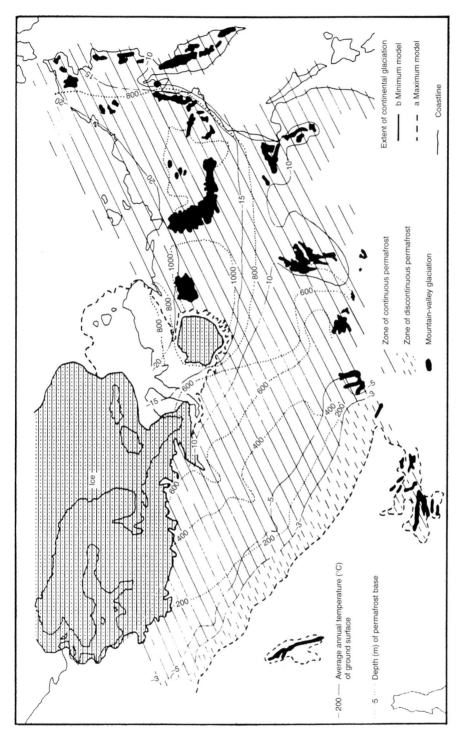

Figure 24.18. Map of permafrost distribution during the last glaciation (upper Valdai) (after Baulin and Danilova, 1984).

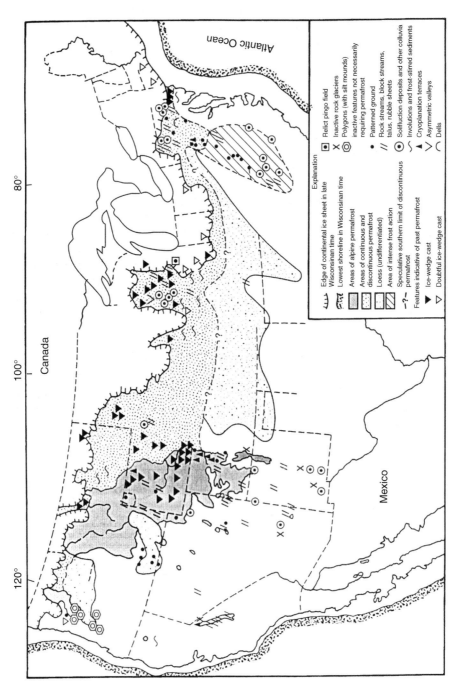

Figure 24.19. Reconstruction of the maximum extent of Late Wisconsinan periglacial conditions in the USA south of the ice sheet limits (based upon several authors, French, 1996, Figure 13.6).

the open system are abundant; all of them were generated in the Holocene. Also the vast cryoplanation terraces in the centre of the territory are significant.

It seems that in the Southern Hemisphere the permafrost did not develop so much, because the described morphologies do not need permafrost conditions. Only ice-wedge casts in southern Patagonia, Argentina, have been mentioned (Grosso and Corte, 1989).

9. Some considerations on global climate change in periglacial areas

The predicted temperature increase due to the greenhouse effect will seriously affect permafrost areas, which at present occupies approximately the 25% of the Earth's surface, as mentioned above. It is reasonable to state that permafrost will decrease in depth and extension. An important increase of the active layer will imply substantial modifications to the hydrologic, geomorphologic and biologic processes of the permafrost regions. Moreover subsidence problems, affecting the infrastructure will happen. By utilizing general circulation models (GCMs) and a scientific paleoreconstruction, assuming an increment in temperature of 2°C, it can be concluded that the continuous permafrost will reduce its extension 29 to 67% during the 21st century in the Northern Hemisphere (Anisimov and Nelson, 1996). Permafrost areas are the most susceptible to global warming (Boer and de Groot, 1990; Smith, 1993). Nevertheless, the results obtained from GCMs do not provide precise forecasts, but indicate the trend of future changes in permafrost regions (Pissart, 1990).

Several studies indicate that permafrost temperatures have been rising in the last 20 to 30 years. Temperature records taken between 1970 and 1990 in arctic Europe and subarctic Russia, show an increment of 0.6 to 0.7°C, at 3 m depth (Pavlov, 1994). Similar values have been obtained in Tibet and in North America. The warming is interpreted, as in other areas of the planet, as the result of the global climate change or due to a natural trend of climate warming.

Chapter 25

Climate change in arid and tropical humid regions

Rainforest, savannah, and semiarid areas, although presenting different morphogenetic processes and being studied separately, have been affected by important climate changes during the Cenozoic period, as is shown by the anomalous paleoforms existing in the different environments. Thus, for example, aeolian deposits in the Congo rainforest or laterites in the arid zones in southern Niger Republic, bear witness to these important climate fluctuations.

The purpose of this chapter is to show the evolution of these regions during late Tertiary and Quaternary, from the analysis of erosive and depositional landforms. Dating results (isotopic, palaeomagnetic, thermoluminescence, and others) make it possible to elaborate a sequence of events, adequately time-ranked, which provide the means for correlations between different areas. From these, an approximation can be obtained of the climate changes that occurred in the recent history of the Earth. Furthermore, it is possible to provide data and valuable interpretations for the prediction of future climate (Petit-Maire, 1999).

1. Arid regions

The study of the landforms in arid regions that are tectonically stable and without significant anthropic activity, reflect that many of the existing landscapes have been generated under geomorphologic conditions that differ from the current ones. Therefore, those landforms should be considered as paleoforms or inherited landforms, originating under morphoclimatic conditions different from the present ones. For instance, the extensive dune fields of the Sahelian zone, covered by grassland vegetation indicate a higher aridity during the time of formation. Similarly, no water currently flows in the complex network of *ouads* (*wadis*) in the central Sahara, indicating periods of higher precipitation in the past. (Tricart, 1969, 1979). These landscapes show significant modifications to the distribution of the arid regions over the surface of the planet.

From this evidence came the concept of the *pluvial period* or *lacustrine phase*, that can be defined as that in which hydrological and biogeographic conditions were more humid than present conditions. In comparison to the one before, an *interpluvial* or *arid period* occurred under dryer conditions (Rognon, 1980). This definition has taken into account water availability, which not only is conditioned by the precipitation, but also by evapotranspiration. Thus, for a given precipitation, a decrease in temperature increases the water availability of soils. Therefore, this double factor of precipitation and temperature makes more difficult a definition of pluvial period. The alternation of pluvial and interpluvial periods can be recognized for deserts by a collection of indirect data of diverse nature: geomorphologic, archaeologic, pedologic, palinologic, hydrologic, and hydrogeologic (Tricart, 1969, 1979; Thomas, 1977c; Demangeot, 1981; Goudie, 1992).

1.1. Fluvial systems

Although precipitation is scarce in arid environments, stream flow paradoxically plays an important role in the construction of the landforms of these areas. This is due to the lack of vegetation cover and to the shortage of soils, so that the geological substrate is unprotected and exposed to rain drops. Precipitation, although scarce, is frequently very intense and with an irregular temporal distribution. Thus, in Saharan semiarid areas there is a flooding per year in average terms, but in more arid regions up to 10 years can happen without any water circulation in the ouads.

   Many authors have dealt with the influence of climate change on the modification of the geomorphic processes in fluvial systems. In Schumm (1977) and Knox (1984) the direct and indirect effects of climate on these systems are presented in detail. In arid regions where vegetal cover is scarce, the variations of the vegetation, as a result of climate change, can produce significant modifications to the hydrology. Knox (1984) established a critical threshold in the vegetation cover that can be of 70 %, corresponding to an annual average precipitation of 400 to 500 mm. For values lower than 70 %, the quantity of soil subject to erosion increases very rapidly. These circumstances can be clearly seen in the curve (Figure 25.1) of Langbein and Schumm (1958), wherein annual average precipitation variations are presented relative to annual sediment yield. In deserts, the higher the precipitation, the greater the runoff and the erosion increases to reach a maximum of about 300 mm in precipitation. Where this precipitation increases the vegetation cover becomes more extensive and the sediment yield decreases rapidly.

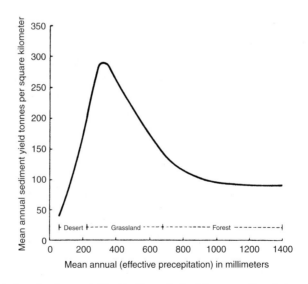

Figure 25.1.   Variation of sediment yield with climate in fluvial systems (Langbein and Schumm, 1958).

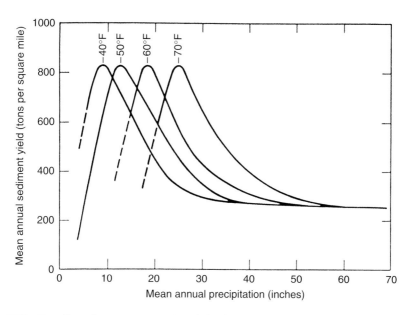

Figure 25.2. The effect of temperature on the relation between mean annual sediment yield and mean annual precipitation (Schumm, 1965).

Meanwhile in very extreme deserts there is little water erosion, but in semiarid regions the stream flow is very frequent due to stormy precipitation, and the sediment yields reach higher values. This graph shows also the importance of vegetation cover as the most effective factor in reducing stream flow and soil erosion. But these maxima of sediment yield vary depending upon the temperature (Schumm, 1965) (Figure 25.2), in such a form that the higher the temperature, the erosion maxima coincide with higher precipitation. That is, for higher temperatures, with a higher evapotranspiration, more precipitation is needed to maintain the percentage of the vegetal cover, there is a lesser stream flow, and the sediment yield maximum moves to the right.

In semiarid regions, which correspond to areas where great erosion is produced, flooding is also much more frequent, because the water from precipitation flows rapidly. In basins with different drainage areas (Figure 25.3), the magnitude of the flooding increases significantly as average annual rainfall decreases from 500 mm to approximately 375 mm (Knox, 1972). This indicates that a climate change from relatively wet to dry will produce a widening of the channel to adapt to bigger floods.

Similarly, the density of the drainage network tends to increase in semiarid regions and normally decrease with the increment in precipitation (Gregory, 1976) (Figure 25.4). The graph indicates a sharp decrease in density from 800 mm precipitation. Obviously, there are variations as a result of the different types of rocks, relief and vegetation.

All these examples illustrate the sensitivity of the landscape of these areas to climate change. In arid regions, the vegetation is more sensitive to climate and a slight modification in the vegetation cover can produce significant hydrologic changes. Therefore, in these regions the intensity of erosion can vary very much as a result of slight climate changes.

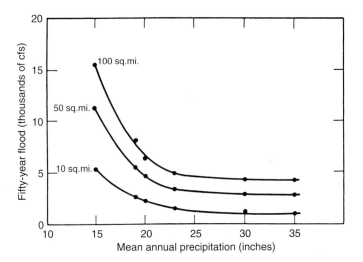

Figure 25.3. Relation between flood magnitude and mean annual precipitation for three sizes of drainage basins (after Knox (1972)).

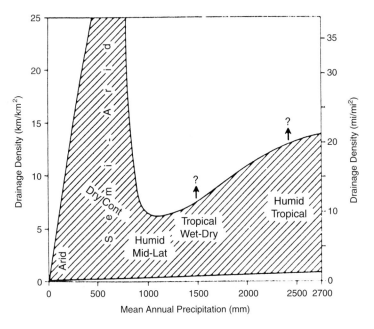

Figure 25.4. Relation between drainage density and mean annual precipitation (modified from Gregory (1976), in Chorley et al. (1984), Figure 18.5).

The oscillations in temperatures and precipitation that occurred during the Holocene (the last 10,000 years) have been of low magnitude ($\pm 2°C$ and 10–20% in precipitation) (Lamb, 1977). Even so, these variations have affected the intensity of many geomorphologic processes, which control the morphologic stability of the landscapes, especially in fluvial systems (Knox, 1984).

1.1.1. Hillslopes and their evolution

The best typical profile of hillslopes in arid regions can be recognized in sequences of stratified rocks. This profile consists of a free face at the top, a slope covered totally or partially by detritus in the middle, and a basal concavity from where a pediment develops (King, 1962).

In a study of the hillslopes of a great part of the territory of the USA, the influence of the climate on the evolution and morphology of the profiles is stated (Toy, 1977). In order to minimize the influence of the geology and the structure, the author carried out the analyses on marine claystones with an angle of dip of less than 5°. According to this work, slopes in arid zones tend to be shorter, steeper and with a smaller radius of curvature than in humid regions (Figure 25.5).

In some places, as in the Judean desert of Israel, hillslopes developed on limestone and marls of the upper Cretaceous, present smooth profiles, with concave–convex shapes, and are covered by a thin layer of detritus (Figure 25.6). This morphology indicates an environment more humid than the existing one (260 mm), when the morphology was generated. The higher precipitation is corroborated by the development of *hohlkarren*, *kluftkarren* (Figure 11.11) and *kamenitzas* on Cretaceous limestones.

1.1.2. Triangular slope facets

The knowledge of the processes affecting slopes in arid and semiarid environments has progressed considerably in the last few decades. Nevertheless, works related to their evolution are scarce, due to the difficulty in dating the different relict hillslopes. In semiarid environments where there is a higher percentage of vegetal cover and there has

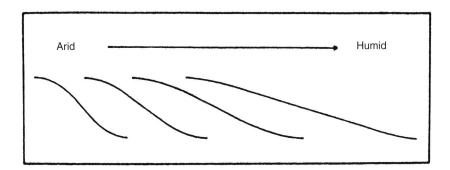

Figure 25.5. Conceptualized hillslope profiles (Toy, 1977).

Figure 25.6.   Regularized slopes on limestones and marls of the upper Cretaceous, outcropping as a result of the incision by gullies. The convex–concave profile seems to indicate a previous stage more humid than present (260 mm). Mishor Adumim, Judean Desert, Israel.

been an old human presence, however, it is possible to find carbonaceous remains and archaeological materials within the deposits, making dating possible.

Slope accumulations can be later affected by a period of incision that can destroy all the accumulation, or can generate relict slopes termed *talus flatirons* (Koons, 1955). They are also known as *tripartite slopes* (Gossmann, 1976) and *triangular slope facets* (Büdel, 1982), although this latter term can be confused with the terminology of eroded fault scarps. These morphologies present a triangular or trapezoidal ground plan and are located around and separated from the scarp (Figure 14.15 and Figure 25.7). Therefore, the generation of the flatirons requires the alternation of accumulative and erosive periods during the slope evolution. The development of several of these stages can produce a *talus flatirons sequence*, with the most recent flatirons being those which are close to the scarp. These relict landforms are not very common due to the fact that they are located on stratified formations with horizontal disposition, in which the cornice overlies easily eroded material (Koons, 1955; Everard, 1963; Büdel, 1970; Blume and Barth, 1972;

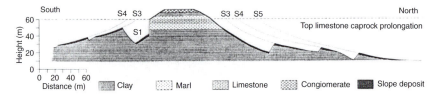

Figure 25.7.   Schematic cross section of the San Pablo Mesa showing a flatiron sequence with five stages of slope evolution.

Gerson, 1982; Gerson and Grossman, 1987; Schmidt, 1989a,b, 1994; Arauzo et al., 1996a). Furthermore, the free face has to be thin to avoid an excessive accumulation of detritus (Schmidt, 1987), allowing the rapid destruction of the cornice and retreat of the slope. On the other hand, the alternation of hard and weak layers below the free face makes difficult the formation of *talus flatirons*.

Once the factors favouring *talus flatirons* are indicated, it is necessary to know their causes. Koons (1955) interpreted the features from the discontinuous balance between sediment yield and removal in the slopes (Figure 14.16). This noncyclic origin is against the interpretation based on the modification of the dominant morphogenetic processes due to climate changes (Everard, 1963; Gerson, 1982; Gerson and Grossman, 1987; Sancho et al., 1988, Schmidt, 1989, 1994; Arauzo et al., 1996a,b; Gutiérrez et al., 1998a,b) (Figure 25.8). In recent periods, the possible influence of anthropic action on the modification of the geomorphic processes has also been pointed out (Everard, 1963; Sancho et al., 1988; Gutiérrez and Peña, 1989, 1992; Arauzo et al., 1996a,b; Gutiérrez et al., 1998a,b).

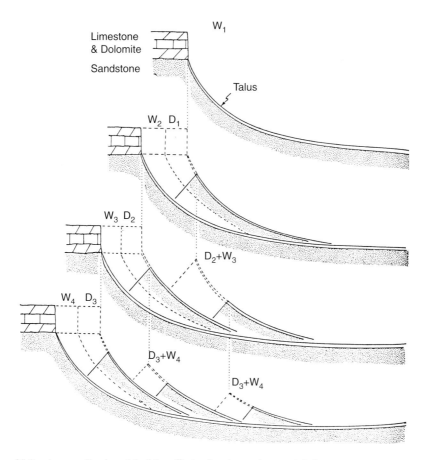

Figure 25.8. A generalized model of the effects of major environmental changes on escarpment retreat. D = dry, arid to extremely arid, climate; W = relatively wet, moderately arid to semiarid, climate (after Gerson, 1982).

The importance of climate changes has been discussed at great length for several types of landscapes. In desert regions with scarce land cover, the variations of vegetation can produce important modifications of the hydrologic conditions. Climate characteristics in the centre of the Ebro Depression in Spain agree with maximum sediment yields of the curve proposed by Schumm (1965). According to this graph, small variations of temperature and annual precipitation can modify the percentage of land cover and, so, the sediment yield.

On the other hand, in periods with more humidity (pluvial) the processes that generate sediments are more active on the free face, mainly undercutting and slumping. This is due to the increase of infiltration and groundwater flow (Ahnert, 1960). When a decrease in the temperature occurs, the number of frost days per year raises and, therefore, the sediment yield in the scarp will be higher due to frost action.

The climatic interpretation of alluviation and trenching stages leads to the establishment of a correlation with known climate sequences. The two most recent accumulation stages are well represented in all the slopes in northern Spain (Gutiérrez and Peña, 1989, 1992, 1998) and have been dated in many places by means of archaeological and radiometric techniques. The most recent, post-medieval accumulation can be correlated with the Little Ice Age (1450 to 1850) (Grove, 1988). The accumulation corresponding to the stage previous to the latter was formed during the Cold Period of the Iron Age (2900 to 2300 BP) (Lamb, 1977; Gribbin and Lamb, 1978; Grove, 1979; Harding, 1982). Gutiérrez and Peña (1989, 1992, 1998) correlated these two accumulation stages with those cold periods for the northeastern Iberian Peninsula (Figure 25.9).

At present there are only two dates for accumulations older than the above mentioned, both obtained by means of C-14 at Mezalocha and Chalamera in Zaragoza province and at Monteagudo, Soria province in Spain (Gutiérrez et al., 1998a; Gutiérrez and Martínez, 2001). In principle, there is a good correlation with the Heinrich events established from the analysis of deep-sea cores in the North Atlantic (Gutiérrez et al., 2005). Ages obtained for the accumulations could correspond to H3 (27 ka BP) and H4 (35 ka BP) (Heinrich, 1988; Bond et al., 1993, 1997). A Heinrich event is represented by glacial sediments coming from the melting of icebergs with scarce foraminifera. The stages of formation of these events are related to movements of a great abundance of icebergs, resulting from the detachment from northern ice sheets. These circumstances produce a widespread cooling of the superficial waters of the ocean. As a result, cool winds from the Atlantic came into the Ebro Depression. This fall in temperature produced a decrease in evapotranspiration and therefore, an increment in the availability of water in soils, causing an increase in land cover. These conditions seem to have been suitable for the accumulation of sediments on slopes. These Heinrich events occur in periods of extreme atmospheric cooling and are followed by net and short interstadials (Dansgaard et al., 1993).

From the above, it may be deduced that accumulation stages coincide with cool-climate epochs, whereas the stages of trenching correspond with warm ones in which talus flatirons are formed. Therefore, the sequences of talus flatirons record climate changes.

In semiarid environments, those changes in slope systems that generate talus flatirons, are produced in relatively small time spans. In dryer areas (Figure 25.10), however, climatic changes on the order of 100 kyr are needed to generate these relict slopes (Gerson, 1982; Gerson and Grossman, 1987; Bull, 1991; Schmidt, 1994, 1996).

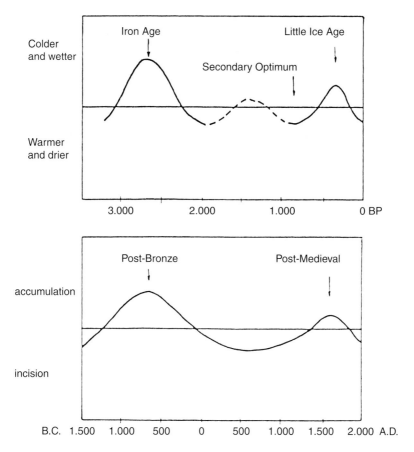

Figure 25.9. Relation between accumulation and erosion stages in slope deposits in NE Spain and climatic epochs of the Upper Holocene (Gutiérrez and Peña, 1989).

Current available chronological data are very scarce and it is necessary to have many more dates available to validate this climate origin. If the research that is being carried out corroborates this hypothesis, it will be possible to obtain a climate curve for this region for the late Pleistocene and Holocene periods. On the other hand, given the position of the Iberian Peninsula between Europe and Africa, these works can be very useful to correlate it with the pluvial periods in northern Africa and climate fluctuations identified in central and northern Europe.

1.1.3. The ramblas or arroyos

Rambla is a term used mainly to name ephemeral fluvial courses in the semiarid areas of the Mediterranean zone of Spain. These courses present a flat-floored channel and vertically cut banks, and are only active when heavy rains occur, producing floods that transport a large amount of sediment load (Segura, 1990). *Arroyo* is the word introduced by the Spanish colonizers of the deserts of the western USA. The term is applied to courses

Figure 25.10. Hillslope triangular facets sequence. Makhtésh Ramon, Negev Desert, Israel.

from 5 to 200 km with deep transverse cross sections and flat floors. They are intermittent courses that can be affected by sporadic storms, incising into unconsolidated material (American Geological Institute, 1972; Bull, 1977). In these alluvial systems it is possible to differentiate several stages of alluviation and trenching (Figure 25.11 and Figure 25.12). Furthermore, the reader will no doubt recognize the strong similarity or even identity of these features with the *ouads, wadis*, or *nullahs* of much of North Africa, the Middle East, and southwest Asia, respectively.

North American deserts constitute the region in which these geomorphologic arroyo systems have been more deeply studied. The most advanced knowledge in the fields of hydrology, paleohydrology, geomorphology, and climate changes come from there. Sedimentation of valley fills by ephemeral intermittent courses exists with precipitation ranges from 100 to 500 mm (Bull, 1977). Temperature regulates discharge, because it controls the evapotranspiration and the type and density of vegetation. These effects are more pronounced in arid regions than in subhumid and humid ones (Langbein, 1949). Sediments in these ephemeral courses come mainly from sand, silt, and clay supply from slope erosion. Later, these particles are easily mobilized, except clay layers, in which infiltration decreases and cohesion rises.

On the other hand, anthropic action is important, because humans have modified land cover in many ways; overgrazing, excessive construction, deforestation, fires, and so forth. The increment in biomass affects the resistance to erosion of the channel (Graf, 1979). In this sense, Cooke and Reeves (1976) showed that the introduction of herbivorous animals into New Mexico and Arizona triggered an important increment in the erosion of soils.

In ramblas or arroyos several stepped levels can be recognized (Figure 13.36, within the sediments filling these landforms, which result from different stages of aggradation and

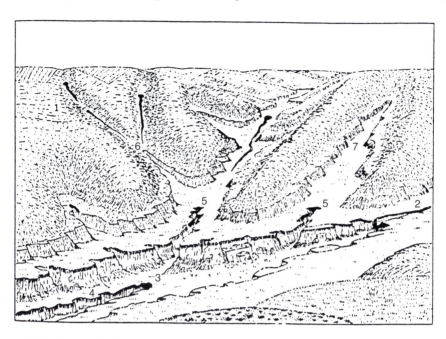

Figure 25.11. Composite sketch, based on field sketches and photographs, showing the different varieties of valley-floor gullies. Actively eroding scarps are indicated by darker shading. Discontinuous gullies (1) form in tributary valleys. They will coalesce when the headcut at 6 reaches them. In this valley there are two knickpoints (2, 3). The incision of the valley floor forms terraces (4, 5). Tributary gullies erode alluvium that fills tributary valleys (7) and they will eventually integrate the drainage network (Brice, 1966, in Schumm et al. (1984), Figure 4.1).

Figure 25.12. Terrace trenched in a valley fill by incision of the Villafranca gully or Villafranca de Ebro, Zaragoza province, Spain. Ebro Depression.

Figure 25.13. Aerial view of a dendritic drainage network consisting of flat floored valleys excavated in Miocene gypsum and marls. Central part of the Ebro Depression of Spain.

incision. In the centre of the Ebro Depression in Spain, an intricate dendritic network of flat valleys that appears incised, showing several levels with sediments from the Holocene, has developed over Miocene marly gypsum material (Figure 25.13), (Peña et al., 1993; Arauzo and Gutiérrez, 1994). In the ephemeral courses, it is also possible to recognize stages with a shortage of sedimentation that are recorded in incipient soil profiles. Therefore, these ephemeral courses are characterized by lengthy periods of sedimentation, alternating with a brief hiatus with shallow incised channels and the development of entisols (Leopold, 1994). The aggradation processes described above are followed by a stage of degradation that corresponds to the incision of the rambla that starts with a rapid incision followed by an exponential decrease, due to the approximation of a new base-level of erosion (Bull, 1997). For southern Arizona, Cooke and Reeves (1976) proposed a model on the different causes of the beginning of the arroyo incision in which many variables interact. Land-use changes, random variations in the frequency and magnitude of the processes and the secular climate changes are the principals. The first and the last seem to be the principal causes of the initiation of the arroyos. Nevertheless, the origin is a classic example of equifinality or convergence of landforms, in which it is very difficult to identify a unique mechanism (Cooke and Reeves, 1976).

In the analysis of the origin of trenching in arroyos it is very complicated to distinguish between anthropic actions and secular or short-term climate changes. Another way to deal with the problem is based on the fact that there is an increase of the erosion produced by waters and a decrease in the resistance to erosion in the sediments of the arroyo. In any case, vegetation seems to be the key to the initiation and development of the arroyos.

Figure 25.14. Rambla or arroyo with the channel covered mainly by shrub-like vegetation. Vertical scarps are composed by erodible materials. Barranco Grande, Las Bárdenas Reales, Navarra, Spain.

Vegetal cover, on slopes and riversides (Figure 25.14), responds rapidly to the changes in temperature, precipitation and anthropic actions. Among the modifications to *land use* are the over-pasturage in arid zones with silty soils that produces a decrease in the vegetal cover, intense trampling, and increments of rainsplash, with collectively a decrease in infiltration capacity. All of these factors, together with construction works (roads, railroads, bridges), deforestation, and fires, are the principal triggering factors of the human actions. Two examples illustrate these anthropic triggering factors.

Thornes (1976), in a work on the Almazán basin of Soria, Spain, indicated that the excessive pasturage due to the in-place livestock and to the en-route transhumance to summer pastures in Urbion was an important factor for intense erosion in these areas, due to the decrease of a great part of the grassland vegetation. As a result, gullies were triggered, which incised into the Tertiary terrains that consisted of easily eroded materials.

A similar problem occurred during the colonization of the southwestern USA. Settlers came in great numbers, building numerous farms and roads and put many thousand head of cattle into this area. All these circumstances produced the disappearance of a great part of the existing vegetation, triggering an intense arroyo action that generated gully trenching of more than 50 m in depth. This erosive process occurred between 1850 and 1920 and especially from 1870 to 1890. Settlers had to migrate to more favourable areas. This anthropic environmental change was known from a long time ago and Cooke and Reeves (1976) studied this in depth in an excellent monograph.

Other authors (Bryan, 1941; Antevs, 1952) held that trenching had a double cause and pointed out that although pasturage also contributed, vegetation was weakened at the end

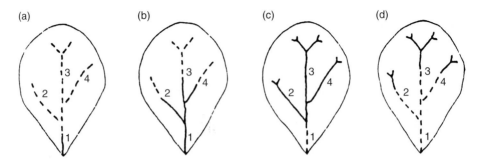

Figure 25.15. The cycle of trenching and alluviation in a semiarid valley. Dotted line indicates alluviation within a channel. Solid line indicates trenching or a well-defined channel (Schumm and Hadley, 1957).

of 19th century due to a climate change. This hypothesis is supported by dendrochronologic studies, which show a period of lower precipitation (Stockton and Fritts, 1971).

Another model proposed by Cooke and Reeves (1976) is the one related to the *random variations in the frequency and magnitude of the processes.* One factor that can contribute to incision is the tectonic fall of the base level (Bull and Pearthree, 1988). Another is related to erosion capacity, which rises exponentially with the increment in discharge (Leopold et al., 1964). This can produce flooding when storms occur. These events produce trenching, while in periods of less discharge alluviation is predominant. Schumm and Hadley (1957) suggested that incision is due to internal adjustments within the fluvial system. For small basins less than 50 km^2, the sedimentation can be produced by flow dissipation, due to infiltration, evaporation, and so forth. These authors generalized those changes and proposed a cycle for semiarid erosion in which the evolution of trenching and alluviation is considered for small basins (Figure 25.15).

Minor and secular climate changes constituted the last model for trenching. For short-term periods there is enough evidence of the fact that erosion and sedimentation result from climate changes, as well as from human activity. Leopold (1976) put instruments for monitoring ephemeral water courses in New Mexico for a 15-year period and deduced that the arroyos that had been eroded during the century changed their trend toward aggradation, from 1961 on. The author ascribed the change to the cooling after 1940. Hereford (1984) described a similar situation in Arizona. On the other hand, the hydrologic and geomorphologic responses to El Niño oscillations can produce droughts (Andrade and Sellers, 1988) or large floods in southern USA. It is also necessary to take into account that this climate perturbation is persistent.

Without any doubt, during the colonization epochs in the semiarid zones that affected the southwestern USA, South Africa, India, and Australia, the trenching of the fluvial courses increased markedly (Cooke et al., 1993). Nevertheless, in prehistoric periods the alternation of aggradation and incision seems to have been due to climate changes (Antevs, 1952). Other researchers, on the basis of dendrochronologic studies with records of thousands of years, indicated that there was an alternation of dry and wet epochs that were related to trenching and sedimentation. Similar results show the studies carried out by means of C-14 and archaeological remains.

Leopold (1994) carried out a synthesis of the climate history following the last Wisconsinan ice retreat. The synthesis was based on much research carried out on the alluvial deposits of southwestern USA. A period of aggradation can be recognized from the last glaciation to the Altithermal Phase or Climate Optimum, some 4000 to 6000 years ago. During dry conditions in this time period intense erosion was triggered, affecting valley fills, which produced terraces up to 10 m above the channel. Later, another aggradation period developed, which finished about 1200 to 1400 yr AD (Medieval Warm Period or Secondary Climate Optimum). In this dry period an incision of the fluvial deposits previously deposited occurred again. During the wet and cold stage corresponding to the Little Ice Age (1450 to 1850 AD) aggradation occurred yet again. From 1880 to 1920 there was a change toward aridity that produced erosion in the valleys.

This general proposal is based on existing data, but is not universal. This sequence of aggradation and erosion stages can be different for tributary headwaters than for principal channels (Balling and Wells, 1990). Furthermore, it is possible that a synchronism exists among degradation stages within a vast region, because each fluvial system has a different sensitivity to climate perturbations (Bull, 1997). This author calculated that a 100 years of time were needed for the trenching, but at least 500 years were necessary to complete the aggradation in incised channels.

In conclusion, it is too simplistic to ascribe channel trenching to only a single factor, as, for instance, an increase in temperature. The changes in the density of vegetation of slopes or riversides play an important role in the incision initiation. Vegetation can be modified by changes in precipitation, temperature, overgrazing, fires, and other human activities. Incision begins when these variables create an imbalance within the system, which can trigger and maintain the process of trenching. Some explanations such as climate change, however, have a regional significance, whereas others can only be locally referenced. There are defenders and detractors of each hypothesis regarding the complicated problems about rambla and arroyos, but it is necessary to take into account that different changes in an area can produce similar geomorphic results.

1.1.4. Alluvial fans

These are semiconical forms that are produced when water flows and confined channel sediments reach the mountain foot, where there is a reduction in the stream capacity, absence of confinement, and sedimentation of the charge, mainly by the larger ephemeral rivers. Although they are located in many climate environments (Rachocki, 1981), here only those actually located in arid regions are considered in order to deduce the climate changes that have occurred.

The study of alluvial fans has progressed a lot through the application of sedimentological techniques (Mayer et al., 1984; Grossman and Gerson, 1987), the analysis of palaeosoils (Bull, 1991), charcoal dating and the rock varnish method (Dorn, 1994a). From that, it is possible to identify the climatic features from deposits and palaeosols, as well as from correlations within the fan and in the region. Thus, the objective is to establish the influence of the climate on the genesis of alluvial fans.

For some authors, the fundamental stages of aggradation in alluvial fans occur in wet periods (Glennie, 1970; Dorn et al., 1987; Maizels, 1987), whereas others think that sedimentation increases in drier stages (Wells et al., 1987; Bull, 1991). An intermediate solution is adopted by other researchers who have indicated that aggradation can be produced in both climate conditions (Williams, 1973; Ponti, 1985). All this shows the complexity of the interpretation and the extreme difficulty in carrying out correlations.

Four conceptual models have been proposed, however, to relate climate changes with alluvial fans (Dorn, 1994b). When a *transition to a drier climate* takes place, there is produced a greater denudation of the accumulated deposits in slopes and floodplains in the upstream basin, previously retained by a greater land cover, creating an aggradation stage (Knox, 1983; Wells et al., 1987; Harvey, 1990; Bull, 1991; Harvey, 1997). The *paraglacial model* is significant due to the great quantity of erodible materials and the large volumes of water released during melting epochs which affect alluvial fans. In this way, glaciations produce great quantities of sediments that supply alluvial fans, lasting for thousands of years after the glacial stage. In the deserts of the western USA and central Asia, many examples can be recognized in which upstream basins of alluvial fans have been affected by glaciers. Another alternative, previously described, is the direct relation between a *wet stage and aggradation*. Some authors relate the latter with glacial conditions in which water availability is higher (Bull, 1991). Finally, the *periglacial model* implies an important production of gelifracts, which when they are eroded, generate a stage of aggradation in the alluvial fan (Williams, 1973; Wasson, 1977). These periglacial processes are very marked in some mountains, in which upstream basins of alluvial fans develop. In the southeastern part of Spain, as indicated, the main aggradation periods coincide with the cold phases of the Quaternary and the incision periods with times of scarce sediments, which correspond to warmer phases (Harvey, 1978, 1990, 1996, 1997).

The construction of an alluvial fan requires a long period of time and, therefore, during its formation there will be changes in the climatic variables. Despite the diversity of the models proposed, it is really difficult to differentiate the climate influence from other "noise," such as the tectonic or the intrinsic factors of the system.

Nevertheless, analysing the paleo-vegetation record of the Mohave Desert, climate correlations between different areas with alluvial fans have been carried out (Spaulding, 1990). In this type of climate correlation it is necessary to take into account that the areas analysed usually have highly variable tectonic histories of fairly recent age.

At present, researchers agree that the construction of the alluvial fans in arid regions is due to storms of high magnitude and low frequency (Beaty, 1974; Baker, 1977; Blair and McPherson, 1994a,b; Dorn, 1994b; Harvey, 1997), which can have occurred in any climatic period of its history. Therefore, it is difficult to forecast the effects of the climate change (Harvey, 1996). As pointed out by this author, threshold conditions and sensitivity to environmental change (Brunsden and Thormes, 1979) vary from one alluvial fan to another one. In the semiarid zone of the southeastern part of Spain, during the period 1985 to 1990, a significant increment in the number of storms occurred. The erosion was limited to the alluvial fan apex, where marked trenches were produced, while eroded sediment was deposited downstream, without any evidence of new incisions in the middle and distal zones of the alluvial fans (Harvey, 1996). Therefore, it is very complicated to reconstruct the evolutionary history or, even ascertain data on the climate changes that can have affected the alluvial fans (Blair and McPherson, 1994a,b).

1.1.5. Desert rivers

In order to deduce environmental changes, the modifications in the form of the channel and in the type of the deposit are analysed. During historic periods many examples of human intervention are known, which trigger important variations in morphology, as well as in fluvial sediments (see Chapter 19). Furthermore, neotectonic activity is another factor that can influence the modification and intensity of the fluvial processes in some areas. Due to this, geomorphological works are carried out in cratonic areas where there is no tectonic activity, and so the "noise" produced by such activity can be avoided. Rivers are relatively insensitive to climate changes if these are not very marked (Reid, 1994). Rivers in arid and wet regions, however, present important differences, however, not only in form but also in the transported solid charge (Wolman and Gerson, 1978). In arid regions few hydrologic records are available, or do not exist, so the recent variations and their relations with climate are not well known. Some researchers, such as Schumm and Lichty (1963), have studied temporal variations of the channel parameters to deduce climate changes. One difficulty in drainage basins is the response of the variables of the fluvial system to climate changes (Schumm and Lichty, 1965). These variables are interrelated and climate affects mainly the weathering, pedogenesis, vegetation and hydrology (Schumm, 1977). Furthermore, the variables susceptible to change can be more active in one basin than in another.

The modifications that the fluvial system can undergo may be the result of changes in the average characteristics of the climate and also of the modifications in the common values of magnitude–frequency of the different events (Knighton, 1998). Similarly, small climate changes have produced modifications in fluvial systems (Knox, 1983). The studies of the floods of the Mississippi River for the last 7000 years reveal that there was a warm and dry stage between 5000 and 3000 yr BP, with minor flooding, and a abrupt climate change of a colder and wetter nature that produced important floods (Knox, 1993). Moreover, this author indicated that the most important floods in the last 9000 years are due to periodic changes in the atmospheric circulation systems (Knox, 1995).

When a channel increases due to a climate change, the width, depth and the inclination increases and vice versa. If the sediment load increases, the erosion will rise in the drainage basin. This increment in sediments can be produced by a climate change toward more aridity or by deforestation. A fall in the solid load can be due to a climate change toward wetter conditions or to soil conservation practices. In general, any modification in the channel is attached to a change in the type of sediment charge. When a change into a more humid climate takes place, an increment of increase is produced in the percentage of the vegetal cover and, therefore, there is a decrease in the loss of soils. This increase in precipitation turns ephemeral courses into perennial ones, the density of drainage grows and erosion in the tributaries rises, which produces an increase in the solid load, mainly during storms. If climate returns to semiarid conditions, soil loss and erosion rise, meanwhile discharge falls. All these variations in discharge and solid load, which in this case can be a climate change from subhumid to semiarid and vice versa, produce important hydrologic modifications and changes in the morphology of the channels, termed *channel metamorphosis* (Schumm, 1977; Chorley et al., 1984).

Many large drainage basins go through different climatic zones. The Nile river rises in central Africa, where it presents an important discharge and transports a small load

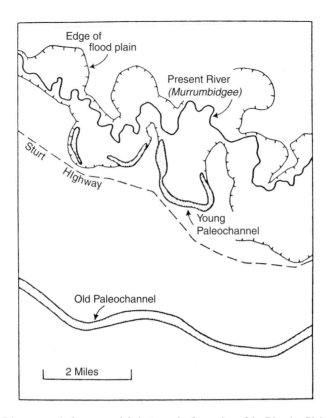

Figure 25.16. Diagram made from an aerial photograph of a portion of the Riverine Plain near Darlington Point, New South Wales. The sinuous Murrumbidgee River, which is about 200 ft wide, flows to the left across the top of the figure. It is confined to an irregular floodplain on which a large oxbow lake (youngest paleochannel) is preserved. The oldest paleochannel crosses the lower part of the figure (after Schumm (1969)).

of sediments. In middle and lower parts, the discharge falls but sediment load increases, because it goes through the arid countries of Sudan and Egypt. The history of the climate variations that occurred during the Quaternary is complex and difficult to study (Williams and Faure, 1980).

One of the best known examples of fluvial metamorphosis is the meandering river Murrumbidgee, which flows through the almost flat riverine plain in New South Wales, Australia (Schumm, 1968, 1969). The channel, 60 m wide, is surrounded by a floodplain in which several abandoned meanders occur (Figure 25.16) indicating that the river had a higher discharge. In the lower part of the figure an old channel of low sinuosity is shown. Different parameters in existing channels were measured, in recent and in old ones (width, depth, sinuosity, slope gradient, meander amplitude), as well as the discharge and the solid charge, to compare the hydrologic variations and the changes in the sediment charge when active. These geomorphological observations, together with the pedologic ones, indicate that the climate was drier when the paleochannels were active, while younger paleochannels occurred under more humid conditions.

In a dry stage the river decreases its discharge, land cover gets reduced and solid transportation rises. The change toward wetter conditions produces an increment in the discharge, width, depth, and in the amplitude of the meanders. The result represents a complete transformation of the fluvial system. An example of fluvial system abandonment and replacement by another system occurs in Oman, in western Arenas Wahiba (Maizels, 1987). There it is possible to recognize a five paleochannels system consisting of cemented gravels, which is part of an alluvial fan, (Figure 25.17). The erosion of the fan produces lowering, emphasized by aeolian deflation, leaving an exhumed landscape in the shape of a cord, which sticks out about 30 m above the plain. Older channels can correspond to humid stages in the early Pliocene–Pleistocene, whereas the exhumation is related to semiarid stages during the Pleistocene, corresponding to lower sea levels.

In arid environments overlaps and interferences between aeolian and fluvial domains are common. Thus, southwestern Egypt close to the frontier with Sudan and Algeria, is constituted by the large sand layer of Selima. That is a hyperarid zone, in which the study of radar images has shown that there is a system of big fluvial channels overlain by the sand layer, having an age that probably corresponds to the early Miocene (McCauley et al., 1982).

The most classic and complex example of interference between alternating aeolian and fluvial stages occurred in the middle course of the river Niger in Mali (Figure 25.18). Fluvial accumulations took place in the humid period of the early Quaternary and totally stopped during the subsequent stage of highest aridity, producing longitudinal dunes, or trade winds ones (ENE-SSW), in which watercourses disappeared. Later, in a period wetter than the present one, they crossed dune fields and filled the existing enclosed depressions, producing the expansion of lakes, as demonstrated by the high quantity of ceramic remains from the Neolithic. Nowadays, the discharge is enough to make the Niger have an exorheic character, although present precipitation is lower than Neolithic and much higher than the one existing during the stage of dune formation (Tricart, 1969; 1979). In the section about dune systems this aspect will be discussed in more detail (Figure 25.18).

1.2. Paleolakes

The study of the shorelines of ancient lakes, together with the analysis of the sediments deposited, provide very valuable data on the hydrologic modifications that have occurred. Many desert lakes present clear geomorphological and stratigraphic evidence of having had, during the late Quaternary in times with a higher effective humidity (Mabbut, 1977), dimensions much larger than the present ones.

Shoreline mapping of lakes requires a detailed photogeomorphologic interpretation and meticulous field work (Sack, 1994). Measures taken at different points around the basin of the lake for a given shoreline normally present different heights. This can be due to coastal geomorphologic processes and in deep lakes to hydroisostatic rebounds (Gilbert, 1890). With these data, it is possible to determine the area and depth of a paleolake for different periods. The existence of subbasins, separated by thresholds, as in Lake Bonneville (Gilbert, 1890), controls the levels of the paleolake (Benson, 1978; Benson and Paillet, 1989) and makes their study difficult, because of the complex dynamics of each subbasin. Another difficulty is related to the paleohydrologic variations that occurred within the

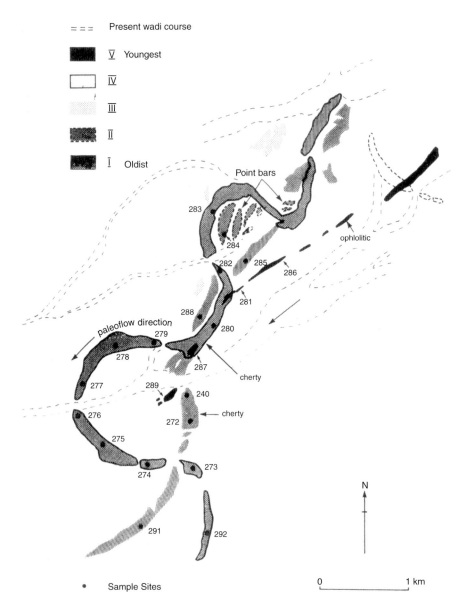

Palaeochannel Generation on Fan I, Area 5

Figure 25.17. Channel pattern and lithology changes associated with successive generations of paleochannels on Fan 1, in the western Wahiba Sands of Oman. Channel sinuosity and mean width decrease progressively with time (after Maizels (1987)).

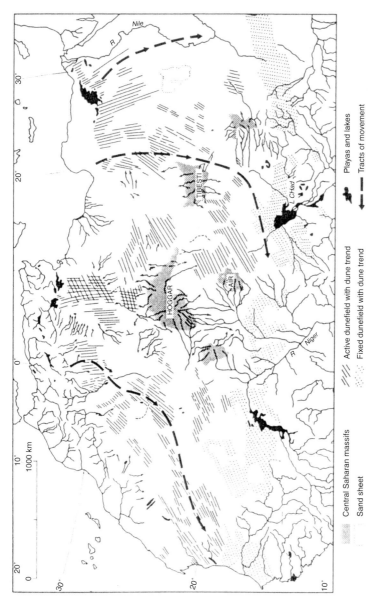

Figure 25.18. Saharan dunefields or ergs tend to occur near the centres of basins of interior drainage beyond the terminals of the present river systems, suggesting an alluvial origin for the sands (Mainguet, 1975).

Figure 25.19. Small trench excavated in the boundary of the sebkha El Melah. The base corresponds to gray rhythmic sediments and the top consists of a fine sand level, Zarzis, southern Tunisia.

fluvial basins that supplied the paleolake. Any important variation that modifies the water discharge can be incorrectly interpreted as a climate change.

The study of the stratigraphic record of a lacustrine depression provides relevant data to know its history (Figure 15.19). Geomorphologic, sedimentologic, mineralogic, geochemical, paleontologic and palynologic research provide keys for obtaining suitable paleo-environmental information on the different events through which the lacustrine system has undergone (Figure 25.20).

Lake expansions are due to a rise in precipitation or to a fall in temperature that produces a decrease in evapotranspiration. These circumstances correspond to the old concept of the pluvial period or lacustrine phase (Rognon, 1980). Nevertheless, conclusions obtained do not have the same verisimilitude that conclusions do based on the analysis of dune systems, because the relationship between precipitation and lake levels get complicated because of the temperature and other nonclimatic factors (Goudie, 1992). Some lakes, as with those related to the Rift Valley in eastern Africa (Figure 25.21), have a marked tectonism during the Quaternary, which can modify the geometry and capacity of the lake. Similarly, volcanic activity, mainly effusive, produces very pronounced modifications. Despite these circumstances, works carried out by numerous authors on lakes in eastern Africa and the southern Sahara indicate that they are excellent paleoclimatic indicators (Figure 25.22). In the deserts on shields and platforms (Mabbutt, 1977) large lakes develop in shallow structural depressions during relatively wet periods. Lake Chad is the largest known endorheic basin (Figure 25.23), whose boundaries are on the Guinea ridge, Congo watershed and in the massifs of central Sahara. This lake has been studied by many researchers (Barbeu, 1961; Grove and Warren, 1968; Pias, 1970; Servant

Figure 25.20. Extensive saline plain corresponding to the sebkha El Melah and, in the foreground, saline polygons, Zarzis, Tunisia.

Figure 25.21. Lake Nakuru, located in the Rift Valley Gregory, and therefore, of tectonic origin, even at present. This lake, of alkaline water, is famous because it contains the highest number of pink flamingos in the world. Nakuru, Kenya.

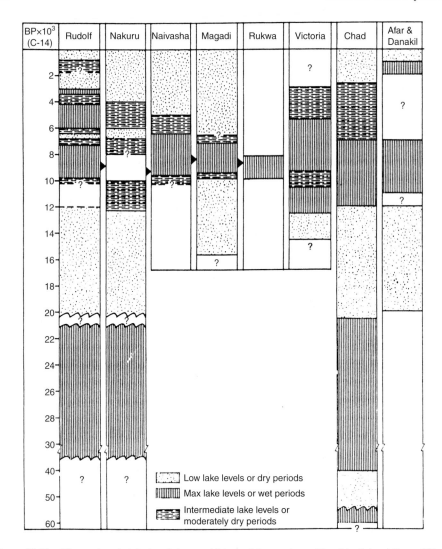

Figure 25.22. Fluctuations in lake levels in east Africa and the southern side of the Sahara. Data: different authors (compiled by Selby (1985)).

and Servan-Vildary, 1980; Nicholson, 1981) who have spatially and temporally analysed dry and wet phases that have occurred in the basin. During several wet periods the extension of the lake was larger than the present one. The flatness of the lacustrine basin is expressed by the depth, 3 to 7 m deep, for an actual area of 20,000 km^2. In one of the expansive stages (7000 to 5400 yr BP) (Pias, 1970), the lake, whose level was 320 m, occupied an area of 350,000 km^2 ("Megachad") and overflowed into the Atlantic Ocean and the river Nile in dry periods the lake was partially covered by dunes of ENE direction (trade wind dune). This type of lake is characterized by overlapping between lacustrine and aeolian landforms, as indicated in Figure 25.23.

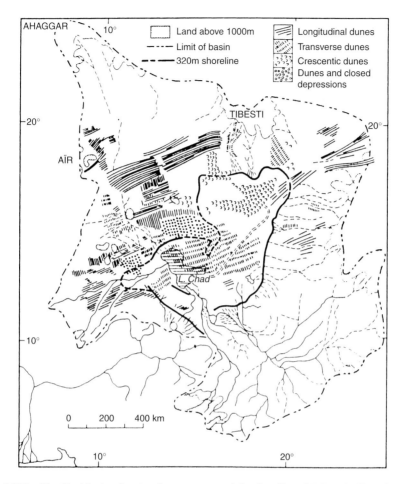

Figure 25.23. The Chad basin, showing dune systems and the shoreline of "Megachad" at about 320 m (Grove and Warren, 1968).

In mountains and desert depressions there are numerous lakes, frequently deep, that originated during stages of intense fracturing. In those lakes, shorelines of different height, beach ridges, spits, small deltas, saline mud plains, and so forth can be recognized (Gracia, 1995). In the Great Basin, located in the northern part of the Basin and Range province in USA, exists a large number of "pluvial" lakes (Morrison, 1965); about 110 to 120 lakes, formed by extensional tectonic block movements during the upper Pliocene and lower Pleistocene (Figure 25.24). In the stage of highest expansion Lake Bonneville was generated, which was studied in detail by Gilbert (1890), one of the founders of the discipline of geomorphology. This lake reached an area of 51,640 km^2 and a depth of 340 m. At present, the Great Salt Lake of Utah, which is the largest lake resulting from its desiccation, has an area ranging from 2600 to 6500 km^2. Toward the west, Lake Lahontan at 22,900 km^2, also developed. Some basins were connected by overflow connections and others overflowed directly in the ocean through the rivers Snake and Colorado.

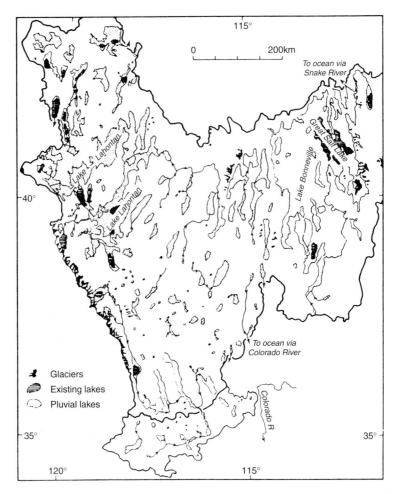

Figure 25.24. Maximum extent of Great Basin lakes during post-Sangamonian time. Arrows indicate overflow connections. The heavy line shows the present drainage divide of Great Basin (after Morrison, 1965).

The aridity of these North American deserts is due to an orographic effect. The high ranges (Coast Ranges, Sierra Nevada and Cascades) are close to the Pacific coast and obstruct the cyclonic circulation coming from the west. As a result, deserts in the rainshadow are generated. In this area, temperatures have been very low, mainly during glacial periods, because the Laurentide ice sheet was close and as in high mountain areas, around the coast, many smaller glaciers also developed. As temperatures fell, evapotranspiration decreased and lowlands in the Great Basin received large amounts of water from glacier melting. Therefore, large lakes were formed within endorheic zones, which nowadays show spectacular lacustrine terraces. This area is an ideal region to recognize the relationships between the mountain glaciations and lake phases. The work of Smith and Street-Perrott (1983) analysed many of these basins and discussed the chronology of the lake phases. They established a period of expansion between 24,000 and 14,000 yr BP; another between 14,000 and 10,000 yr BP with wide fluctuations that can be synchronous or not. Finally, between 10,000 and 5000 yr BP, an arid or dry phase can be recognized, with a fall of the lake levels. Recently, Oviatt (1997) pointed out that the different stages of lake-level fall have ages of 21, 18.5 to 19, 17.5, 16 to 15.5, 14 to 13, and 10 kyr, all of them obtained by means of C-14. They are synchronous with the end of the Heinrich H1 and H2 events and with other smaller stages of iceberg drift (a, b, c, and Younger Dryas) in the northern Atlantic, suggesting a global change.

The study of shorelines and their interpretation in deep lakes is complicated due to the fact that they have been deformed because of the hydrostatic unloading. The study of lake levels is crucial, however, to determine the magnitude of the subsidence in particular areas of the Earth's surface, as a result of the load due to the water volume and the subsequent rise resulting from the removal of lacustrine water. Figure 25.25 on the left shows isobaths

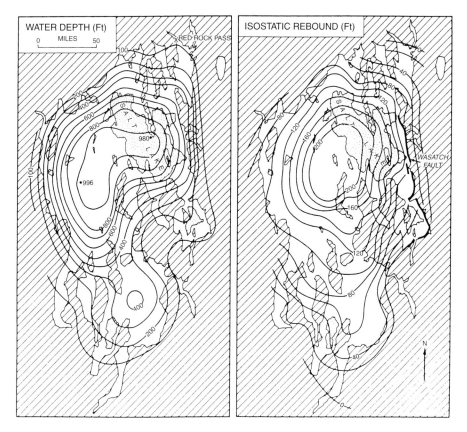

Figure 25.25. Pleistocene Lake Bonneville, Utah. Left: Maximum water depth (in feet) averaged over circles of 25-mile radius. The white area shows the extent of the lake at the Bonneville shoreline and the stippled areas indicate the present lakes. Drainage of the Pleistocene lake occurred as the result of an outlet being cut at Red Rock Pass. Right: Isostatic rebound (in feet) resulting from the removal of lake water, as measured by the deformation of the Bonneville shoreline. Recent displacement along the Wasatch Fault is shown. (Crittenden, 1963).

of the lake Bonneville during the period 25,000 to 11,000 yr BP and the one on the right indicates the isostatic rebound produced by lake desiccation, which is 20 % of the column of water (Crittenden, 1963).

The Aral-Caspian Sea system is one of the most spectacular cases of lake expansion during the Pleistocene era, which is formed by wide shallow deformations. Huge amounts of glacial melt water went into this system from Syr Darya and Amu Darya rivers, to the Aral Sea, and from Volga and Ural to the Caspian Sea. When the level of the lake reached 76 m above the Caspian Sea, it met the Aral Sea, generating the largest "pluvial" lake in the world, flooding an area of 1,100,000 km^2 and penetrating 1300 km into the river Volga (Goudie, 1992). The Caspian Sea also reached the Black Sea through the Mantych depression.

Despite the difficulties of interpreting climate on the basis of geomorphologic studies of paleolakes, these are crucial for paleoclimatic reconstructions, especially if they are based on multidisciplinary paleoenvironmental studies.

1.3. Dune systems

1.3.1. Introduction

Large aeolian accumulations or *ergs*, together with the erosive landforms existing in intertropical deserts, occupy 20 to 25% of the Earth's surface (Livingstone and Warren, 1996). The extension of these dune fields is well known in most of the deserts in the world (Figure 18.2) (Thomas, 1997b), from aerialphotos and satellite images. Dunes have been generated in periods of 1000 to 100,000 years (Lancaster, 1995). Expansions and contractions of the principal ice sheets, which considerably affected lower latitudes, changing climate and sea level, occurred during dune-field formation. The most significant variations become evident through changes of the boundaries of great sand deserts, in sand-particle supply and in the mobility of the dunes. These sand deserts are not exclusively Pleistocene, but many of them also became larger during the global climate cooling (Figure 25.26) (Shackleton and Kennet, 1975; Goudie, 1992); this is what Williams (1994) call "upper Cenozoic desiccation."

Many periods of dryness can be identified in the course of the Quaternary. The study of the ice cores dated in the upper part of the Greenland ice sheet (GISP2: Greenland Ice Sheet Project 2), show fluctuations in the electrical conductivity of the ice, at scales of <5 to 20 years. These reveal rapid oscillations in the atmospheric dust content (Figure 25.27) in periods in which the atmosphere was very dusty, separated by others without dust. The analyses of continental sediments within deep oceanic cores can also be used to interpret the extension, intensity and duration of the arid zones during the Quaternary. Continents in intertropical regions provide fluvial and aeolian sediments to the marine bottom. Furthermore, there is a relatively constant supply of organic carbonates, and thus, those cores with a high content of carbonates indicate a scarce supply from continents. Aeolian supplies are represented by high percentage of silt-sized quartz grains, which can be differentiated from fluvial ones because of their tendency for reddish staining (Diester-Haas, 1996). Increments of aeolian dust took place during Quaternary cold periods, in which the quartz content at the bottom of the Atlantic Ocean moved toward the south,

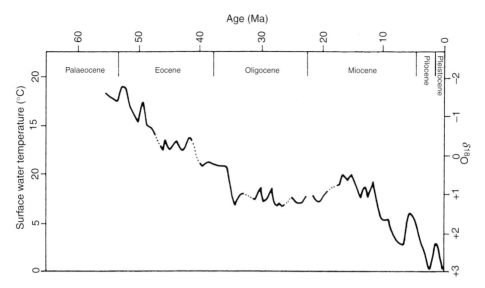

Figure 25.26. Cenozoic sea surface temperatures in the Southern Ocean deduced from changes in oxygen isotopic composition of planktonic foraminifera at DSDP sites 277, 279, and 281 (Shackleton and Kennett, 1975, in, Abrahams and Parsons, (1994), Figure 26.4).

at about 8°N, indicating an expansion of the aridity and of the Saharan sand desert (Figure 25.28) (Kolla et al., 1979). The existence of fresh water diatoms in cores corresponding to the maximum of the last glaciation can be interpreted as particles deflated from desiccated desert lakes, which also supports the idea of this expansion (Parmenter and Folger, 1974; Pokras and Mix, 1985). Finally, important data can be obtained from the analysis of deep-sea drilling cores on the basis of the study of feldspar weathering, pollen and phytoliths, and salinity and temperature of the marine water from oxygen isotopic composition in foraminifera.

This research provides indirect data on the important dry periods that can be distinguished in the recent geologic record. The detailed study of the different active or inactive dune fields,

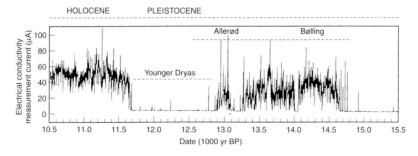

Figure 25.27. Yearly averages of electrical conductivity (related to dust content) of a core at the summit of the Greenland icecap (Taylor et al., 1993). The core shows remarkable (and sudden) alternations between the dustiness of cold periods and the relatively clear skies of interglacial periods.

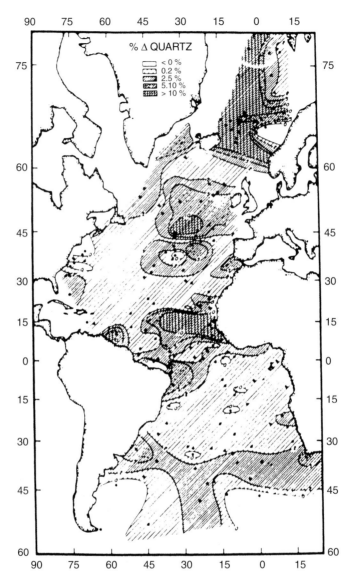

Figure 25.28. Difference in quartz percentage in late-glacial and Holocene sections of Atlantic sediment cores. Enhanced late-glacial concentrations in the Northern Hemisphere tropical zone are interpreted as the product of dust inputs deflated from an expanded Sahara and/or due to stronger circulation systems (after Kolla et al. (1979)).

within the different deserts in the world, however, provides valuable information on the extension, duration and distribution of the periods of marked aridity, as well as on the changes in wind directions and circulation systems through time (Lancaster, 1995).

The dating of sandy accumulations has been one of the most important problems. Until a short time ago, dating was difficult and dates were relative. Thus, for example, C-14

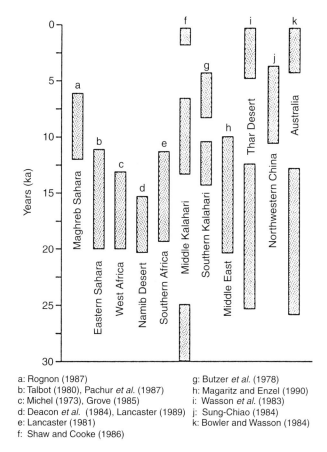

Figure 25.29. A summary of late Quaternary dry periods for the tropical and subtropical continental deserts (Africa, Asia, and Australia) for the last 30,000 years. Dune building episodes most likely took place during the above periods (compiled from several authors by Tchakerian (1994)).

dating of the organic matter existing in paleosols developed on dunes reveals an age pre- or post-. Another difficulty is to have available reliable dates, because many dune fields are formed during several epochs. There was extraordinary progress at the beginning of 1980s, with the use of many new techniques such as thermoluminescence (TL), and optically simulated luminescence (OSL), which permitted the dating of sand in dunes and loess (Wintle, 1993). Ages up to 500,000 years have been obtained. These dates made possible correlations with paleoclimatic data by using other techniques. With all these data, it was possible to make paleoclimatic correlations and interpretations. As a result of this progress, actually it is now possible to know the different dry periods that occurred in all deserts in the world (Figure 25.29), for which different chronologic histories can be obtained.

Dunes are sensitive to variations in atmospheric parameters, such as wind direction and intensity, changes in precipitation that influence evapotranspiration, soil humidity, vegetation percentage, and particle mobility. The areas of dunes tend to emphasize

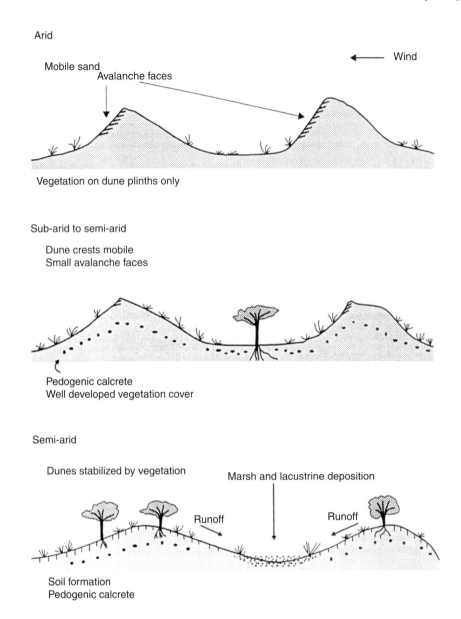

Figure 25.30. Possible responses of dunes to changes in climate (after Rognon (1982)).

the effects of dry and wet phases (Rognon, 1980) and rapidly respond to climate changes. If precipitation increases a little, vegetation grows up mainly in the lower parts of the dunes, whereas in the upper parts sand mobilizes producing leeward avalanches (Figure 25.30). When precipitation increases, vegetation colonizes the entire dune, except the crest, and calcification starts in sandy accumulations. Finally, when periods of important and durable rainfall occur, the groundwater table appears on the surface generating marsh deposits in the interdune areas; furthermore, vegetation of large size settles, stabilising the dunes and generating soils. In the cases in which precipitation is very intense, dunes can be eroded (Talbot and Williams, 1978). On the other hand, in dry periods the aquifers are low and, therefore, water availability is virtually zero, hampering the growth of the vegetation and favouring the aeolian activity during these periods.

1.3.2. Active and inactive dunes

The knowledge of the degree of activity of dunes is problematic and imprecise. From numerous data, it is estimated that when average precipitation exceeds 100 to 300 mm (Goudie, 1992), vegetation is effective enough to restrict the movement of the dunes. The degree of activity can be obtained from climatic indexes collected by Cooke et al. (1993), and elaborated by Chepil et al. (1962), Wasson (1984), and Lancaster (1988). Climate indexes are imprecise due to limited meteorological stations and to great year-on-year variability in semiarid climates (Livingstone and Warren, 1996), in such a way that in some years the dunes are active and in others with less precipitation, they lose most of the mobility or even become stable.

The transport and sedimentation of *active dunes* (Figure 25.31) become apparent in surfaces with ripples and leeward avalanches. Nowadays, overgrazing and other human activities in deserts can trigger the reactivation of the dunes, as occurs in the very populated area of Rajasthan, India, in the desert of Thar (Goudie, 1992). *Dormant dunes* (Lancaster, 1995) or *episodically active dunes* (Livingstone and Warren, 1996) are those in which the movement of the sand is slow or does not exist for long time spans

Figure 25.31. Large barchanoid dunes. Moving sand can be observed in the foreground. Al Liwa, Rub al Khali Desert, United Arab Emirates.

Figure 25.32. Dormant longitudinal dunes with well-developed shrub-like vegetation. The mobility of the sand is limited to the crests of dunes, while flanks are stabilized by biocrusts and shrubs. Experimental Estation of Nizzana, western Negev Desert, Israel.

(Figure 25.32). In general, they are located in areas with 100 to 300 mm of precipitation, great variability in year-on-year precipitation, in such a form that in dry periods land cover decreases and sand mobilizes. The effect is the opposite in wet years. In these dunes, primary sedimentary structures are affected by bioturbation. Vegetation shows a good state of development with high percentages of perennial plants. *Inactive, fixed or relict*

Figure 25.33.   Inactive dunes with shrubs and trees cover. Beer Sheva region, Negev Desert, Israel.

dunes are remains from past drier climates and, therefore, are indicators of paleodeserts (Figure 25.33). These dune systems are located in areas that nowadays have rainfalls of 250 mm, and even 2000 mm, as in Zaire (Thomas and Goudie, 1984), the Orinoco Basin (Tricart, 1985), Pantanal of Brazil (Klammer, 1982b), and elsewhere. They consist of dunes and sand sheets that are stabilized by structured soils. Within these soils, caliches and ferralites develop due to the precipitation increase (Tricart, 1979; Völkel and Grunert, 1990). Alternatively, dunes can be cemented, generally variably, which produces aeolianites that commonly have weathering alveoles and karstic micromorphologies (Figure 25.34). On the other hand, dunes become modified by nonaeolian processes and elluviation. Crests become rounded and reduced, flanks eroded by water erosion and the resulting sediments fill up interdune depressions (Mabbutt, 1977). Nowadays, flanks of longitudinal dunes in the southern Sahara do not reach 5°. During the initial degradation stages of a dune, rilling, gullying, and piping processes can be triggered in wet seasons, which can destroy former aeolian structures. Flooding in wetter stages is another cause of inactivity, as has occurred in the desert of Thar (Singh, 1971) and in the Chad Basin (Grove, 1958), where dunes are situated below lacustrine deposits. In all these relict dunes the return to an active state is only possible through important environmental changes.

An aspect-causing controversy and, on the other hand, also common in other branches of geomorphology, is the estimation of the time needed for the development of a landform such as a dune. Some researchers have pointed out that actual dunes are probably relict (Glennie, 1970; Besler, 1982), because only very strong winds, as the ones that occurred in upper Pleistocene, are capable of generating these landforms. It is also said, alternatively, that dunes require long periods of time to be formed. Thus, on the basis of empirical observations, Wilson (1972) estimated that the 100-m high dunes in eastern Great Erg

Figure 25.34. Aeolianite with cross-stratification affected by alveolinization controlled laminae. Sila, United Arab Emirates.

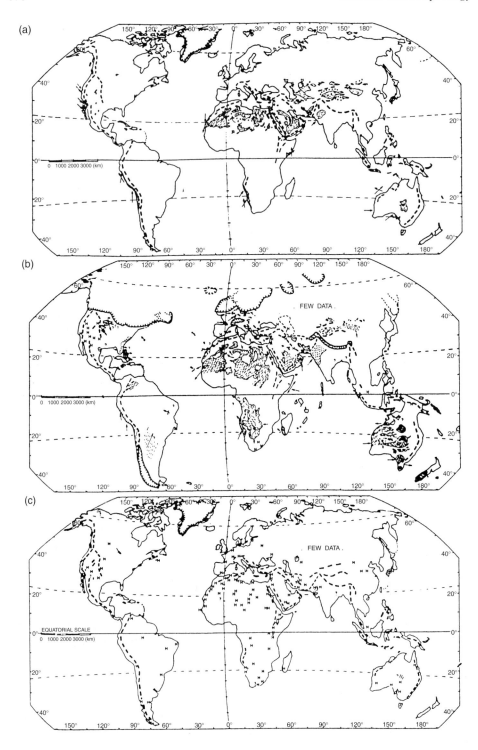

(a)

(b)

(c)

have an age of 10,000 years, and Lancaster (1989a) estimated in the age of the large dunes of the Namib Desert to be some 42,000 years. Dunes of 70 m height exist on the coast of the Huelva province, in El Asperillo of southern Spain, showing an age, by means of archaeological and historical data, from the 18th century (Borja et al., 1999; Zazo et al., 1999).

Today it is well known that there are extensive areas of stabilized dunes on all the continents, in tropical zones as well as at high latitudes, in relation to periglacial climates (Sarnthein, 1978) (Figure 25.35). Dunes formed, during the last glacial maximum, 2.5 million km^2 in Africa, which means the largest desert in the world (Thomas and Shaw, 1991). Comparing the extension of these dune fields with the current area with active dunes, it can be deduced that great changes in have occurred precipitation and in the percentage of land cover. At present, active dune deserts, located between 30°N and 30°S, occupy only 10%, while 18,000 years ago their extension was 50% (Goudie, 1992).

In the boundaries of most zonal deserts there are stabilized longitudinal dunes. The most extensive ones are located in the Sahara, over 5°N (Figure 25.18) and stretch from the Atlantic coast of Senegal to Sudan. They extend toward the south over 500 km, in areas that nowadays have a precipitation of 500 mm. Three generations of dunes can be recognized (Grove and Warren, 1968; Talbot, 1980). The oldest group is older than 20,000 years, the second between 20,000 and 13,000 years and the most recent has an age of about 5000 years. Most of them were stabilized between 11,000 and 7000 years ago. The last droughts and deforestation have reactivated some dunes and have generated new ones (Nickling and Gillies, 1993). Other examples can be found in the Kalahari, northern Arabia, Thar Desert, Australia, Venezuela, and elsewhere. All these dunes are completely colonized by vegetation and have soils developed on them.

1.3.3. Paleoclimatic information provided by dune systems

Active dunes that developed on continents are located around and below the isohyet of 150 mm (Mainguet et al., 1980), with the southern Sahara as an excellent example of that spatial distribution (Figure 25.36). The existence of these relict dunes represents an irrefutable evidence of previous dryer stages. The situation of these dunes in areas of high precipitation, as for example the selva of the Congo, reveals the drastic change that occurred. Moreover, it is necessary to take into account that the reactivation or formation of dunes can be the result of changes in the velocity of the wind and their interactions with humidity (Thomas, 1992).

Comparing wind directions that mobilize sand with dune alignments, valuable paleoclimate information can be obtained. There are wind systems that at present blow in the same direction as during the late Pleistocene, as in the southern Sahara (trade winds).

Figure 25.35. Active dune fields and long-term resultant effective wind regimes for the present, 18,000 yr BP (glacial maximum), and 6000 yr BP Climatic Optimum. Continental outlines at 18,000 BP represent sea-level lowering of 120 m. Hatched areas indicate dune fields and major dune contours. Arrows are long-term resultant surface-wind directions. H signs stand for fossil wet conditions excluding dune formation (after Sarnthein (1978)).

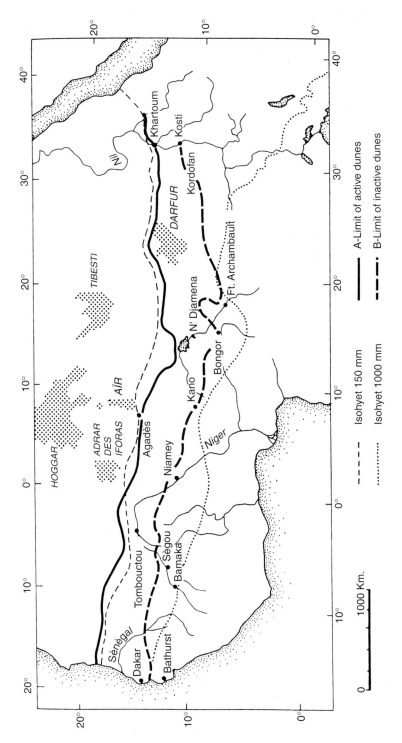

Figure 25.36. Active and relict dune field limits at present in southern regions of Sahara (Mainguet et al., 1980).

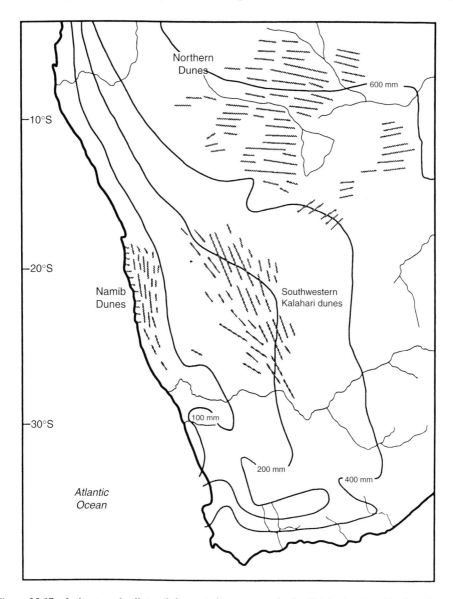

Figure 25.37. Isohyets and relict and dormant dune systems in the Kalahari region. Northern Dunes: <20,000–30,000 years ago; Southwestern Kalahari Dunes:18,000 years and southern dunes started in the early Holocene (after Lancaster, 1981).

On the other hand, there are regions in which present and past directions are very different. In the desert of the Kalahari the dunes are principally inactive and in the main are covered by acacia forest, shrubs and grass. Three dune systems can be recognized on it, generated by winds with different paleodirections (Figure 25.37) (Lancaster, 1981).

Figure 25.38. Star-shaped dunes overlaying laminated marsh deposits. Maharaga, Rub al Khali Desert, United Arab Emirates.

The majority of present sand deserts have been formed slowly during the climate changes of the Quaternary and subsequent variations of the sea level. As a result, different surfaces of regional extension can be identified (Kocurek, 1988), which separate different stages of accumulation. Over them, pedogenic processes, water erosion, marsh and lacustrine sedimentation, and so forth can be recognized to have occurred.

A great part of the sand deserts are fed by deflation of marine, lacustrine (Figure 25.38) and fluvial sands. The more important and constant through time is the source area, the major transport, the accumulation, and the migration of dunes. Due to the sea level lowering during glacial epochs, the Persian Gulf was mostly emergent (Kassler, 1973) and acted as a source area, supplying the construction of the Rub al Khali desert, as indicated by the directions of the dune systems (Glennie et al., 1994). These dunes were formed between 20,000 and 9000 years ago (McClure, 1978). In the Mojave Desert of California, the sand sources came from the bottom of desiccated lakes and from fluvial systems. At the Pleistocene–Holocene boundary there was an important increase in the aridity and the sand supply fell sharply. The aeolian action was limited to the remobilization of dunes (Lancaster, 1995).

In many sand deserts, interdune deposits are common, as in the Al Liwa region in the Rub al Khali Desert of Arabia (Figure 25.39). Shallow lakes develop, which are supplied by springs, rises of the phreatic level, or by water flowage. The ephemeral character, sometimes seasonal, with an alternation of expansion and desiccation stages, is the main characteristic. The importance lies in the fossil content, which makes possible dating and paleoenvironmental interpretations. Thus, interdune lacustrine deposits of the Rub al Khali desert show wet conditions between 2400 and 3200 yr BP and also between 5000 and 8500 yr BP (Lancaster, 1995).

Figure 25.39.   Interdune playa with shrubs in the back. These playas can become flooded by precipitation or water emanation. Al Liwa, Rub al Khali Desert, United Arab Emirates.

The areas in which there are successions consisting of alternations of aeolian accumulations corresponding to arid stages, and marsh and lacustrine sediments produced in wetter stages, are locations ideal for the study of climate changes. One of those places is located in the Erg Akchar of western Mauritania, which constitutes an excellent example of periodicity produced by climate changes. The stratigraphic work and dating, by means of C-14 and prehistoric remains, were carried out by Kocurek et al. (1991) (Figure 25.40). The substrate is formed by longitudinal dunes with an age ranging from 13,000 to 20,000 yr BP. It is also possible to recognize two other stages of aeolian reactivation, separated by surfaces that can be identified on the basis of specific stratigraphic characteristics. Finally, the other periods correspond to phases of aeolian inactivity because of the growth of the land cover. Marsh and lacustrine systems also develop between dunes, with sedimentation of sand, carbonates and gypsum. In interdune areas close to the ocean, marine transgressions turn these depressions into coastal *sebkhas*.

The southeastern part of the River Niger was supplied during the late Pleistocene, by tributaries coming from the Hoggar massif (Figure 25.18), which are currently inactive. Downstream the discharge increases with other tributaries, flowing into its delta. The upper part of the Niger, which flows toward the northeast and is fed by Guinean mountains, flowed into the Senegal Gulf during the late Pliocene and early Pleistocene (Goudie, 1992). Later, during a dry phase, vast dune fields were formed, which acted as a barrier blocking the former watercourse (Tricart, 1979). As a result, many ephemeral lakes were generated. After that, by capture or overflowing, this part of the river flowed into the Niger, forming a wide bend in the region of Timbuktu, Mali. This capture occurred 5000 to 6000 years ago. The River Niger, during dry phases, was incapable of going through the extensive dunes and disappeared inside and into the zone of lakes located in southern Timbuktu. In "pluvial"

(a) Ogolian (13,000–20,000 yr BP)

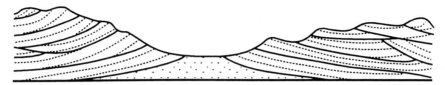

(b) Tchadian–Nouakchottian (4,000–11,000 yr BP)

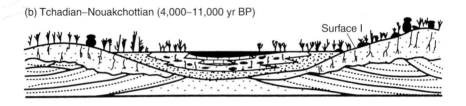

Surface I

(c) Tafolian–Modern ? (?–4,000 yr BP)

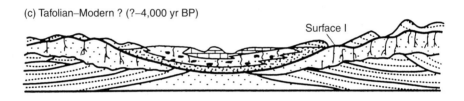

Surface I

(d) Modern (?)

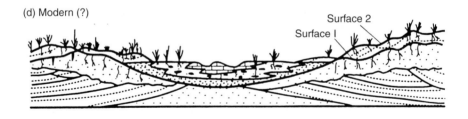

Surface 2
Surface I

(e) Modern (1990)

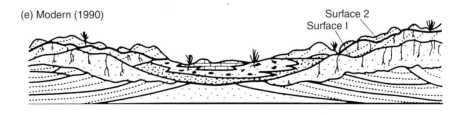

Surface 2
Surface I

stages, the higher supply of the watercourse made it possible to cross the aeolian accumulations, transforming the fluvial basin from endorheic into exorheic. All lakes in this area expand during "pluvial" stages, while in dry periods increase the salinity by evaporation, becoming even desiccated, as occurred in 1957 to Lake Fagnibine (Tricart, 1979). The present climate allows the Niger to flow into the Gulf of Guinea, which indicates that actual precipitation is higher than that of the past, when the river was endorheic.

1.4. Causes of the paleoclimate fluctuations during the late Quaternary period

This is one of the subjects that has most fascinated experts on arid zones and has been restricted mainly to the African continent, probably because it is the most well known by many geomorphologists in different countries. According to Tricart (1969), however, the reconstructions of the mechanisms that produce paleoclimate oscillations are different. Areas located within tropical high pressure zones generate zonal deserts, whereas extrazonal deserts are formed under specific geographic configurations, such as continentality, orographic effect, and cold oceanic currents. The latter are normally more sensitive to variations in the heat balance of the Earth and their evolution is determined by local factors.

Theories on general circulation during the Pleistocene are presented in the excellent work of Nicholson and Flohn (1980), in which the numerous authors theorize about the trajectory of cyclones in middle latitudes, on the heat contrast imposed by the presence of large ice masses with respect to the glacial–pluvial synchronism, and so forth. Nicholson and Flohn (1980) have carried out the most relevant and synthetic work with regard to climate and environmental changes in Africa and the general atmospheric circulation in the Pleistocene and Holocene periods. During the biggest extension of the ice (18,000 yr BP) in the Northern Hemisphere, the seasonal variations were minimal. Therefore, there was a displacement of the cyclonic zones toward the south and an increase in the temperature gradient. On the other hand, the Southern Hemisphere showed characteristics close to present ones.

For these authors, in this epoch and during the greater part of the interval from 20,000 to 12,000 yr BP, the aridity increased in the regions located in the southern Sahara and eastern Africa. This contradicts the data of Sarnthein (1978), who held that the period of aridity was about 2000 to 3000 years, around 18,000 years ago. But this supposition does not explain that, from that date to 12,000 years, wet conditions occurred in northern Africa, in the massifs of the interior Sahara and in southern Egypt. Nicholson and Flohn (1980) pointed out that the biggest environmental changes in tropical Africa coincided with the accelerated growth of the ice sheets (previous to 22,000 yr BP). Circulation systems meant, for most of the period, aridity in tropical Africa and relatively wet conditions in northern Africa (Figure 25.41), as a result of the displacement toward the south of the atmospheric circulation systems.

Figure 25.40. Evolution sequence of the aeolian activity and stabilization periods in the erg Akchar, Mauritania. (a) Longitudinal dunes formed in the Ogolian phase. (b) Dune stabilized by vegetation and interdune lake formation during Tchandian and Novakchottian. (c) Dune reactivation and lake desiccation in the Tafolian. (d) Formation of a new revegeted surface during the most recent period. (e) Modern reactivation of dunes, which remobilize former ones (after Kocurek et al. (1991)).

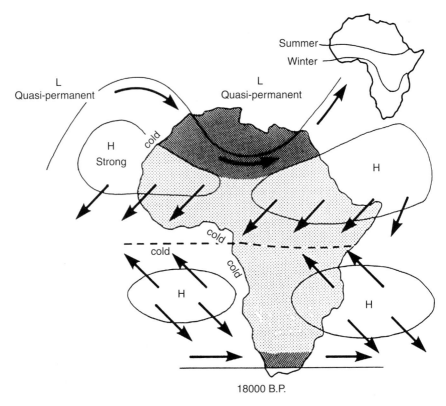

Figure 25.41. Inferred pattern of atmospheric circulation over Africa during the last glacial maximum. Dark shading indicates those areas more humid than today; lighter shading indicates areas considered drier than today. The inset shows the present position of the ITCZ in summer and winter months (slightly modified after Nicholson and Flohn (1980)).

Approximately 10,000 to 8000 years ago, there was a change in the circulation system that produced the first subtropical lacustrine event, as a result of an increase in precipitation that caused the expansion of the lakes in the southern Sahara, from eastern Africa to Senegal and Mauritania. Meanwhile, aridity took place in northeastern Africa. These circumstances are related to a gradual warming in the north of the Atlantic Ocean, a fall in the gradient of temperature and a maximum drift toward north that displaced the aridity toward the northeast, as indicated by the advances of dunes in Morocco and western Algeria.

Latter, about 7000 years ago, the Fennoscandian ice sheet disappeared and northern Europe reached the Climatic Optimum over 6000 yr BP, while the Southern Hemisphere remained cold. During the period 6500 to 4500 yr BP, both northern and southern Sahara were more humid than at present, but with an intensity smaller than in the period 10,000 to 8000 yr BP, perhaps due to the fact that the Northern Hemisphere was warmer during this third period.

The model of Nicholson and Flohn (1980) is the most complete among existing ones, but presents several contradictions, mainly regarding the interpretation of the Southern

Hemisphere in Africa (Thomas, 1997c). The problem reaches high degrees of complexity when interpretations are made globally and do not only refer to the African continent. Without a doubt, a long and hard job must be done, because it is necessary to obtain new data, reinterpret the past, and to have at one's disposal the highest number of accurate datas. It is also essential to know the Quaternary climate evolution in the Mediterranean areas, because this represent the transition zone between Europe and hot Africa. Therefore, a detailed exploration in arid zones is needed.

2. Tropical regions

2.1. Introduction

Tropical climates include a wide variety of precipitation regimes and vegetation associations, from selva to savanna and the transitions to semiarid zones. On the other hand, the activity and intensity of the geomorphological processes change depending on the zone. The variations in time of these morphoclimatic limits produce the modification of geomorphologic processes and, therefore, the existing landscapes. Nevertheless, according to Douglas (1969), humid tropical cores are stable ecosystems in which the energy is used by vegetation and do not present much water erosion in the rainforest plains.

The most significant studies on climate change in the humid tropics have been carried out since 1950. Brückner (1955) in Ghana, and De Ploey (1965) in Zaire, on the basis of the analysis of superficial deposits, found clear evidence of stages of aridity in the Pleistocene period. In South America there are ancient dunes in the Llanos (*plains*) of Venezuela, in Bahia state of Brazil and in central Amazon (Tricart, 1974a, 1975, 1985), which show arid conditions in these humid tropical areas. Bigarella and Andrade (1965) and Bigarella et al. (1969) developed the hypothesis of peneplanation in southeastern Brazil. The works of Aubreville (1962) and Butzer (1978) analysed the spatial variations of the limits of the savanna and rainforest because of climate changes. During the last 20 years there has been an important increment in the number of publications on this subject, but there are not too many long-term chronological records (Thomas, 1994), which force us to use temporal records of the lakes in the arid zones of Africa and Central America (Stret-Perrott et al., 1985), because other long-term records are very rare.

For the study of climate change, there are available, apart from the record of semiarid lakes, data from ice cores in ice sheets and from deep-sea drilling cores. There is an important limitation, however, due to the C-14 and AMS C-14 techniques because they can only be applied back to $\sim 60,000$ years ago.

Works carried out from cores in the equatorial Atlantic have analysed the content of freshwater diatoms, opal phytoliths, pollen, and sapropel muds for the last 150,000 years. From these analyses, up to seven different arid phases can be identified, depending on the methodology used (Jausen and Van Iperen, 1991).

2.2. Biological modifications

Variations in the configuration of seas and lands, atmospheric circulation, and physiography can produce important changes in the humidity, in such a form that the rainforest can be replaced by the savanna and vice versa, and thus, the morphogenesic changes (Tricart, 1974b). In the lower Congo and in the southern Ivory Coast, the forest trends toward savanna (Tricart, 1974b). The opposite effect can be observed in Senegal. On the basis of a review of the forest and savanna zones, Aubréville (1962) found anomalous ecologic evidence in African humid tropics, as well as in the Guyana. These anomalies are interpreted by this author as a lack of recuperation of the forest, after the drier conditions occurred in the last glacial period. He pointed out that forest regions drifted toward the south (the principal core in Angola), with some forest traces remaining in the coastal areas of Guinea Gulf and Congo. This interpretation contradicts the idea of a stable forest. A similar study was carried out by Butzer (1978), in which, during the glacial optimum (18,000 BP) a large reduction of the extension of tropical forest occurred, resulting in three bastions: Guinea–Sierra Leona–Liberia, Nigeria–Cameroon, and Zaire. The work on the distribution of flora and fauna in the forest areas by Hamilton (1976), provided important data about relevant variations of the central Africa forest in the last 20,000 year (Figure 25.42).

In the South American forest there are curious systems of speciation probably due to changes in the extension of the forest in the upper Pleistocene era. The forest has been fragmented into isolated sections in which speciation occurred (Figure 25.43) (Goudie, 1983a; Bradley, 1985; Crowley and North, 1991). In Brazil there is paleogeographic evidence that supports the existence of xeromorphic plants in *cerrados* and *tabuleiros* in central Brazil and in the western Amazon. The Barreiras Formation, consisting of quartziferous sand and kaolin, which crops out in the area, was intensively incised during the last glacial period as a result of the lowering of the base level resulting from the marine regression. This erosion was favoured by the existence of a dry climate that produced the disappearance of the forest in the lowlands along the Amazon River. This implies a reduction of the area of forest during dry stages; there are forest refuges only in high areas of the Guyana and Brazilian shield, where precipitation is higher. In the Holocene, climate turned more humid and the Flandrian transgression raised the base level (Dias de Ávila-Pires, 1974; Tricart, 1985).

2.3. Geomorphologic evidences

2.3.1. Regolith and relict weathering profiles

The greatest weathering depths occur in the humid tropics and, normally, these profiles were developed in the past, although actual climate conditions in the tropics can increase weathering effects (Ollier and Pain, 1996). Some authors point out that the regolith thickness in high areas is much smaller than in areas of gentle relief, while others, to the contrary, indicate that in some cases the greatest depths are located in high areas, and in the bottom of the valleys the bedrock crops out, because of the fluvial incision (Pain and Ollier, 1981). From this, it can be deduced that the weathering depth depends on the weathering intensity and erosion. On the other hand, very thick profiles need deep groundwater to maintain an active weathering front (Ollier, 1988a,b).

In many known cases, weathering has been developed during the Tertiary and Cretaceous periods, reaching thicknesses higher than 100 m. Most of these profiles are

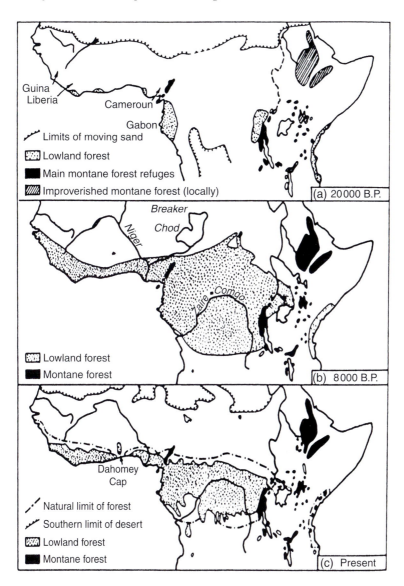

Guina
Liberia
Cameroun
Gabon
Limits of moving sand
Lowland forest
Main montane forest refuges
Improverished montane forest (locally)
(a) 20000 B.P.

Breaker
Niger
Chod
Zaïre–Congo
Lowland forest
Montane forest
(b) 8000 B.P.

Dahomey
Cap
Natural limit of forest
Southern limit of desert
Lowland forest
Montane forest
(c) Present

Figure 25.42. The probable distribution of forest in central Africa at 20,000 and 8000 BP compared with the present (after Hamilton, 1976). Note that the existence of Greater Lake Chad at 8000 BP is now questioned.

balanced with the present climate conditions, but have registered many climate changes in the past, which makes the interpretation difficult. In Chapter 20, examples of different deep weathering profiles, corresponding to humid tropics and extratropical regions, are shown. The latter are located in Europe, Australia, India and USA; weathering can even be Jurassic. From all these works, it can be stated that the principal period of deep weathering took place in Mesozoic-lower Tertiary, when thick regoliths developed and later eroded as

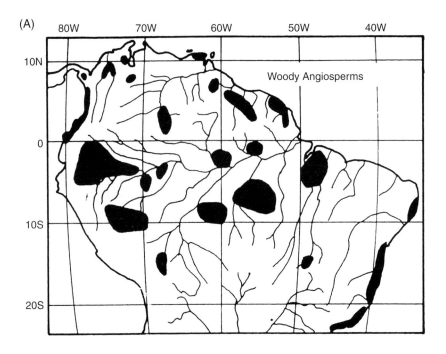

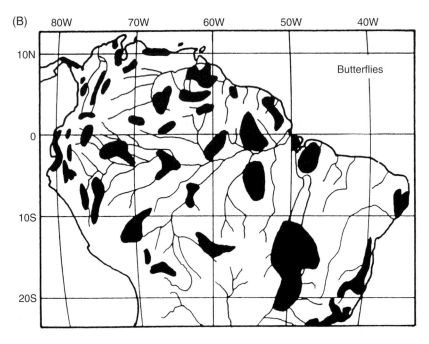

Figure 25.43. Proposed refuge areas for certain species of (A) woody angiosperms and (B) butterflies in the Amazon Basin during dry climatic phases of the Pleistocene (after Bradley (1985), in Crowley and North (1991), Figure 3.8).

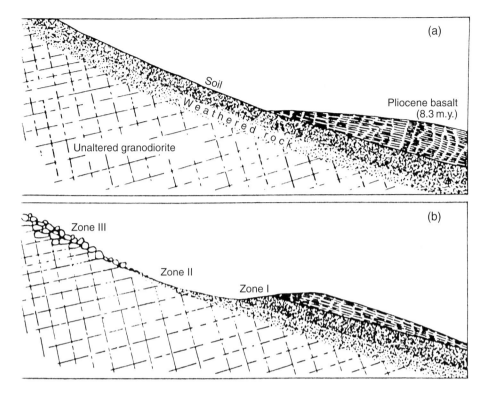

Figure 25.44. Diagrammatic representation of conditions at one site in the Mojave Desert of California: (a) Immediately after emplacement of the basalt. (b) Present conditions, showing the relict weathering preserved beneath the basalt (Zone I); a partly-stripped surface (Zone II), and exposed relict core stones forming the modern boulder surface (Zone III) (after Oberlander (1972)).

weathering decreased in intensity (Valeton, 1994). During the Eocene and several stages in the Cretaceous, Earth was warmer and more humid than at present. Poles were warmer and temperatures in the tropics were colder (Walker and Sloan, 1992).

The dating of weathering profiles is generally difficult, although there are some illustrative examples. In the Mojave Desert of California, basalts of the upper Miocene (8 to 9 million years) overlap weathering profiles developed in granites. At present there exists a later stage, 9 more arid, eroded part of the basalt and weathered rock (Oberlander, 1972) (Figure 25.44). The analysis of oxygen isotopes in deep regoliths of Australia also makes it possible to differentiate between regoliths generated in upper Mesozoic and lower Tertiary and those formed later than middle Tertiary. These differences in the profiles can be interpreted as the result of the drift of the Australia continent from high to low latitudes, along the Mesozoic and Tertiary (Bird and Chivas, 1988).

2.3.2. Inselbergs

These landforms are magnificent and difficult to interpret. The main controversy is caused by the genesis of these landforms. On the one hand, some authors such as King (1948,

1975) and Kesel (1973) consider that inselbergs as subaerial remains are generated from scarp retreat. On the other, several researchers who stand out from the others; Büdel (1957), Ollier (1960), Thomas (1965, 1994), Twidale (1964, 1982a,b), and Vidal and Twidale (1998) maintained that inselbergs are landforms exhumed from thick chemically-weathered mantles.

Pediplanation, which commonly occurs together with these landforms, in the first hypothesis is due to the intrinsic development of the landform and, therefore, weathering is superimposed on the inselberg–pediment system once this generated. In the exhumation hypothesis, the development of the regolith is the dominant process in the genesis of the inselbergs, playing in the accompanying pediment process a secondary and belated role.

Obviously, the development of inselberg landscapes, according to the theory on slope evolution, occurs in semiarid morphoclimatic zones. The genesis of these landscapes by exhumation occurs in morphoclimatic systems of humid tropics. The latter explains the small domal inselbergs (*bornhardts*), up to 50 m high, simply by denudation of a thick saprolite. When this hypothesis is applied to the origin of large *bornhardts* many problems rise. These difficulties can be solved if a continuous weathering is supposed at the foot of the slope, as the weathering allows the surrounding plains to be lowered.

In savanna areas, the history is not monoclimatic (Birot, 1978), at least not in Quaternary times, and, precisely, in those areas is where inselbergs proliferated (Kesel, 1973). The alternation of wet and dry climate stages is what facilitates the development of these landforms. Furthermore, it is necessary to take into account that once the bornhardt-plain was established, the contrast in the velocity of the weathering between both stages assures the persistence of the system (Thomas, 1978).

Inselbergs, in some cases, can have been formed in quite old periods of the geologic past and have remained until present. This interpretation has been proposed by Twidale (1978) for the famous Ayers Rock (Uluru), in the desert plains of central Australia. The author, on the basis of correlative sediment analyses and geomorphologic interpretations, concluded that the development of that monolith started at the boundary of the Mesozoic–Cenozoic period.

2.3.3. Fluvial systems

The morphology of the channels and fluvial sediments in the humid tropics records the effects of environmental fluctuations, data on fluvial regimes as well as the evolution of the system, because many of these sediments can be dated (Thomas, 1994; Thomas and Thorp, 1995).

The Amazon River, which represents 15% of the total discharge of all the rivers in the world, has a slope gradient of 0.1 m/km near the Peruvian Andes, decreasing drastically to 0.03 m/km from Manaus, in the centre of the basin, to the mouth, 1400 km distant. These contrasts in altimetry and gradient clearly affect the morphology of the channels. Rivers that plough through the Andes, as with the Solimoes, the principal course of the Amazon River, transport coarse-grained material but mainly sediments held in suspension (Figure 21.4) and in the alluvial plains channels present low sinuosity. On the other hand, in rivers in the centre of the fluvial basin, as with the river Juruá, the sediment load transportation is almost zero and the channels have a high sinuosity. All that can be interpreted is a function of the

Figure 25.45. Meander of the auriferous river Nzi crossing the woody savanna. The river transports a great quantity of sediments held in suspension. Eastern Bouaké, Ivory Coast.

capacity of rivers to mobilize sediments in dry periods. These periods coincide with the glacial periods, in which channels present low sinuosity (Baker, 1978).

In the past, it was thought that tropical rivers did not have terraces, but the exploration for alluvial placers ended this opinion (Tricart, 1974b). In the majority of the rivers in Ivory Coast, terraces occur that consist of rounded quartz alluvium, covered by about 2 m of sticky red clays. These deposits are situated 3 to 4 m over the actual river, and there gold and a higher proportion of diamonds are exploited (Figure 25.45) (Vogt, 1959).

In the diamond exploitations of the alluvial placers in the centre of Sierra Leone, alluvial plains and channel fills have been studied by means of stratigraphic profiles and drill cores. Materials have been dated by C-14, revealing a long history over the last 30,000 years. Moreover, older ferruginated terraces without dating occur in large rivers. In the subsequent period, black clays of swamp environments can be identified in the margins of the flood plains (36,000 and 20,500 BP) and gravel channel fills in latter periods. The last two groups seem to have been formed in arid phases, while the excavation and filling are related to wet periods in the Holocene (Thomas and Thorp, 1985). In southern Ghana, a fluvial substrate consisting of weathered slates with filled polygonal cracks of 2 m depth can be identified. These cracks indicate an intense contraction due to desiccation in dry environments with almost no vegetation. Fluvial deposits with diamonds were deposited in the channels, which have been dated at many levels, making possible the identification of different dry and wet stages (Hall et al., 1985).

Two terraces at 5 to 6 m and 12 m above water level can be identified in the river Juruá, in the southwestern part of the Amazon basin. This deposit consists of sands and clays on which a brownish-red soil has developed. The bottom of the valley is also formed by sandy

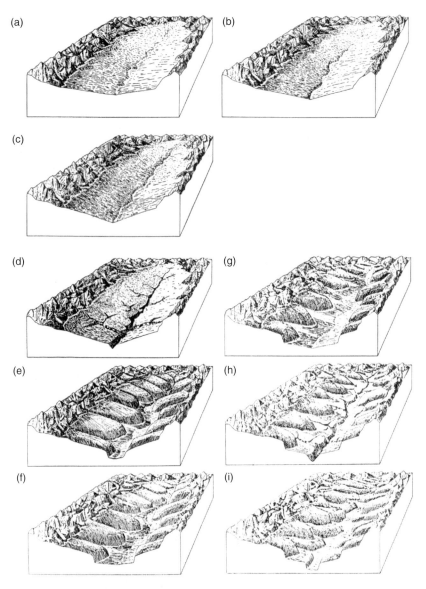

Figure 25.46. Basic scheme of slope evolution (Bigarella and Andrade, 1965). Stages depicted are: (a) extensive intermontane surface formed by pediplanation under arid climate; (b) and (c) slight erosion of the planated surface caused by a slight lowering of the local base level of erosion due to small climatic fluctuations toward humid conditions within the arid epoch. (d) Generalized dissection of the planated surface due to onset of humidity. (e) Valley-widening, alluviation, and colluviation accelerated by short arid episodes within the humid epoch. (f) Escarpment retreat and formation of a pediment surface under arid conditions. (g) Slight erosion pediment slopes during slight humid fluctuations within the arid epoch. (h) Generalized incision under a new humid epoch. (i) Widening and alluviation of valleys caused by episodic climatic fluctuations trending toward increasing aridity.

clays. As a whole, the data indicate that nowadays there is a phase of incision and, previously, two stages of alluviation occurred. This alternation seems to be related to climate changes. In the river Solimoes toward the northwest, the same sequence of accumulation and incision can be observed and there are some abandoned meanders on the terrace. Here, the important regression that occurred during the glacial period triggered the dry climate in which an intense incision developed. Therefore, the origin of these terraces seems to be eustatic and not climatic (Tricart, 1985).

The classic works of Bigarella and Andrade (1965) and Bigarella et al. (1969) deduced dry stages in the Pliocene on the basis of analyses on pediments. The study area is located in the Serra do Mar and its continuation toward southern coastal areas of Paraná and Santa

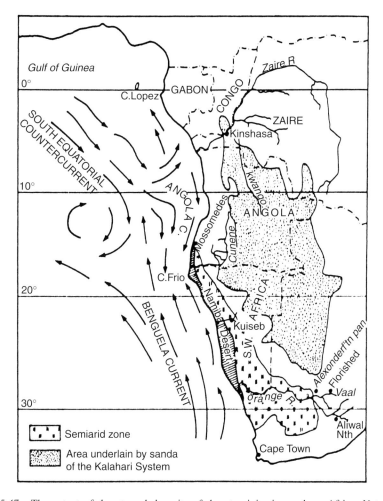

Figure 25.47. The extent of desert, and deposits of desert origin, in southern Africa. Note that the northward extent of the Namib Desert is controlled by the extent of the Benguela Current, and that the Kalahari Sands reach across the Zaire River although they are now covered by forest and savanna woodlands in the northern area (after Selby et al. (1979)).

Catarina states in Brazil. The authors pointed out that the margins of the valleys are terraced due to an alternation of dry stages, in which pediments are formed, and humid stages in which weathering and incision predominate (Figure 25.46). Dry phases corresponded to glacial periods, whereas wet ones can be correlated with interglacial stages.

2.3.4. Aeolian action

Aeolian accumulations and landforms generated by deflation in the humid tropics are excellent indicators of the occurrence of ancient arid stages. Most works on this field have been carried out in central-southern Africa and in northern and central South America.

In southwestern Africa the aridity is controlled by the cold current of Benguela, which reached its highest energy during the glacial periods, producing more aridity. It is stated also that this current produced the drift toward north in the Angola basin, as the cores from deep Atlantic Ocean confirm. Therefore, the ideas of De Ploey (1965) are reaffirmed, regarding the aeolian sands of the Kalahari in the western Congo (Figure 25.47). This extension can be probably explained by the expansion of the Benguela current (Selby et al., 1979).

The work by Tricart (1985) is a good state-of-the-art review of the information about dry climates in the centre and north of South America. In the margin of the Llanos del Orinoco (Orinoco Plains) extensive dune fields have been described (Tricart, 1974a). In Landsat images it is observed that in those areas are formed by longitudinal dunes with a NE–SW direction, parabolic dunes also occur. These aeolian accumulations come from the alluvial cover deposited by piedmont Andean rivers. Their age can be related with the last glacial period, in which a glacioeustatic lowering occurred. During the Holocene the climate was warmer and more humid. Dunes have been fossilized by alluvium and become flooded in rainy epochs. Dunes and landforms produced by deflation, which are partially fossilized by sediments corresponding to the Flandrian transgression, can be also recognized near the Orinoco Delta (Tricart and Alfonsi, 1981).

The depression of the Mato Grosso marsh of southwestern Brazil, where current precipitation is about 1200 mm, have many big alluvial fans coming from Paleozoic mountains. These landforms seem to indicate a period of dry climate, with scarce vegetation and dune formation. In the fan Taquari, the greatest, it is possible to identify deflation basins of up to 3 km diameter. At present, El Pantanal, as indicated by its name, is flooded or colonized by marsh vegetation (Tricart, 1982).

References

Aber, J.S., Croot, D.G., Fenton, M.M., 1989. Glaciotectonic Landforms and Structures. Kluwer, Dordrecht, p. 200.

Abrahams, A.D., Parsons, A.J., 1991a. Relation between infiltration and stone cover on a semiarid hillslope, southern Arizona. J. Hydrol. 122, 49–59.

Abrahams, A.D., Parsons, A.J., 1991b. Relation between sediment yield and gradient on debris-covered hillslopes, Walnut Gulch, Arizona. Bull. Geol. Soc. Am. 103, 1109–1113.

Abrahams, A.D., Parsons, A.J., (Eds), 1994. Geomorphology of Desert Environments. Chapman & Hall, London, p. 674.

Abrahams, A.D., Howard, A.D., Parsons, A.J., 1994. Rock mantled slopes. In: Abrahams, A.D., Parsons, A.J. (Eds), Geomorphology of Desert Environments. Chapman & Hall, London, pp. 173–212.

Acres, B.D., Blaire Rain, A., King, R.B., Lawton, R.M., Mitchell, A.J.B., Rackham, L.J., 1985. African dambos: their distribution, characteristics and use. Z. Geomorphol. Suppl.bd. 52, 63–86.

Adams, F.D., 1938. The Birth and Development of the Geological Sciences. Dover, New York, p. 506.

Adams, J., Maslin, M., Thomas, E., 1999. Sudden climate transitions during the quaternary. Prog. Phys. Geog. 23, 1–36.

Aghassy, J., 1973. Man-induced badlands topography. In: Coates, D.R. (Ed.), Environmental Geomorphology and Landscape Conservation. Volume III. Non-Urban. Dowden, Hutchinson and Ross, Stroudsburg, pp. 124–136.

Ahlmann, H.W., 1935. Contribution to the physics of glaciers. Geogr. J. 86, 97–113.

Ahnert, F., 1960. The influence of Pleistocene climates upon the morphology of cuesta scarps on the Colorado Plateau. Ann. Assoc. Am. Geogr. 50, 139–156.

Ahnert, F., 1970. Functional relationships between denudation relief and uplift in large, mid-latitude drainage basins. Am. J. Sci. 268, 243–263.

Ahnert, F., 1987a. Approaches to dynamic equilibrium in theoretical simulations of slope development. Earth Surf. Processes Landforms 12, 3–15.

Ahnert, F., 1987b. An approach to identification of morphoclimates. In: Gardiner, V. (Ed.), International Geomorphology 1986. Wiley, Chichester, pp. 159–188, Part II.

Ahnert, F., 1996. Introduction to Geomorphology. Arnold, London, p. 352.

Aires-Barros, L., 1975. Dry and wet laboratory tests and thermal fatigue of rocks. Eng. Geol. 9, 249–265.

Aitchison, G.D., Wood, C.C., 1965. Some interactions of compaction, permeability, and post-construction deflocculation affecting the probability of piping failure in small earth dams, Proceedings 6th International Conference on Soil Mechanics and Foundation Engineering, Montreal, Vol. 2, pp. 442–446.

Alberto, F., 1989. La desertización por salinización en el Valle del Ebro. Azara 1, 45–53.

Albritton, C.C., Brooks, J.E., Issawi, B., Swedan, A., 1990. Origin of the Qattara depression, Egypt. Bull. Geol. Soc. Am. 102, 952–960.

Al-Durrah, M., Bradford, J.M., 1982. New methods of studying soil detachment due to raindrop impact. J. Soil Sci. Soc. Am. 45, 949–953.

Aleva, G.J.J., 1985. Indonesian fluvial casiterite placers and their genetic environments. J. Geol. Soc. London 142, 815–836.

Aleva, G.J.J., 1991. Tropical weathering, denudation and mineral accumulation. Geol. Mijnbouw 70, 35–38.

Alexander, R.W., Calvo, A., 1990. The influence of lichens on slope processes in some Spanish badlands. In: Thornes, J.B. (Ed.), Vegetation and Erosion. Wiley, Chichester, pp. 385–398.

Alexandre, L.T., Cady, J.G., 1962. Genesis and hardening of laterite in soils. Tech. Bull. 1282, 90, United States Department of Agriculture, Soil Conservation Service.

Allen, J.G., 1972. Aluminium ore deposits. In: Fairbridge, R.W. (Ed.), The Encyclopedia of Geochemistry and Environmental Sciences. Van Nostrand Reinhold, New York, pp. 23–27.

Allen, J.R.L., 1965. A review of the origin and characteristics of recent alluvial sediments. Sedimentology 5, 89–191.

Allen, J.R.L., 1971. Transverse erosional marks of mud and rock: their physical basis and geological significance. Sediment. Geol. 5, 1–38.

Allison, R.J., 1997. Middle East and Arabia. In: Thomas, D.S.G. (Ed.), Arid Zone Geomorphology: Process, Form and Change in Drylands. Wiley, Chichester, pp. 507–521.

Allison, R.J., Thomas, D.G., 1993. The sensivity of landscapes. In: Thomas, D.S.G., Allison, R.J. (Eds), Landscape Sensivity. Wiley, Chichester, pp. 1–5.

Ambrose, J.W., 1964. Exhumed paleoplains of the Precambrian shield of North America. Am. J. Sci. 262, 815–857.

American Geological Institute, 1972. Glossary of Geology. American Geological Institute, Washington, p. 805.

Anadón, P., 1989. Lagos. In: Arche, A. (Ed.), Sedimentología. Consejo Superior de Investigaciones Científicas, Madrid, pp. 219–276.

Anderson, B., 1961. The Rufiji Basin, Tanganyka. 7. Soils of main irrigatable areas. Rept. Govt. Tanganyka, Prelim. Reconnaissance Surv. Rufiji Basin, 125 pp.

Anderson, R.S., 1986. Erosion profiles due to particles entrained by wind: application of an eolian sediment transport model. Bull. Geol. Soc. Am. 97, 1270–1278.

Anderson, R.S., 1987. A theoretical model for aeolian impact ripples. Sedimentology 34, 943–956.

Anderson, R.S., 1989. Saltation of sand: a qualitative review with biological analogy. In: Gimminghang, C.H., Richtie, W., Willets, B.B., Willis, A.J. (Eds), Symposium: Coastal Sand Dunes, Royal Society of Edinburgh, Proceeding, Vol. B96, pp. 149–165.

Anderson, R.S., Bunas, K.L., 1993. Grain size segregation and stratigraphy in aeolian ripples modelled with a cellular automaton. Nature 365, 740–743.

Andersson, J.G., 1906. Solifuction; a component of subaerial denudation. J. Geol. 14, 91–112.

Andersson, R.S., Haff, P.K., 1988. Simulation of aeolian saltation. Science 241, 820–823.

Andrade, E.R., Sellers, W.D., 1988. El Niño and its effect on precipitation in Arizona and western New Mexico. J. Climatol. 8, 403–410.

André, M.F., 1993. Les Versants du Spitsberg. Presses Universitaires de Nancy, Nancy, p. 361.

Andrews, J.T., 1970. A geomorphological study of post-glacial uplift with particular reference to Arctic Canada. Inst. Br. Geogr. Spec. Publ. 2, 156.

Andrews, J.T., 1975. Glacial Systems. An Approach to Glaciers and their Environments. Duxbury Press, North Scituate, MA, p. 191.

Andrews, J.T., Dugdale, R.E., 1971. Quaternary history of Northern Cumberland Peninsula, Baffin Island, N.W.T.: Part V: factors affecting corrie glacierization in Okoa Bay. Quaternary Res. 1, 532–551.

Andrews, J.T., Smithson, B.B., 1966. Till fabric of the cross-valley moraines of north-central Baffin Island. Bull. Geol. Soc. Am. 77, 271–290.

Anisimov, D.A., Nelson, F.E., 1996. Permafrost distribution in the Northern Hemisphere under scenarios of climatic change. Global Planet. Change 14, 59–72.

Antevs, E., 1952. Arroyo cutting and filling. J. Geol. 60, 375–385.

Arauzo, T., Gutiérrez, M., 1994. Evolución de los valles de fondo plano del centro de la Depresión del Ebro. In: Arnáez, J., García-Ruiz, J.M., Gómez Villar, A. (Eds), Geomorfología en España, Logroño, Vol. 1, pp. 277–290.

Arauzo, T., Gutiérrez, M., Sancho, C., 1996a. Facetas triangulares de ladera como indicadores paleoclimáticos en ambientes semiáridos (Depresión del Ebro). Geogaceta 20, 1093–1095.

Arauzo, T., Gutiérrez-Elorza, M., Sancho, C., 1996b. Retroceso de escarpes en ambientes semiáridos a partir de facetas triangulares de ladera (Depresión del Ebro). Cadernos do Laboratorio Xeolóxico de Laxe 21, 405–416, Sada (A Coruña).

Aristarain, L.F., 1970. Chemical analysis of caliche profiles from high plains, New Mexico. J. Geol. 78, 201–212.

Aristarain, L.F., 1971. On the definition of caliche deposits. Z. Geomorphol. 15, 274–289.

Arnborg, L.B., Walker, H.J., Peippo, J., 1967. Suspended load in the Colville River, Alaska, 1962. Geogr. Ann. 49, 131–144.

Artieda, O., 1993. Factores geológicos que inciden en el desarrollo de los suelos en un medio semiárido. El Caso de Quinto (Zaragoza). Tesis de Licenciatura, p. 305, Facultad de Ciencias. Zaragoza (Inédita).

Ash, J.E., Wasson, R.J., 1983. Vegetation and sand mobility in the Australian desert dunefield. Z. Geomorphol. Suppl.bd. 45, 7–25.

Ashida, K., 1985. Debris Flow. Disaster on Alluvial Fans. Kokon-shoin, Tokyo, p. 224.

Aubreville, A., 1962. Savanisation tropicale et glaciation quaternaires. Adansonia 2, 16–84.

Aufrére, M.L., 1928. L'orientation des dunes et la direction des vents. C. R. Acad. Sci. Paris 187, 833–835.

Aufrère, M.L., 1935. Essai sur les dunes du Sahara algérien. Geogr. Ann. 18, 481–550.

Augustinus, P.C., 1992. Outlet glaciar trough size-drainage area relationships. Fiordland, New Zealand. Geomorphology 4, 347–361.

Aune, Q.A., 1983. Quick clays and California's clays: no quick solution. In: Tank, R.W. (Ed.), Environmental Geology. Oxford University Press, New York, pp. 145–150.

Aylsworth, J.M., Shilts, W.W., 1989. Bedforms of the keewatin ice sheet, Canada. Sediment. Geol. 62, 407–428.

Bagnold, R.A., 1941. The Physics of Blown Sand and Desert Dunes. Methuen, London, p. 265.

Bagnold, R.A., 1953. The surface movement of blown sand in relation to meteorology. Res. Counc. Israel, Spec. Publ. 2, 89–96.

Baker, A.A., 1936. Geology of the Monument Valley–Navajo Mountain region, San Juan County, Utah. US Geol. Surv. 865, 106.

Baker, V.R., 1977. Stream channel response to floods with examples from Central Texas. Bull. Geol. Soc. Am. 88, 1057–1071.

Baker, V.R., 1978. Adjustment of fluvial system to climate and source terrain in tropical and subtropical environments. In: Miall, A.D. (Ed.), Fluvial Sedimentology, Canadian Society of Petroleum Geologists, Memoir 5, pp. 211–230.

Baker, V.R., (Ed.), 1981b. Catastrophic Flooding. The Origin of the Chanelled Scabland. Dowden, Hutchinson and Ross, Stroudsburg, p. 360.

Baker, V.R., 1993. Extraterrestrial geomorphology: science and philosophy of Earthlike planetary landscapes. Geomorphology 7, 9–35.

Baker, V.R., Bunker, R.C., 1985. Cataclysmic late Pleistocene flooding from glacial Lake Missoula: a review. Quaternary Sci. Rev. 4, 1–41.

Baker, V.R., Twidale, C.R., 1991. The reenchantment of geomorphology. Geomorphology 4, 73–100.

Baker, V.R., Kochel, R.C., Laity, J.E., Howard, A.D., 1990. Spring sapping and valley network development. In: Higgins, C.G., Coates, D.R. (Eds), Groundwater Geomorphology, Geological Society of America Special Paper, 252, pp. 235–265.

Baker, V.R., Benito, G., Rudey, A.N., 1993. Paleohydrology of late Pleistocene superflooding, Altay Mountains, Siberia. Science 259, 348–350.

Bakker, J.P., 1965. A forgotten factor in the development of glacial stairways. Z. Geomorphol. 9, 18–34.

Bakker, J.P., 1967. Weathering of granites in different climates. In: Macar, P. (Ed.), L'Evolution des Versant, Congrès et Colloque Université Liége, Vol. 40, pp. 51–68.

Bakker, J.P., Levelt, Th.W.M., 1964. An inquiry into the probability of a polyclimatic development of peneplains and pediments (etchplains) in Europe during the Senonian and Tertiary period. Publication Service Carte Géologique, Luxembourg 14, 27–75.

Balazs, D., 1973. Relief types of tropical karst areas. In: Jacucs, L. (Ed.), Symposium on Karst-Morphogenesis. Attila Joseph University, Szeged, pp. 16–32.

Ballantyne, C.K., Kirkbride, M.P., 1987. Rockfall activity in upland Britain during the Loch Lomond stadial. Geogr. J. 153, 86–92.

Balling, R.C., Wells, S.G., 1990. Historical rainfall patterns and arroyo activity within the Zuni River drainage basin, New Mexico. Ann. Assoc. Am. Geogr. 80, 603–617.

Barbary, J.P., Maire, R., Shouyuz, Z., 1991. Grottes et Karsts Tropicaux de Chine Méridionale, Vol. 4. Karstologia, Mémoire n°, Venissieux, p. 232.

Barbeau, J., 1961. Morphologie du Quaternaire des abords orientaux du Lac Tchad. Bulletin Institute Equatoriale Recherches Geologiques et Minières 14, 73–82.

Barbier, R., 1967. Nouvelles réflexions sur le problème des "pains de sucre" à propos d'observations dans le Tassili N'Ajjer (Algérie). Travaux Laboratoire de Géologie. Faculté des Sciences de Grenoble 43, 15–21.

Bardossy, G., 1981. Karst Bauxites. Akadémiai Kiadó, Budapest, p. 480.

Bardossy, J., Aleva, G.J.J., 1990. Lateritic Bauxites. Elsevier, Amsterdam, p. 624.

Barry, R.G., 1997. Paleoclimatology, climate system processes and the geomorphic record. In: Stoddart, D.R. (Ed.), Process and Form in Geomorphology. Routledge, London, pp. 187–214.

Barry, R.G., Chorley, R.J., 1987. Atmosphere, Weather and Climate. Methuen, London.

Barsch, D., 1977a. Eine abschätaung von schottproduktion und schutttransport in bereich activer blockgletscher der schiveizer Alpen. Z. Geomorphol. Suppl.bd. 28, 148–160.

Barsch, D., 1977b. Nature and importance of mass-wasting by rock glaciers in the alpine permafrost environments. Earth Surf. Processes 2, 231–245.

Barsch, D., 1988. Rockglaciers. In: Clark, M.J. (Ed.), Advances in Periglacial Geomorphology. Wiley, Chichester, pp. 69–90.

Barsch, D., 1990. Geomorphology and geoecology. Z. Geomorphol. Suppl.bd. 79, 39–49.

Barsch, D., 1996. Rock Glaciers. Springer, Berlin, p. 331.

Barsch, D., Caine, N., 1984. The nature of mountain geomorphology. Mt. Res. Dev. 4, 287–298.

Barsch, D., Hell, G., 1975. Photogrammetrische bewegungsmessungen am blockgletscher Murtèl Oberengadin, Schweizer Alpen. Zeitschrift für Gletscherkunde und Glazialgeologie 9, 111–142.

Bateman, A.M., 1950. Economic Mineral Deposits. Wiley, New York.

Baulig, H., 1957. Peneplains and pediplains. Bull. Geol. Soc. Am. 68, 913–920.

Baulin, V.V., Danilova, N.S., 1984. Dynamics of late quaternary permafrost in Siberia. In: Velichko, A.A. (Ed.), Late Quaternary Environments of the Soviet Union. Longman, London, pp. 69–78.

Baumer, M., 1990. Agroforestry and Desertification. Technical Centre for Agricultural and Rural Cooperation, p. 249, Wageningen.

Beaty, C.B., 1974. Debris flow, alluvial fans and a revitalized catastrophism. Z. Geomorphol. Suppl.bd. 21, 39–51.

Beaumont, P., 1993. Drylands. Environmental Management and Development. Routledge, London, p. 536.

Beckinsale, R.F., Chorley, R.J., 1968. Geomorphology, history of. In: Fairbridge, R.W. (Ed.), The Encyclopedia of Geomorphology. Dowden, Hutchinson and Ross, Stroudsburg, pp. 410–415.

Beckinsale, R.F., Chorley, R.J., 1991. The History of the Study of Landforms on the Development of Geomorphology. Vol. 3: Historical and Regional Geomorphology 1890–1950. Routledge, London, p. 496.

Begin, A.B., Schumm, S.A., 1984. Gradational thresholds and landforms singularity: significance for quaternary studies. Quaternary Res. 21, 267–274.

Bender, M., Sowers, T., Dickson, M.L., Orchado, J., Grooters, P., Mayewski, P.A., Meese, D.A., 1994. Climate connection between Greenland and Antarctica during the last 100,000 yr. Nature 372, 663–666.

Benedict, J.B., 1970. Frost cracking in the Colorado Front Range. Geogr. Ann. 52, 87–93.

Benito, G., 1987. Karstificación y colapsos kársticos en los yesos del sector de la Depresión del Ebro. Cuaternario y Geomorfología 1, 61–76.

Benito, G., 1997. Energy expenditure and geomorphic work of the cataclysmic Missoula flooding in the Columbia River George, USA. Earth Surf. Processes Landforms 22, 457–472.

Benito, G., Gutiérrez, M., 1988. Karst in gypsum and its environmental impact on the Middle Ebro Basin, Spain. Environ. Geol. Water Sci. 12, 107–111.

Benito, G., Pérez del Campo, P., 1991. Sinkhole evolution in alluvial deposits within the Central Ebro Basin, Northeast Spain. In: Johnson, A.I. (Ed.), Land Subsidence, International Association of Hydrological Sciences, 200, pp. 323–331.

Benito, G., Gutiérrez, M., Sancho, C., 1993. The influence of physico-chemical properties on erosion processes in badland areas, Ebro basin, NE-Spain. Z. Geomorphol. 37, 199–214.

Benito, G., Machado, M.J., Sancho, C., 1993. Sandstone weathering processes damaging prehistoric rock paintings at the Albarracín Cultural Park, NE Spain. Environ. Geol. 22, 71–79.

Benito, G., Pérez del Campo, P., Gutiérrez, M., Sancho, C., 1995. Natural and human-induced sinkholes in gypsum terrain and associated environmental problems in NE Spain. Environ. Geol. 25, 156–164.

Benn, D.I., 1995. Fabric signature of subglacial till deformation Breidamerkurjökull Iceland. Sedimentology 42, 735–747.

Bennet, M.R., Glasser, N.F., 1996. Glacial Geology: Ice Sheets and Landforms. Wiley, Chichester, p. 364.

Bennet, M.R., Hambrey, M.J., Huddart, D., Glasser, N.F., 2000. Resedimentation of debris on ice-cored lateral moraine in the high-Artic (Konsvegen, Svalbard). Geomorphology 35, 21–40.

Benson, L.V., 1978. Fluctuation in the level of pluvial Lake Lahotan during the last 40,000 years. Quaternary Res. 9, 300–318.

Benson, L.V., Paillet, F.L., 1989. The use of total lake-surface area as an indicator of climatic change: examples from the Lahotan Basin. Quaternary Res. 32, 262–275.

Berry, L., Ruxton, B.P., 1959. Notes on weathering zones and soils on granite rocks in two tropical regions. J. Soil Sci. 10, 54–73.

Beskow, G., 1930. Erdfliessen und Structurböden der Hochgebirge im Licht der Frostheburg. Geol. Fören. Stockholm, 52, 622-638.

Besler, H., 1982. The north-eastern Rub' al Khäli within the borders of the United Arab Emirates. Z. Geomorphol. NF 26, 495–505.

Bickerton, R.W., Matthews, J.A., 1992. On the accuracy of lichenometric dates: an assessment based on the "little ice age" moraine sequence of Nigardsbreen, southern Norway. Holocene 2, 227–237.

Bigarella, J.J., Andrade, G.O. de, 1965. Contribution to the study of the Brazilian quaternary. Geol. Soc. Am. Spec. Paper 84, 435–451.

Bigarella, J.J., Becker, R.D., (Eds), 1975. International symposium on the quaternary, Boletim Paranaense de Geociências, p. 33.

Bigarella, J.J., Mousinho, M.M., Da Silva, J.X., 1969. Processes environments of the Brazilian quaternary. In: Péwé, T.L. (Ed.), The Periglacial Environment. McGill-Queen's University Press, Montreal, pp. 417–487.

Bird, M.I., Chivas, A.R., 1988. Oxygen isotope dating of the Australian regolith. Nature 331, 513–516.

Birkeland, P.W., 1974. Pedology, Weathering and Geomorphology. Oxford University Press, New York, p. 285.

Birkeland, P.W., 1984. Soils and Geomorphology. Oxford University Press, New York, p. 372.

Birot, P., 1949a. Essai sur Quelques Problèmes de Morphologie Génerale, Instituto para a Alta Cultura. Centro de Estudos Geográficos, Lisboa, p. 176.

Birot, P., 1949b. Les Surfaces d'erosion du Portugal Central et Septentrional, Rapport de la Commission pour la Cartographie des surfaces d'aplanissement, Union Geographique International. Congrès International de Géographie, Lisboa, pp. 9–116.

Birot, P., 1954. Problèmes de morphologie karstique. Ann. Géogr. 63, 161–192.

Birot, P., 1955. Les Méthodes de la Morphologie. Presses Universitaires de France, Paris, p. 177.

Birot, P., 1958. Les dômes crystallines. Mem. Doc., Cent. Natl Rech. Sci. 6, 8–34.

Birot, P., 1960. Le Cycle d'Erosion sous les Différents Climats, Centro de Pesquisas de Geografía do Brasil. Faculdade Nacional de Filosofia, Rio de Janeiro, p. 137.

Birot, P., 1966. Le Relief Calcaire. Centre de Documentation Universitaire, Paris, p. 238.

Birot, P., 1968. Contribution a l'Étude de la Désagrégation des Roches. Centre de Documentation Universitaire, Paris, p. 232.

Birot, P., 1973. Géographie Physique Générale de la Zone Intertropicale. Centre de Documentation Universitaire, Paris, p. 280.

Birot, P., 1978. Evolution des conceptions sur la genèse des inselbergs. Z. Geomorphol., Suppl.bd. 31, 42–63.

Birot, P., Dresch, J., 1966. Pédiments et glacis l'Ouest des États-Unis. Ann. Géogr. 411, 513–552.

Birot, P., Solé, L., 1954a. Recherches morphologiques dans le Nord-Ouest de la Peninsule Iberique. Memoires et Documents du Centre National de la Recherche Scientifique 4, 9–61, Paris.

Birot, P., Solé, L., 1954b. Investigaciones sobre la Morfología de la Cordillera Central Española, Instituto Juan Sebastián Elcano. Consejo Superior de Investigaciones Científicas, Madrid, p. 87.

Bjornsson, H., 1979. Glaciers in Iceland. Jökull 29, 74–80.

Björnsson, H., 1992. Jökulhlaups in Iceland: prediction, characteristics and simulation. Ann. Glaciol. 18, 95–106.

Black, R.F., Barksdale, W.L., 1949. Oriented lakes of Northern Alaska. J. Geol. 57, 105–118.

Blackwelder, E., 1925. Exfoliation as a phase of rock weathering. J. Geol. 33, 793–806.

Blackwelder, E., 1928. Mudflow as a geologic agent in semiarid mountains. Bull. Geol. Soc. Am. 39, 465–484.

Blackwelder, E., 1931. Desert plains. J. Geol. 39, 133–140.

Blair, R.W., 1986. Karst landforms and lakes. In: Short, N.M., Blair, R.W. (Eds), Geomorphology from Space. A Global Overview of Regional Landforms. NASA, Washington, pp. 407–446.

Blair, T.C., 1987. Sedimentary processes, vertical stratification sequences, and geomorphology of the Roaring River alluvial fan, Rocky Mountain National Park, Colorado. J. Sediment. Petrol. 57, 1–18.

Blair, T.C., McPherson, J.G., 1994a. Alluvial fans and their natural distinction from rivers based on morphology, hydraulic processes, sedimentary processes, and facies assemblages. J. Sediment. Res. A64, 450–489.

Blair, T.C., McPherson, J.G., 1994b. Alluvial fan processes and forms. In: Abrahams, A.D., Parsons, A.J. (Eds), Geomorphology of Desert Environments. Chapman & Hall, London, pp. 354–402.

Blake, D.H., Paijmans, J.K., 1973. Landforms types of eastern Papua and their associated characteristics. Land Res. Ser. CSIRO Aust. 32, 11–21.

Bland, W., Rolls, D., 1998. Weathering: An Introduction to the Scientific Principles. Arnold, London, p. 271.

Blissenbach, E., 1952. Relation of surface angle distribution to particle size distribution on alluvial fans. J. Sediment. Petrol. 22, 25–28.

Blissenbach, E., 1954. Geology of alluvial fans in semiarid regions. Bull. Geol. Soc. Am. 65, 175–190.

Blume, H., Barth, H.K., 1972. Rampenstufen und Schuttrampen als Abtragungsformen in ariden Schichtstufenlandschaften. Erkunde 26, 108–116.

Blümel, W.D., 1981. Pedologische und geomorphologische aspekte der kalkkrustenbildung in südwesafrika und Südtospanien, Karlsruher Geographische Hefte, Vol. 10, p. 228.

Blümel, W.D., 1982. Calcretes in Namibia and SE Spain – relations to substratum, soil formation and geomorphic actors. Catena Suppl. 1, 67–95.

Boast, R., 1990. Dambos: a review. Prog. Phys. Geog. 14, 153–177.

Boer, M.M., de Groot, R.S., 1990. Landscape-Ecological Impact of Climatic Change. IOS Press, Amsterdam, p. 429.

Bögli, A., 1980. Karst Hydrology and Physical Speleology. Springer, Berlin, p. 270.

Bolt, B.A., Horn, W.L., Macdonald, G.A., Scott, R.F., 1975. Geological Hazards. Springer, Berlin, p. 328.

Bond, G., Broecker, W., Johnsen, S., McManus, J., Labeyrie, L/, Jozuel, J., Bonani, G., 1993. Correlations between climate records from North Atlantic sediments and Greenland ice. Nature 365, 143–147.

Bond, G., Showers, W., Cheseby, M., Lotti, R., Almasi, P., deMenocal, P., Priore, P., Cullen, H., Hajdas, I., Bonani, G., 1997. A pervasive millenial-scale cycle in North Atlantic Holocene and glacial climates. Science 278, 1257–1265.

Bondurant, D.C., 1951. Sedimentation studies at Conchas Reservoir in New Mexico. Trans. Am. Soc. Civ. Eng. 116, 1292–1295.

Bonython, C.W., Mason, R., 1953. The filling and drying of Lake Eyre. Geogr. J. 119, 321–330.

Boothroyd, J.C., Nummedal, D., 1978. Proglacial braided outwash: a model for humid alluvial-fan deposits. In: Miall, A.D. (Ed.), Fluvial Sedimentology. Canada Society Petroleum Geology, Calgary, pp. 641–668, Memoire 5.

Bordonau, J., 1993. The upper Pleistocene ice-lateral till complex of Cerler (Esera Valley, central Southern Pyrenees: Spain). Quaternary Int. 18, 5–14.

Borja, F., Zazo, C., Dabrio, C.J., Diaz del Olmo, F., Goy, J.L., Lario, J., 1999. Holocene eolian facies and human settlements along the Atlantic coast of southern Spain. Holocene 9, 333–339.

Bornhardt, W., 1900. Zur Oberflachengestaltung und Geologie Deutsch-Ostafrikas. Reimer, Berlin.

Boulangé, B., Millot, G., 1988. La distribution des bauxites sur le craton Ouest-Africain. Bull. Sci. Géolog. 41, 113–123, Strasbourg.

Boulton, G.S., 1970a. On the origin and transport of englacial debris in Svalbard glaciers. J. Glaciol. 9, 213–229.

Boulton, G.S., 1970b. On the deposition of subglacial and melt-out tills at the margins of certain Svalbard glaciers. J. Glaciol. 9, 231–245.

Boulton, G.S., 1971. Till genesis and fabric in Svalbarg, Spitsbergen. In: Goldthwait, R.P. (Ed.), Till: A Symposium. Ohio State University Press, Columbus, pp. 41–72.

Boulton, G.S., 1972. Modern arctic glaciers as depositional models for former ice sheets. J. Geol. Soc. London 128, 361–393.

Boulton, G.S., 1974. Processes and patterns of glacial erosion. In: Coates, D.R. (Ed.), Glacial geomorphology. State University of New York, Binghamton, pp. 41–87.

Boulton, G.S., 1975. Processes and patterns of subglacial sedimentation: a theoretical approach. In: Wright, A.E., Moseley, E. (Eds), Ice Ages: Ancient and Modern. Seel House Press, Liverpool, pp. 7–42.

Boulton, G.S., 1976. The origin of glacially fluted surfaces-observation and theory. J. Glaciol. 17, 287–309.

Boulton, G.S., 1978. Boulder shapes and grain-size distributions of debris as indicators of transport paths through a glacier and till genesis. Sedimentology 23, 15–37.

Boulton, G.S., 1982. Subglacial processes and the development of glacial bedforms. In: Davidson-Arnott, R., Nickling, W., Fahey, B.D. (Eds), Research in Glacial, Glacio-Fluvial and Glacio-Lacustrine Systems. Geobooks, Norwich, pp. 1–31.

Boulton, G.S., 1987. A theory of drumlin formation by subglacial sediment deformation. In: Menzies, J., Rose, J. (Eds), Drumlin Symposium. Balkema, Rotterdam, pp. 25–80.

Boulton, G.S., 1993. Ice ages and climatic change. In: Duff, P.McL.D. (Ed.), Holmes' Principles of Physical Geology. Chapman & Hall, London, pp. 439–469.

Boulton, G.S., Vivian, R., 1973. Underneath the glaciers. Geogr. Mag. 45, 311–319.

Bowen, N.L., 1928. The Evolution of Igneous Rocks. Dover Publications, New York, p. 332.

Bowler, J.M., 1973. Clay dunes: their occurrence, formation and environmental significance. Earth Sci. Rev. 9, 315–338.

Bowler, J.M., 1986. Spatial variability and hydrologic evolution of Australian lake basins: analogue for Pleistocene hydrologic change and evaporate formation. Palaeogeogr, Palaeocl. Palaeoecol. 54, 21–41.

Bowler, J.M., McGee, J.W., 1978. Geomorphology of the Malle region in semiarid northern Victoria and Western New South Wales. Proc. R. Soc. Victoria 90, 5–20.

Bowley, W.W., Burghardt, M.D., 1971. Thermodynamics and stones, EOS. Am. Geophys. Union Trans. 52, 4–7.

Bowler, J. M., Wasson, R. J., 1984. Glacial ages environments of inland Australia. In: Vogel, J. C. (Ed.). Late Cainozoic Palaeoclimates of the Southern Hemisphere, Balkema, Rotterdam, pp. 183–208.

Bowyer-Bower, T.A.S., Bryan, R.B., 1986. Rill initiation: concepts and evaluation on badland slopes. Z. Geomorphol., Suppl.bd. 59, 161–175.

Boyé, M., 1949. Importance du défongage périglaciaire dans l'elaboration des formes glaciaires. Compte Rendus de l'Académie des Sciences de Paris 229, 723–724.

Boyé, M., Frisch, P., 1973. Dégagement artificiel d'un dóme crystallin au Sud-Cameroun. Travaux et Documents du Géographie Tropicale, Bordeaux 8, 31–62.

Boyle, R.W., 1979. The geochemistry of gold and its deposits. Bull. Geol. Surv. Can. 280, 584.

Bradley, R.S., 1985. Quaternary Paleoclimatology: Methods of Paleoclimatic Reconstruction. Allen & Unwin, Boston, MA.

Bradley, W.C., 1963. Large-scale exfoliation in massive sandstones of the Colorado Plateau. Bull. Geol. Soc. Am. 74, 519–528.

Bradley, W.C., Hutton, J.T., Twidale, C.R., 1978. Role of salts in development of granitic tafoni, South Australia. J. Geol. 86, 647–654.

Brakenridge, G.R., Shuster, J., 1986. Late quaternary geology and geomorphology in relation to archaeological site locations, southern Arizona. J. Arid Environ. 10, 225–239.

Brammer, H., 1990. Floods in Bangladesh: I. Geographical background to the 1987 and 1988 floods. Geogr. J. 156, 12–22.

Brand, E.W., Premchitt, J., Phillipson, H.B., 1985. Relationship between rainfall and landslides in Hong Kong, IV International Symposium on Landslides, Toronto, 1984, Vol. 1, pp. 377–384.

Breed, C.S., Grow, T., 1979. In: McKee, E.D. (Ed.), Morphology and distribution of dunes in sand seas observed by remote sensing. United States Geological Survey, Washington, pp. 253–303, Professional Paper, 1052.

Breed, C.S., McCauley, J.F., Davis, P.A., 1987. Sand sheets of the eastern Sahara and ripple blankets on Mars. In: Frostick, L.E., Reid, I. (Eds), Desert Sediments, Ancient and Modern. Blackwell, Oxford, pp. 337–360.

Breed, C.S., McCauley, J.F., Whitney, M.I., 1989. Wind erosion forms. In: Thomas, D.S.G. (Ed.), Arid Zone Geomorphology. Belhaven Press, London, pp. 284–307.

Breed, C.S., McCauley, J.F., Whitney, M.I., Tchakarian, V.P., Laity, J.E., 1997. Wind erosion in drylands. In: Thomas, D.S.G. (Ed.), Arid Zone Geomorphology: Process, Form and Change in Drylands. Wiley, Chichester, pp. 437–464.

Bremer, H., 1971. Flüsse, Flächen- und Stufenbildung in den feuchten Tropen. Wuzburger Geographische Arbeiten 35, 194.

Bremer, H., 1983. Albrecht Penck (1858–1945) and Walther Penck (1888–1923), two German geomorphologists. Z. Geomorphol. 27, 129–138.

Bremer, H., 1985. Soil and slope development in the wet zone of Sri Lanka. In: Douglas, I., Spencer, T. (Eds), Environmental Change and Tropical Geomorphology. Allen & Unwin, London, pp. 295–302.

Bremer, H., 1996. Climatic, climatogenetic and tectonic geomorphology. In: Mc Cann, S.B., Ford, D.C. (Eds), Geomorphology sans Frontières. Wiley, Chichester, pp. 34–39.

Bretz, J.H., 1923. The channeled scablands of the Columbia Plateau. J. Geol. 31, 617–649.

Bretz, J.H., 1969. The Lake Missoula floods and the channeled scablands. J. Geol. 72, 505–543.

Brice, J.L., 1966. Erosion and deposition in the loess mantled great plains, medicine creek drainage basin, Nebraska. US Geol. Surv., Prof. Paper 350 H, 255–339.

Briere, P.R., 2000. Playa, playa lake, sebkha. Proposed definitions for old terms. J. Arid Environ. 45, 1–7.

Brimhall, G.H., Lewis, C.J., Ague, J.J., Dietrich, W.E., Hampel, J., Teague, T., Rix, C., 1988. Metal enrichment in bauxites by deposition of chemically nature aeolian dust. Nature 333, 819–824.

Brodzikowski, K., Van Loon, A.J., 1991. Glacigenic Sediments. Elsevier, Amsterdam, p. 674.

Broecker, W.S., 1994. Massive iceberg discharges as triggers for global climate change. Nature 372, 421–424.

Broecker, W.S., Denton, G.H., 1990. What drives glacial cycles? Sci. Am. January, 43–50.

Brooks, G.A., 1978. A new approach to the study of inselberg landscapes. Z. Geomorphol. Suppl.bd. 31, 138–160.

Brosche, K.U., 1971. Neue Beobacktungen zu vorzeitslichen periglaziaterscheinungen in Ebrobecken. Z. Geomorphol. 15, 107–114.

Brosche, K.U., 1972. Vorzeitliche Periglazialerscheinunger im Ebrobecken in der Ungebung voz Zaragoza sowie ein Beitrag zur Ausdehbung von Schuttund Bloakdecken in Zentral-und-W-Teil der Iberische Halbinsel. Göttingen Geographische Abh. 60, 293–316.

Brown, R.J.E., 1970. Permafrost in Canada; its Influence on Northern Development. University of Toronto Press, Toronto, p. 234.

Brückner, W.D., 1955. The mantle rock (laterite) of the Gold Coast and its origin. Geol. Rundsch. 43, 307–327.

Brum Ferreira, A. de, 1978. Planaltos e Montanhas do Norte da Beira. Estudo de Geomorfología, Vol. 4. Memórias do Centro de Estudos Geográficos, Lisboa, p. 374.

Brunner, F.K., Scheidegger, A.E., 1974. Kinematics of a scree slope. Rev. Ital. di Geofis. 23, 89–94.

Brunsden, D., 1990. Tablets of stone: towards the ten commandments of geomorphology. Z. Geomorphol. Suppl.bd. 79, 1–37.

Brunsden, D., Thornes, J.B., 1979. Landscape sensivity and change. Trans. Inst. Br. Geogr. 4, 463–484.

Bryan, K., 1922. Erosion and sedimentation in the Papago country, Arizona. US Geol. Surv. Bull. 730, 19–90.

Bryan, K., 1927. Pedestal rocks formed by differential erosion. US Geol. Surv. Bull. 790A, 1–15.

Bryan, K., 1941. Pre-Columbian agriculture in the southwest as conditioned by periods of alluviation. Assoc. Am. Geogr. Ann. 31, 219–242.

Bryan, K., 1946. Cryopedology-the study of frozen ground and intensive frost-action with suggestions on nomenclature. Am. J. Sci. 244, 622–642.

Bryan, R.B., 1987. Processes and significance of rill development. Catena Suppl. 8, 1–15.

Bryan, R.B., Poesen, J., 1989. Laboratory experiments on the influence of slope length on runoff, percolation and rill development. Earth Surf. Processes Landforms 14, 211–231.

Bryan, R.B., Yair, A., 1982. Perspectives on studies of badland geomorphology. In: Bryan, R.B., Yair, A. (Eds), Badland geomorphology and piping. Geobooks, Norwich, pp. 1–12.

Bryan, R.B., Yair, A., Hodges, W.K., 1978. Factors controlling the initiation of runoff and piping in Dinosaur Provincial Park badlands, Alberta, Canada. Z. Geomorphol. Suppl.bd. 29, 151–168.

Bryant, E., 1991. Natural Hazards. Cambridge University Press, Cambridge, p. 294.

Bryant, E., 1997. Climate, Process and Change. Cambridge University Press, Cambridge, p. 209.

Bryant, R.G., Sellwood, B.W., Millington, A.C., Drake, N.A., 1994. Marine-like potash evaporate formation on a continental playa: case study from Chott el Djerid, southern Tunisia. Sediment. Geol. 90, 269–291.

Buchanan, F., 1807. A Journey from Madras through the Countries of Mysore, Kahara and Malabar, Vol. 2. East Indian Company, London, see also Vol. 3, pp. 436–461; pages 66, 89, 251, 258, 378, 559.

Budd, W.F., Jenssen, D., Radok, U., 1970. The extent of basal melting in Antarctica. Polarforschung 6, 293–306.

Büdel, J., 1944. Die morphologischen wirkungen des eiszeitclimas im glitocherfreien gebiet. Geol. Rundsch. 34, 482–519.

Büdel, J., 1948. Das system der klimatischen Geomorphologie. Verhandlungen der Deutschen Geographentag 27, 69–100, trans. Derbyshire, E. (Ed.) 1973. Climatic Geomorphology. Macmillan, London, pp. 104–130.

Büdel, J., 1951. Die klimazonen des Eiszeitalters. Eiszeitalter und Gevenwart 1, 16–26, Trad. Inglesa, 1959. Int. Geol. Rev., 1(9), 72–79.

Büdel, J., 1957. Die "doppelten Einebnungsflachen" in den feuchten tropen. Z. Geomorphol. 1, 201–288.

Büdel, J., 1960. Die Frostschott zone Südorst Spitsbergen, Colloquium Geographica, Vol. 6, p. 105, Bonn.

Büdel, J., 1963. Klima-genetische geomorphologie. Geogr. Rundsch. 15, 269–285, trans. Derbyshire, E. (Ed.) 1973. Climatic Geomorphology. Macmillan, London, pp. 202–227.

Büdel, J., 1965. Die Relieftypen de Flachenspül-Zone Sud-Indiens am Ostabfall Dekans gegen Madras, Colloquium Geographicum, Vol. 8, p. 100, Bonn.

Büdel, J., 1968. Geomorphology-principles. In: Fairbridge, R.W. (Ed.), The Encyclopedia of Geomorphology. Dowden, Hutchinson and Ross, Stroudsburg, pp. 416–422.

Büdel, J., 1970. Pedimente, Rumpfflächen und Rückland-Steilhänge. Z. Geomorphol. 14, 1–57.

Büdel, J., 1977. Klima-Geomorphologie. Borntraeger, Berlín, p. 304.

Büdel, J., 1978. Das Inselberg-Rumpfflächenrelief des heutigen Tropen und das Schicksal seiner fossilen Altformen in anderen Klimazonen. Z. Geomorphol. Suppl.bd. 31, 79–110.

Büdel, J., 1980. Climatic and climatomorphic geomorphology. Z. Geomorphol. Suppl.bd. 36, 1–8.

Büdel, J, 1982. Climatic Geomorphology. Princeton University Press, Princeton, p. 443, transl. Fischer, L. and Busche, D.

Budyko, M.I., 1982. The Earth's Climate: Past and Future. Academic Press, Orlando, p. 307.

Bull, W.B., 1963. Alluvial fan deposits in Western Fresno County, California. J. Geol. 71, 243–251.

Bull, W.B., 1964a. Geomorphology of segmented alluvial fans in Western Fresno County, California. US Geol. Surv. Prof. Paper 352E, 89–125.

Bull, W.B., 1964b. History and causes of channel trenching in Western Fresno County, California. Am. J. Sci. 262, 249–258.

Bull, W.B., 1968. Alluvial fan cone. In: Fairbridge, R.W. (Ed.), Enclycopedia of Geomorphology. Dowden, Hutchinson and Ross, Stroudsburg, pp. 7–10.

Bull, W.B., 1972. Recognition of alluvial fan deposits in the stratigraphic record. In: Rigby, J.K., Hamblin, W.K. (Eds), Recognition of Ancient Sedimentary Environments, Society of Economic Palaeontologist and Mineralogists. Special Publication, 16, pp. 63–83.

Bull, W.B., 1977. The alluvial fan environments. Prog. Phys. Geog. 1, 222–270.

Bull, W.B., 1979. The threshold of critical power in stream. Bull. Geol. Soc. Am. 90, 453–464.

Bull, W.B., 1991. Geomorphic Responses to Climate Change. Oxford University Press, New York, p. 326.

Bull, W.B., 1997. Discontinous ephemeral streams. Geomorphology 19, 227–276.

Bull, W.B., McFadden, L.D., 1977. Tectonic geomorphology north and south of the Garlock fault, California. In: Doehring, D.C. (Ed.), Geomorphology in Arid Regions. Publications in Geomorphology, Binghamton, NY, pp. 115–137.

Bull, W.B., Pearthree, P.A., 1988. Frequency and size of quaternary surface ruptures of the Pilaycachi fault, northeastern Sonora, Mexico. Bull. Seismol. Soc. Am. 78, 956–978.

Burton, I., Kates, R.W., White, G.F., 1978. The Environment as Hazard. Oxford University Press, New York, p. 240.

Butzer, K.W., 1976. Geomorphology from the Earth. Harper & Row, New York, p. 463.

Butzer, K.W., 1978. Climate patterns in an unglaciated continents. Geogr. Mag. 51, 201–208.

Butzer, K. W., Stuckenrath, A. J., Bruzewicz, Helgrem, D. M., 1978. Late Cenozoic Palaeoclimates of the Face Escarpment, Kalahari margin, South Africa. Quaternary Research, 10, 310–339.

Cailleux, A., 1952. Polissage et surcreusement glaciaires dans l'hypothèse de Boyé. Rev. Géomorphol. Dyn. 5, 247–257.

Cailleux, A., 1961. Histoire de la Géologie. Presses Universitaires de France, Paris, p. 128.

Caine, N., 1972. Air photo analysis of blockfield fabrics in Talus Valley, Tasmania. J. Sediment. Petrol. 42, 33–48.

Calaforra, J.M., 1996. Contribución al Conocimiento de la Karstología de Yesos. Tesis Doctoral, p. 350, Facultad de Ciencias. Granada (Inédita).

Calvin, P., Cailleux, A., 1962. A quantitative study of cavernous weathering (taffonis) and its application to glacial chronology in Victoria Valley, Antarctica. Z. Geomorphol. 6, 317–324.

Campbell, I.A., 1974. Measurements of erosion on badlands surfaces. Z. Geomorphol. Suppl.bd. 21, 122–137.

Campbell, R.H., 1975. Soil slips, debris flows and rainstorms in the Santa Monica Mountains and vicinity, southern California. US Geol. Surv. Prof. Paper 851, 51.

Campbell, I.A., 1982. Surface morphology and rates of change during a ten-year period in the Alberta badlands. In: Bryan, R., Yair, A. (Eds), Badland Geomorphology and Piping. GeoBooks, Norwich, pp. 221–236.

Campbell, I.A., 1989. Badlands and badland gullies. In: Thomas, D.S.G. (Ed.), Arid Zone Geomorphology. Halstead Press, New York, pp. 159–193.

Campbell, I.A., Honsaker, J.L., 1982. Variability in badlands erosion; problems of scale and threshold identification. In: Thorn, C.E. (Ed.), Space and Time in Geomorphology. George Allen & Unwin, London, pp. 59–79.

Campbell, S.E., Seeler, J.S., Golubic, S., 1989. Desert crust formation and soil stabilization. Arid Soil Res. Rehab. 3, 217–228.

Capot-Rey, R., 1957. Le vent et le modelé éolien au Borkou. Travaux de l'Institut de Recherches Sahariennes 15, 155–157.

Capot-Rey, R., 1963. Contribution à l'ètude et la reprèsentation des barkhanes. Travaux de l'Institut de Recherches Sahariennes 22, 37–60.

Carol, H., 1947. The formation of roches moutonnées. J. Glaciol. 1, 57–59.

Carr, M.H., 1981. The Surface of Mars. Yale University Press, New Haven.

Carroll, D., 1970. Rock Weathering. Plenum Press, New York, p. 203.

Carson, C.E., Hussey, K.M., 1962. The oriented lakes of Arctic Alaska. J. Geol. 70, 417–439.

Carson, M.A., 1971. An application of the concept of threshold slopes to the Laramie Mountains, Wyoming. In: Brunsden, D. (Ed.), Slope Form and Process, Institute of British Geographers. Special Publication, 3, pp. 31–47.

Carson, M.A., Kirkby, M.J., 1972. Hillslope Form and Process. Cambridge University Press, Cambridge, p. 475.

Carter, R.W.G., 1988. Coastal Environments. An Introduction to the Physical, Ecological and Cultural Systems of Coastlines. Academic Press, London, p. 617.

Chaput, J.L., 1971. Aspects morphologiques du sud-est de la Sierra Morena (Espagne). Rev. Géogr. Phys. Géol. Dyn. 13, 55–66.

Charlesworth, J.K., 1957. The Quaternary Era, with Special Reference to its Glaciation. Arnold, London, p. 1700.

Chepil, W.S., Siddoway, F.H., Armbrust, D.V., 1962. Climatic factor form estimating wind erodibility of farm fields. J. Soil Water Conserv. 17, 162–165.

Chester, D., 1993. Volcanoes and Society. Arnold, London, p. 351.

Chico, R.J., 1963. Playa mud cracks: regular and king size. Geol. Soc. Am., Spec. Paper, 76.

Chorley, R.J., 1964. The nodal position and anomalous character of slope studies in geomorphological research. Geogr. J. 130, 503–506.

Chorley, R.J., 1978. Bases for theory in geomophology. In: Embleton, C., Brunsden, D., Jones, D.K.C. (Eds), Geomorphology: Present Problems and Future Prospects. Oxford University Press, Oxford, p. 113.

Chorley, R.J., Beckinsale, R.P., 1980. G.K. Gilbert's geomorphology. Geol. Soc. Am. Spec. Paper 183, 129–142.

Chorley, R.J., Beckinsale, R.P., Dunn, A.J, 1964. The History of the Study of Landforms or the Development of Geomorphology. Vol. 1: Geomorphology Before Davis. Methuen, London, p. 678.

Chorley, R.J., Beckinsale, R.P., Dunn, A.J., 1973. The History of the Study of Landforms or the Development of Geomorphology. Vol. 2: The Life and Work of William Morris Davis. Methuen, London, p. 874.

Chorley, R.J., Schumm, S.A., Sugden, D.E., 1984. Geomorphology. Methuen, London, p. 605.

Chueca, J., Julián, A., López, I., 2003. Variations of glacier coronas, Pyrenees, Spain, during the 20th century. J. Glaciol. 49 (166), 449–455.

Church, M., 1972. Baffin Island Sandurs; a Study of Arctic Fluvial Processes, Geological Survey of Canada, Bulletin, 216, p. 208.

Church, M., 1988. Floods in cold climates. In: Baker, V.R., Craig Kochel, R., Patton, P.C. (Eds), Flood Geomorphology. Wiley, New York, pp. 205–229.

Chylek, P., Box, J.E., Lesins, G., 2004. Global warming and the Greenland ice sheet. Climatic Change 63, 201–221.

Clare, K.E., 1960. Roadmaking gravel and soil in Central Africa. Rd. Res. Overseas Bull., 12.

Clark, D.A., 1974. The Kankoma elzy deposit. Economic Report No. 49. Geological Survey, Zambia.

Clark, J.A., 1980. A numerical model of worldwide sea level changes on a viscoelastic earth. In: Mörner, N.A. (Ed.), Earth Rheology, Isostasy and Eustasy. Wiley, Chichester, pp. 525–534.

Clark, J.A., Farrell, W.E., Peltier, W.R., 1978. Global changes in postglacial sea level, a numerical calculation. Quaternary Res. 9, 265–287.

Clark, M.J., 1988. Periglacial hydrology. In: Clark, M.J. (Ed.), Advances in Periglacial Geomorphology. Wiley, Chichester, pp. 415–462.

Clark, P.U., 1994. Unstable behavior of the Laurentide ice sheet over deforming sediment and its implications for climate change. Quaternary Res. 41, 19–25.

Clark, P.U., Bartlein, C.J., 1995. Correlation of the late Pleistocene glaciation in the western United States with North Atlantic Heinrich events. Geology 23, 483–486.

Clarke, G.K.C., 1991. Length, width and slope influences on glacier surging. J. Glaciol. 37, 236–246.

Clayton, L., 1964. Karst topography on stangnant glaciers. J. Glaciol. 5, 107–112.

Cloos, H., 1939. Hebung, spaltung, vulkanismus. Geol. Rundsch. 30, 3A.

Clos-Arceduc, A., 1969. Essai d'Explication des Formes Dunaires Sahariennes, Études de Photo-Interprétation. Institut Géographique National, Paris, p. 66.

Coates, D.R., (Ed.), 1974. Glacial Geomorphology. State University of New York, Binghamton, p. 398.

Coates, D.R., 1983. Large-scale land subsidence. In: Gardner, R., Scoging, H. (Eds), Mega-Geomorphology. Clarendon Press, Oxford, pp. 212–240.

Coates, D.R., 1987. Subsurface impacts. In: Gregory, K.J., Walling, D.E. (Eds), Human Activity and Environmental Processes. Wiley, Chichester, pp. 271–304.

Cole, K.L., Mayer, L., 1982. Use of packrat middens to determine rates of cliff retreat in the Eastern Grand Canyon, Arizona. Geology 10, 597–599.

Coleman, J.M., 1969. Brahmaputra river: channel processes and sedimentation. Sediment. Geol. 3, 129–239.

Colombo, F., 1979. Introducció a l'Estudi Sedimentologic dels Cons de Deyeccó (Alluvial Fans) i dels Sediments Associats, Vol. 5. Seminaris d'Estudis Universitaris, Barcelona, p. 163.

Colombo, F., 1989. Abanicos aluviales. In: Arche, A. (Ed.), Sedimentología, Vol. 1. Consejo Superior de Investigaciones Científicas, Madrid, pp. 143–218.

Coltrinari, L., Nogueira, F., 1989. Dambo-like landforms in southeastern Brazil, Abstracts of Papers, Second International Conference on Geomorphology, Frankfurt/Main, Vol. 1, p. 58, Geöko-plus.

Combes, P.J., 1969. Recherches sur la Genèse des Bauxites dans le Nord-Est de l'Espagne, le Languedoc et l'Ariège (France), Mémoires du Centre d'Études et de Recherches Géologiques et Hydrogéologiques, p. 342, Montpellier.

Common, R., 1966. Slope failure and morphogenetic regions. In: Dury, G.H. (Ed.), Essays in Geomorphology. Heineman, London, pp. 53–81.

Cooke, R.U., 1970a. Stone pavements in deserts. Ann. AAG 60, 560–577.

Cooke, R.U., 1970b. Morphometric analysis of pediments and associated landforms in the western Mojave Desert, California. Am. J. Sci. 269, 26–38.

Cooke, R.U., 1977. Applied geomorphological studies in deserts: a review of examples. In: Hails, J.R. (Ed.), Applied Geomorphology. Elsevier, Amsterdam, pp. 183–225.

Cooke, R.U., 1979. Laboratory simulation of salt weathering processes in arid environments. Earth Surf. Processes 4, 347–359.

Cooke, R.U., 1981. Salt weathering in deserts. Proc. Geol. Assoc. 92, 1–16.

Cooke, R.U., 1984. Geomorphological Hazards in Los Angeles. Allen & Unwin, London, p. 206.

Cooke, R.U., 1986. Hot drylands. In: Fookes, P.G., Vaughan, P.R. (Eds), A Handbook of Engineering Geomorphology. Surrey University Press, New York, pp. 109–124.

Cooke, R.U., Doornkamp, J.C., 1974. Geomorphology in Environmental Management. An Introduction. Clarendon Press, Oxford, p. 413.

Cooke, R.V., Doornkamp, J.C., 1990. Geomorphology in Environmental Management. An Introduccion. Clarendon Press, Oxford, p. 410.

Cooke, R.U., Mason, P., 1973. Desert knolls pediment and associated landforms in the western Mojave Desert, California. Rev. Geomorphol. Dyn. 20, 71–78.

Cooke, R.U., Reeves, R.W., 1976. Arroyos and Environmental Change in the American Southwest. Clarendon Press, Oxford, p. 213.

Cooke, R.U., Smalley, I.J., 1968. Salt weathering in deserts. Nature 220, 1226–1227.

Cooke, R.U., Warren, A., 1973. Geomorphology in Deserts. Batsford, London, p. 394.

Cooke, R.U., Brunsden, D., Doornkamp, J.C., 1982. Urban Geomorphology in Drylands. Oxford University Press, Oxford, p. 324.

Cooke, R.U., Warren, A., Goudie, A., 1993. Deserts Geomorphology. UCL Press, London, p. 526.

Coque, R., 1955. Morphologie et croût dans le sud-Tunesien. Ann. Géogr. 64, 359–370.

Coque, R., 1962. La Tunisie présaharienne, Etude Géomorphologique, Imprimerie Oberthur, p. 476.

Corbel, J., 1963. Pédiments d'Arizona. Centre de Documentation Cartographique et Géographique. Memoires et Documents 9, 33–95.

Corbel, J., 1964. L'érosion terrestre, étude quantitative (Méthodes-techniques-résultats). Ann. Géogr. 73, 385–412.

Corte, A.E., 1969. Formación en el laboratorio de estructuras como pliegues por congelamiento y descogelamiento múltiple. Cuartas Jornadas Geológicas Argentinas 1, 215–227.

Corte, A.E., 1971. Particle sorting by repeated freezing and thawing. Biuletyn Peryglacjalny 15, 175–240.

Corte, A.E., 1976. Rock glaciers. Biuletyn Peryglacjalny 26, 157–197.

Corte, A.E., Higashi, A., 1964. Experimental research on desiccation cracks in soil. U.S. Army Material Command Cold Region Research and Engineering Laboratory, Research Report, 66, p. 72.

Costa, J.E., 1984. Physical geomorphology of debris flows. In: Costa, J.E., Fleisher, P.J. (Eds), Developments and Applications of Geomorphology. Springer, Berlin, pp. 268–317.

Costa, J.E., 1988a. Floods from dam failures. In: Baker, V.R., Craig Kochel, R., Patton, P.C. (Eds), Flood Geomorphology. Wiley, New York, pp. 439–463.

Costa, J.E., 1988b. Rheologic, geomorphologic and sedimentologic differentiation of water floods, hyperconcentrated flows and debris flows. In: Baker, V.R., Cochel, R.C., Patton, P.C. (Eds), Flood Geomorphology. Wiley, New York, pp. 113–122.

Costa, J.E., Baker, V.R., 1981. Surficial Geology: Building with the Earth. Wiley, New York, p. 498.

Costard, F., Dupeyrat, L., Gautier, E., Carey-Gailhardis, E., 2003. Fluvial thermal erosion investigations along a rapidly eroding river bank: application to the Lena River (Central Siberia). Earth Surf. Processes Landforms 28, 1349–1359.

Cotton, C.A., 1942. Climatic Accidents in Landscape Making. Whitcomb & Tombs, Wellington.

Cotton, C.A., 1961. Theory of savanna planation. Geography 46, 89–96.

Cotton, C.A., 1962. Plains and inselbergs of the humid tropics. Roy. Soc. New Zeal., Trans. (Geol.) 1, 269–277.

Coudé-Gaussen, G., 1981. Les Serras da Peneda et do Gerês. Etude Géomorphologique. Memórias do Centro de Estudos Geográficos 5, 254.

Coudé-Gaussen, G., 1991. Les Poussières Sahariennes: Cycle Sédimentaire et Place dans les Environments et Paléoenvironments Désertiques. John Libby Eurotext, Montrouge, p. 485.

Coudé-Gaussen, G., Rognon, P., 1988. Charactérisation sédimentologique et conditions paléoclimatiques de la mise en place de loess and Nord du Sahara á partir de l'example du sud-Tunesien. Bulletin de la Societé Géologique de France 4, 1081–1090.

Coussot, P., Meunier, M., 1996. Recognition, classification and mechanical description of debris flow. Earth-Sci. Rev. 40, 209–227.

Cowardin, L. M., Carter, V., Golet, F. C., La Roe, E. T., 1979. Classification of wetlands and deepwater habitats of the United States: U.S. Fish and Wildlife Service FWS/OBS - 79/31.

Crandell, D.R., Mullineaux, D.R., Rubin, M., 1975. Mount St. Helens Volcano: recent and future behaviour. Science 187, 438–441.

Crittenden, M.D., 1963. New data on the isostatic deformation of Lake Bonneville. US Geol. Surv. Prof. Paper, 454-E.

Croot, D.G., 1987. Glacio-tectonic structures: a mesoscale model of thin-skinned thrust sheets? J. Struct. Geol. 9, 797–808.

Crory, E. E., 1985. Long-term foundation studies of three bridges in the Fairbanks Area. U.S. Army Cold Regions Res. And Engineering Lab. Technical Report.

Crouch, R.J., 1983. The role of tunnel erosion in gully head progression. J. Soil Conser., N.S.W. 39, 148–155.

Crouch, R.J., 1990. Rates and mechanisms of discontinuous gully erosion in a red-brown earth catchment, New South Wales, Australia. Earth Surf. Processes Landforms 15, 277–282.

Crowley, T.J., North, G.R., 1991. Paleoclimatology. Oxford University Press, New York, p. 339.

Crozier, M.J., 1986. Landslides: Causes, Consequences and Environment. Croom Helm, London, p. 252.

Currey, D.R., 1994. Hemiarid lake basins: hydrographic patterns. In: Abrahams, A.D., Parsons, A.J. (Eds), Geomorphology of Desert Environments. Chapman and Hall, London, pp. 405–421.

Curtis, C.D., 1976. Chemistry of rock weathering: fundamental reactions and controls. In: Derbyshire, E. (Ed.), Geomorphology and Climate. Wiley, New York, pp. 25–57.

Cutts, J.A., Blasius, K.R., Roberts, W.J., 1979. Evolution of Martian polar landscapes: interplay of long-term variations in perennial ice cover and dust storm intensity. J. Geophys. Res. 84, 2975–2994.

Czeppe, Z., 1964. Exfoliation in a periglacial climate. Geographia Polonica 2, 5–10.

Czudek, T., 1964. Periglacial slope development in the area of the Bohemian Massif in Northern Moravia. Biuletyn Peryglacjalny 14, 169–194.

Czudek, T., 1973. Zur Klimatischen Talasymmetrie des Westeiles der Tschechoslowakei. Z. Geomorphol. 17, 49–57.

Czudek, T., Demek, J., 1970a. Pleistocene cryopedimentation in Czechoslovakia. Acta Geographica Lodziensia 24, 101–108.

Czudek, T., Demek, J., 1970b. Thermokarst in Siberia and its influence on the development of lowland relief. Quaternary Res. 1, 103–120.

Czudek, T., Demek, J., 1971. Pleistocene cryoplanation in the Ceska vysocina highlands, Czechoslovakia. Trans. Inst. Br. Geogr. 52, 95–112.

D'Almeida, G.A., 1989. Desert aerosol: characteristics and effects of climate. In: Leinen, M., Sarnthein, M. (Eds), Paleoclimatology and Palaeometereology: Modern and Past Patterns of Global Atmospheric Transport. Kluwer Academic, Dordrecht, pp. 311–337.

D'Ercole, R., 1994. Natural hazards in the French West Indies. An overall view, Third European Intensive Course on Applied Geomorphology. Tropical Regions. Université Louis Pasteur, Strasbourg, p. 10.

Dahl, R., 1965. Plastically sculptured detail forms on rock surfaces in Northern Nordland, Norway. Geogr. Ann. 47, 83–140.

Dahl, R., 1966. Blockfields and other weathering forms in the Narvik Mountains. Geogr. Ann. 48A, 55–85.

Danin, A., 1983. Weathering of limestone in Jerusalem by cyanobacteria. Z. Geomorphol. 27, 413–421.

Dansgaard, W., Johnsen, S.J., Möller, J., Langway, C.C. Jr., 1969. One thousand centuries of climatic record from camp century on the Greenland ice sheet. Science 166, 377–381.

Dansgaard, W., Johnsen, S.J., Clausen, H.B., Dahl-Jensen, D., Gundestrup, N.S., Hammer, C.U., Hvidberg, C.S., Steffensen, J.P., Sveinbjörnsdottir, A.E., Jouzel, J., Bond, G., 1993. Evidence for general instability of past climate from a 250-ky ice-core record. Nature 364, 218–220.

Davies, J.L., 1969. Landforms of Cold Climates. The M.I.T. Press, Cambridge, p. 200.

Davis, L., 1992. Encyclopedia of Natural Disasters. Headline, London, p. 433.

Davis, W.M., 1899. The geographical cycle. Geogr. J. 14, 481–504.

Davis, W.M., 1905. The geographical cycle in arid climate. J. Geol. 13, 381–407.

Davis, W.M., 1906. The sculpture of mountains by glaciers. Scot. Geogr. Mag. 22, 76–89.

Davis, W.M., 1932. Piedmont benchlands and Primärrumpfe. Bull. Geol. Soc. Am. 43, 399–440.

Davis, W.M., 1954. Geographical Essays. Dover, New York, p. 777.

Davison, C., 1889. On the creeping on the soilcap through the action of frost. Geol. Mag. 6, 255–261.

Dawson, A.G., 1992. Ice age earth. Late Quaternary Geology and Climate. Routledge, London, p. 293.

Deacon, J., Lancaster, N., Scott, L., 1984. Evidence for Pate Quaternary climatic change in Southern Africa. Late Cainozoic palaeoclimates of the Southern Hemisphere. In: Vogel, J. C., (Ed.) Balkema, Rotterdam, pp. 21–34.

Dearman, W.R., Shibakova, V. S., 1989. The tropical environment. In: Dearman, W.R., Sergeev, E.M., Shibakoya, V.S. (Eds), Engineering Geology of the Earth. Nauka Publishers, Moscow, pp. 128–141.

Dearman, W.R., Baynes, F.J., Irfan, T.Y., 1978. Engineering grading of weathered granite. Eng. Geol. 12, 345–374.

De Chetelat, E., 1938. Le modelé latérique de l'ouest de la Guinée francaise. Rev. Géogr. Phys. Géol. Dyn. 11, 5–12.

De Dapper, M., 1978. Couvertures limono-sableuses, stone-line, indurations ferrugineuses et action des termites sur le plateau de La Manika (Kolwezi, Shaba, Zaire). Géo-Eco-Trop 2, 265–278.

De Dapper, M., 1989. Pedisediments and stone-line complexes in Peninsular Malaysia. In: Alexandre, J., Symoens, J.J. (Eds), Stone-Lines, Geo-Eco-Trop, Vol. 11. Académie Royale des Sciences D'Outre-Mer, Bruxelles, pp. 37–59.

De Gans, W., 1988. Pingo scars and their identification. In: Clark, M.J. (Ed.), Advances in Periglacial Geomorphology. Wiley, Chichester, pp. 299–322.

DeGraff, J.V., 1978. Regional landslide evaluation: two Utah examples. Environ. Geol. 2, 203–214.

Demangeot, J., 1981. Les Milieux Natural Désertique. SEDES, Paris, p. 261.

Demangeot, J., 1996. Certitudes et incertitudes du "global change". Bull. Assoc. Géogr. Fr. 4, 278–282.

Demek, J., 1964. Castle koppies and tors in the Bohemian highland (Czechoslovakia). Biuletyn Peryglacjalny 14, 195–216.

Demek, J., 1965. Slope development in granite areas of Bohemian massif (Czechoslovakia). Z. Geomorphol. Suppl.bd. 5, 82–106.

Demek, J., 1969. Cryogene processes and the development of cryoplanation terraces. Biuletyn Peryglacjalny 18, 115–125.

Dendy, F.E., Bolton, G.C., 1976. Sediment yield-runoff drainage area relationships in the United States. J. Soil Water Conserv. 31, 264–266.

Denny, C.S., 1967. Fans and pediments. Am. J. Sci. 265, 81–105.

De Ploey, J., 1964. Stone-lines and clayey–sandy mantles in Lower Congo: their formation and the effect of termites. In: Bouillon, A. (Ed.), Etudes sur les Termites Africains. Université de Louvanium, Leopoldville, pp. 399–414.

De Ploey, J., 1965. Position géomorphologique, génèse et chronologie de certains dépots superficiels au Congo occidentale. Quaternaria 7, 131–154.

De Ploey, J., 1980. Some field measurements and experimental data on wind-blown sands. In: Boodt, M. de., Gabriels, D. (Eds), Assessment of Erosion. Wiley, New York, pp. 541–552.

De Ploey, J., Gabriels, D., 1980. Measuring soil loss and experimental studies. In: Kirkby, M.J., Morgan, R.P.C. (Eds), Soil Erosion. Wiley, New York, pp. 63–108.

De Ploey, J., Poesen, J., 1989. Experimental geomorphology and the interpretation of stone-lines. In: Alexandre, J., Symoens, J.J. (Eds), Stone Lines, Vol. 11. Académie Royale des Sciences D'Outre-MerGeo, Geo-Eco-Trop, Bruxelles, pp. 75–82.

De Ploey, J., Savat, J., Moeyerons, D., 1976. The differential impact of some soil loss factors on flow, runoff, creep, and rainwash. Earth Surf. Process. 1, 151–161.

De Ploey, J., Kirkby, M.J., Ahnert, F., 1991. Hillslope erosion by rainstorms – a magnitude – frequency analysis. Earth Surf. Processes Landforms 16, 399–409.

Derbyshire, E., 1973. Introduction. In: Derbyshire, E. (Ed.), Climatic Geomorphology. Macmillan, London, pp. 11–18.

Derbyshire, E., (Ed.), 1976a. Geomorphology and Climate. Wiley, London, p. 512.

Derbyshire, E., (Ed.), 1976b. Geomorphology and climate: background. Geomorphology and Climate. Wiley, London, pp. 1–24.

Derbyshire, E., 1977. Periglacial environments. In: Hails, J.R. (Ed.), Applied Geomorphology. Elsevier, Amsterdam, pp. 227–276.

Derbyshire, E., 1983a. On the morphology, sediments and origin of the Loess Plateau of central China. In: Gardner, R.A.M., Scoging, H. (Eds), Megageomorphology. Oxford University Press, Oxford, pp. 172–194.

Derbyshire, E., 1983b. Origin and characteristics of some Chinese loess at two locations in China. In: Brookfield, M.E., Ahlbrandt, T.S. (Eds), Eolian Sediments and Processes. Elsevier, Amsterdam, pp. 69–90.

Derbyshire, E., Evans, I.S., 1976. The climatic factor in cirque variation. In: Derbyshire, E. (Ed.), Geomorphology and Climate. Wiley, New York, pp. 447–494.

Derbyshire, E., Love, M.A., 1986. Glacial environments. In: Fookes, P.G., Vaughan (Eds), A Handbook of Engineering Geomorphology. Chapman & Hall, New York, pp. 66–81.

Derruau, M., 1965. Précis de Géomorphologie. Masson, Paris.

DeWolf, Y., 1988. Stratified slope deposits. In: Clark, M.J. (Ed.), Advances in Periglacial Geomorphology. Wiley, Chichester, pp. 91–110.

Dias de Avila-Pires, F., 1974. Caracterizao zoogeográfica de Provincia Amazónica. Ann. Acad. Cienc. Brasil 46, 133–181.

Dickinson, R.E., 1986. Impact of human activities on climate - a framework. In: Clark, W.C., Munn, R.E. (Eds), Sustainable Development of the Biosphere. Cambridge University Press, Cambridge, pp. 252–289.

Dickinson, R.E., Meleshko, V., Randall, D., Sarashik, E., Silva-Dias, P., Slingo, A., 1996. Climate Processes. In: Houghton, J.T. et al. (Eds), Climate Change. The Science of Climate Change. Cambridge University Press, Cambridge, pp. 193–227.

Diester-Haas, L., 1976. Late Quaternary climatic variations in Northwest Africa deduced from east Atlantic sediment cores. Quaternary Res. 6, 299–314.

Dietrich, R.V., 1977. Impact abrasion of harder by softer materials. J. Geol. 85, 242–246.

Dionne, J.P., 2004. Pot-hole. In: Goudie, A.S. (Ed.), Encyclopedia of Geomorphology. Routledge, London, p. 807.

Dixon, J.C., 1994. Duricruts. In: Abrahams, A.D., Parson, A.J. (Eds), Geomorphology of Desert Environments. Chapman & Hall, London, pp. 82–105.

Doelling, H.H., 1985. Geology of Arches National Park, Utah Geological and Mineral Survey, Map 74.

Dohrenwend, J.C., 1994. Pediments in arid environments. In: Abrahams, A.D., Parsons, A.J. (Eds), Geomorphology of Desert Environments. Chapman & Hall, London, pp. 321–353.

Dokuchayev, V.V., 1883. The Russian Chernozem. St. Petersburg.

Donn, W.L., Farrand, W.R., Ewing, M., 1962. Pleistocene ice volumes and sea lowering. J. Geol. 70, 206–214.

Doornkamp, J.C., Ibrahim, H.A.M., 1990. Salt weathering. Prog. Phys. Geogr. 14, 335–348.

D'Orefice, M., Pecci, M., Smiraglia, C., Ventura, R., 2000. Retreat of Mediterranean Glaciers since the Little Ice Age: Case Study of Ghiacciaio del Calderone, Central Apennines, Italy. Arct. Antarct. Alp. Res. 32, 197–201.

Dorn, R.I., 1994a. Rock varnish as evidence of climatic change. In: Abrahams, A.D., Parsons, A.J. (Eds), Geomorphology of Desert Environments. Chapman & Hall, London, pp. 539–552.

Dorn, R.I., 1994b. The role of climatic change in alluvial fan development. In: Abrahams, A.D., Parsons, A.J. (Eds), Geomorphology of Desert Environments. Chapman & Hall, London, pp. 593–615.

Dorn, R.I., 1995. Alterations of ventifat surfaces at the glacier/desert interface. In: Tchakerian, V.P. (Ed.), Desert Aeolian Processes. Chapman & Hall, London, pp. 199–217.

Dorn, R.I., 1998. Rock Coatings. Elsevier, Amsterdam, p. 429.

Dorn, R.I., Oberlander, T.M., 1981. Rock varnish origin, characteristics, and usage. Z. Geomorphol. 25, 420–436.

Dorn, R.I., Oberlander, T.M., 1982. Rock varnish. Prog. Phys. Geogr. 6, 317–367.

Dorn, R.J., Tanner, D.L., Turrin, B.D., Dohrenwend, J.C., 1987. Cation-ratio dating of Quaternary materials in the east-central Mojave Desert, California. Phys. Geogr. 8, 72–81.

Douglas, I., 1967. Man, vegetation and the sediment yield of rivers. Nature 215, 925–928.

Douglas, I., 1969. The efficiency of humid tropical denudation systems. Inst. Br. Geogr. Trans. 40, 1–16.

Douglas, I., 1977. Humid Landforms. The MIT Press, Cambridge, MA, p. 288.

Douglas, I., 1978. Tropical geomorphology: present problems and future prospects. In: Embleton, C., Brunsden, D., Jones, D.K.C. (Eds), Geomorphology: Present Problems and Future Prospects. Oxford University Press, Oxford, pp. 162–184.

Douglas, I., 1986. Hot wetlands. In: Fookes, P.G., Vaughan, P.R. (Eds), A Handbook of Engineering Geomorphology. Surrey University Press, Glasgow, pp. 137–149.

Douglas, I., Spencer, T., 1982. Introduction: applied geomorphology in the tropics. Z. Geomorphol. Suppl.bd. 44, 1–3.

Douglas, I., Spencer, T., 1985a. The history of geomorphology in low latitudes. In: Douglas, I., Spencer, T. (Eds), Environmental Change and Tropical Geomorphology. Allen and Unwin, London, pp. 3–11.

Douglas, I., Spencer, T., 1985b. Present day processes as a key to the effects of environmental change. In: Douglas, I., Spencer, T. (Eds), Environmental Change and Tropical Geomorphology. Allen and Unwin, London, pp. 39–73.

Douglas, I., Spencer, T., (Eds), 1985c. Environmental Change and Tropical Geomorphology. Allen and Unwin, London, p. 378.

Dov Nir, 1983. Man, a Geomorphological Agent. An Introduction to Anthropic Geomorphology. Keter Publication, Jerusalem, p. 165.

Dowdeswell, J.A., Hamilton, G.S., Hagen, J.O., 1991. The duration of the active phase on surge-type glaciers: contrasts between Svalbard and other regions. J. Glaciol. 37, 388–400.

Dragovich, D., 1969. The origin of cavernous surfaces (tafoni) in granitic rocks of Southern Australia. Z. Geomorphol. 13, 163–181.

Dregne, H.E., 1999. Pan. In: Mares, M.A. (Ed.), Encyclopaedia of Deserts. University of Oklahoma Press, Norman, OK, pp. 499–500.

Dreimanis, A., 1953. Studies of friction cracks along shores of Cirrus Lake and Kasakokwog, Ontario. Am. J. Sci. 251, 769–783.

Dreimanis, A., 1988. Tills: their genetic terminology and classification. In: Goldthwait, R.P., Matsch, C.L. (Eds), Classification of Glacigenic Deposits. Balkema, Rotterdam, pp. 17–83.

Dreimanis, A., Vagners, J.J., 1971. Bimodal distribution of rock and mineral fragments in basal tills. In: Goldthwait, R.P. (Ed.), Till: A Symposium. Ohio State University Press, Columbus, pp. 237–250.

Dresch, J., (1938). Les surfaces de piémont dans les Djebilet et le Massif Central du Grand Atlas. Compte Rendu Congress Internationale de Géographie, Vol. 2, section 2a, p. 135. Amsterdam.

Dresch, J., 1957. Pédiments et glacis d'erosion, pediplains et inselbergs. Inf. Géogr. 21, 183–196.

Dresch, J., 1970. A propos du désert de Chihuahua. Ann. Géogr. 79, 51–57.

Dresch, J., 1982. Géographie des Régions Arides. Presses Universitaires de France, Paris, p. 277.

Drewry, D., 1986. Glacial Geologic Processes. Edward Arnold, London, p. 276.

Duchaufour, P., 1977. Pédologie. 1. Pédogènese et Classification. Masson, Paris, p. 477.

Dumas, B., 1967. Place et signification des glacis dan le Quaternaire. Bull. Assoc. Fr. Etude Quatern. 3, 223–244.

Dunne, T., 1980. Formation and control of channel networks. Prog. Phys. Geogr. 4, 211–239.

Dunne, T., Aubry, B.F., 1986. Evaluation of Horton's Theory of sheetwash and rill erosion on basis of field experiments. In: Abrahams, A.D. (Ed.), Hillslope Processes. Allen and Unwin, Boston, pp. 31–53.

Dunne, T., Dietrich, W.E., 1980. Experimental study of Horton overland flow on tropical hillslopes. 1. Soil conditions, infiltration and frequency of runoff. Z. Geomorphol. Supplement Band 35, 40–59.

Durgin, P.B., 1977. Landslides and the weathering of granite rocks. In: Coates, D.R. (Ed.), Landslide Perspectives, Reviews in Engineering Geology, Vol. III. Geological Society of America, Salt Lake City, Utah, pp. 127–131.

Dury, G.H., 1969. Rational descriptive classification of duricrusts. Earth Sci. J. 3, 77–86.

Dury, G.H., 1971. Relict deep weathering and duricrusting in relation to the palaeoenvironments of middle latitudes. Geogr. J. 137, 511–522.

Dury, G.H., 1978. The future of geomorphology. In: Embleton, C., Brunsden, D., Jones, D.K.C. (Eds), Geomorphology: Present Problems and Future Prospects. Oxford University Press, Oxford, pp. 263–274.

Dyke, A.S., Prest, V.K., 1987. Late Wisconsinan and Holocene history of the Laurentide Ice Sheet. Géogr. Phys. Quatern. 41, 237–264.

Eakin, H.M., 1916. The Yukon-Koyukuk region, Alaska. US Geol. Surv. Bull. 631, 88.

Ehlers, J., 1996. Quaternary and Glacial Geology. Wiley, Chichester, p. 578.

Eissmann, L., (1981). Periglaciäre prozesse und permafroststrukturen aus sechs kaltzeiten des quartäs. Ein beitrag zur periglazialgeologie aus des sicht des Saale-Elbe-Gebiets. Altenburger Naturwissenschaftliche Forschunge, 1, p. 171.

Elson, J.A., 1989. Comment on glacitectonite, deformation till and comminution till. In: Goldthwait, R.P., Matsch, C.L. (Eds), Genetic Classification of Glacigenic Deposits. Balkema, Rotterdam, pp. 85–88.

Elvidge, C.D., Moore, C.B., 1979. A model for desert varnish formation. Geol. Soc. Am. Abstr. Prog. 11, 271.

Ellenberger, F., (1988). Histoire de la Géologie. Tome I. Des Anciens à la Première Moitié du XVIIe Siècle. Tecnique et Documentation-Lavoisier, p. 352 Paris.

Ellison, W.D., 1944. Studies of raindrop erosion. Agric. Eng. 25, 131–136, see also 181–182.

Embleton, C., 1979. Glacial Processes. In: Embleton, C., Thornes, J. (Eds), Process in Geomorphology. Edward Arnold, London, pp. 272–306.

Embleton, C., 1989. Natural hazards and global change. ITC J. (1989), 169–178.

Embleton, C., King, C.A.M., 1975a. Glacial Geomorphology. Edward Arnold, London, p. 573.

Embleton, C., King, C.A.M., 1975b. Periglacial Geomorphology. Edward Arnold, London, p. 203.

Emery, K.O., 1960. Weathering of the Great Pyramid. J. Sediment. Petrol. 30, 140–143.

Emiliani, C., 1955. Pleistocene paleotemperatures. J. Geol. 63, 539–578.

Emmett, W.W., 1978. Overland flow. In: Kirkby, M.J. (Ed.), Hillslope Hydrology. Wiley, Chichester, pp. 145–176.

Engle, C.E., Sharp, R.P., 1958. Chemical data on desert varnish. Bull. Geol. Soc. Am. 69, 487–518.

EPICA community members, 2004. Eight glacial cycles from an Antarctic ice core. Nature 429, 623–628.

Erhardt, H., 1951. Sur l'importance des phénomènes biologiques dans la formation des cuirasses ferrugineuses en zone tropicale. Compte Rendus de la Académie des Sciences de Paris 223, 805–806.

Erhart, H., (1967). La Genèse des Sols en tant que Phénomène Géologique. Esquisse d'une Théorie Géologique et Géochimique. Biostasie et Rhexistasie. Masson, p. 177 Paris.

Eronen, M., 1983. Late Weichselian and Holowing shore displacement in Finland. In: Smith, D.E., Dawson, K.G., (Eds). Shorelines and Isostary. Academic Press, London, 183–208.

Eugster, H.P., Hardie, L.A., 1975. Sedimentation in an ancient playa-lake complex: the William Peak Member of the Gran River Formation of Wyoming. Bull. Geol. Soc. Am. 86, 319–334.

Eugster, H.P., Kelts, K., 1983. Lacustrine chemical sediments. In: Goudie, A.S., Pye, K. (Eds), Chemical Sediments and Geomorphology. Academic Press, London, pp. 321–367.

Euster, H.P., Hardie, L.A., 1978. Saline Lakes. In: Lerman, A. (Ed.), Lakes: Chemistry, Geology, Physics. Springer, New York, pp. 237–293.

Evans, I.S., 1969. The geomorphology and morphometry of glacial and nival areas. In: Chorley, R.J. (Ed.), Water, Earth and Man. Methuen, London, pp. 369–380.

Evans, D.J.A., 2004a. Surging glaciers. In: Goudie, A.S. (Ed.), Encyclopedia of Geomorphology. Routledge, London, pp. 1028–1029.

Evans, D.J.A., 2004b. Fjord. In: Goudie, A.S. (Ed.), Encyclopedia of Geomorphology. Routledge, London, pp. 374–376.

Evans, I.S., 2004c. Cirque, glacial. In: Goudie, A.S. (Ed.), Encyclopedia of Geomorphology. Routledge, London, pp. 154–158.

Evans, I.S., 1969–1970. Salt crystallization and rock weathering: a review. Rev. Géomorphol. Dyn. 70, 153–177.

Evenari, M., Shanan, L., Tadmor, N.H., 1971a. The Negev: the Challenge of a Desert. Harvard University Press, Cambridge, MA, p. 437.

Evenari, M.L., Shanan, N., Tadmor, N.H., 1971b. The Negev: the Challenge of a Desert. Harvard University Press, Cambridge, MA.

Evenson, E.B., 1971. The relationship of macro and micro-fabric of till and the genesis of glacial landforms in Jefferson County, Wisconsin. In: Goldthwait, R.P. (Ed.), Till: A Symposium. Ohio State University Press, Columbus, OH, pp. 345–364.

Everard, C.E., 1963. Contrats in the form and evolution of hill-side slopes in central Cyprus. Trans. Inst. Br. Geogr. 32, 31–47.

Eyles, N., (Ed.), 1983. Glacial Geology: An Introduction for Engineers and Earth Scientists. Pergamon Press, Oxford, p. 409.

Eyles, N., 1993. Earth's glacial record and its tectonic setting. Earth-Sci. Rev. 35, 1–248.

Fahey, B.D., 1986. A comparative laboratory study of salt crystallisation and salt hydration as potential weathering agents in deserts. Geogr. Ann. 68A, 107–111.

Fair, T.J.D., 1947. Slope form and development in the interior of Natal, South Africa. Trans. Geol. Soc. S. Afr. 50, 105–120.

Fair, T.J.D., 1948. Hill-slopes and pediments of the semi-arid Karroo. S. Afr. Geogr. J. 30, 71–79.

Fairbanks, R.G., 1989. A 17,000-year glacio-eustatic sea level record: influence of glacial melting rates on the Younger Dryas event and deep-ocean circulation. Nature 342, 637–642.

Fairbridge, R.W., 1968a. Fjord, fiord. In: Fairbridge, R.W. (Ed.), The Encyclopedia of Geomorphology. Dowden, Hutchinson and Ross, Stroudsburg, pp. 358–359.

Fairbridge, R.W., 1968b. The Encyclopedia of Geomorphology. Dowden, Hutchinson and Ross, Stroudsburg, p. 1295.

Faniran, A., Jeje, L.K., 1983. Humid Tropical Geomorphology. Longman, London, p. 414.

Farr, T.G., Adams, J.B., 1984. Rock coatings in Hawai. Bull. Geol. Soc. Am. 95, 1077–1083.

Fauque, L., Strecker, M.R., 1988. Large rock avalanche deposits. (Strurzstrome, sturzstroms) at Sierra Aconquija, northern Sierras Pampeanas, Argentina. Eclogae Geol. Helv. 81, 579–592.

Fayzullin, R.M., 1969. Lithological-facies characteristics and the gold contents of a series of conglomerates of the Kamensk deposit. Chem. Abstr. 70, 108202b.

Feininger, T., 1971. Chemical weathering and glacial erosion of crystalline rocks and the origin of till. US Geol. Surv. Prof. Paper 750, C65–C81.

Ferguson, R.I., 1973. Sinuosity of supraglacial streams. Bull. Geol. Soc. Am. 84, 251–256.

Fernández-Nieto, C., Fernández;, R., Gutiérrez Elorza, M., Arrese, F., 1981. Papel de la karstificación en la génesis de los yacimientos de Sierra Menera (Teruel y Guadalajara). Bol. Geol. Minero 92, 127–140.

Finkel, H.J., 1959. The barchans of southern Peru. J. Geol. 67, 614–647.

Fitzpatrick, E.A., 1963. Deeply weathered rock in northeast Scotland, its occurrence, age and contribution to the soils. J. Soil Sci. 14, 33–42.

Fletcher, J.E., Carroll, P.H., 1948. Some properties of soils associated with piping in southeastern Arizona. Soil Sci. Soc. Am. Proc. 13, 545–547.

Flint, R.F., 1971. Glacial and Quaternary Geology. Wiley, New York, p. 892.

Fölster, H., 1969. Slope development in SW Nigeria during the late Pleistocene and Holocene. Göttinger Bodenkundliche Berichte 10, 3–56.

Fookes, P.G., French, W.J., 1977. Soluble salt damage to surface roads in the Middle East. J. Inst. Highway Eng. 24 (12), 10–20.

Fookes, P.G., Vaughan, P.R., 1986. A Handbook of Engineering Geomorphology. Surrey University Press, Glasgow, p. 343.

Ford, D.C., Williams, D.W., 1989. Karst Geomorphology and Hydrology. Unwin and Hyman, London, p. 601.

Fosberg, F.R., Garnier, B.J., Kuchler, A.W., 1961. Delimitation of the humid tropics. Geogr. Rev. 51, 333–347.

Foster, G.R., Meyer, L.D., 1975. Mathematical Simulation of Upland Erosion Using Fundamental Erosion Mechanics, Present and Prospective Technology for Predicting Sediment Yields and Sources. US Department of Agriculture, Mississippi, US Agricultural Research Service Report ARS-S-40, pp. 190–207.

Fournier, F., 1960. Climat et Érosion: La Relation entre l'Érosion du Sol par l'Eau et les Précipitations Atmosphériques. Presses Universitaires de France, Paris, p. 201.

Francis, C., 1994. Plants on desert hillslopes. In: Abrahams, A.D., Parsons, A.J. (Eds), Geomorphology of Desert Enviroments. Chapman & Hall, London, pp. 243–254.

Francis, C.F., Thornes, J.B., 1990. Runoff hydrographs from three Mediterranean vegetation cover types. In: Thornes, J.B. (Ed.), Vegetation and Erosion. Wiley, Chichester, pp. 363–384.

Francis, P., 1993. Volcanoes: A Planetary Perspective. Clarendon Press, Oxford, p. 443.

Francou, B., 1988. Eboulis stratifiés dans les Hautes Andes Centrales du Pérou. Z. Geomorphol. 32, 47–76.

French, H.M., 1971. Slope asymmetry of the Beaufort Plain, northwest Banks Island, N.W.T, Canada. Can. J. Earth Sci. 8, 717–731.

French, H.M., 1987. Permafrost and ground ice. In: Gregory, K.J., Walling, D.E. (Eds), Human Activity and Environmental Processes. Wiley, Chichester, pp. 237–269.

French, H.M., 1996. The Periglacial Environment. Longman, Essex, p. 341.

French, H.M., Egginton, P., 1973. Thermokarst development, Banks Island, Western Canadian Arctic, Permafrost: North American Contribution, Second International Permafrost Conference, Yakutsk, USSR. National Academy of Science, Washington, DC, Publication 2115, pp. 203–212.

French, H.M., Karte, J., 1988. A periglacial overview. In: Clark, M.J. (Ed.), Advances in Periglacial Geomorphology. Wiley, Chichester, pp. 463–473.

French, H.M., Harris, S.A., Van Everdingen, R.O., 1983. The Klondike and Dawson. In: French, H.M., Heginbottom, J.A. (Eds), Guidebook to Permafrost and Related Features of the Northern Yukon Territory and Mackenzie Delta, Canada, Fourth International Conference of Permafrost. Division of Geological and Geophysical Surveys, State of Alaska, Fairbanks, pp. 35–63.

Friedman, J.D. et al., 1971. Observations on icelandic polygon and palsa areas. Photointerpretation and field studies. Geogr. Ann. 53, 115–145.

Frostick, L.E., Reid, I., Layman, J.T., 1983. Changing size distribution of suspended sediment in arid-zone flash floods. Spec. Publ. Int. Assoc. Sediment. 6, 97–106.

Frumkin, A., 1995. Morphology and development of salt caves. Bull. Natl Speleol. Soc. 56, 82–95.

Frumkin, A., Ford, D.C., 1995. Rapid entrenchment of stream profiles in the salt caves of Mount Sedom, Israel. Earth Surf. Processes Landforms 20, 139–152.

Fryberger, S.G., Ahlbrandt, T.S., 1979. Mechanism for the formation of aeolian sand seas. Z. Geomorphol. 23, 440–460.

Fryberger, S.G., Goudie, A.S., 1981. Arid geomorphology. Prog. Phys. Geogr. 5, 420–428.

Frye, J.C., Willman, H.B., 1973. Wisconsinian climatic history interpreted from Lake Michigan lobe deposits and soils. Geol. Soc. Am. Mem. 136, 135–152.

Furdada, G., 1996. Estudi de les allams al Pirineu occidental de Catalunya: predicaio espaniol I aplicanions de la cartografia. Geoforma Ediciones, Logrono, p. 315.

Furrer, G., 1968. Unterswchungen an Strukturböden in Ostspitbergen, ihre Bedentung für die Erforchung rezenter und fossiler Frostmusterformen in den Alpen baw im Alpen vorland. Polarforchung 6, 202–206.

Furrer, G., 1972. Bewegungsmessungen auf solifluktionsdecken. Z. Geomorphol. Suppl.bd. 13, 87–101.

Gallart, F., Solé, A., Puigdefábregas, J., Lázaro, R., 2002. Badlands systems in the Mediterranean. In: Bull, L.J., Kirby, M.J. (Eds), Drylands Rivers: Hydrology and Geomorphology of Semi-arid Channels. Willey, New York, pp. 299–326.

García Salmerón, J., 1967. Erosión Eólica. Ministerio de Agricultura, Madrid, p. 579.

Gardner, J.S., 1987. Evidence for headwall weathering zones, Boundary Glacier, Canadian Rocky Mountains. J. Glaciol. 33, 60–67.

Gardner, L.R., 1972. Origin of the Mormon Mesa caliche, Clark County, Nevada. Bull. Geol. Soc. Am. 83, 143–156.

Gardner, R., 1983. Introduction. In: Gardner, R., y Scoging, H. (Eds), Mega-Geomorphology. Clarendon Press, Oxford, pp. X–XIII.

Gaylord, D.R., Dawson, P.J., 1987. Airflow-terrain interactions through a mountain gap, with an example of eolian activity beneath an atmospheric hydraulic jump. Geology 15, 789–792.

Gellert, J.F., 1970. Climatomorphology and palaeoclimates of the central European Tertiary. In: Pecsi, M. (Ed.), Problems of Relief Planation. Akadémiai Kaidó, Budapest, pp. 107–112.

Geological Society, 1990. Engineering group working party report: tropical residual soils. Q. J. Eng. Geol. 23, 1–101.

Georges, C., 2004. 20th Century Glacier Fluctuations in the Tropical Cordillera Blanca, Perú. Arct. Antarct. Alp. Res. 36, 100–107.

Gerits, J., Imeson, A.C., Verstraten, J.M., Bryan, R.B., 1987. Rill development and badland regolith properties. Catena Suppl. 8, 141–160.

Gerrard, A.J., 1988. Rocks and Landforms. Unwin Hyman, London, p. 319.

Gerrard, A.J., 1990. Mountain Environments: An Examination of the Physical Geography of Mountains. Belhaven Press, London, p. 317.

Gerson, R., 1982. Talus relicts in deserts: a key to major climatic fluctuations. Israel J. Earth Sci. 31, 123–132.

Gerson, R., Grossman, S., 1987. Geomorphic activity on escarpments and associated fluvial systems in hot deserts. In: Rampino Sanders, J.E., Newman, W.S., Königsson, L.K. (Eds), Climate, History, Periodicity, Predictability. Van Nostrand Reinhold, New York, pp. 300–322.

Gerstenhauer, A., 1960. Der tropische kelgelkarst in Tabasco (Mexico). Z. Geomorphol. Suppl.bd. 2, 22–48.

Ghose, B., Kar, A., Hussain, Z., 1979. The lost courses of the Sarawasti river in the Great Indian Desert – new evidence from Landsat imagery. Geogr. J. 145, 446–451.

Giardino, J.R., 1983. Movement of ice-cemented rock glaciers by hydrostatic pressure: an example from Mt. Mestas, Colorado. Z. Geomorphol. 27, 297–310.

Giardino, J.R., Vick, S.G., 1987. Geologic engineering aspects of rock glaciers. In: Giardino, J.R., Schroeder, J.F. Jr., Vitek, J.D. (Eds), Rock Glaciers. Allen and Unwin, London, pp. 265–287.

Gibbs, R.J., 1967. The geochemistry of Amazon River System, Part I. The factors that control the salinity and composition of the suspended solids. Bull. Geol. Soc. Am. 78, 203–1232.

Gilbert, G.K., 1875. Report on The Geology of Portions of Nevada, Utah, California and Arizona, 1871–1872. Report upon Geographical and Geological Explorations and Surveys West of the One Hundredth Meridian 3, pp. 21–187.

Gilbert, G.K., 1877. Report on the geology of the Henry Mountains. United States Geographical and Geological Survey of the Rocky Mountain Region. US Department of the Interior, Washington, DC.

Gilbert, G.K., 1890. Lake Bonneville. US Geol. Surv. Monogr. 1.

Gilbert, G.K., 1904. Domes and dome structure of the High Sierra. Bull. Geol. Soc. Am. 15, 29–36.

Gilbert, G.K., 1906. Crescentic gouges on glaciated surfaces. Bull. Geol. Soc. Am. 17, 303–313.

Gile, L., Hawley, J.W., 1966. Periodic sedimentation and soil formation on an alluvial-fan piedmont in southern New Mexico. Soil Sci. Soc. Am. Proc. 30, 261–268.

Gile, L.H., 1975. Holocene soils and soil-geomorphic relations in an arid region of southern New Mexico. Quaternary Res. 5, 321–360.

Gile, L.H., Peterson, F.F., Grossman, R.B., 1966. Morphological and genetic sequences of carbonate accumulation in desert soils. Soil Sci. 101, 347–360.

Gilley, J.E., Kottwitz, E.R., Simanton, J.R., 1990. Hydraulic characteristics of rills. Trans. Am. Soc. Agr. Eng. 33, 1900–1906.

Gjessing, J., 1965. On "plastic scouring" and "subglacial erosion". Norsk Geogr. Tidskr. 20, 1–37.

Glennie, K.W., 1970. Desert Sedimentary Environments. Elsevier, Amsterdam, p. 222.

Glennie, K.W., 1972. Permian Rotliegendes of northwest Europe interpreted in the light of modern desert sedimentation studies. Bull. Am. Assoc. Petr. Geol. 56, 1048–1071.

Glennie, K.W., Pugh, J.M., Goodall, T.M., 1994. Late Quaternary Arabian desert models of Permian Rotliegend reservoirs. Shell Explor. Bull. 274, 1–19.

Goldich, S.S., 1938. A study of rock weathering. J. Geol. 46, 17–58.

Goldsmith, V., 1985. Coastal dunes. In: Davis, R.A. (Ed.), Coastal Sedimentary Environments. Springer, New York, pp. 171–236.

Goldthwait, R.P., 1951. Development of end moraines in east-central Baffin Island. J. Geol. 59, 567–577.

Gómez Ortiz, A., 1996. El Relleu d'Andorra. Morfologia glacial i periglacial. Govern d'Andorra. Ministeri d'Educació, Joventut i Esports, p. 129, Andorra.

Gómez Ortiz, A., Salvador Franch, F., 1997. Procesos geomórficos periglaciares en el Pandero del Mulhacén (Sierra Nevada). Experimentación de campo sobre la efectividad de la geliturbación y del hielo en el suelo. Cuaternario y Geomorfología 11, 81–97.

Gómez Ortiz, A., Salvador Franch, F., 1998. Procesos periglaciares actuales en montaña mediterránea. Ideas clave, trabajos de campo y resultados en Sierra Nevada. In: Gómez Ortiz, A., Salvador Franch, F., Schulte, L., García Navarro, A. (Eds), Procesos biofísicos actuales en medios fríos. Publicacions de la Universitat de Barcelona, Barcelona, pp. 217–234.

Gómez Villar, A., 1996. Abanicos aluviales: aportación teórica a sus aspectos más significativos. Cuaternario y Geomorfología 10, 77–124.

Goodess, C.M., Palulikof, J.P., Davies, T.D., 1992. The Nature and Causes of Climate Change. Assessing the Long Term Future. Belhaven Press, London, p. 248.

Goodsell, B., Hambrey, M.J., Glasser, N.F., 2002. Formation of band ogives and associated structures at Bas Glacier d'Arolla, Valais, Switzerland. J. Glaciol. 48, 207–300.

Goosens, D., Offer, Z.I., 1990. A wind-tunnel simulation and field verification of desert dust deposition. Sedimentology 37, 7–22.

Gordon, J.E., 1981. Ice-scoured topography and its relationships to bedrock structure and ice movement in parts of northern Scotland and west Greenland. Geogr. Ann. 63A, 55–65.

Gordon, J.E., Whalley, W.B., Gellatly, A.F., Vere, D.M., 1992. The formation of glacial flutes: assessment of models with evidence from Lyngsdalen, north Norway. Quaternary Sci. Rev. 11, 709–731.

Gossmann, H., 1976. L'importance des processus se déroulant à la ligne de partage locale des eaux por l'évolution des versants sous la dominance du ruissellement pluvial (á l'aide de formules mathématiques élémentaires). Actes du Symposium sur les versants en pays méditerranées, Aix en Provence, 28–30 Avril 1975. Aix en Provence: Centre d'Etudes Géographiques et de Recherches Méditerranéennes Vol. V, 139–143.

Goudie, A.S., 1972a. On the definition of calcrete deposits. Z. Geomorphol. 16, 464–468.

Goudie, A.S., 1972b. The chemistry of world calcrete deposits. J. Geol. 80, 449–463.

Goudie, A.S., 1973. Duricrust in Tropical and Subtropical Landscapes. Clarendon Press, Oxford, p. 174.

Goudie, A.S., 1974. Further experimental investigation of rock weathering by salt and other mechanical processes. Z. Geomorphol. 21, 1–12.

Goudie, A.S., 1977. Sodium sulphate weathering and the disintegration of Mohenjo-Daro, Pakistan. Earth Surf. Processes 2, 75–86.

Goudie, A.S., 1978. Dust storms and their geomorphological implications. J. Arid Environ. 1, 291–310.

Goudie, A.S., 1981a. Geomorphological Techniques. Allen and Unwin, London, p. 395.

Goudie, A.S., 1981b. The Human Impact. Man's Role in Environmental Change. Blackwell, Oxford, p. 326.

Goudie, A.S., 1983a. The arid earth. In: Gardner, R., Scoging, H. (Eds), Megageomorphology. Clarendon Press, Oxford, pp. 152–171.

Goudie, A.S., 1983b. Calcrete. In: Goudie, A.S., Pye, K. (Eds), Chemical Sediments and Geomorphology: Precipitates and Residua in the Near Surface Environment. Academic Press, New York, pp. 93–131.

Goudie, A.S., 1986. The Human Impact on the Natural Environment. Blackwell, Oxford, p. 338.

Goudie, A.S., 1988. The geomorphological role of termites and earthworms in the tropics. In: Viles, H. (Ed.), Biogeomorphology. Blackwell, Oxford, p. 365.

Goudie, A.S., 1989a. Weathering processes. In: Thomas, D.S.G. (Ed.), Arid Zone Geomorphology. Belhaven Press, London, pp. 11–24.

Goudie, A.S., 1989b. Wind erosion in deserts. Proc. Geologist. Assoc. 100, 83–92.

Goudie, A.S., 1990a. Desert degradation. In: Goudie, A.S. (Ed.), Techniques for Desert Reclamation. Wiley, Chichester, pp. 1–33.

Goudie, A.S., 1990b. The global geomorphological future. Z. Geomorphol. Suppl.bd. 79, 51–62.

Goudie, A.S., (Ed.), 1990c. Techniques for Desert Reclamation. Wiley, Chichester, p. 271.

Goudie, A.S., 1991. Pans. Prog. Phys. Geog. 15, 221–237.

Goudie, A.S., 1992. Environmental Change. Clarendon Press, Oxford, p. 329.

Goudie, A.S., 1994a. Salt attack on buildings and other structures in arid lands. In: Fookes, P.G., Parry, R.H.G. (Eds), Engineering Characteristics of Arid Soils. Balkema, Rotterdam, pp. 15–28.

Goudie, A.S., 1994b. Deserts in a warmer world. In: Millington, A.C., Pye, K. (Eds), Environmental Change in Drylands: Biogeographical and Biogeomorphological Perspectives. Wiley, Chichester, pp. 1–24.

Goudie, A.S., 1995. The Changing Earth. Rates of Geomorphological Processes. Blackwell, Oxford, p. 302.

Goudie, A.S., 1998. The salt weathering hazards in deserts. In: Kalvoda, J., Rosenfeld, C.L. (Eds), Geomorphological Hazards in High Mountain Areas. Kluwer, Dordrecht, pp. 107–120.

Goudie, A.S., 2004a. Glacial erosion. In: Goudie, A.S. (Ed.), Encyclopedia of Geomorphology. Routledge, pp. 447–448.

Goudie, A.S., 2004b. Pan. In: Goudie, A.S. (Ed.), Encyclopaedia of Geomorphology. Routledge, London, 758–759.

Goudie, A.S., Pye, K., (Eds), 1983. Chemical Sediments and Geomorphology. Academic Press, London, p. 439.

Goudie, A.S., Thomas, D.S.G., 1985. Pans in southern Africa with particular reference to South Africa and Zimbabwe. Z. Geomorphol. NF 29, 1–19.

Goudie, A.S., Viles, H., 1995. The nature and pattern of debris liberation by salt weathering: a laboratory study. Earth Surf. Processes Landforms 20, 437–449.

Goudie, A.S., Viles, H., 1997. Salt Weathering Hazards. Wiley, Chichester, p. 241.

Goudie, A.S., Wells, G.L., 1995. The nature, distribution and formation of pans in arid zones. Earth Sci. Rev. 38, 1–69.

Goudie, A.S., Wilkinson, J., 1977. The Warm Desert Environment. Cambridge University Press, Cambridge, MA, p. 88.

Goudie, A.S., Cooke, R.U., Evans, I., 1970. Experimental investigation of rock weathering by salts. Area, 42–48.

Goudie, A.S., Warren, A., Jones, D.K.C., Cooke, R.U., 1987. The character and possible origins of the aeolian sediments of the Wahiba Sand Sea, Oman. Geogr. J. 153, 231–256.

Gow, A.J., Epstein, S., Stteeby, W., 1979. On the origin of stratified debris in ice cores from the bottom of the Antarctic ice sheet. J. Glaciol. 23, 185–192.

Gracia, F.J., 1985. Geomorfología de las Bardenas Orientales, Tesis de Licenciatura. Facultad de Ciencias, Zaragoza, p. 172.

Gracia, F.J., 1995. Shoreline forms and deposits in Gallocanta Lake (NE Spain). Geomorphology 11, 323–335.

Graf, W.L., 1979. The development of montane arroyos and gullies. Earth Surf. Processes 4, 1–14.

Graf, W.L., 1980. The effect of dam closure on downstream rapids. Water Resour. Res. 16, 129–136.

Graf, W.L., 1988. Fluvial Processes in Dryland Rivers. Springer, Berlin, p. 346.

Grant, W.H., 1969. Abrasion pH, and index of weathering. Clays Clay Miner. 17, 151–155.

Gravis, G.F., 1969. Fossil slope deposits in the northern Arctic asymmetrical valleys. Biuletyn Peryglacjalny 20, 239–257.

Gray, J.M., 1981. P-forms from the Isle of Mull. Scot. J. Geol. 17, 39–47.

Gray, J.M., 1991. Glaciofluvial landforms. In: Ehlers, J., Gibbard, P.L., Rose, J. (Eds), Glacial Deposits in Britain and Ireland. Balkema, Rotterdam, pp. 443–453.

Greeley, R., Iversen, J.D., 1985. Wind as a Geological Process on Earth, Mars, Venus and Titan. Cambridge University Press, Cambridge, MA, p. 333.

Greenway, D.R., 1987. Vegetation and slope stability. In: Anderson, M.G., Richards, K.S. (Eds), Slope Stability. Wiley, Chichester, pp. 187–230.

Gregory, K.J., 1976. Drainage networks and climate. In: Derbyshire, E. (Ed.), Geomorphology and Climate. Wiley, London, pp. 289–315.

Gribbin, J., Lamb, H.H., 1978. Climatic change in historical times. In: Gribbin, J. (Ed.), Climatic Change. Cambridge University Press, Cambridge, MA, pp. 68–82.

Griggs, D.T., 1936. The factor of fatigue in rock exfoliation. J. Geol. 44, 783–796.

Grimm, E.C., Jacobson, G.L., Watts, W.A., Hansen, B.C.S., Maasch, K., 1993. A 50,000 year record of climate oscillations from Florida and its temporal correlation with the Heinrich events. Science 261, 198–200.

Grolier, M.J., McCauley, J.F., Breed, C.S., Embabi, N.S., 1980. Yardangs of the Western desert. Geogr. J. 135, 191–212.

Grossman, S., Gerson, R., 1987. Fluviatile deposits and morphology of alluvial surfaces as indicators of Quaternary environmental changes in the southern Negev, Israel. In: Frostick, L., Reid, Y. (Eds), Desert Sediments: Ancient and Modern, Geological Society of London Special Publication, pp. 17–29.

Grosso, S.A., Corte, A.E., 1989. Pleistocene ice-wedge coasts at 34°S, eastern Andes piedmont, south-west of South America. Geogr. Ann. 71A, 125–136.

Grove, A.T., 1958. The ancient ergs of Hausaland and similar formations of the south side of the Sahara. Geogr. J. 124, 528–533.

Grove, A.T., 1977. The geography of semi-arid lands. Philos. Trans. R. Soc. London Ser. B 278, 457–475.

Grove, A.T., 1985. The physical evolution of the river basins. In: Grove, A.T. (Ed.). The Niger and its neighbours. Balkema, Rotterdam, pp. 2–60.

Grove, A.T., Warren, A., 1968. Quaternary landforms and climate on the south side of the Sahara. Geogr. J. 134, 194–208.

Grove, J.M., 1979. The glacial history of the Holocene. Prog. Phys. Geog. 3, 1–54.

Grove, J.M., 1988. The Little Ice Age. Routledge, London, p. 498.

Grund, A., 1914. Der geographische zyclus im karst. Z. Gesselchaft Erdkund 52, 621–740.

Grünert, J., Busche, D., 1980. Large-scale fossil landslides at the Msak Mallat and Hamadat Manghini escarpement. In: Salem, M.J., Busrewil, M.T. (Eds), Geology of Libya. Academic Press, London, pp. 849–860.

Guillén, J., Diaz, J.I., Palanques, A., 1992. Cuantificación y evolución durante el Siglo XX de los aportes de sedimentos transportados como carga de fondo por el río Ebro al medio marino. Revista de la Sociedad Geológica de España 5, 27–37.

Guillien, Y., 1951. Les grèzes litèes de Charente. Revue Geographique de Pyrénées et de Sud-Ouest 22, 154–162.

Guillien, Y., Lautridou, J.P., 1970. Recherches de gélifraccion experimentale du Centre de Géomorphologic. I. Calcaires de Charentes. Bull. Cent. Géomorphol. Caen 5, 45, C.N.R.S.

Günster, A., Hadley, N.F., Mitchell, D., Nobel, P.S., Seely, M., 1993. Form and function. In: Seely, M. (Ed.), Deserts. Weldon Ower Limited, Australia, pp. 66–81.

Gupta, A., 1984. Urban hydrology and sedimentation in the humid tropics. In: Costa, J.E., Fleisher, P.J. (Eds), Developments and Applications of Geomorphology. Springer, Berlín, pp. 240–267.

Gupta, A., 1988. Large floods as geomorphic events in the humid tropics. In: Baker, V.R., Craig Kockel, R., Patton, P.C. (Eds), Flood Geomorphology. Wiley, New York, pp. 301–315.

Gupta, A., 1993. The changing geomorphology of the humid tropics. Geomorphology 7, 165–186.

Gupta, A., Dutt, A., 1989. The Auranga: description of a tropical monsoon river. Z. Geomorphol. 33, 73–92.

Gupta, A., Rahman, A., Poh Poh, W., Pitts, J., 1987. The old alluvium of Singapore and the extinct drainage system of the South China Sea. Earth Surf. Processes Landforms 12, 259–275.

Gutiérrez, F., 1995. Synsedimentary and postsedimentary subsidence due to gypsum karstification and derived hazards (Calatayud Graben, Iberian Range, Spain). In: Barends, F.B.J., Brouwer, F.J.J., Schröder, F.H. (Eds), Land Subsidence, Natural Causes Measuring Techniques and Groningen Gasfields. Balkema, The Hague, pp. 47–57.

Gutiérrez, F., 1996. Gypsum karstification induced subsidence: effects on alluvial systems and derived geohazards, Calatayud Graben, Iberian Range, Spain. Geomorphology 16, 277–293.

Gutiérrez, F., 1998. Fenómenos de subsidencia por disolución de formaciones evaporíticas en las fosas neógenas de Teruel y Calatayud (Cordillera Ibérica). Tesis Doctoral, Facultad de Ciencias de Zaragoza.

Gutiérrez, F., Gutiérrez, M., Sancho, C., 1998. Geomorphological and sedimentological analysis of a catastrophic flash flood in the Arás drainage basin (Central Pyrenees, Spain). Geomorphology 22, 265–283.

Gutiérrez, M., 1986. Some remarks on the problem of climatic changes and geomorphological processes in arid zones. In: López-Vera, F. (Ed.), Quaternary Climate in Western Mediterranean, pp. 3–29, Madrid.

Gutiérrez, M., 1990. Historia de la Geomorfología, Historia de la Geología. Real Academia de Ciencias Exactas, Físicas y Naturales, Madrid, pp. 115–131.

Gutiérrez, M., 2001. Geomorfologia Climatica Omega, Barcelona, p. 642.

Gutiérrez, M., Ibáñez, M.J., 1979. Las "gnammas" de la región de Alcañiz. Estudios Geológicos 35, 193–198.

Gutiérrez, M., Gracia, F.J., 1997. Environmental interpretation and evolution of the Tertiary erosion surfaces in the Iberian Range (Spain). In: Widdowson, M. (Ed.), Paleosurfaces: Recognition, Reconstruction and Palaeoenvironmental Interpretation, Vol. 120. The Geological Society Special Publication, London, pp. 147–158.

Gutiérrez, M., Gutiérrez, F., 1998. Geomorphology of the Tertiary gypsum formations in the Ebro Depression. Geoderma, 1–29.

Gutiérrez, M., Martínez, V.H., 2001. Multiple talus flatirons, variations of scarp retreat rates and the evolution of the slopes in Almazán Basin (semi-arid central Spain). Geomorphology 38, 19–29.

Gutiérrez, M., Peña, J.L., 1976. Glacis y terrazas en el curso medio del río Alfambra. Boletín Geológico y Minero 87, 561–570, Madrid.

Gutiérrez, M., Peña, J.L., 1977. Las acumulaciones periglaciares del Macizo del Tremedal (Sierra de Albarracín). Boletín Geológico y Minero 92, 101–110.

Gutiérrez, M., Peña, J.L., 1981. Los glaciares rocosos y el modelado acompañante en el área de la Bonaigua (Pirineo de Lérida). Boletín Geológico y Minero 92, 101–110.

Gutiérrez, M., Peña, J.L., 1989. Upper Holocene climatic change and geomorphological processes on slopes and infilled valleys from archaeological dating (NE-Spain), European Conference on Landscape Ecological Impact of Climate Change Lunteren, Netherlands, 3-7 December 1989. Dutch Ministry of the Environment, Lunteren, p. 21.

Gutiérrez, M., Peña, J.L., 1992. Evolución climática y geomorfológica de Holoceno Superior (Cordillera Ibérica, Depresión del Ebro y Pre-Pirineo). In: Cearreta, A., Ugarte, F.M. (Eds), The Late Quaternary in the Western Pyrenean Region. Servicio Editorial de la Universidad del País Vasco, Bilbao, pp. 109–124.

Gutiérrez, M, Peña, J.L., 1994. Depresión del Ebro. In: Gutiérrez, M. (Coord.), Geomorfología de España. Rueda, Madrid, pp. 305–349.

Gutiérrez, M., Peña, J.L., 1998. Geomorphology and Upper Holocene climatic change in northeastern Spain. Geomorphology 23, 205–217.

Gutiérrez, M., Rodríguez, J., 1978. Consideraciones sobre la morfogénesis del Sistema Central. Boletín Geológico y Minero 89, 109–113.

Gutiérrez, M., Sancho, C., 1993. Applied geomorphology in arid and semiarid regions, Second Intensive Course of Applied Geomorphology: Arid Regions. Universidad de Zaragoza-Erasmus, Zaragoza, pp. 3–31.

Gutiérrez, M., Ibáñez, M.J., Peña, J.L., Rodríguez, J., Soriano, M.A., 1985. Quelques exemples de karst sur gypse dans la Dépression de l'Ebre. Karstología 6, 29–36.

Gutiérrez, M., Benito, G., Rodríguez, J., 1988. Piping in badlands areas of the middle Ebro basin, Spain. Catena Suppl. 13, 49–60.

Gutiérrez, M., Sancho, C., Arauzo, T., 1998a. Scarp retreat in semiarid environments from talus flatirons (Ebro Basin). Geomorphology 25, 111–121.

Gutiérrez, M., Sancho, C., Arauzo, T., Peña, J.L., 1998b. Evolution and paleoclimatic meaning of the talus flatirons in the Ebro Basin (NE of Spain). In: Alsharhan, A.S., Glennie, K.W., Whittle, G.L. (Eds), Quaternary Deserts and Climatic Change. Balkema, Rotterdam, pp. 593–599.

Gutiérrez, M., Sancho, C., Desir, G., Sirvent, J., Benito, G., Calvo, A., 1995. Erosión hídrica en terrenos arcillosos y yesíferos de la Depresión del Ebro, Proyecto Lucdeme. Ministerio de Agricultura, Pesca y Alimentación, Zaragoza, p. 389.

Gutiérrez, M., Sancho, C., Benito, G., Sirvent, J., Desir, G., 1997. Quantitative study of piping processes in badland areas of the Ebro Basin. Geomorphology 20, 237–253.

Gutiérrez, M., Desir, G., Gutiérrez, F., 2002. Yardangs in the semiarid central sector of the Ebro Depression (NE Spain). Geomorphology 44, 155–170.

Gutiérrez, M., Gutiérrez, F., Desir, G., 2005. Considerations on the chronological and causal relationships between tapus flatiron and palaeoclimatic changes in central and northeastern Spain. Geomorphology (accepted).

Gutiérrez, M., Desir G., Gutiérrez, F., Marin C., 2005. Origin and evolution of playas and blowouts in the semi arid zone of Tierra de Pinares (Duero Basin, Spain). Geomorphology (accepted).

Hack, J.T., 1960. Interpretation of erosional topography in humid temperate regions. Am. J. Sci. 258A, 80–97.

Hack, J.T., 1975. Dynamic equilibrium and landscape evolution. In: Melhorn, W.C., Flemal, R.C. (Eds), Theories of Landform Development. George Allen and Unwin, London, pp. 87–102.

Hackman, R.J., 1965. Interpretation of Alaskan post-earthquake photographs. Photogramm., 31, 604–620.

Hadley, R.F., Schumm, S.A., 1961. Sediment sources and drainage basin characteristics in upper Cheyenne River basin. US Geol. Surv. Water Supply Paper 1531, 137–198.

Hagedorn, H., 1968. Über äolische Abtragung und Formung in der Südest-Sahara. Erdekunde 22, 257–269.

Haigh, M.J., 1978. Microrills and dessication cracks: some observations. Z. Geomorphol. 12, 457–461.

Halimov, M., Fezer, F., 1989. Eight yardang types in central Asia. Z. Geomorphol. NF 33, 205–217.

Hall, A.M., 1986. Deep weathering patterns in north-east Scotland and their geomorphological significance. Z. Geomorphol. 30, 407–422.

Hall, A.M., 1988. The characteristics and significance of deep weathering in the Gaick area, Grampian Highlands, Scotland. Geogr. Ann. 70A, 309–314.

Hall, A.M., Thomas, M.F., Thorp, M.B., 1985. Late Quaternary alluvial placer development in the humid tropics: the case of the Birim Diamond Placer, Ghana. J. Geol. Soc. 142, 777–787.

Hall, A.M., Mellor, A.M., Wilson, M.J., 1989. The clay mineralogy and age of deeply weathered rock in north-east Scotland. Z. Geomorphol. Suppl.bd. 72, 97–108.

Hall, K., Thorn, C.E., Matsuoka, N., Prick, A., 2002. Weathering in cold regions: some thoughts and perspectives. Prog. Phys. Geog. 26, 577–603.

Hallet, B., 1979. Subglacial regelation water film. J. Glaciol. 23, 321–334.

Hallsworth, E.G., Robertson, G.K., Gibbons, F.R., 1955. Studies of pedogenesis in New South Wales, VII: the "gilgai" soils. J. Soil Sci. 6, 1–31.

Hambrey, M.J., 1977. Foliation, minor folds and strain in glacier ice. Tectonophysics 39, 397–416.

Hambrey, M., 1994. Glacial Environments. UCL Press, London, p. 296.

Hambrey, M., Alean, J.C., 1992. Glaciers. Cambridge University Press, Cambridge, MA, p. 207.

Hamilton, A., 1976. The significance of patterns of distribution shown by forest plants and animals in tropical Africa for the reconstruction of Upper Pleistocene palaeoenvironments: a review. In: Van Zinderen Baker, E.M. (Ed.), Palaeoecology of Africa, Vol. 9. Balkema, Rotterdam, pp. 63–97.

Hansen, W.R., 1965. Effects of the earthquake of March 27, 1964, at Anchorage, Alaska. US Geol. Surv. Prof. Paper 542-A, 64.

Hardie, L.A., Smoot, J.P., Eugster, H.P., 1978. Saline lakes and their deposits: a sedimentological approach. In: Matter, A., Tucker, M. (Eds), Modern and Ancient Lake Sediments, International Association of Sedimentologists, Special Publication, Vol. 2, pp. 7–41.

Harding, A.F., 1982. Introduction: climate change and archaeology. In: Harding, A.F. (Ed.), Climatic change in Later Prehistory. Edinburgh University Press, Edinburgh, pp. 1–10.

Harris, C., Davies, M.C.R., 2000. Gelifluction: observations from large-scale laboratory simulations. Arct. Antarct. Alp. Res. 32, 202–207.

Harris, C., Davies, M.C.R., Rea, B.R., 2003. Gelifluction: viscous flow or plastic creep? Earth Surf. Processes Landforms 28, 1289–1301.

Harris, S.A., 1968. Gilgai. In: Fairbridge, R.W. (Ed.), The Encyclopedia of Geomorphology. Dowden, Hutchinson and Ross, Stroudsburg, PA, pp. 425–426.

Harris, S.A., 1981. Climatic Relationships of Permafrost Zones in areas of low winter snow-cover. Biuletyn Periyglacjalny, 28, 227–240.

Harris, S.A., 1986. The Permafrost Environment. Croom Helm, London, p. 276.

Harris, S.A., 1988. The alpine periglacial zone. In: Clark, M.J. (Ed.), Advances in Periglacial Geomorphology. Wiley, Chichester, pp. 369–413.

Harris, S.A., 1998. A genetic classification of the palsa-like mounds in Western Canada. Biuletyn Peryglacjalny 37, 115–130.

Harris, S.E., 1943. Friction cracks and the direction of glacial movement. J. Geol. 51, 244–258.

Harry, D.G., 1988. Ground ice and permafrost. In: Clark, M.J. (Ed.), Advances in Periglacial Geomorphology. Wiley, Chichester, pp. 113–149.

Hart, M.G., 1986. Geomorphology Pure and Applied. Allen and Unwin, London, p. 228.

Harvey, A.M., 1978. Dissected alluvial fans in southeast Spain. Catena 5, 177–211.

Harvey, A.M., 1984a. Debris flows and fluvial deposits in Spanish Quaternary alluvial fans: implications for fan geomorphology. In: Koster, E.H., Steel, R.J. (Eds), Sedimentology of Gravels and Conglomerates, Canadian Society Petroleum Geologist, Memoir, Vol. 10, pp. 123–132.

Harvey, A.M., 1984b. Aggradation and dissection sequences on spanish alluvial fans: influence on morphological development. Catena 11, 289–304.

Harvey, A.M., 1987a. Alluvial fans dissection: relationship between morphology and sedimentation. In: Frostik, L., Reid, I. (Eds), Desert Sediments, Ancient and Modern, Geological Society of London, 35, pp. 87–103.

Harvey, A.M., 1987b. Patterns of Quaternary aggradational and dissectional landform development in the Almeria region, southeast Spain: a dry region, tectonically active landscape. Die Erde 118, 193–215.

Harvey, A.M., 1990. Factors influencing Quaternary alluvial fans development in southeast Spain. In: Rachocki, A., Church, M. (Eds), Alluvial fans. A Field Approach. Wiley, Chichester, pp. 247–269.

Harvey, A.M., 1996. The role of alluvial fans in the mountain fluvial systems of southeast Spain: implications of climatic change. Earth Surf. Processes Landforms 21, 543–553.

Harvey, A.M., 1997. The role of alluvial fans in arid zone fluvial systems. In: Thomas, D.G.S. (Ed.), Arid Zone Geomorphology. Process, Form and Change in Drylands. Wiley, Chichester, pp. 231–259.

Haugen, R.K., Brown, J., 1971. Natural and man-induced disturbances of permafrost terrane. In: Coates, D.R. (Ed.), Environmental Geomorphology. Geom. State University, Binghamton, Nueva York, pp. 139–149.

Hay, R.L., Wiggins, B., 1980. Pellets, ooids, sepiolite and silica in three calcretes of the southwest United States. Sedimentology 27, 559–576.

Hayden, B.P., 1988. Flood climates. In: Baker, V.R., Craig Kockel, R., Patton, P.C. (Eds), Flood Geomorphology. Wiley, New York, pp. 13–26.

Hays, J., 1967. Land surfaces and laterites of the Northern Territory. In: Jennings, J.N., Mabbutt, J.A. (Eds), Landform Studies from Australasia and New Guinea. Cambridge University Press, Cambridge, MA, pp. 182–210.

Heathcote, R.L., 1983. The Arid Lands: Their Use and Abuse. Longman, London.

Heede, B.H., 1971. Characteristics and processes of soil piping in gullies. United States Department of Agriculture Forest Service, Rocky Mountain Forest and Range Experiment Station Research Paper RM-68, p. 15.

Heginbottom, J.A., 2002. Permafrost mapping: a review. Prog. Phys. Geog. 26, 623–642.

Heine, K., Walter, R., 1996. Gypcretes of the central Namib Desert, Namibia. In: Heine, K. (Ed.), Palaecology of Africa and the Surroundings Islands, Vol. 24. Balkema, Rotterdam, pp. 173–201.

Heinrich, H., 1988. Origin and consequences of cyclic ice rafting in the Northeast Atlantic ocean during the past 130,000 years. Quaternary Res. 29, 142–152.

Henderson, E.P., 1956. Large nivation hollows near Knob Lake, Quebec. J. Geol. 64, 607–616.

Henderson-Sellers, A., 1992. Continental cloudiness changes this century. GeoJournal 27, 255–262.

Hereford, R., 1984. Climate and ephemeral-stream processes: twentieth-century geomorphology in alluvial stratigraphy of the Little Colorado River, Arizona. Bull. Geol. Soc. Am. 95, 654–668.

Herrero, J., Aragüés, R., 1988. Suelos afectados por salinidad en Aragón. Surcos de Aragón 9, 5–10.

Herrero, J., Aragüés, R., Amezketa, E., 1993. Salt-affected soils and agriculture in the Ebro Basin. In: Gutiérrez, M., Sancho, C., Benito, G. (Eds), Second Intensive Course of Applied Geomorphology: Arid Regions. Universidad de Zaragoza-Erasmus, Zaragoza, pp. 139–150.

Herron, S., Langway, C.C., 1979. The debris-laden ice at the bottom of the Greenland ice sheet. J. Glaciol. 23, 193–207.

Hestners, E., 1985. A contribution to the prediction of slush avalanches. Ann. Glaciol. 6, 1–4.

Hewitt, K.J., 1967. Ice-front deposition and the seasonal affect, a Himalayan example. Trans. Inst. Br. Geogr. 42, 93–106.

Hickson, C.J., Peterson, D.W., (Eds), 1990. Special Symposium Commemorating the 10th Anniversary of the Eruption of Mount St Helens, May 18, 1980, Geosci. Can., 17, pp. 125–187.

Higgins, C.G., 1956. Formation of small ventifacts. J. Geol. 64, 506–516.

Higgins, C.G., 1975. Theories of landscape development: a perspective. In: Melhorn, W.N., Flemal, R.C. (Eds), Theories of Landform Development. Allen and Unwin, Boston, pp. 1–28.

Higgins, C.G., 1984. Piping and sapping: development of landforms by groundwater flow. In: LaFleur, R.G. (Ed.), Groundwater as a Geomorphic Agent. Allen and Unwin, Boston, pp. 18–58.

Higgins, C.G., Osterkamp, W.R., 1990. Seepage-induced cliff recession and regional denudation. In: Higgins, C.G., Coates, D.R. (Eds), Groundwater Geomorphology, Geological Society of America Special Paper, Vol. 252, pp. 291–317.

Hills, E.S., 1940. The lunette, a new landform of aeolian origin. Aust. Geogr. 3, 15–21.

Hill, I.D., Rackham, I.J., 1978. Indications of mass movement on the Jos Plateau, Nigeria. Z. Geomorphol. 22, 258–274.

Hodges, W.K., 1982. Hydrologic characteristics of a badland pseudopediment slope system during simulated rainstorm experiments. In: Bryan, R.B., Yair, A. (Eds), Badland Geomorphology and Piping. Geobooks, Norwich, pp. 127–152.

Hodges, W.K., Bryan, R., 1982. The influence of material behaviour on runoff initiation in the Dinosaur Badlands, Canada. In: Bryan, R., Yair, A. (Eds), Badland Geomorphology and Piping. GeoBooks, Norwich, pp. 13–46.

Holmes, A., 1944. Principles of Physical Geology. Thomas Nelson, London.

Holmes, A., 1965. Principles of Physical Geology. Nelson, London, p. 1288.

Holmes, G.W., Hopkins, D.M., Foster, H.L., 1968. Pingos in central Alaska. Geogr. Surv. Bull. 1341 H, 40, Washington.

Holzer, T.L., 1984. Ground failure induced by ground-water withdrawal from unconsolidated sediment. In: Holzer, T.L. (Ed.), Man-Induced Land Subsidence, The Geological Society of America. Reviews in Engineering Geology, Vol. 6. Boulder, Colorado, pp. 67–105.

Holzer, T.L., Davis, S.N., Lofgren, B.E., 1979. Faulting caused by groundwater extraction in south-central Arizona. J. Geophys. Res. 84, 603–612.

Höllermann, P., 1983. Blockgletscher als mesoformen der periglazialstufe. Bonner Geogr. Abh. 67, 1–73.

Hooke, R., Le, B., 1967. Processes on arid-region alluvial fans. J. Geol. 75, 438–460.

Hooke, R., Le, B., 1972. Geomorphic evidence for late-Wisconsin and Holocene tectonic deformation, Death Valley, California. Bull. Geol. Soc. Am. 83, 2073–2098.

Hooke, R., Le, B., 1987. Mass movement in semi-arid environments and the morphology of alluvial fans. In: Anderson, M.G., Richards, K.S. (Eds) Slope Stability. Wiley, Chichester, pp. 505–529.

Hooke, R., Le, B., Rohrer, W.L., 1979. Geometry of alluvial fans: effect of discharge and sediment size. Earth Surf. Processes 4, 147–166.

Hooke, R., Le, B., Miller, S.B., Kohler, J., 1988. Character of englacial and subglacial drainage system in the upper part of the ablation area of Storglaciären, Sweden. J. Glaciol. 34, 228–231.

Hoppe, G., 1959. Glacial morphology and inland ice recession in northern Sweden. Geogr. Ann. 41, 193–212.

Horton, R.E., 1933. The role of infiltration in the hydrologic cycle. Trans. Am. Geophys. Union 14, 446–460.

Horton, R.E., 1945. Erosional development of streams and their drainage basins: hydrophysical approach to quantitative morphology. Bull. Geol. Soc. Am. 56, 275–370.

Houghton, J.T., Jenkins, G.C., Ephraums, J.J., 1990. Climatic Change, The IPCC Scientific Assessment. Cambridge University Press, Cambridge, MA, p. 356.

Houghton, J.T., Meira Filho, L.G., Callander, B.A., Harris, N., Kattenberg, A., Maskell, K., 1996. Climate Change 1995, The Science of Climatic Change. Published for the IPCC. Cambridge University Press, Cambridge, MA, p. 572.

Howard, A.D., 1942. Pediment passes and the pediment problem. J. Geomorphol. 5, 3–32, 95–136.

Howard, A.D., Dolan, R., 1981. Geomorphology of the Colorado River in the Grand Canyon. J. Geol. 89, 269–298.

Howard, A.D., Selby, M.J., 1994. Rock slopes. In: Abrahams, A.D., Parsons, A.J. (Eds), Geomorphology of Desert Environments. Chapman and Hall, London, pp. 123–172.

Howard, A.D., Morton, J.B., Gad-el-Hak, M., Pierce, D.B., 1978. Sand transport model of barchan dune equilibrium. Sedimentology 25, 307–338.

Hubble, G.D., Isbell, R.F., Northcote, K.H., 1983. Features of Australian soils, Soils: an Australian viewpoint. Academic Press, Melbourne, pp. 17–47.

Hudson, N., 1981. Soil Conservation. Batsford, London, p. 324.

Hughes, T.J., 1992. Abrupt climatic change related to unstable ice-sheet dynamics: toward a new paradigm. Palaeogeogr. Palaeoclimatol. Palaeoecol. 97, 203–234.

Humlum, O., 1982. Rock glacier types on Disko, Central West Greenland. Jesgr. Tidgskr., 82, 59–66.

Hunt, C.B., y Mabey, D.R., 1966. Stratigraphy and Structure. Death Valley, California. US Geol. Surv. Prof. Paper 494A, 162.

Hunt, C.B., Washburn, A.L., 1960. Salt features that simulate ground patterns formed in cold climates. US Geol. Surv. Prof. Paper 494-B, 104–133.

Hunt, C.B., Washburn, A.L., 1966. Patterned ground. In: Hunt, C.B. et al. (Eds), Hydrologic basin Death Valley, California, US Geological Survey Professional Paper, 494B, pp. 104–133.

Hunter, R.E., 1985. A kinematic model for the structure of lee-side deposits. Sedimentology 32, 409–422.

Hunter, R.E., Richmond, B.M., Alpha, T.R., 1983. Storm-controlled oblique dunes of the Oregon coast. Bull. Geol. Soc. Am. 94, 1450–1465.

Hutchinson, G.E., 1957. A Treatise on Limnology: Geography, Physics and Chemistry, Vol. 1. Wiley, New York, p. 1015.

Hutchinson, J.N., 1968. Mass movement. In: Fairbridge, R.W. (Ed.), The Encyclopedia of Geomorphology. Reinhold, New York, pp. 688–695.

Hutchinson, J.N., 1988. General report: morphological and geotechnical parameters of landslides in relation to geology and hydrogeology, Proceedings Fifth International Symposium on Landslides (Laussanne). Balkema, Rotterdam, pp. 3–35.

Hutton, J.T., Dixon, J.C., 1981. The chemistry and mineralogy of some South Australian calcretes and associated soft carbonates and their dolomitization. J. Geol. Soc. Aust. 28, 71–79.

Idso, S.B., 1974. Tornado or dust devil: enigma of desert whirlwinds. Am. Sci. 62, 530–541.

Imeson, A.C., Kwaad, F.J.P.M., Verstratten, J.M., 1982. The relationship of soil physical and chemical properties to the development of badlands in Morocco. In: Bryan, R.B., Yair, A. (Eds), Badland Geomorphology and Piping. Geobooks, Norwich, England, pp. 47–70.

Inbar, M., 1972. A geomorphic analysis of a catastrophic flood in a Mediterranean basaltic watershed. University of Haifa, 22 International Geographical Congress Publications.

Ingles, O.G., Aitchinson, G.D., 1970. Soil water disequilibrium as a cause of subsidence in natural soils and earth embankments, International Symposium on Land Subsidence, International Association of Hydrological Sciences, Publication 89, pp. 342–353.

Innes, J., 1983. Debris flows. Prog. Phys. Geogr. 7, 469–501.

Isarin, R.F.B., 1997. Permafrost distribution and temperatures in Europe during the Younger Dryas. Permafrost Periglac. Processes 8, 313–333.

Ishihara, K., 1989. Liquefaction-induced landslide and debris flow in Tajikistan, URSS. Landslide News 3, 6–7.

International Society of Soil Mechanics and Foundation Engineering, 1985. Progress Report: peculiarities of Geotechnical behaviour of tropical lateritic and saprolitic soils, First International Conference and Saprolitic Soils. Committee on Tropical Soils of International Society of Soil Mechanics and Foundation Engineering, Sao Paulo, p. 449.

Iverson, R.N., 1980. Processes of accelerated pluvial erosion on desert hillslopes modified by vehicular activity. Earth Surf. Processes 5, 369–388.

Iverson, N.R., 1995. Processes of erosion. In: Menzies, J. (Ed.), Modern Glacial Environments: Processes, Dynamics and Sedimets. Butterworth-Heineman, Oxford, pp. 241–260.

Iversen, J.D., Greeley, R., Marshall, J.R., Pollack, J., 1987. Aeolian saltation threshold: effect of density ratio. Sedimentology 34, 699–706.

Jahn, A., 1972. Tundra polygons in the Mackenzie Delta. Göttinger Geogr. Abh. 60, 285–292.

Jahn, A., 1975. Problems of the Periglacial Zone. National Technical Information Service. Panstowe Wydawnistwo Naukowe, Warsaw, Poland, p. 223.

Jahn, A., 1976. Geomorphological modeling and nature protection in Arctic and Subarctic environments. Geoforum 7, 121–137.

Jahns, R.H., 1943. Sheet structure in granites: its origin and use as a measure of glacial erosion in New England. J. Geol. 60, 7–98.

Jakcus, L., 1977. Morphogenetics of Karst Regions. Adam Hilger, Bristol, p. 284.

Jausen, J.H.F., Van Iperen, J.M., 1991. A 220,000 years climatic record for the east equatorial Atlantic Ocean and equatorial Africa: evidence from diatoms and opal phytoliths in the Zaire (Congo) deep sea fan. Palaeoceanography 6, 573–591.

Jauzein, A., 1974. Les données sur le système $CaSO_4 \cdot H_2O$ et leurs implications géologiques. Rev. Géogr. Phys. Géol. Dyn. 16, 151–159.

Jeje, L.K., 1973. Inselberg evolution in a humid tropical environment: the example of south-western Nigeria. Z. Geomorphol. 17, 194–225.

Jennings, J.N., 1971. Karst. The MIT Press, Cambridge, MA, p. 252.

Jennings, J.N., 1985. Karst Geomorphology. Blackwell, Oxford, p. 293.

Jenny, H., 1941. Factors of Soil Formation. McGraw-Hill, New York, p. 281.

Jenny, H., 1950. Origin of soils. In: Trask, P.D. (Ed.), Applied Sedimentation. Wiley, New York, pp. 41–61.

Jiskoot, H., Murray, T., Boyle, P., 2000. Controls on the distribution of surge-type glaciers in Svalbard. J. Glaciol. 46, 412–422.

Johannessen, C.L., Feireisen, J.J., Wells, A.K., 1982. Weathering of ocean cliffs by salt expansion in mid-latitude coastal environment. Shore Beach 50, 26–34.

Johnson, A.M., 1970. Physical Processes in Geology. Freeman Cooper and Co., San Francisco, p. 577.

Johnson, A.M., Rodine, J.R., 1984. Debris flow. In: Brunsden, D., Prior, D.B. (Eds), Slope Instability. Wiley, Chichester, pp. 257–361.

Johnson, D.W., 1932a. Rock planes of arid regions. Geogr. Rev. 22, 656–665.

Johnson, D.W., 1932b. Rock fans of arid regions. Am. J. Sci. 23, 389–416.

Johnson, G., 1960. Cryoturbartion at Zaragoza, Northern Spain. Z. Geomorphol. 4, 74–80.

Johnson, W.H., Menzies, J., 1996. Pleistocene supraglacial and ice-marginal deposits and landforms. In: Menzies, J. (Ed.), Past Glacial Environments. Sediments, Forms and Techniques. Butterworth-Heinemann, Oxford, pp. 137–160.

Joly, F., 1950. Pédiments et glacis d'érosion dans le Sud-Est du Maroc. Congres Internationale de Géographie 2, 110–125, Lisboa.

Jones, D.E., Holtz, W.G., 1973. Expansive soils – the hidden disaster: civil engineering. Am. Soc. Civil Eng. 43 (8), 49–51.

Jones, D.K.C., Brunsden, D., Goudie, A.S., 1983. A preliminary geomorphological assessment of the Karakoram Highway. Q. J. Eng. Geol. 16, 331–355.

Jones, F.O., 1973. Landslides of Rio de Janeiro and the Serra das Araras Scarpment, Brazil. US Geol. Surv. Prof. Paper 697, p. 42.

Jones, J.A.A., 1981. The Nature of Soil Piping — A Review of Research. Geobooks, Norwich, p. 301.

Jones, J.A.A., 1990. Piping effects in humid lands. Geol. Soc. Am. Spec. Paper 252, 111–138.

Jouzel, J., Lorius, C., 1999. Evolution du climat: du passé récent vers le futur. C. R. Acad. Sci. Paris 328, 229–239.

Jouzel, J., Lorius, C., Petit, J.R., Barkev, N.I., Kotlyakov, V.M., Petrov, V.M., 1987. Vostok ice core. A continuous isotopic temperature record over the last climatic cycle (160,000 years). Nature 329, 403–408.

Jungerius, P.D., 1984. A simulation model of blowout development. Earth Surf. Processes Landforms 9, 509–512.

Jungerius, P.D., van der Meulen, F., 1989. The development of dune blowouts, as measured with erosion pins and sequential air photos. Catena 16, 369–376.

Kääb, A., Haeberli, W., 2001. Evolution of a high-mountain thermokarst lake in the Swiss Alps. Arct. Antarct. Alp. Res. 33, 385–390.

Kamb, B., LaChapelle, E.R., 1964. Direct observations of the mechanism of glacier sliding over bedrock. J. Glaciol. 6, 159–172.

Kapitsa, A.P., Ridley, J.K., Robin, G.de Q., Siegert, M.J., Zotikov, I.A., 1996. A large deep freshwater lake beneath the ice of central East Antarctica. Nature 381, 684–686.

Kaplar, C.W., 1965. Stone migration by freezing of soils. Science 149, 1520–1521.

Kasse, C., 1997. Cold-climate sand sheet formation in North-Western Europe (c.14–12.4 ka); a response to permafrost degradation and increased aridity. Permafrost Periglac. Processes 8, 295–311.

Kasse, C., 2002. Sandy aeolian deposits and environments and their relation to climate during the Last Glacial Maximum and Late glacial in northwest and central Europe. Prog. Phys. Geog. 26, 507–532.

Kassler, P., 1973. The structural and geomorphic evolution of the Persian Gulf. In: Purser, B.H. (Ed.), The Persian Gulf. Springer, New York, pp. 11–32.

Kaye, C.A., 1967. Kaolinisation of bedrock of the Boston, Massachusetts area. US Geol. Surv. Prof. Paper 575-C, C165–C172.

Keller, C., 1946. El Departamento de Arica. Ministerio de Economía y Comercio, Santiago de Chile.

Keller, W.D., 1957. The Principles of Chemical Weathering. Lucas Brothers, Columbia, MO, p. 111.

Kemmis, T.L., 1989. Importance of the vegetation process to certain properties of basal tills deposited by the Laurentide ice sheet in Iowa and Illinois, USA. Ann. Glaciol. 2, 147–152.

Kerr, R.C., Nigra, J.O., 1952. Eolian sand control. Bull. Am. Assoc. Petrol. Geol. 36, 1541–1573.

Kershaw, G.P., Gill, D., 1979. Growth and decay of palsas and peat plateaux in the Macmillan Pass – Tsichu River area, Northwest Territories, Canada. Can. J. Earth Sci. 16, 1362–1367.

Kesel, R.H., 1973. Inselberg landform elements: definition and synthesis. Rev. Géomorphol. Dyn. 22, 97–108.

Kesel, R.H., 1976. The use of the refraction-seismic techniques in geomorphology. Catena 3, 91–98.

Kesel, R.H., 1977a. Slope runoff and denudation in the Rupununi savanna, Guyana. J. Trop. Geogr. 44, 33–42.

Kesel, R.H., 1977b. Some aspects of the geomorphology of inselbergs in central Arizona, USA. Z. Geomorphol. 21, 119–146.

Kesseli, J.E., Beaty, C.B., 1959. Desert flat conditions in the White mountains of California and Nevada. US Army Quartermaster Research and Engineering Center. Technical Report, EP-108.

Keylock, C., 1997. Snow avalanches. Prog. Phys. Geogr. 21, 481–500.

Khalil, G.M., 1990. Floods in Bangladesh: a question of disciplining the rivers. Nat. Hazards 3, 379–401.

Kiewietdejonge, C.J., 1984. Büdel's geomorphology II. Prog. Phys. Geogr. 8, 365–397.

Killigrew, L.P., Gilkes, R.J., 1974. Development of playa lakes in south western Australia. Nature 247, 454–455.

King, C.A.M., 1976a. Introduction. In: King, C.A.M. (Ed.), Periglacial Processes. Dowden, Hutchinson and Ross, Stroudsburg, pp. 1–6.

King, C.A.M., 1976b. Landforms and Geomorphology. Concepts and History. Dowden, Hutchinson and Ross, Stroudsburg, p. 404.

King, L.C., 1942. South African Scenery. A Textbook of Geomorphology. Oliver and Boyd, Edinburgh, p. 308.

King, L.C., 1948. A theory of bornhardts. Geogr. J. 112, 83–87.

King, L.C., 1949. The pediment problem: some current problems. Geol. Mag. 86, 245–250.

King, L.C., 1953. Canons of landscape evolution. Bull. Geol. Soc. Am. 64, 721–752.

King, L.C., 1957a. Landscape study in southern Africa. Proc. Geol. Soc. S. Afr. 50, 22–52.

King, L.C., 1957b. The uniformitarian nature of hillslopes. Trans. Edin. Geol. Soc. 17, 81–102.

King, L.C., 1962. The Morphology of the Earth. A Study and Synthesis of World Scenery. Oliver and Boyd, Edinburgh, p. 726.

King, L.C., 1975. Bornhardt landforms and what they teach. Z. Geomorphol. 19, 299–318.

King, L.C., 1976. Planation remnants upon high lands. Z. Geomorphol. N.F. 20, 133–148.

King, L.C., 1983. Wandering Continents and Spreading Sea Floors on an Expanding Earth. Wiley, Chichester, p. 232.

King, P.B., Schumm, S.A., 1980. The Physical Geography (Geomorphology) of William Morris Davis. Geo Books, Norwich, p. 217.

Kirkbride, M.P., 1995. Processes of transportation. In: Menzies, J. (Ed.), Modern Glacial Environments: Processes, Dynamics and Sediments. Butterworth-Heineman, Oxford, pp. 261–292.

Kirkby, A.T.V., Kirkby, M.J., 1974. Surface wash at the semiarid break in slope. Z. Geomorphol. Suppl.bd. 21, 151–171.

Kirkby, M.J., 1994. Thresholds and instability in stream head hollow: a model of magnitude and frequency for wash processes. In: Kirkby, M.J. (Ed.), Process, Models and Theoretical Geomorphology. Wiley, Chichester, pp. 215–314.

Klammer, C., 1982. Die Palaeowüste des Pantanal von Mato Grosso und die Pleistozane Klimageschichte der Brazilianischen Randtropen. Z. Geomorphol. 26, 393–416.

Klappa, C.F., 1979. Lichen stromatolites: criterion for subaerial exposure and a mechanism for the formation of laminar calcretes (caliche). J. Sediment. Petrol. 49, 387–400.

Klein, M., 1984. Weathering rates of limestones tombstones measured in Haifa, Israel. Z. Geomorphol. 28, 105–111.

Klimchouk, A., Andrejchuk, V., 1996. Environmental problems in gypsum karst terrains. Int. J. Speleol. 25, 145–156.

Klimchouk, A., Forti, P., Cooper, A., 1996. Gypsum karst of the world: a brief overview. Int. J. Speleol. 25, 159–181.

Knight, M.J., 1980. Structural analysis and mechanical origins of gilgai at Boorook, Victoria, Australia. Geoderma 23, 245–283.

Knight, P.G., Waller, R.I., Patterson, C.J., Jones, A.P., Robinson, Z.P., 2002. Discharge of debris from ice at the margin of the Greenland ice sheet. J. Glaciol. 48, 192–200.

Knighton, D., 1998. Fluvial Form and Processes. A New Perspective. Arnold, London, p. 383.

Knox, J.C., 1972. Valley alluviation in southwestern Wisconsin. Ann. Assoc. Am. Geogr. 62, 401–410.

Knox, J.C., 1983. Responses of river systems to Holocene climates. In: Wright, H.E. Jr. (Ed.), Late Quaternary Environments of the United States, Holocene, Vol. 2. University of Minnesota Press, Minneapolis, pp. 26–41.

Knox, J.C., 1984. Fluvial responses to small scale climate changes. In: Costa, J.E., Fleisher, P.J. (Eds), Developments and Applications of Geomorphology. Springer, Berlin, pp. 318–342.

Knox, J.C., 1993. Large increases in flood magnitude in response to modest changes in climate. Nature 361, 430–432.

Knox, J.C., 1995. Fluvial systems since 20,000 years BP. In: Gregory, K.J., Starkel, L., Baker, V.R. (Eds), Global Continental Palaeohydrology. Wiley, Chichester, pp. 87–108.

Kocurek, G., 1988. First-order and super bounding surfaces in eolian sequences. Bounding surfaces revisited. In: Kocurek, G. (Ed.), Late Paleozoic and Mesozoic eolian deposits of the western interior of the United States, Sediment. Geol., 56, pp. 193–206.

Kocurek, G., Nielson, J., 1986. Conditions favourable for the formation of warm-climate aeolian sand sheets. Sedimentology 33, 795–816.

Kocurek, G., Harholm, K.G., Deynoux, M., Blakey, R.C., 1991. Amalgamated accumulations resulting from climatic and eustatic changes, Akchar Erg, Mauritania. Sedimentology 38, 751–772.

Kocurek, G., Townsley, M., Yeh, E., Sweet, M., Havholm, K., 1992. Dune and dune-field development stages on Padre Island, Texas: effects of lee airflow and sand saturation levels and implications for interdune deposition. J. Sediment. Petrol. 62, 622–635.

Kolla, V., Biscaye, P.E., Hanley, A.F., 1979. Distribution of quartz in Late Quaternary Atlantic sediments in relation to climate. Quaternary Res. 11, 261–277.

Koons, D., 1955. Cliff retreat in the southwestern United States. Am. J. Sci. 253, 44–52.

Köppen, W., 1936. Das geographische system der klimate, Handbuch der Klimatologie, Vol. 1. Gerbruder Borntraeger, Berlin.

Kor, P.S.G., Shaw, J., Sharpe, D.R., 1991. Erosion of bedrock by subglacial meltwater, Georgian Bay, Ontario: a regional review. Can. J. Earth Sci. 28, 623–642.

Koster, E.A., Dijkmans, J.W.A., 1988. Niveo-aeolian deposits and denivation forms, with special reference to the Great Kobuk Sand Dunes, Northwestern Alaska. Earth Surf. Processes Landforms 13, 153–170.

Krahn, J., Johnson, R.F., Fredlund, D.G., Clifton, A.W., 1979. A highway failure in Cretaceous sediments at Maymount, Saskatchewan. Can. Geotech. J. 16, 703–715.

Krainer, K., Mostler, W., 2002. Hydrology of active rock glaciers: examples from the Austrian Alps. Arct. Antarct. Alp. Res. 34, 142–149.

Krauskopf, K.B., 1956. Dissolution and precipitation of silica at low temperatures. Acta Geochim. Cosmochim. 10, 1–26.

Krinsley, D.B., 1970. A geomorphological and paleoclimatological study of the playas of Iran. United States Geological Survey, Washington, 2 Volumes.

Krinsley, D., Wellendorf, W., 1980. Wind velocities determined from the surface textures of sand grains. Nature 283, 372–373.

Krumbein, W.E., Jens, K., 1981. Biogenic rock varnishes of the Negev Desert (Israel): an ecological study of iron and manganese transformation by cyano-bacteria and fungi. Oecologia 50, 25–28.

Kuenen, Ph., 1960. Experimental abrasion, 4: eolian action. J. Geol. 68, 427–449.

Kühnel, R.A., 1987. The role of cationic and anionic scavengers in laterites. Chem. Geol. 60, 31–40.

Lachenbruch, A., 1962. Mechanics of thermal contraction cracks and ice-wedge polygons in permafrost. Geol. Soc. Am. Spec. Paper 70, p. 69.

Lachenbruch, A.H., 1970. Thermal considerations in permafrost. In: Adkison, W.L., Borsge, M.M. (Eds), Geological Seminar on the North Slope of Alaska, Proc. Am. Assoc. Petr. Geol., J1-2 and discussion J2-5. Los Angeles.

Laity, J.E., 1994. Landforms of aeolian erosion. In: Abrahams, A.D., Parsons, A.J. (Eds), Geomorphology of Desert Environments. Chapman & Hall, London, pp. 506–537.

Laity, J.E., 1995. Wind abrasion and ventifact formation in California. In: Tchakerian, V.P. (Ed.), Desert Aeolian Processes. Chapman & Hall, London, pp. 295–321.

Laity, J.E., Malin, M.C., 1985. Sapping processes and the development of theater-headed valley networks in the Colorado Plateau. Bull. Geol. Soc. Am. 96, 203–217.

Lal, R., 1986. Deforestation and soil erosion. In: Lal, R., Sánchez, P.A., Cummings, R.W. (Eds), Land Clearing and Development in the Tropics. Balkema, Rotterdam, pp. 299–315.

Lal, R., 1990. Water erosion and conservation: an assessment of the water erosion problem and the techniques available for soil conservation. In: Goudie, A. (Ed.), Techniques for Desert Reclamation. Wiley, Chichester, pp. 161–198.

Lal, R., Lawson, T.L., Anastase, A.H., 1981. Erosivity of tropical rains. In: De Broodt, M., Gabriels, D. (Eds), Assessment of Erosion. Wiley, Chichester, pp. 143–151.

Lamb, H.H., 1972. Climate: Past, Present and Future, Vol. 1. Methuen, London, p. 613.

Lamb, H.H., 1977. Climates: Present, Past and Future, Vol. II. Methuen, London, p. 835.

Lancaster, I.N., 1978. The pans of the Southern Kalahari, Botswana. Geogr. J. 144, 80–98.

Lancaster, N., 1981. Palaeoenvironmental implications of fixed dune systems in southern Africa. Palaeogeogr. Palaeoclimatol. Palaeoecol. 33, 327–346.

Lancaster, N., 1982. Dunes on the Skeleton Coast, Namibia (South West Africa): geomorphology and grain size relationships. Earth Surf. Processes Landforms 7, 575–587.

Lancaster, N., 1983. Linear dunes of the Namib sand sea. Z. Geomorphol. Suppl.bd. 45, 27–49.

Lancaster, N., 1985. Variations in wind velocity and sand transport on the windward flanks of desert sand dunes. Sedimentology 32, 581–593.

Lancaster, N., 1987. Variations in wind velocity and sand transport on the windward flanks of desert sand dunes-reply. Sedimentology 43, 516–520.

Lancaster, N., 1988. Development of linear dunes in the southwestern Kalahari, southern Africa. J. Arid Environ. 14, 233–244.

Lancaster, N., 1989a. The Namib Sand Sea: Dune Forms, Processes and Sediments. Balkema, Rotterdam, p. 180.

Lancaster, N., 1989b. The dynamics of star dunes: an example from the Gran Desierto. Mexico. Sedimentology 36, 273–289.

Lancaster, N., 1989c. Star dunes. Prog. Phys. Geogr., 67–91.

Lancaster, N., 1994. Dune morphology and dynamics. In: Abrahams, A.D., Parsons, A.J. (Eds), Geomorphology of Desert Environments. Chapman & Hall, London, pp. 474–505.

Lancaster, N., 1995. Geomorphology of Desert Dunes. Routledge, London, p. 290.

Lancaster, N., Ollier, C.D., 1983. Sources of sand for the Namib Sand Sea. Z. Geomorphol. Suppl.bd. 45, 71–83.

Land, L.S., 1970. Phreatic versus vadose meteoric diagenesis of limestones: evidence from a fossil water table. Sedimentology 14, 175–185.

Langbein, W.B., 1949. Annual runoff in the United States. US Geol. Surv. Circ. 52, p. 14.

Langbein, W.B., 1961. Salinity and hydrology of closed lakes. US Geol. Surv. Prof. Paper 412, 1–20.

Langbein, W.B., Schumm, S.A., 1958. Yield of sediment in relation to mean annual precipitation. Trans. Am. Geophys. Union 39, 1076–1084.

Lautridou, J.P., 1988. Recent advances in cryogenic weathering. In: Clark, M.J. (Ed.), Advances in Periglacial Geomorphology. Wiley, Chichester, pp. 33–47.

Lawson, A.C., 1915. The epigene profiles of the desert. Univ. Calif. Dep. Geol. Bull. 9, 23–48.

Lawson, D.E., 1979. A comparison of the pebble orientations in ice and deposits of the Matanuska glacier, Alaska. J. Geol. 87, 629–645.

Lawson, M.P., Thomas, D.S.G., 2002. Late Quaternary lunette dune sedimentation in the southwestern Kalahari Desert, South Africa: luminescence based chronologies of aeolian activity. Quaternary Sci. Rev. 21, 825–836.

Lecce, S.A., 1990. The alluvial fan problem. In: Rachocki, A.H., Church, M. (Eds), Alluvial Fans. A Field Approach. Wiley, Chichester, pp. 3–24.

Lee, J.A., Wigner, K.A., Gregory, J.M., 1993. Drought, wind and blowing dust on the southern High Plains of the United States. Phys. Geogr. 14, 56–67.

Lee, S.W., 1981. Landslides in Taiwan. Proceedings South East Asian regional symposium on Problems of soil erosion and sedimentation, Bangkok, pp. 195–206.

Legget, R.F., 1974. Glacial landforms and civil engineering. In: Coates, D.R. (Ed.), Glacial Geomorphology. State University of New York, Binghamton, pp. 351–374.

Lehmann, H., 1936. Morphologische Studien auf Java. Geogr. Abhandl III Heft 9, Stuttgart, p. 114.

Lehmann, H., 1955. Der tropische kegelkarst in westindien. Tagungsbericht Wiss. Abhandl., 126–131, Wiesbaden.

Lehmann, H., Roglic, J., Rathjens, C., Lasserre, G., Harrassowitz, H., Corbel, J., Birot, P., 1954. Karstphänomen in den verschiedenen klimazonen. Erdkunde 8, 112–122.

Lelong, F., 1966. Régimes des nappes phreatiques contenues dans las formations d'altération tropicale. Conséquences pour la pédogenèse. Sciences de la Terre 11, 203–244.

Leopold, L.B., 1968. Hydrology for urban planning – a guidebook on the hydrologic effects of urban land use. US Geol. Surv. Circ. 554, p. 18.

Leopold, L.B., 1976. Reversal of erosion cycle and climatic change. Quaternary Res. 6, 557–562.

Leopold, L.B., 1994. A View of the River. Harvard University Press, Cambridge, MA, p. 298.

Leopold, L.B., Maddock, T., 1954. The Flood Control Controversy. Ronald Press, New York, p. 178.

Leopold, L.B., Wolman, M.G., Miller, G.P., 1964. Fluvial Processes in Geomorphology. W.H. Freeman, San Francisco, p. 522.

Leopold, L.B., Emmet, W.W., Myrick, R.M., 1966. Channel and hillslope processes in a semiarid area of New Mexico, US Geol. Surv. Prof. Paper, 352G.

Létolle, R., Mainguet, M., 1993. Aral. Springer, Paris, p. 357.

Levson, V.M., Morison, S.R., 1996. Geology of placer deposits in glaciated environments. In: Menzies, J. (Ed.), Past Glacial Environments: Sediments, Forms and Techniques. Butterworth-Heinemann, Oxford, pp. 441–478.

Linell, K.A., Johnston, G.H., 1973. Engineering design and construction in permafrost regions, Proceedings of Permafrost 2st International Conference. Natural Research Council, Yakutsk, URSS, pp. 553–575.

Linell, K.A., Lobacz, E.F., 1978. Some experiences with tunnel entrances in permafrost, Proceedings Third International Conference of Permafrost, pp. 813–819. Ottawa.

Linton, D.L., 1955. The problem of tors. Geogr. J. 121, 470–487.

Linton, D.L., 1963. The forms of glacial erosion. Trans. Inst. Br. Geogr. 33, 1–28.

Lipman, P.W., Mullineaux, D.R., (Eds), 1981. The 1980 Eruptions of Mount St. Helens, Washington, US Geol. Surv. Prof. Paper, 1250.

Lister, L.A., Secrest, C.D., 1985. Giant desiccation cracks and differential surface subsidence, Red Lake Playa, Mohave County, Arizona. Bull. Assoc. Eng. Geol. 22, 299–314.

Livingstone, I., Warren, A., 1996. Aeolian Geomorphology. An Introduction. Longman, Essex, p. 211.

Llamas, M.R., 1962. Estudio geológico-técnico de los terrenos yesíferos de la Cuenca del Ebro y de los problemas que plantean en los canales. Ministerio de Obras Públicas, Boletín 12, Madrid, p. 192.

Lliboutry, L., 1965. Traité de Glaciologie. Masson, Paris, 2 Volumes, p. 1040.

Lliboutry, L., Briat, M., Cresoveur, M., Fourchet, M., 1976. 15 m deep temperatures in the glaciers of Mont Blanc (French Alps). J. Glaciol. 16, 197–203.

Lloyd, J.W., 1983. Hydrogeology in glaciated terrains. In: Eyles, N. (Ed.), Glacial Geology. An Introduction for Engineers and Earth Scientists. Pergamon Press, Oxford, pp. 349–368.

Lobacz, E.F., Quinn, W.F., 1966. Thermal regime beneath buildings constructed on permafrost, Proceedings First International Conference of Permafrost. National Academy of Sciences, Washington, Publication 1287, pp. 447–491.

Löffler, E., 1974. Piping and pseudokarst features in the tropical lowlands of New Guinea. Erdkunde 28, 13–18.

López-Bermúdez, F., 1973. La Vega Alta del Segura. Departamento de Geografía. Universidad de Murcia, Murcia, p. 288.

López Bermúdez, F., 1986. Evaluación de la erosión hídrica en las áreas receptoras de los embalses de la Cuenca del Segura. Aplicación de la USLE. In: López Bermúdez, F., Thornes, J. (Eds), Estudios sobre Geomorfología del Sur de España. Universidad de Murcia, Murcia, pp. 93–101.

López Bermúdez, F., 1988. Desertificación: magnitud del problema y estado actual de las investigaciones. In: Gutiérrez, M., Peña, J.L. (Eds), Perspectivas en Geomorfología, Vol. 2. Monografía de la Sociedad Española de Geomorfología, Zaragoza, pp. 155–170.

López-Bermúdez, F., Conesa-García, C., Alonso-Sarriá, F., 2002. Floods: magnitude and frequency in ephemeral streams of the Spanish Mediterranean region. In: Bull, L.J., Kirby, M.J. (Eds), Drylands Rivers: Hydrology and Geomorphology of Semi-arid Channels. Wiley, New York, pp. 329–350.

López-Martínez, J., 1988. El riesgo debido a los aludes. In: Ayala, F.J., Durán, J.J., v Peinado, T. (Eds), Riesgos geológicos. Instituto Geológico y Minero de España, Madrid, pp. 215–225.

Lorius, C., Barkov, N.I., Jouzel, J., Korotkevich, Y.S., Kotlyakov, V.M., Raynaud, D., 1988. Antarctic ice core: CO_2 and change over the last climatic cycle. EOS 69, 681–684.

Loughnan, F.C., 1969. Chemical Weathering of the Silicate Minerals. Elsevier, New York, p. 154.

Loughnan, F.C., Bayliss, P., 1961. The mineralogy of the bauxitic deposits near Weipa, Queensland. Am. Mineral. 46, 209–217.

Loughnan, F.C., Grim, R.W., Vernet, J., 1962. Weathering of some Triassic shales in the Sydney area. J. Geol. Soc. Aust. 8, 245–257.

Lovegrove, G.W., Fookes, P.G., 1972. The planning and implementation of a site investigation for a highway in tropical conditions in Fiji. Q. J. Eng. Geol. 5, 43–68.

Lozinski, W.Von., 1909. Über die mechanische Verwitterung der Sandsteine im gemässigten Klima. Acad. Sci. Cracovie Bull. Int. Cl. Sci. Math. Nat. 1, 1–25.

Lucchitta, I., 1975. Application of ERTS images and image processing to regional geologic problems and geologic mapping in northern Arizona – Part IV B, the Shivwits Plateau. National Aeronautics and Space Administration Technical Report 32–1597, 41–72.

Lumb, P., 1962. The properties of decomposed granite. Géotechnique 12, 226–243.

Lumb, P., 1983. Engineering properties of fresh and decomposed igneous rocks from Hong-Kong. Eng. Geol. 27, 287–298.

Lundquist, G., 1969. Earth and ice mounds: a terminological discussion. In: Péwé, T.L. (Ed.), The Periglacial Environment. McGill-Queen's University Press, Montreal, pp. 203–215.

Lundqvist, J., 1989. Rogen (ribbed) moraine: identification and possible origin. Sediment. Geol. 62, 281–292.

Lvovitch, M.I., 1967. Water resources of the world and their future. General Assembly of Bern. Int. Assoc. Sci. Hydrol., 317–322.

Lyford, F.P., Qashu, H.K., 1969. Infiltration rates as affected by desert vegetation. Water Resour. Res. 5, 1373–1376.

Mabbutt, J.A., 1961a. Basal surface or weathering front. Proc. Geol. Assoc. 72, 357–358.

Mabbutt, J.A., 1961b. A stripped landsurface in Western Australia. Trans. Inst. Br. Geogr. 29, 101–114.

Mabbutt, J.A., 1965a. Stone distribution in a Stony Table-land soil. Aust. J. Soil Res. 3, 131–142.

Mabbutt, J.A., 1965b. The weathered landsurface of central Australia. Z. Geomorphol. 9, 82–114.

Mabbutt, J.A., 1966. Mantle-controlled planation of pediments. Am. J. Sci. 264, 78–91.

Mabbutt, J.A., 1977. Desert landforms. The MIT Press, Cambridge, MA, p. 340.

MacAyeal, D.R., 1993. A low-order model of the Heinrich event cycle. Paleoceanography 8, 767–773.

Mackay, J.R., 1973a. Problems in the origin of glassif Icy Beds, Western Arctic, Canada. North American Contribution. Permafrost. 2nd Int. Conf., Yakutsk, USSR. Nat. Acad. Sciences, Washington, 10, 223–228.

Mackay, J.R., 1973b. The growth of pingos, western Arctic coast, Canada. Can. J. Earth Sci. 10, 979–1004.

Mackay, J.R., 1979. Pingos of the Tuktoyaktuk Peninsula area, Northwest Territories. Géogr. Phys. Quatern. 33, 3–61.

Mäckel, R., 1974. Dambos. A study in morphodynamic activity on the plateau regions of Zambia. Catena 1, 327–365.

Mäckel, R., 1985. Dambos and related landforms in Africa – an example for the ecological approach to tropical geomorphology. Z. Geomorphol. Suppl.bd. 52, 1–23.

Machette, M.N., 1985. Calcic soils of the southwestern United States. In: Weide, D. (Ed.), Soils and Quaternary geology of the Southwestern United States, Geological Society of America Special Paper, Vol. 203, pp. 1–21.

Magarerz, M., Enzel, Y., 1990. Standing-water deposits as indicators of Late Quaternary dune migration in the northwestern Neger, Israel. Climatic Change, 336, 367–318.

Maignien, R., 1966. Review of Research on Laterites, Vol. 4. UNESCO Natural Resources Research, Paris.

Mainguet, M., 1972. Le Modelé des Grés. Problemes Généraux. Institut Géographique National, Paris, p. 657.

Mainguet, M., 1975. Etude comparée des ergs à l'échelle continentale (Sahara et déserts d'Australie). Bull. Assoc. Géogr. Françaises 52, 135–140.

Mainguet, M., 1978. The influence of trade winds, local air masses and topographic obstacles on the aeolian movement of sand particles and the origin and distribution of ergs in the Sahara and Australia. Geoforum 9, 17–28.

Mainguet, M., 1983. Dunes vives, dunes fixées, dunes vêtues: une classification selon le bilan d'alimentation, le régime èolien et la dynamique des édifices sableux. Z. Geomorphol. Suppl.bd. 45, 265–285.

Mainguet, M., 1984. A classification of dunes based on aeolian dynamics and the sand budget. In: El Baz, F. (Ed.), Deserts and Arid Lands. Martinus Nijhof, The Hague, pp. 59–77.

Mainguet, M., 1991. Desertification. Natural Background and Human Mismanagement. Springer, Berlin, p. 306.

Mainguet, M., Canon, N., Chemin, M., 1980. Le Sahara: géomorphologie et paléogéomorphologie éoliennes. In: Williams, M.A.J., Faure, H. (Eds), The Sahara and the Nile. Balkema, Rotterdam, pp. 17–35.

Maizels, J.K., 1981. Freeze/thaw experiments in the simulation of sediment cracking patterns. Bedford College London, Papers in Geography 13.

Maizels, J.K., 1987. Plio-Pleistocene raised channel systems of the western Sharqiya (Wahiba), Oman. In: Frostick, L.E., Reid, I. (Eds), Desert Sediments: Ancient and Modern. Blackwell Scientific, Oxford, pp. 31–50.

Maizels, J., 1995. Sediments and landforms of modern proglacial terrestrial environments. In: Menzies, J. (Ed.), Modern Glacial Environments. Processes, Dynamics and Sediments. Butterworth-Heinemann, Oxford, pp. 365–416.

Maizels, J., 1997. Jökulhlaup deposits in proglacial areas. Quaternary Sci. Rev. 16, 783–819.

Mammerickx, J., 1964. Quantitative observations on pediments in the Mojave and Sonoran deserts (Southwestern United States). Am. J. Sci. 262, 417–435.

Mann, A.W., 1984. Mobility of gold and silver in lateritic weathering profiles: some observations from western Australia. Econ. Geol. 79, 38–49.

Mannion, A.M., 1997. Global Environmental Change. Longman, Essex, p. 387.

Marko, J.R., Fissel, D.B., Miller, J.D., 1988. Iceberg movement prediction off the Canadian east coast. In: El-Sabh, M.I., Murty, T.S. (Eds), Natural and Man-Made Hazards. D. Reidel Publishers, Dordrecht, pp. 435–462.

Marsden, S.S., Davis, S.N., 1967. Geological subsidence. Sci. Am. 216, 93–100.

Marshall, A.F., Workman, J.P., 1977. Identification of dispersive clays in the Texas Gulf Coast Area. In: Sherard, J.L., Decker, R.S. (Eds), Dispersive Clays, Related Piping, and Erosion in Geotechnical Projects, pp. 274–286.

Martínez, J., Garzón, M.G., Arche, A., 1987. Avenidas e inundaciones. Ministerio de Obras Públicas y Urbanismo, Madrid, p. 67.

Martini, J.P., 1978. Tafoni weathering, with examples from Toscany, Italy. Z. Geomorphol. 22, 44–67.

Martín-Serrano, A., 1988. El relieve de la región occidental zamorana. La evolución geomorfológica de un borde del Macizo Hespérico. Instituto de Estudios Zamoranos Florián de Ocampo, Zamora, p. 311.

Martonne, E. de, Aufrère, L., 1928. L'extension des régions privées d'écoulement vers l'ocean. Ann. de Géogr., 38, 1–24.

Mason, B., 1958. Principles of Geochemistry. Wiley, New York, p. 310.

Matthes, F.E., 1900. Glacial sculpture of the Bighorn Mountains, Wyoming, United States Geological Survey 21st Annual Report 1899–1900, pp. 167–190.

Matthes, F.E., 1930. Geologic history of the Yosemite valley. US Geol. Surv. Prof. Paper 160, 54–103.

Maxson, J.H., 1940. Fluting and faceting of rock fragments. J. Geol. 48, 717–751.

Mayer, L., Gerson, R., Bull, W.B., 1984. Alluvial gravel production and deposition - a useful indicator of Quaternary climatic changes in deserts (a case study in south-western Arizona). Catena Suppl. 5, 137–151.

McBean, G.A., Liss, P.S., Schneider, S.H., 1996. Advancing our understanding. In: Houghton, J.T. (Ed.), Climate Change 1995. The Science of Climatic Change. Cambridge University Press, Cambridge, MA, pp. 517–531.

McCall, J.G., 1960. The flow characteristics of a cirque glacier and their effect on cirque formation. In: Lewis, W.V. (Ed.), Investigations on Norvegian Cirque Glaciers, Royal Geographers Society, Research Series, Vol. 4, pp. 39–62.

McCauley, J.F., Grolier, M.J., Breed, C.S., 1977. Yardangs. In: Doehring, D.O. (Ed.), Geomorphology in Arid Regions. Allen and Unwin, London, pp. 233–269.

McCauley, J.F., Breed, C.S., El-Baz, F., Whitney, M.I., Grolier, M.J., Ward, A.W., 1979. Pitted and fluted rocks in the Western Desert of Egypt–Viking comparisons. J. Geophys. Res. 84, 8222–8232.

McCauley, J.F. et al., 1982. Subsurface valleys and geoarchaeology of the eastern Sahara revealed by Shuttle Radar. Science 218, 1004–1020.

McClung, D.M., 2003. Magnitude and frequency of avalanches in relation to terrain and forest cover. Arct. Antarc. Alp. Res. 35, 82–90.

McClung, D., Schaerer, P., 1993. The Avalanche Handbook, The Mountaineers, p. 272.

McClure, H.A., 1978. ArRub' Alkhali. In: Al-Sayari, S.S., Zötl, J.G. (Eds), Quaternary Period in Saudi Arabia. Springer, Viena, pp. 252–263.

McFarlane, M.J., 1976. Laterite and Landscape. Academic Press, London, p. 151.

McFarlane, M.J., 1983. Laterites. In: Goudie, A.S., Pye, K. (Eds), Chemical Sediments and Geomorphology. Academic Press, London, pp. 7–58.

McFarlane, M.J., 1987. The key role of microorganisms in the process of bauxitisation. Mod. Geol. 11, 325–344.

McFarlane, M.J., 1989. Dambos - their characteristics and geomorphological evolution in parts of Malawi and Zimbabwe, with particular reference to their role in the hydrogeological regime of surviving areas of African surface. Proceedings of the groundwater exploration. Harare, Zimbabwe, 15-24 June 1987. Commonwealth Science Council, Vol. 1, 254–308.

McFarlane, M.J., Pollard, S., 1989. Some aspects of stone-lines and dissolution fronts associated with regolith and dambo profiles in parts of Malawi and Zimbabwe. In: Alexandre, J., Symoens, J.J. (Eds), Stone-Lines, Geo-Eco-Trop, Vol. 11. Académie Royale des Sciences D'Outre-Mer, Bruxelles, pp. 23–35.

McFarlane, M.J., Whitlow, R., 1990. Key factors affecting the initiation and progress of gullying in dambos in parts of Zimbabwe and Malawi. Land Degrad. Rehabil. 2, 215–235.

McGee, W.J., 1897. Sheetflood erosion. Bull. Geol. Soc. Am. 8, 87–112.

McGowen, J.H., 1979. Alluvial fan systems. In: Galloway, W.E., Kreitler, C.V., McGowen, J.H. (Eds), Depositional and Ground Water Flow Systems in the Exploration for Uranium, Texas Bureau of Economic Geology Research Colloquium, pp. 43–79.

McGreevy, J.P., 1982. Frost and salt weathering: further experimental results. Earth Surf. Processes Landforms 7, 475–488.

McGregor, G.R., Nieuwolt, S., 1998. Tropical climatology, An Introduction to the Climates of the Low Latitudes. Wiley, Chichester, p. 339.

McIntyre, D.S., 1979. Exchangeable sodium, subplasticity, and hydraulic conductivity of some Australian soils. Aust. J. Soil Res. 17, 115–120.

McKee, E.D., 1979a. A study of global sand seas. US Geol. Soc. Prof. Paper 1052, Washington, p. 429.

McKee, E.D., 1979b. Introduction to a study of global sand seas. In: McKee, E.D. (Ed.), A Study of Global Sand Seas, US Geological Survey Professional Paper, Vol. 1052, pp. 1–19, Washington.

McKee, E.D., Bigarella, J.J., 1979. Sedimentary structures in dunes. In: McKee, E.D. (Ed.), A Study of Global Sand Seas, US Geol. Surv. Prof. Paper, 1052, pp. 83–134, Washington.

McManus, J.F., 2004. A great grand-daddy of ice cores. Nature 429, 611–612.

McTainsh, G.H., 1987. Desert loess in northern Nigeria. Z. Geomorphol. NF 31, 145–165.

Meadows, M.E., 1985. Dambos and environmental change in Malawi, central Africa. Z. Geomorphol. Suppl.bd. 52, 147–169.

Mehta, P.K., 1983. Mechanism of sulphate attack on Portland cement concrete-another look. Cem. Concr. Res. 13, 401–406.

Meigs, P., 1953. The world distribution of arid and semiarid homoclimates, Reviews of Research on Arid Zone Hydrology. UNESCO, Paris, pp. 203–209.

Meir, M.F., Post, A., 1969. What are glacier surges? Can. J. Earth Sci. 6, 807–817.

Mellet, J.S., 1990. Ground-penetrating radar enhances knowledge of Earth's surface layer. Geotimes 35, 12–14.

Mensching, H., 1958. Glacis-Fussfläche-Pediment. Z. Geomorphol. N.F. 2, 165–186.

Menzies, J., 1979. A review of the literature on the formation and location of drumlins. Earth Sci. Rev. 14, 315–359.

Menzies, J., 1995a. Glaciers and ice sheets. In: Menzies, J. (Ed.), Modern Glacial Environments. Processes, Dynamics and Sediments. Butterworth-Heinemann, Oxford, pp. 101–138.

Menzies, J., (Ed.), 1995b. Modern Glacial Environments: Processes, Dynamics and Sediments. Butterwoth-Heinemann, Oxford, p. 621.

Menzies, J., 1995c. Hydrology of glaciers. In: Menzies, J. (Ed.), Modern Glacial Environments. Processes, Dynamics and Sediments. Butterworth-Heinemann, Oxford, pp. 197–239.

Menzies, J., (Ed.), 1996. Past Glacial Environments: Sediments, Forms and Techniques. Butterwoth-Heinemann, Oxford, p. 598.

Menzies, J., Shilts, W.W., 1996. Subglacial environments. In: Menzies, J. (Ed.), Past Glacial Environments. Sediments, Forms and Techniques. Butterworth-Heinemann, Oxford, pp. 15–136.

Meybeck, M., 1979. Concentrations des eaux fluviales en éléments majeurs et apports en solution aux océans. Rev. Géol. Dyn. Géogr. Phys. 21, 215–246.

Meyer, L.D., 1986. Erosion processes and sediment properties for agricultural cropland. In: Abrahams, A.D. (Ed.), Hillslope Processes. Allen and Unwin, Boston, pp. 55–76.

Miall, A.D., 1978. Lithofacies types and vertical profile models in braided river deposits: a summary. In: Miall, A.D. (Ed.), Fluvial Sedimentology, Canadian Society Petroleum Geologists Memoir, 5, pp. 597–604, Calgary.

Michel, P., 1973. Les basins des fleuves Sénégal et Gambie. Etude géomorphologique. Memoirs ORSTROM 63, (3 Volumes).

Middleton, N., 1997. Desert dust. In: Thomas, D.S.G. (Ed.), Arid Zone Geomorphology. Process, Form in Change in Drylands. Wiley, Chichester, pp. 413–436.

Middleton, N., Thomas, D., (Eds), 1997. World Atlas of Desertification. Arnold, London, p. 182.

Middleton, N.J., 1990. Wind erosion and dust-storm control. In: Goudie, A. (Ed.), Techniques for Desert Reclamation. Wiley, Chichester, pp. 87–108.

Middleton, N.J., Goudie, A.S., Wells, G.L., 1986. The frequency and source areas of dust storms. In: Nickling, W.G. (Ed.), Aeolian Geomorphology. Allen and Unwin, Boston, pp. 237–268.

Milnes, A.R., Thiry, M., Wright, V.P., 1991. Silica accumulations in saprolite and soils in South Australia. In: Nettleton, W.D. (Ed.), Occurrence, characteristics, and genesis of carbonate, gypsum and silica accumulation in soils, Soil Science Society of America, Special Publication, 26, pp. 121–149.

Milliman, J.D., Meade, R.H., 1983. World-wide delivery of river sediment to the oceans. J. Geol. 91, 1–21.

Millington, A.C., Drake, N.A., White, K., Bryant, R.G., 1995. Salt ramps: wind-induced depositional features on Tunisian playas. Earth Surf. Processes Landforms 20, 105–113.

Millot, G., 1964. Géologie des Argiles. Masson, Paris, p. 499.

Mills, H.H., 1990. Thickness and character of regolith on mountain slopes in the vicinity of Mountain Lake, Virginia, as indicated by seismic refraction, and implications for hillslope evolution. Geomorphology 3, 143–157.

Moeyersons, J., 1989. The concentration of stones into a stone-line, as a result from subsurface movements in fine and loose soils in the tropics. In: Alexandre, J., Symoens, J.J. (Eds), Stone Lines, Académie Royale des Sciences D'Outre-Mer, Geo-Eco-Trop, Vol. 11, pp. 11–22.

Mol, J., Vandenberghe, J., Kasse, C., 2000. River response of variations of periglacial climate. Geomorphology 33, 131–148.

Molina, E., Blanco, J.A., 1980. Quelques précisions sur l'altération du Massif Hercynien espagnol. C. R. Acad. Sci. Paris 280, 1293–1296.

Molina, E., Pellitero, E., 1982. Formas periglaciares actuales en la Paramera de Avila. Hipótesis sobre su origen. Botelín de la Real Sociedad Española de Historia Natural 80, 49–56.

Monroe, W.H., 1964. Lithological control in the formation of the towers of Puerto Rico, 20th International Geographical Conference Karst Symposium.

Monroe, W.H., 1966. Formation of tropical karst topography by limestone solution and reprecipitation. Caribb. J. Sci. 6, 1–7.

Mook, W., Woillard, G., 1982. Carbon-14 dates at Gran Pile. Correlation of land and sea chronologies. Science 215, 159–161.

Ministerio de Obras Públicas y Urbanismo, 1987. Medio Ambiente en España, 1986, Dirección General de Medio Ambiente. Ministerio de Obras Públicas y Urbanismo, Madrid, p. 368.

Moran, S., 1971. Glaciotectonic structures in drift. In: Goldthwait, R.P. (Ed.), Till: A Symposium. Ohio State University Press, Columbus, pp. 127–148.

Morgan, R.P.C., 1986. Soil Erosion and Conservation. Longman, Harlow, p. 298.

Morgan, R.P.C., Finney, H.J., Lavee, H., Merrit, E., Noble, C.A., 1986. Plant cover effects on hillslope runoff and erosion: evidence from two laboratory experiments. In: Abrahams, A.D. (Ed.), Hillslope Processes, Binghampton Symposia in Geomorphology. Unwin, London, pp. 77–90.

Morin, J., Benyami, Y., 1977. Rainfall infiltration into bare soils. Water Resour. Res. 13, 813–817.

Mörner, N.A., 1980. The Fennoscandian uplift: geological data and their geodynamical implication. In: Mörner, N.A. (Ed.), Earth Rheology, Isostasy and Eustasy. Wiley, Chichester, pp. 251–284.

Morrison, R.B., 1965. Quaternary geology of the Great Basin. In: Wright, H.E., Frey, D.G. (Eds), The Quaternary of the United States. Princeton University Press, Princeton, NJ, pp. 265–285.

Moss, R.P., 1965. Slope development and soil morphology in a part of south west Nigeria. J. Soil Sci. 16, 192–209.

Motts, W.S., 1970 (Ed.). Geology and Hydrology of Selected Playas in Western United States. Geol. Dept. Unit of Massachussets, Amherst.

Muller, S.W., 1945. Permafrost or permanently frozen ground and related engineering problems. US Geol. Surv. Spec. Paper 62, p. 231.

Müller, F., 1959. Beobachtung uber pingos. Meddeleser om Gronland 153, p. 127.

Müller, F., 1962. Zonation in the accumulation area of the glaciers of Axel Heiberg Island, N.W.T., Canada. J. Glaciol. 4, 302–318.

Murchinson, S.B., 1989. Fluctuation history of Great Salt Lake, Utah, during the last 13,000 years. Ph. Dissertation, University of Utah, Salt Lake City.

Murty, T.S., El-Sabh, M.I., 1988. Edge waves and ice ride-up on shorelines. In: El-Sabh, M.I., Murty, T.S. (Eds), Natural and Man-Made Hazards. D. Reidel Publication, The Netherlands, pp. 429–434.

Mustoe, G.E., 1982. Origin of honeycomb weathering. Bull. Geol. Soc. Am. 93, 108–115.

Mustoe, G.E., 1983. Cavernous weathering in the Capitol Reef Desert, Utah. Earth Surf. Processes Landforms 8, 517–526.

Muxart, T., Birot, P., 1977. L'Alteration Meteorique des Roches, Publications du Département de Géographie de l'Université de Paris-Sorbonne, p. 279.

Nadler, C.T., Schumm, S.A., 1981. Metamorphosis of South Platte and Arkansas rivers, eastern Colorado. Phys. Geogr. 2, 95–115.

Nagel, R.H., 1962. Geology of the Serra do Navio manganese district. Econ. Geol. 57, 481–498.

Nahon, D., 1986. Evolution of iron crusts in tropical landscapes. In: Colman, S.M., Dethier, D.P. (Eds), Rates of Chemical Weathering of Rocks and Minerals. Academic Press, New York, pp. 169–191.

Nahon, D., Beauvales, A., Trescases, J.J., 1985. Manganese concentration through chemical weathering of metamorphic rocks under lateritic conditions. In: Drever, J.I. (Ed.), The Chemistry of Weathering. Reidel, Boston, pp. 277–291.

Nahon, D.B., 1991. Introduction to the Petrology of Soils and Chemical Weathering. Wiley, New York, p. 313.

Nalpanis, P., 1985. Saltating and suspended particles over flat and sloping surfaces. II. Experiments and numerical simulations. In: Barndorff-Nielsen, O.E., Moller, J.T., Romer-Rasmussen, K., Willets, B.B. (Eds), Proceedings of the International Workshop on the Physics of Blown Sand, 8. Institute of Mathematics, University of Aarhus, Denmark, pp. 37–66.

Návar, J., Bryan, R., 1990. Interception loss and rainfall redistribution by three semiarid growing shrubs in northeastern Mexico. J. Hydrol. 115, 51–63.

Neal, J.T., 1965. Environmental setting and general surface characteristics of playas. In: Neal, J.T. (Ed.), Geology, Mineralogy and Hydrology of United States Playas, Air Force Cambridge Research Laboratory, Environmental Research Paper, Vol. 96, pp. 1–29.

Neal, J.T., (Ed.), 1975. Playas and Dried Lakes. Dowden, Hutchinson and Ross, Stroudsburg, p. 411.

Netterberg, F., 1969. Ages of calcretes in southern Africa. S. Afr. Archeol. Bull. 24, 117–122.

Newman, J.C., Philips, J.R.H., 1957. Tunnel erosion in the Riverina. J. Soil Conserv. Serv. NSW 13, 159–169.

Nickling, W.G., 1983. Grain size characteristics of sediment transported during dust storms. J. Sediment. Petrol. 53, 1011–1024.

Nickling, W.G., 1988. The initiation of particle movement by wind. Sedimentology 35, 499–511.

Nickling, W.G., Gillies, J.A., 1993. Dust emission and transport in Mali, West Africa. Sedimentology 40, 859–868.

Nicholas, R.M., Dixon, J.C., 1986. Sandstone scarp form and retreat in the Land of Standing Rocks, Canyonlands National Park, Utah. Z. Geomorphol. 30, 167–187.

Nicholson, S.E., 1981. Saharan climates in historical times. In: Allan, J.A. (Ed.), The Sahara: Ecological Change and Early Economic History. Menas Press, London, pp. 35–59.

Nicholson, S.E., 1993. What is a desert? In: Seely, M. (Ed.), Deserts. Weldon Ower Limited, Australia, pp. 14–25.

Nicholson, S.E., Flohn, H., 1980. African environmental and climatic changes and the general atmospheric circulation in Late Pleistocene and Holocene. Climatic Change 2, 313–348.

Nicod, J., 1992. Formes d'aplanissement et de regularisation des versants dans les roches carbonatées: position des problèmes et eléments de discussion. Tübinger Geogr. Stud. 109, 1–22, Tübingen.

Nicoletti, P.G., Sorriso-Valvo, M., 1991. Geomorphic controls of the shape and mobility of rock avalanches. Bull. Geol. Soc. Am. 103, 1365–1373.

Nielson, J., Kocurek, G., 1987. Surface processes, deposits, and development of star dunes, Dumont dune field, California. Bull. Geol. Soc. Am. 99, 177–186.

Nilsen, T.H., 1993. What is an alluvial fan? Int. Work. Group Fan Deltas Newslett. 1, 3–4.

Nilsen, T.H., Moore, T.E., 1984. Bibliography of Alluvial Fan Deposits. Geo Books, Norwich, p. 96.

Nortcliff, S., Ross, S.M., Thornes, J.B., 1990. Soil moisture, runoff and sediment yield from differentially cleared tropical rainforest plots. In: Thornes, J.B. (Ed.), Vegetation and Erosion. Wiley, Chichester, pp. 420–436.

Norton, S.A., 1973. Laterite and bauxite formation. Econ. Geol. 68, 353–361.

Nyamweru, C., 1980. Rifts and Volcanoes, A Study of the East African Rift System. Nelson Africa, Nairobi, Kenya, p. 128.

Nye, J.F., 1952. The mechanics of glacier flow. J. Glaciol. 2, 82–93.

Nye, J.F., 1958. Surges in glaciers. Nature 181, 1450–1451.

O'Hara, S.L., 1997. Human impacts on drylands geomorphic processes. In: Thomas, D.S.G. (Ed.), Arid Zone Geomorphology. Process, Form and Change in Drylands. Wiley, Chichester, pp. 639–658.

Oberlander, T.M., 1972. Morphogenesis of granitic boulder slopes in the Mojave Desert California. J. Geol. 80, 1–20.

Oberlander, T.M., 1977a. Origin of segmented cliffs in massive sandstones of southeastern Utah. In: Doehring, D.O. (Ed.), Geomorphology in Arid Regions. George Allen and Unwin, London, pp. 79–114.

Oberlander, T.M., 1997b. Slope and pediment systems. In: Thomas, D.S.G. (Ed.), Arid Zone Geomorphology. Wiley, Chichester, pp. 135–163.

Oberlander, T.M., 1994. Rock varnish in deserts. In: Abrahams, A.D., Parsons, A.J. (Eds), Geomorphology of Desert Environments. Chapman & Hall, London, pp. 106–119.

Oerlemans, J., 1989a. On the response of valley glaciers to climatic change. In: Oerlemans, J. (Ed.), Glacier Fluctuations and Climatic Change. Kluwer, Dordrecht, pp. 353–371.

Oerlemans, J., (Ed.), 1989b. Glacier Fluctuations and Climatic Change. Kluwer, Dordrecht, p. 417.

Ogunkoya, O.O., Jeje, L.K., 1987. Sediment yield from some third order basins of the basement complex rocks in southwestern Nigeria. Catena 14, 383–397.

Ollier, C., 1977. Applications of weathering studies. In: Hails, J.R. (Ed.), Applied Geomorphology. Elsevier, Amsterdam, pp. 9–50.

Ollier, C., 1984. Weathering. Longman, London, p. 304.

Ollier, C.D., 1960. The inselbergs of Uganda. Z. Geomorphol. 4, 43–52.

Ollier, C.D., 1965. Some features of granite weathering in Australia. Z. Geomorphol. 9, 285–304.

Ollier, C.D., 1966. Desert gilgai. Nature 212, 581–583.

Ollier, C.D., 1978. Inselbergs of the Namib Desert, processes and history. Z. Geomorphol. Suppl.bd. 31, 161–176.

Ollier, C.D., 1981. Tectonics and Landforms. Longman, London, p. 324.

Ollier, C.D., 1984. Weathering. Longman, London, p. 270.

Ollier, C.D., 1988a. Deep weathering, groundwater and climate. Geogr. Ann. 70A, 285–290.

Ollier, C.D., 1988b. The regolith in Australia. Earth Sci. Rev. 25, 355–362.

Ollier, C.D., 1991. Ancient Landforms. Belhaven Press, London, p. 233.

Ollier, C.D., Ash, J.E., 1983. Fire and rock breakdown. Z. Geomorphol. 27, 363–374.

Ollier, C.D., Pain, C., 1996. Regolith, Soils and Landforms. Wiley, Chichester, p. 316.

Oostwoud, D.J.W., Bryan, R.R., 1994. Gully head cuts as sediment sources on the Njemps Flats and initial low-cost gully control measures. In: Bryan, R.R. (Ed.), Soil erosion, land degradation and social transition, Advances in GeoEcology, Vol. 27, pp. 205–229.

Ortí, F., 1989. Evaporitas marinas. In: Arche, A. (Coor.), Sedimentología, Vol. II. Consejo Superior de Investigaciones Científicas, Madrid, pp. 89–177.

Ortiz, A., 1993. Hydrology information and voice communication automatic system in the Ebro basin. In: Gutiérrez, M., Sancho, C., Benito, G. (Eds), Second Intensive Course on Applied Geomorphology: Arid Regions, Zaragoza, pp. 89–105.

Osterkamp, W.R., Wood, W.W., 1987. Playa-lake basins on the Southern High Plains and New Mexico: Part 1 – hydrologic, geomorphic and geologic evidence for their development. Bull. Geol. Soc. Am. 99, 215–223.

Ostrem, G., Haakensen, N., Melander, O., 1973. Atlas over Breer i Nord-Skandinavia. Norges Veerdrags of Elektrisitetsvesen on Stockholm Univ., 315 p.

Oviatt, C.G., 1997. Lake Bonneville fluctuations and global climate change. Geology 25, 155–158.

Owens, I., 2004. Avalanche, snow. In: Goudie, A.S. (Ed.), Encyclopedia of Geomorphology, Routledge, pp. 41–44.

Owens, L.B., Watson, J.P., 1979. Landscape reduction by weathering in small Rhodesian watersheds. Geology 7, 281–284.

Pachur, H.J., Röper, H.P., Kroplin, S., Goschin, M., 1987. Late Quarternary hydrography of the eastern Sahara. Berliner Geowissenschaften Abhandlungen, 75, 331–384.

Pain, C.F., Ollier, C.D., 1981. Geomorphology of a Pliocene granite in Papua New Guinea. Z. Geomorphol. 25, 249–258.

Palmer, R.S., 1963. The influence of a thin water layer on water drop impact forces. Int. Assoc. Sci. Hydrol. 65, 141–148.

Pallister, J.W., 1956. Slope development in Buganda. Geogr. J. 122, 80–87.

Parker, G.G., 1963. Piping, a geomorphic agent in landform development of the drylands. Int. Assoc. Sci. Hydrol. Publ. 65, 103–113.

Parker, G.G., Jenne, E.A., 1967. Structural failure of Western US highways caused by piping, United States Geological Survey. Water Resources Division, p. 27.

Parker, G.G., Higgins, G.C., Parker, G.G., Wood, W.W., 1990. Piping and pseudokarst in drylands. In: Higgins, G.C., Coates, D.R. (Eds), Groundwater Geomorphology: The Role of Subsurface Water in Earth-surface Processes and Landforms. Geological Society of America, Boulder, pp. 77–110, Special Paper 252.

Parmenter, C., Folger, D.W., 1974. Eolian biogenic detritus in deep sea sediments: a possible index of equatorial Ice Age aridity. Science 185, 695–698.

Parra, E., Capeda, H., 1990. Volcanic hazards maps of the Nevado del Ruiz volcano, Columbia. J. Volcanol. Geotherm. Res. 42, 117–127.

Passarge, S., 1904. Die Kalahari: Versuch einer physisch-geographischen darstellung der sandfelden des Südafrikanischen beckens. Reiner, Berlin.

Passarge, S., 1931. Geomorfología. Ed. Labor, Barcelona, p. 189.

Paté-Cornell, E., 1996. Uncertainties in global climate change estimates. An editorial essay. Climatic Change 33, 145–149.

Paterson, W.S.B., 1994. The Physics of Glaciers. Pergamon, New York, p. 480.

Pavich, M.J., 1985. Appalachian piedmont morphogenesis: weathering, erosion, and Cenozoic uplift. In: Morisawa, M., Hack, J.T. (Eds), Tectonic Geomorphology. George Allen and Unwin, London, pp. 27–51.

Pavlov, A.V., 1994. Current changes of climate and permafrost in the Arctic and Sub-Arctic of Russia. Permafrost Periglac. Processes 5, 101–110.

Pearce, F., 1994. Not warming, but cooling. New Sci. 9, 37–41, July.

Pécsi, M., 1968. Loess. In: Fairbridge, R.W. (Ed.), The Encyclopedia of Geomorphology. Reinold, New York, pp. 674–678.

Pécsi, M., Sailárd, J., 1970. Planated surfaces: principal problems of research and terminology. In: Pécsi, M. (Ed.), Problems of Relief Planation. Akadémiai Kiadó, Budapest, pp. 13–27.

Pedraza, J., Sanz, M.A., Martín, A., 1989. Formas Graníticas de La Pedriza. Agencia de Medio Ambiente, Madrid, p. 205.

Pedro, G., 1966. Essai sur le caractérisation géochimique des différents processus zonaux résultant de l'altération superficielle. C. R. Acad. Sci. Paris 262, 1828–1831.

Pedro, G., 1968. Distributions des principaux types d'altération chimique à la surface du globe. Rev. Géogr. Phys. Géol. Dyn. 10, 457–470.

Pedro, G., Jamague, M., Bejon, J.C., 1969. Mineral interactions and transformations in relation to pedogenesis during the Quaternary. Soil Sci. 107, 462–469.

Peel, R.F., 1974. Insolation weathering: some measurements of diurnal temperature changes in exposed rocks in the Tibesti region, central Sahara. Z. Geomorph. - Suppl., 21, 19–28.

Peltier, L.C., 1950. The geographic cycle in periglacial regions as it is related to climatic geomorphology. Ann. Assoc. Am. Geogr. 40, 214–236.

Peltier, W.R., Andrews, J.T., 1983. Glacial geology and glacial isostasy of the Hudson Bay region. In: Smith, D.E., Dawson, A.G. (Eds), Shorelines and Isostasy. Academic Press, London, pp. 285–320.

Peña, J.L., Chueca, J., Julián, A., 1998. Los derrubios estratificados del sector central pirenaico: cronología y límites altitudinales. In: Gómez Ortiz, A., Salvador Franch, F., Schulte, L., García Navarro, A. (Eds), Procesos biofísicos actuales en medios fríos. Publicaciones de la Universitat de Barcelona, Barcelona, pp. 205–216.

Peña, J.L., Echeverría, M.T., Petit-Maire, N., Lafont, R., 1993. Cronología e interpretación de las acumulaciones holocenas de la Val de las Lenas (Depresión del Ebro, Zaragoza). Geographicalia, Zaragoza, 30, pp. 321–332.

Penck, A., 1905. Glacial features in the surface of the Alps. J. Geol. 13, 1–19.

Penck, W., 1924. Die Morphologische Analyse. Geogr. Abh. 2, Stuttgart, 1–283.

Penck, W., 1953. Morphological Analysis of Landforms. Macmillan, London, p. 429.

Pérez del Campo, P., 1989. Nuevos datos sobre colapsos kársticos en la Depresión del Ebro. Dirección e incidencia en la nueva línea de alta velocidad Madrid-Barcelona. II Reunión del Cuaternario Ibérico, Madrid, pp. 905–911.

Persons, B.S., 1970. Laterite: Genesis, Location and Use. Plenum Press, New York, p. 103.

Persons, B.S., 1984. Laterite, Engineering Geology. In: Finkl, C.W. (Ed.), The Encyclopedia of Applied Geology. Van Nostrand Reinhold, New York, pp. 320–326.

Perthuisot, J.P., Jauzein, A., 1975. Sebkha et dunes d'argile: l'enclave endoreique de Pont du Fars, Tunisie. Rev. Geógr. Phys. Géol. Dyn. 17, 295–306.

Petit-Maire, N., 1999. Variabilité naturelle des environments terrestres: les deux dernieres extrêmes climatiques (18,000 + 1000 ans BP). C. R. Acad. Sci. Paris 328, 273–279.

Petri, S., Fúlgaro, V.F., 1983. Geología do Brasil. Editora da Universidade de Sao Paulo, Sao Paulo, p. 631.

Petts, G., Foster, I., 1985. Rivers and Landscape. Arnold, London, p. 274.

Péwé, T.L., 1955. Origin of the upland silt near Fairbanks, Alaska. Bull. Geol. Soc. Am. 66, 699–724.

Péwé, T.L., 1959. Sand wedge polygons (tesselations) in the McMurdo Sound region, Antarctica. Am. J. Sci. 257, 545–552.

Péwé, T.L., 1966. Paleoclimatic significance of fossil wedges. Biuletyn Peryglacjalny 15, 65–73.

Péwé, T.L., 1981. Desert dust: an overview. In: Pewé, T.L. (Ed.), Desert Dust: Origin, Characteristics and Effect on Man. Geological Society of America, Boulder, pp. 1–10, Special Paper 186.

Péwé, T.L., 1983. The periglacial environment in North America during Wisconsin time. In: Porter, S.C. (Ed.), Late Quaternary of the United States, The Late Pleistocene, Vol. I. Longman, London, pp. 157–189.

Péwé, T.L., Tungsheng, L., Slatt, R.M., Bingyuan, L., 1995. Origin and Character of Loesslike Silt in the Southern Qinghai-Xizang (Tibet) Plateau, China, United States Geological Survey Professional Paper, Vol. 1549, p. 55.

Pfeffer, W.-T., 1992. Stress-induced foliation in the terminus of variegated glacier, Alaska, USA, formed during the 1982–1983 surge. J. Glaciol. 38, 213–222.

Philbrick, S.S., 1970. Horizontal configuration and the rate of erosion of Niagara Falls. Bull. Geol. Soc. Am. 81, 3723–3732.

Pias, J., 1970. Les Formations Sédimentaires Tertiaires et Quaternaires de la Cuvette Tchadienne et les Sols qui en Dérivent. ORSTOM Mem. 43.

Pichler, E., 1958. Aspectos geológicos dos escorregamentos de Santos. Noticia Geomorfológica 2, 40–44.

Pierson, T.C., 1983. Soil pipes and slope stability. Q. J. Eng. Geol. 16, 1–11, London.

Pilgrim, D., Cordery, I., Baron, B., 1982. Effects of catchment size on runoff relationships. J. Hydrol. 58, 205–221.

Pissart, A., 1965. Les pingos des Hautes Fagnes: le problème de leur genèse. Ann. Soc. Géol. Bel. 89, 377–402.

Pissart, A., 1970. Les phénoménes physiques essentiels liés au gel; les structures périglaciaires qui en résultent et leur signification climatique. Ann. Soc. Géol. Belg. 93, 7–49.

Pissart, A., 1988. Pingos: an overview of the present state of knowledge. In: Clark, M.J. (Ed.), Advances in Periglacial Geomorphology. Wiley, Chichester, pp. 279–297.

Pissart, A., 1990. Advances in periglacial geomorphology. Z. Geomorphol. Suppl.bd. 79, 119–131.

Pissart, A., 2000. Remnants of lithalsas of the Hautes Fagnes, Belgium: a summary of present-day knowledge. Permafrost Periglac. Processes 11, 327–355.

Pissart, A., 2004. Thermokarst. In: Goudie, A.S. (Ed.), Encyclopedia of Geomorphology. Routledge, London, pp. 1049–1051.

Plafker, G., Ericksen, G.E., 1978. Nevados Huascarán avalanches, Perú. In: Voight, B. (Ed.), Rockslides and Avalanches, Vol. 1. Elsevier, Amsterdam, pp. 277–314.

Ploey, J. de., 1980. Some field measurements and experimental data on wind-blown sands. In: Boodt, M. de y Gabriels, D., (Eds), Assessment of erosion. John Wiley, New York, pp. 541–552.

Poessen, J., 1986. Surface sealing as influenced by slope angle and position of simulated stones in the top layer of loose sediments. Earth Surf. Processes Landforms 11, 1–10.

Poessen, J., Lavee, H., 1991. Effects of size and incorporation of synthetic mulch on runoff and sediment yield from interrills in a laboratory study with simulated rainfall. Soil Till. Res. 21, 209–223.

Poessen, J., Ingelmo-Sánchez, F., Mucher, H., 1990. The hydrological response of soil surfaces to rainfall as affected by cover and position of rock fragments in the top layer. Earth Surf. Processes Landforms 15, 653–672.

Pohjola, V.A., 1994. TV-video observations of bed and basal sliding on Storglaciären, Sweden. J. Glaciol. 39, 111–118.

Pokras, E.M., Mix, A.C., 1985. Eolian evidence for spatial variability of Late Quaternary climates in tropical Africa. Quaternary Res. 24, 137–149.

Poland, J.F., Davis, G.H., 1969. Land subsidence due to withdrawal of fluids. In: Varnes, D.J., Kiersch, G. (Eds), Reviews in Engineering Geology, Geological Society of America, Vol. 2, pp. 187–269.

Polynov, B., 1937. The Cycle of Weathering. Murby, London, p. 220.

Pollard, W.E., 1988. Seasonal frost mounds. In: Clark, M.J. (Ed.), Advances in Periglacial Geomorphology. Wiley, Chichester, pp. 201–229.

Ponti, D.J., 1985. The Quaternary alluvial sequence of the Antelope Valley, California. Geol. Soc. Am. Spec. Paper 203, 79–96.

Porath, A., Shick, A.P., 1974. The use of remote sensing systems in monitoring desert floods. Int. Assoc. Hydrol. Sci. Publ. 112, 133–138.

Porter, S.C., Zhisheng, An., 1995. Correlation between climate events in the North Atlantic and China during the last glaciation. Nature 375, 305–308.

Posamentier, H.W., 1978. Thoughts on ogive formation. J. Glaciol. 20, 218–220.

Poser, H., 1948. Boden-und klimaverhältnisse in Mittel und Westeuropa während der Würmmeiszeit. Erdkunde 2, 53–68.

Poser, H., 1950. Zur rekonstruktion der spätglazialen luftdruckverhältnisse in Mittel-und Westeuropa auf grund der vorzeitlichen binnendünen. Erdkunde 4, 81–88.

Potter, N., 1972. Ice-cored rock glacier, Galena Creek, Northern Absaroka Mountains, Wyoming. Bull. Geol. Soc. Am. 83, 3025–3057.

Potter, N., Moss, J.H., 1968. Origin of the Blue Rocks block field and adjacent deposits, Berks County, Pennsylvania. Bull. Geol. Soc. Am. 79, 255–262.

Powell, R.D., 1981. A model for sedimentation by tidewater glaciers. Ann. Glaciol. 2, 129–134.

Powell, R.D., 2003. Subaquatic landsystems: fjords. In: Evans, D.J.A. (Ed.), Glacial Landsystems. Arnold, London, pp. 313–347.

Prest, N.K., 1968. Nomenclature of moraines and ice-flow features as applied to the glacial map of Canada. Geol. Surv. Can., Paper 66-57.

Price, R.J., 1966. Eskers near the Casement, Alaska glacier. Geogr. Ann. 48, 111–125.

Price, R.J., 1969. Moraines, sandar, kames and eskers near Breidamerkurjökull, Iceland. Trans. Inst. Br. Geogr. 46, 17–43.

Price, R.J., 1970. Moraines at Fjallsjökull, Iceland. Arctic Alpine Res. 2, 27–42.

Price, R.J., 1973. Glacial and Fluvioglacial Landforms. Longman, Edinburgh, p. 242.

Price, R.J., 1980. Rates of geomorphological changes in proglacial areas. In: Cullinford, R.A., Davidson, D.A., Lewin, J. (Eds), Timescales in Geomorphology. Wiley, Chichester, pp. 79–93.

Price, W.A., 1963. Physico-chemical and environmental factors in clay-dune genesis. J. Sediment. Petrol. 33, 766–778.

Price, W.A., 1968. Carolina bays. In: Fairbridge, R.W. (Ed.), The Encyclopaedia of Geomorphology. Reinhold, New York, pp. 102–109.

Priesnitz, K., 1988. Cryoplanation. In: Clark, M.J. (Ed.), Advances in Periglacial Geomorphology. Wiley, Chichester, pp. 49–67.

Prospero, J.M., Bonatti, E., Schubert, C., Carlson, T.N., 1970. Dust in the Caribbean atmosphere traced to an African dust storm. Earth Planet. Sci. Lett. 9, 287–293.

Prouty, W.F., 1952. Carolina bays and their origin. Bull. Geol. Soc. Am. 63, 167–224.

Pueyo, J.J., 1978–1979. La precipitación evaporítica actual en las lagunas del área: Bujaraloz, Sástago, Caspe, Alcañiz y Calanda (Provs. de Zaragoza y Teruel). Revista del Instituto de Investigaciones Geológicas 33, 5–56, Barcelona.

Pugh, J.C., 1956. Fringing pediments and marginal depressions in the inselberg landscape of Nigeria. Trans. Inst. Br. Geogr. 22, 15–31.

Pugh, J.C., 1966. Landforms in low latitudes. In: Dury, G.H. (Ed.), Essays in Geomorphology. American Elsevier, New York, pp. 121–138.

Pulido-Bosch, A., 1986. Le karst dans les gypses de Sorbas (Almería). Aspects morphologiques et hydrogéologiques. Karstologie, 27–35, Memoire 1.

Punkari, M., 1997. Subglacial processes of the Scandinavian Ice Sheet in Fennoscandia inferred from flow-parallel features and lithostratigraphy. Sediment. Geol. 111, 263–283.

Pye, K., 1982. Morphological development of coastal dunes in a humid tropical environment, Cape Bedford and Cape Flattery, North Queensland. Geogr. Ann. A64, 212–227.

Pye, K., 1987. Aeolian Dust and Dust Deposits. Academic Press, London, p. 334.

Pye, K., 1995. The nature, origin and accumulation of loess. Quaternary Sci. Rev. 14, 653–667.

Pye, K., Tsoar, H., 1987. The mechanics and geological implications of dust transport and deposition in deserts, with particular reference to loess formation and dune sand diagenesis in the northern Negev, Israel. In: Frostick, L.E., Reid, I. (Eds), Desert Sediments, Ancient and Modern, Special Publication 35, Geological Society of London. Blackwell, Oxford, pp. 139–156.

Pye, K., Tsoar, H., 1990. Aeolian Sand and Sand Deposits. Unwin Hyman, London, p. 396.

Quirantes, J., 1965. Notas sobre las lagunas de Bujalaroz-Sástago. Geographica 12, 30–34.

Rachocki, A., 1981. Alluvial Fans. An Attempt at an Empirical Approach. Wiley, Chichester, p. 161.

Rachocki, A.H., Church, M., (Eds), 1990. Alluvial Fans. A Field Approach. Wiley, Chichester, p. 391.

Ramírez, E., Francou, B., Ribstein, P., Descloitres, M., Guérin, R., Mendoza, J., Gallaire, R., Poyaud, B., Jordan, E., 2001. Small glacier disappearing in the tropical Andes: a case-study in Bolivia: Glaciar Chacaltaya (16°S). J. Glaciol. 47 (157), 187–194.

Rantz, S.E., 1970. Urban sprawl and flooding in Southern California. US Geol. Surv. Circ. 601B.

Rapp, A., 1957. Studien über Schutthalden in Lappland und auf Spitzbergen. Z. Geomorphol. 1, 179–200.

Rapp, A., 1984. Are terra rossa soils in Europe eolian deposits from Africa? Geologiska Föreningens I Stocklom Förhendlingar 105, 161–168.

Rapp, A., 1986. Slope processes in high latitude mountains. Prog. Phys. Geogr. 10, 53–68.

Rasid, H., Pramanik, M.A.H., 1993. Areal extent of the 1988 flood in Bangladesh: how much did the satellite imagery show? Nat. Hazards 8, 189–200.

Raso, J.M., García Loureiro, M.C., 1998. Oscillacions termomètiques al'entorn de 08 C a Andorra. In: Gómez Ortiz, A., Salvador Franch, F., Schulte, L., García Navarro, A. (Eds), Procesos biofísicos actuales en medios fríos. Publicacions de la Universitat de Barcelona, Barcelona, pp. 217–234.

Raunet, M., 1985. Les bas-fonds en Afrique et à Madagascar. Z. Geomorphol. Suppl.bd. 52, 25–62.

Raymo, M.E., Ruddiman, W.F., Shackleton, N.J., Oppo, D.W., 1990. Evolution of Atlantic–Pacific d^{13}C gradients over the last 2.5 m.y. Earth Planet. Sci. Lett. 97, 353–368.

Rea, B.R., Evans, D.J.A., 1996. Landscapes of areal scouring in NW Scotland. Scot. Geogr. Mag. 112, 47–50.

Rea, B.R., Evans, D.J.A., Dixon, T.S., Whalley, W.B., 2000. Contemporaneous, localized, basal ice-flow variations: implications for bedrock erosion and the origin of p-forms. J. Glaciol. 46, 470–476.

Reading, A.J., Thompson, R.D., Millington, A.C., 1995. Humid Tropical Environments. Blackwell, Oxford, p. 429.

Reeves, C.C., 1964. Gas rings from Terry County, Texas. J. Sediment. Petrol. 34, 190–193.

Reeves, C.C., 1968. Introduction to Paleolimnology. Elsevier, Amsterdam, p. 228.

Reeves, C.C., 1976. Caliche: Origin, Classification, Morphology and Uses. Estacado Books, Lubbock, TX, p. 233.

Regües, D., Pardini, G., Pini, R., 1992. Estudio del comportamiento de regolitos arcillosos en zonas abarrancadas frente a variaciones de la temperatura y la humedad. In: López, F., Conesa, C., Romero, M.A. (Eds), Estudios de Geomorfología en España, pp. 171–182, Murcia.

Reiche, P., 1950. A Survey of Weathering Processes and Products, Publications in Geology n° 3. New Mexico University, p. 95.

Reid, I., 1994. River landforms and sediments: evidence of climatic change. In: Abrahams, A.D., Parsons, A.J. (Eds), Geomorphology of Desert Environments. Chapman & Hall, London, pp. 571–592.

Renard, K.G., 1969. Evaporation from an ephemeral stream bed: discussion. J. Hydr. Div. Proc. Am. Soc. Civ. Eng. 95, 2200–2204.

Renard, K.G., 1970. The hydrology of semiarid rangeland watersheds. US Dept. Agr. Res. Ser. Pub., 41–162.

Renard, K.G., Keppel, R.V., 1966. Hydrographs of ephemeral streams in the Southwest. Proc. Am. Soc. Civ. Eng. J. Hydrol. Div. 92, 33–52.

Rettig, S.L., Jones, B.F., Risacher, F., 1980. Geochemical evolution of brines in the Salar of Uyuni, Bolivia. Chem. Geol. 30, 57–79.

Rhoades, J.D., 1990. Soil salinity–causes and controls. In: Goudie, A.S. (Ed.), Techniques for Desert Reclamation. Wiley, Chichester, pp. 109–134.

Rhoads, B.L., 1986. Flood hazard assessment for land-use planning near desert mountains. Environ. Manage. 10, 97–106.

Riba, O., Llamas, M.R., 1962. Canales de La Violada, Monegros y Flumen, Libro-Guía del Viaje de Estudio n° 5. I Coloquio Internacional sobre las Obras Públicas en Terrenos Yesíferos. Servicio Geológico de Obras Públicas, Madrid, p. 28.

Richardson, C., Holmund, P., 1996. Glacial cirque formation in northern Scandinavia. Ann. Glaciol. 22, 102–106.

Ritter, D.F., 1978. Process Geomorphology. WM. C. Brown, Dubuque, Iowa, p. 603.

Roberts, C.R., Mitchell, C.W., 1987. Spring mounds in southern Tunisia. In: Frostick, L.E., Reid, I. (Eds), Desert Sediments: Ancient and Modern, Geological Society of London, Special Publication 35, pp. 321–336.

Robin, G., De, Q., Drewry, D.J., Meldrum, D.T., 1977. International studies of ice sheet and bedrock. Philos. Trans. R. Soc. London 279, 185–196.

Robinson, D.A., Williams, R.B.G., 1992. Sandstone weathering in the High Atlas, Morocco. Z. Geomorphol. 4, 413–429.

Rodríguez, J., Navascués, L., 1982. La tafonización de las areniscas miocenas en los alrededores de Huesca. Tecniterrae 19, 7–12.

Rogers, R.D., Schumm, S.A., 1991. The effect of sparse vegetative cover on erosion and sediment yield. J. Hydrol. 123, 19–24.

Rognon, P., 1980. Pluvial and arid phases in the Sahara: the role of nonclimatic factors. In: Sarnthein, M., Seibold, E., Rognon, P. (Eds), Palaeoecology of Africa. Sahara and Surrounding Seas. Balkema, Rotterdam, pp. 45–62.

Rognon, P., 1982. Pluvial and arid phases in the Sahara: the role of nonclimatic factors. Palaeoecology of Africa, 12, 45–62.

Rognon, P., 1987. Late Quarternary climatic reconstruction for the Maghreb (North Africa). Palaeogeography, Palaeoclimatology, Palaeoecology, 58, 11–34.

Rogowski, A.S., Khanbilvardi, R.M., Deangelis, R.J., 1985. Estimating erosion on plot, field and watershed scales. In: El-Swaifi, S.A., Molldenhaner, W.C., Andrew, L. (Eds), Soil Erosion and Conservation. Soil Conservation Society of America, pp. 149–166.

Rojo, L., Sánchez, M.C., 1996. Red de Estaciones Experimentales de Seguimiento y Evaluación de la Erosión y Desertificación (RESEL), Catálogo de Estaciones, Proyecto Lucdeme. Dirección General de Conservación de la Naturaleza. Ministerio de Medio Ambiente, Madrid, p. 121.

Roose, E.J., Lelong, F., 1976. Les facteurs de l'érosion hydrique en Afrique tropicale. Etudes sur petites parcelles expérimentales du sol. Rev. Géogr. Phys. Géol. Dyn. 18, 365–374.

Roque, C., Pallí, L., 1998. Geología de l'Ardenya i Formes Granítiques Associades. Publicacions de l'Institut d'Estudis del Baix Empordá 17, 5–42.

Rosen, M.R., 1994. The importance of groundwater in playas: a review of playa classifications and the sedimentology and hydrology of playas. In: Rosen, M.R. (Ed.), Paleoclimate and basin evolution of playa systems, Geological Society of America, Special Paper 289, pp. 1–18.

Rust, B.R., 1979. Facies models, coarse alluvial deposits. In: Walker, R.G. (Ed.), Facies models, Geoscience Canada Reprint Series, Vol. 1, pp. 9–21.

Rutte, E., 1958. Kalkruste in Spanien. Neues Jahrb. Geol. Paläeontol. 106, 52–138.

Rutter, N.W., 1965. Foliation pattern of Gulkana glacier, Alaska Range, Alaska. J. Glaciol. 5, 711–718.

Ruxton, B.P., 1958. Weathering and sub-surface erosion in granites at the piedmont angle, Balos, Sudan. Geol. Mag. 95, 353–377.

Ruxton, B.P., 1967. Slope wash under nature primary rainforest in northern Papua. In: Jennings, J.N., Mabbutt, J.A. (Eds), Landform Studies from Australia and New Guinea. Cambridge University Press, London, pp. 85–94.

Ruxton, B.P., Berry, L., 1957. Weathering of granite and associated erosional features in Hong Kong. Bull. Geol. Soc. 68, 1263–1292.

Ruxton, B.P., Berry, L., 1961. Weathering profiles and geomorphic position on granite in two tropical regions. Rev. Géomorphol. Dyn. 12, 16–31.

Sack, D., 1994. Geomorphic evidence of climate change from desert-basin paleolakes. In: Abrahams, A.D., Parsons, A.J. (Eds), Geomorphology of Desert Environments. Chapman & Hall, London, pp. 616–630.

Sakamoto-Arnold, C.M., 1981. Eolian features produced by the December, 1977 windstorm, southern San Joaquin Valley, California. J. Geol. 89, 129–137.

Salinas, J.L., 1988. Riesgos ligados a arcillas expansivas. In: Ayala, F.J., Durán, J.J., Peinado, T. (Eds), Riesgos Geológicos. Instituto Geológico y Minero de España, Madrid, pp. 295–304.

Sanches Furtado, A.F.A., 1968. Altération des granites dans les régions intertropicales sous différents climats. Ninth International Congress of Soil Science, Adelaide 4, 403–409.

Sánchez, J.A., Martínez, F.J., Floría, E., Schumann, S., 1993. Hydrogeological characterization of the Monegros Lakes. In: Gutiérrez, M., Sancho, C., Benito, G. (Eds), Second Intensive Course on Applied Geomorphology: Arid Regions. Universidad de Zaragoza, Zaragoza, pp. 245–252.

Sancho, C., Benito, G., 1990. Factors controlling tafoni weathering in the Ebro Basin (NE Spain). Zeitscrift für Geomorphol. 34, 165–177.

Sancho, C., Gutiérrez, M., 1990. Análisis morfométrico de la tafonización de la arenisca de Peraltilla (Anticlinal de Barbastro. Depresión del Ebro): Influencia de los factores mineralógicos-texturales. Cuaternario y Geomorfología 4, 131–145.

Sancho, C., Gutiérrez, M., 1993. Geomorphological features of the Bujalaroz salt lakes, Second Intensive Course of Applied Geomorphology: Arid Regions. Universidad de Zaragoza, Zaragoza, pp. 241–243.

Sancho, C., Meléndez, A., 1992. Génesis y significado ambiental de los caliches pleistocenos de la región del Cinca (Depresión del Ebro). Revista de la Sociedad Geológica de España 5, 81–93.

Sancho, C., Gutiérrez, M., Peña, J.L., Burillo, F., 1988. A quantitative approach to scarp retreat starting from triangular slope facets (Central Ebro Basin Spain). Catena Suppl. 13, 139–146.

Sancho, C., Meléndez, A., Signes, M., Bastida, J., 1992. Chemical and mineralogical characteristics of Pleistocene caliche deposits from the Central Ebro basin. Clay Miner. 27, 293–308.

Sanjaume, E., 1985. Las Costas Valencianas. Sedimentología y Geomorfología. Sección de Geografía. Universidad de Valencia, Valencia, p. 505.

Santer, B.D., Wigley, T.M.L., Barnett, T.P., Anyamba, B., 1996. Detection of climatic change and attribution of causes. In: Houghton, J.T et al. (Eds), Climate Change 1995. The Science of Climatic Change. Cambridge University Press, Cambridge, MA, pp. 406–443.

Sarin, M.M., Krishnaswami, S., Dilli, K., Somayajulu, B.L.K., Moore, W.S., 1989. Major ion chemistry of the Ganga–Brahmaputra river system: weathering processes and fluxes to the Bay of Bengal. Acta Geochim. Cosmochim. 53, 997–1009.

Sarnthein, M., 1978. Sand deserts during the last glacial maximum and climatic optimum. Nature 272, 43–46.

Sarnthein, M., Walger, K., 1974. Der äolische Sandstrom aus der W-Sahara zur Atlantikküste. Geologische Rundschau 63, 1065–1087.

Saunders, I., Young, A., 1983. Rate of surface processes on slopes, slope retreat and denudation. Earth Surf. Processes Landforms 8, 473–501.

Savat, J., Poesen, J., 1981. Detachment and transportation of loose sediments by raindrop splash. Part I. The calculation of absolute data on detachability and transportability. Catena 8, 1–17.

Savigear, R.A.G., 1952. Some observations on slope development in South Wales. Trans. Inst. Br. Geogr. 18, 31–51.

Savigear, R.A.G., 1960. Slopes and hills in West Africa. Z. Geomorphol. Suppl.bd. 1, 156–171.

Scheidegger, A.E., Schumm, S.A., Fairbridge, R.W., 1968. Badlands. In: Fairbridge, R.W. (Ed.), The Encyclopedia of Geomorphology. Hutchinson and Ross, Stroudsburg, PA, pp. 43–48.

Schellmann, W., 1977. The formation of lateritic silicate bauxites and criteria for their exploration and assessment of deposits. Nat. Resour. Dev. 5, 119–134.

Schellmann, W., 1981. Considerations on the definition and classification of laterites. *Proceedings International Seminar on Lateritisation Processes, Trivandrum. Lateritisation Processes*. Balkema, Rotterdam, pp. 1–10.

Schick, A.P., Lekach, J., Hassan, M.A., 1987. Vertical exchange of coarse bedload in desert streams. In: Frostick, L.E., Reid, I. (Eds), Desert Sediments: Ancient and Modern, Geological Society of London, Special Publication 35. Blackwell, Oxford, pp. 7–16.

Schipull, K., 1980. Die Cedar Mesa – Schichtstufe aud dem Colorado Plateau – ein Beispiel für die Morphodynamik arider Schichtstufen. Z. Geomorphol. 24, 318–331.

Schmidt, K.H., 1980. Eine neue Metode zur Ermittlung von Stufenruckwanderungsrate, dargestellt am Beispiel der Black Mesa Schichtstufen, Colorado Plateau, USA. Z. Geomorphol. 24, 180–191.

Schmidt, K.H., 1987. Factors influencing structural landform dynamics on the Colorado Plateau about the necessity of calibrating theoretical models by empirical data. Catena Suppl. 10, 51–66.

Schmidt, K.H., 1988. Rates of scarp retreat: a means of dating neotectonic activity. In: Jacobshagen, V.H. (Ed.), The Atlas System of Morocco Studies on its Geodynamic Evolution, Lecture Notes in Earth Science 15, Berlin, pp. 445–462.

Schmidt, K.H., 1989a. Talus and pediment flatirons – erosional and depositional features of dryland cuesta scarps. Catena Suppl. 14, 107–118.

Schmidt, K.H., 1989b. The significance of scarp retreat for Cenozoic landform evolution on the Colorado Plateau, USA. Earth Surf. Processes Landforms 14, 93–105.

Schmidt, K.H., 1994. Hillslopes as evidence of climatic change. In: Abrahams, A.D., Parsons, J. (Eds), Geomorphology of Deserts Environments. Chapman & Hall, London, pp. 553–570.

Schmidt, K.H., 1996. Talus and pediment flatirons-indicators of climatic change on scarp slope on the Colorado Plateau, USA. Z. Geomórphol. NF Suppl.bd. 103, 135–158.

Schumm, S.A., 1956a. Evolution of drainage systems and slopes in badlands at Perth Amboy, New Jersey. Bull. Geol. Soc. Am. 67, 597–646.

Schumm, S.A., 1956b. The role of creep and rainwash on the retreat of badland slopes. Am. J. Sci. 254, 693–706.

Schumm, S.A., 1962. Erosion on miniature pediments in Badlands National Monument, South Dakota. Bull. Geol. Soc. Am. 73, 719–724.

Schumm, S.A., 1964. Seasonal variations of erosion rates and processes on hillslopes in western Colorado. Z. Geomorphol. Suppl.bd. 5, 215–238.

Schumm, S.A., 1965. Quaternary paleohydrology. In: Wright, H.E., Frey, D.G. (Eds), The Quaternary of the United States. Princeton University Press, Princeton, NJ, pp. 783–794.

Schumm, S.A., 1968. River adjustment to altered hydrologic regime. Murrumbidgee River and paleochannels. US Geol. Surv. Prof. Paper 598.

Schumm, S.A., 1969. Geomorphic implications of climatic change. In: Chorley, R.J. (Ed.), Water, Earth and Man. Methuen, London, pp. 525–534.

Schumm, S.A., 1973. Geomorphic thresholds and complex response of drainage systems. In: Morisawa, M. (Ed.), Fluvial Geomorphology. George Allen and Unwin, London, pp. 299–310.

Schumm, S.A., 1977. The Fluvial System. Wiley, New York, p. 338.

Schumm, S.A., 1979. Geomorphic thresholds: the concept and its applications. Inst. Br. Geogr. Trans. 4, 485–515.

Schumm, S.A., 1981. Evolution and response of the fluvial system, sedimentological implications. Soc. Econ. Paleontol. Mineral. 31, 18–29, Special Publication.

Schumm, S.A., Chorley, R.J., 1966. Talus weathering and scarp recession in the Colorado Plateau. Z. Geomorphol. 10, 11–36.

Schumm, S.A., Hadley, R.F., 1957. Arroyos and the semiarid cycle of erosion. Am. J. Sci. 255, 161–174.

Schumm, S.A., Lichty, R.W., 1963. Channel widening and flood plain construction along Cimarron river in south-western Kansas. US Geol. Surv. Prof. Paper 352D, 71–88.

Schumm, S.A., Lichty, R.W., 1965. Time, space and causality in geomorphology. Am. J. Sci. 263, 110–119.

Schumm, S.A., Lusby, G.C., 1963. Seasonal variation of infiltration capacity and runoff on hillslopes in western Colorado. J. Geophys. Res. 68, 3655–3666.

Schumm, S.A., Harvey, M.D., Watson, C.C., 1984. Incised Channels: Morphology, Dynamics and Control. Water Resources Publications, Littleton, CO, p. 200.

Schumm, S.A., Mosley, M.P., Weaver, W.E., 1987. Experimental Fluvial Geomorphology. Wiley, Chichester, p. 413.

Schunke, E., 1974. Formungsvorgäuge an schneeflecken im isländischen Hochland. In: Poser, H. (Ed.), Geomorphologische Prozesse und Prozesskombinationen in der Gegenwart unter Verschiedenen Klimabedingungen, Akademie Wissenschaften zu Göttingen Abh. Math.-Phys., 29, pp. 274–286.

Schunke, E., 1975. Die periglazialerscheinungen Islands in Abhängigkeit von Klima und substrat. Akademie Wissenchaften zu Göttingen Math.-Phys., Kl. Folge 3 (30), 273.

Schunke, E., Zoltai, S.C., 1988. Earth hummocks (thufur). In: Clark, M.J. (Ed.), Advances in Periglacial Geomorphology. Wiley, Chichester, pp. 231–245.

Schwarzbach, M., 1967. Islandische wasserfalle und eire genetische systematik des wasserfalles überhaupt. Z. Geomorphol. 11, 377–417.

Scoging, H., 1989. Run-off generation and sediment mobilization by water. In: Thomas, D.S.G. (Ed.), Arid Zone Geomorphology. Belhaven Press, London, pp. 87–116.

Segura, F.S., 1990. Las Ramblas Valencianas. Departamento de Geografía, Universidad de Valencia, p. 229.

Selby, M.J., 1972. Antarctic tors. Z. Geomorphol. Suppl.bd. 13, 73–86.

Selby, M.J., 1977. Bornhardts of the Namib Desert. Z. Geomorphol. 21, 1–13.

Selby, M.J., 1985. Earth's Changing Surface. Clarendon Press, Oxford, p. 607.

Selby, M.J., 1993. Hillslope Materials and Processes. Oxford University Press, Oxford, p. 451.

Selby, M.J., Hendy, C.H., Seely, M.K., 1979. A late Quaternary lake in the central Namib Desert, southern Africa, and some implications. Palaeogeogr. Palaeoclimatol. Palaeoecol. 26, 37–41.

Seppälä, M., 1988. Palsas and related forms. In: Clark, M.J. (Ed.), Advances in Periglacial Geomorphology. Wiley, Chichester, pp. 247–278.

Seppälä, M., 2004. Wind as a Geomorphic Agent in Cold Climates. Cambridge University Press, Cambridge, MA, p. 358.

Seppälä, M., Lindé, K., 1978. Wind tunnel studies of ripple formation. Geografiska Annaler 60A, 29–40.

Serrano, E., Agudo, C., 1998. Glaciares rocosos activos de los Pirineos. In: Gómez Ortiz, A., Salvador Franch, L., Schulte, A., García Navarro, A. (Eds), Procesos biofísicos actuales en medios fríos. Publicaciones de la Universitat de Barcelona, Barcelona, pp. 133–154.

Serrat, D., 1979. Rock glacier morainic deposits in the eastern Pyrenees. In: Schlüchter, Ch. (Ed.), Moraines and Varves. Origin, Genesis, Classification. Balkema, Rotterdam, pp. 93–100.

Servant, M., Servant-Vildary, S., 1980. L'environnement Quaternaire du bassin du Tchad. In: Williams, M.A.J., Faure, H. (Eds), The Sahara and the Nile. Balkema, Rotterdam, pp. 133–162.

Shackleton, N.J., Kennett, J.P., 1975. Paleotemperature history of the Cenozoic and initiation of Antarctic glaciation: oxygen and carbon isotope analyses in DSDP Sites 277, 279 and 281. In: Kennett, J.P., Houtz, R.E., Andrews, P.B., Edwards, A.R. et al. (Eds), Initial Reports of the Deep Sea Drilling Project No. 29. US Government Printing Office, Washington, DC.

Shackleton, N.J., Opdyke, N.D., 1973. Oxygen isotope and paleomagnetic stratigraphy of equatorial Pacific core V28-238: oxygen isotope temperatures and ice volumes on a 10^5 and 10^6 year scale. Quaternary Res. 3, 39–55.

Shackleton, N.J., Opdyke, N.D., 1976. Oxygen isotope and paleomagnetic stratigraphy of equatorial Pacific core V28-239, late Pliocene to latest Pleistocene. In: Cline, R.M., Hays, R.D. (Eds), Investigation of Late Quaternary Paleoceanography and Paleoclimatology, Memoirs of the Geological Society of America, Vol. 145, pp. 449–464.

Sharon, D., 1962. On the nature of hamadas in Israel. Z. Geomorphol. 6, 129–147.

Sharp, M.J., 1982. Modification of clasts in lodgement tills by glacial erosion. J. Glaciol. 28, 475–481.

Sharp, R.P., 1940. Geomorphology of the Ruby-East Humboldt Range, Nevada. Bull. Geol. Soc. Am. 51, 337–372.

Sharp, R.P., 1942a. Soil structures in the St Elias Range, Yukon Territory. J. Geomorphol. 5, 274–304.

Sharp, R.P., 1942b. Periglacial involutions in northeastern Illinois. J. Geol. 50, 113–133.

Sharp, R.P., 1949. Pleistocene ventifacts east of the Bighorn Mountains, Wyoming. J. Geol. 57, 175–195.

Sharp, R.P., 1958. Malaspina Glacier, Alaska. Bull. Geol. Soc. Am. 69, 617–646.

Sharp, R.P., 1960. Glaciers. Condon Lecture Publishing, University Oregon Press, Eugene, p. 78.

Sharp, R.P., 1963. Wind ripples. J. Geol. 71, 617–636.

Sharp, R.P., 1964. Wind-driven sand in the Coachella valley, California. Bull. Geol. Soc. Am. 75, 785–804.

Sharp, R.P., 1980. Wind-driven sand in Coachella Valley, California: further data. Bull. Geol. Soc. Am. 91, 724–730.

Sharp, R.P., 1988. Living Ice. Cambridge University Press, Cambridge, MA, p. 225.

Sharpe, C.F.S., 1938. Landslides and Related Phenomena. Columbia University Press, New York, p. 137.

Sharpe, D.R., 1967. The stratified nature of drumlins from Victoria Island and southern Ontario. In: Menzies, J., Rose, J. (Eds), Drumlin Symposium. Balkema, Rotterdam, pp. 195–213.

Shaw, P. A., Cooke, H. J., 1986. Geomorphic evidence for the Late Quaternary paleoclimate of the Middle Kalahari of Northern Botswana. Catena, 13, 349–359.

Shaw, P.A., Thomas, D.S.G., 1993. Geomorphological processes, environmental change and landscape sensivity in the Kalahari Region of Southern Africa. In: Thomas, D.S.G., Allison, R.J. (Eds), Landscape Sensivity. Wiley, Chichester, pp. 83–96.

Shaw, P.A., Thomas, D.S.G., 1997. Pans, playas and salt lakes. In: Thomas, D.S.G. (Ed.), Arid Zone Geomorphology. Process, Form and Change in Drylands. Wiley, Chichester, pp. 293–317.

Sherard, J.L., Decker, R.S., 1977. Summary–Evaluation of Symposium on Dispersive Clays. In: Sherard, J.L., Decker, R.S. (Eds), Dispersive Clays, Related Piping and Erosion in Geotechnical Projects, Vol. 623. American Society for Testing and Materials, Philadelphia, PA, pp. 467–479.

Sherard, J.L., Decker, R.S., Ryker, N.L., 1972. Piping in earth dams of dispersive clays. Proc. ASCE Spec. Conf. Perform. Earth Earth-Support. Struct. 1, 589–626.

Sherard, J. L., Dunnigan, L., Decker, R. S., Steel, E. F., 1976. Journal of the Geotechnical Engineering Division, 1, 69–85.

Sherard, J.L., Dunningan, L.P., Decker, R.S., 1977. In: Sherard, J.L., Decker, R.S. (Eds), Dispersive Clays, Related Piping and Erosion in Geotechnical Projects. American Society for Testing and Materials, Philadelphia, PA, pp. 3–12.

Sheridan, D., 1981. Problems of Desertification of the United States, Council Environmental Quality, p. 142.

Sherman, G.D., 1952. The genesis and morphology of the alumina-rich laterite clays, Problems of Clay and Laterite Genesis. American Institute of Mining and Metallurgical, New York, pp. 154–161.

Shroder, J.E., 1973. Erosional residual in Malawi near the edge of the African Rift. Z. Geomorphol. Suppl.bd 18, 121–143.

Simanton, J.R., Rawitz, E., Shirley, E., 1984. Efects of rock fragments on erosion of semiarid rangeland soils. Soil Sci. Soc. Am. Spec. Publ. 13, 65–72.

Simonett, D.S., 1967. Landslide distribution and earth-quakes in the Bewanni and Toricelli Mountains, New Guinea, a statistical analysis. In: Jennings, J.N., Mabbutt, J.A. (Eds), Landform Studies from Australia and New Guinea. Cambridge University Press, London, pp. 64–84.

Simons, F.S., Eriksen, G.E., 1953. Some desert features of northwest central Peru. Boletín de la Sociedad Geológica del Perú 26, 229–245.

Singer, A., 1979/1980. The paleoclimatic interpretation of clay minerals in soils and weathering profiles. Earth-Sci. Rev. 15, 303–326.

Singh, G., 1971. The Indus valley culture seen in context of post-glacial climate and ecological studies in northwest India. Archaeol. Anthropol. Oceania 6, 177–189.

Sirvent, J., Desir, G., Gutiérrez, M., Sancho, C., Benito, G., 1997. Erosion rates in badland areas recorded by collectors, erosion pins and profilometer techniques (Ebro Basin, NE-Spain). Geomorphology 18, 61–75.

Sivarajasingham, S., Alexander, L.T., Cady, J.G., Cline, M.G., 1962. Laterite. Adv. Agron. 14, 1–60.

Small, R.J., 1970. The Study of Landforms. A Textbook of Geomorphology. Cambridge University Press, Cambridge, MA, p. 502.

Small, R.J., 1987. Englacial and supraglacial sediment: transport and deposition. In: Gurnell, A.M., Clark, M.J. (Eds), Glacio-Fluvial Sediment Transfer-an Alpine Perspective. Wiley, Chichester, pp. 111–145.

Smalley, I.J., 1972. The interaction of great rivers and large deposits of primary loess. Trans. NY Acad. Sci. 34, 534–542.

Smalley, I.J., 1990. Possible formation mechanisms for the modal coarse-silt quartz particles in loess. Quaternary Int. 7/8, 23–27.

Smalley, I.J., 1995. Maxing the material: the formation of silt-sized primary mineral particles for loess deposits. Quaternary Sci. Rev. 14, 645–651.

Smalley, I.J., Unwin, D.J., 1968. The formation and shape of drumlins and their distribution and orientation in drumlin fields. J. Glaciol. 7, 377–390.

Smith, B., Warke, P., 1997. Controls and uncertainties in the weathering environment. In: Thomas, D.S.G. (Ed.), Arid Zone Geomorphology. Wiley, Chichester, pp. 41–54.

Smith, B.J., 1978. The origin and geomorphic implications of cliff foot recesses and tafoni on limestone hamadas in the northwest Sahara. Zeitschrift für Geomorphol. 22, 21–43.

Smith, B.J., 1994. Weathering processes and forms. In: Abrahams, A.D., Parson, J.P. (Eds), Geomorphology of Desert Environments. Chapman & Hall, London, pp. 39–63.

Smith, B.J., de Sánchez, B.A., 1992. Erosion hazards in a Brazilian suburb. Geogr. Rev. 6, 37–41.

Smith, G.I., Streat-Perrott, F.A., 1983. Pluvial lakes of the Western United States. In: Porter, S.C. (Ed.), Late-Quaternary Environments of the United States. University of Minnesota Press, Minneapolis, pp. 190–212.

Smith, H.T.U., 1948. Giant glacial groves in Northwest Canada. Am. J. Sci. 246, 503–514.

Smith, H.T.U., 1953. The Hickory Run boulderfield, Carbon County, Pennsylvania. Am. J. Sci. 25, 625–642.

Smith, H.T.U., Smith, A.P., 1945. Periglacial rock streams in the Blue Ridge area. Bull. Geol. Soc. Am. 56, 1198.

Smith, K., Ward, R., 1998. Floods: Physical Processes and Human Impacts. Wiley, Chichester, p. 382.

Smith, M.W., 1993. Climatic change and permafrost. In: French, H.M., Slaymaker, O. (Eds), Canada's Cold Environments. McGill-Queen's University Press, Montreal, pp. 291–311.

Smith, N.D., Minter, W.E.C., 1980. Sedimentologic central of gold and uranium and two Witwatersrand palaeoplacers. Econ. Geol. 75, 1–14.

Smith, N.D., Vendl, M.A., Kennedy, S.K., 1982. Comparison of sedimentation regimes in four glacier-fed lakes of western Alberta. In: Davidson-Arnott, R., Nickling, W., Fahey, B.D. (Eds), Research in Glacial, Glaciofluvial and Glaciolacustrine Systems. Geo Books, Norwich, pp. 203–238.

So, C.L., 1971. Mass movements associated with the rainstorm of June 1966 in Hong Kong. Trans. Inst. Br. Geogr. 53, 55–66.

Söderman, G., 1985. Planation and weathering in eastern Fennoscandia. Fennia 163, 347–352.

Soloviev, P.A., 1973. Thermokarst phenomena and landforms due to frost heaving in Central Yakutia. Biuletyn Peryglacjalny 23, 135–155.

Soriano, M.A., Simón, J.L., 1995. Alluvial dolines in the central Ebro basin, Spain: a spatial and developmental hazard analysis. Geomorphology 11, 295–309.

Soriano, M.A., Simón, J.L., Gracia, J., Salvador, T., 1994. Alluvial sinkholes over gypsum in the Ebro basin (Spain): genesis and environmental impact. Hydrol. Sci. 39, 257–268.

Sorriso-Valvo, M., 1988. Landslide-related fans in Calabria. Catena Suppl. 13, 109–121.

Souchez, P., 1966. Gélivation et évolution des versants en bordure de l'Islandsis d'Antartides orientale. In: Macar, P. (Ed.), L'Evolution des Versants. Les Congrès et Colloques de L'Univesité de Liége, Vol. 40, pp. 291–298.

Spaulding, W.G., 1990. Vegetation and climatic development of the Mojave Desert: The last glacial maximum to the present. In: Betancourt, J.L., Van Devender, T.R., Martin, P.S. (Eds), Packrat Middens, the Last 40,000 Years of Biotic Change. University of Arizona Press, Tucson, pp. 166–199.

Sperling, C.H.B., Cooke, R.U., 1985. Laboratory simulation of rock weathering by salt crystallization and hydration processes in hot, arid environments. Earth Surf. Processes Landforms 10, 541–555.

Srivastava, P., Juyal, N., Singhve, A.K., Wasson, R.J., Bateman, M.D., 2001. Luminiscence chronology of river-adjustment and incision of Quaternary sediments in the alluvial plain, of the Sabarmati River, north Gujarat, India. Geomorphology 36, 217–229.

Stager, J.K., 1956. Progress report on the analysis on the characteristics and distribution of pingos east of the Mackenzie Delta. Can. Geogr. 7, 13–20.

Stanistreet, I.G., McCarthy, T.S., 1993. The Okavango Fan and the classification of subaerial fan systems. Sediment. Geol. 85, 115–133.

Starkel, L., 1976. The role of extreme (catastrophic) meteorological events in contemporary evolution of slopes. In: Derbyshire, E. (Ed.), Geomorphology and Climate. Wiley, London, pp. 203–246.

Steijn, H.van, Boelhouwers, J., Harris, S., Hétu, B., 2002. Recent research on the nature, origin and climatic relations of blocky and stratified slope deposits. Prog. Phys. Geogr. 26, 551–575.

Stenborg, T., 1970. Delay of run-off from a glacier basin. Geogr. Ann. 52, 1–30.

Stocking, M., 1976. Tunnel erosion. Rhod. Agric. J. 73, 35–39.

Stockton, C.W., Fritts, H.C., 1971. Conditional probability of occurrence for variations in climate based on width of annual tree-rings in Arizona. Tree-Ring Bull. 31, 3–24.

Stoddart, D.R., 1969a. Climatic geomorphology: review and re-assessment. Prog. Geogr. 1, 159–222.

Stoddart, D.R., 1969b. Climatic geomorphology. In: Chorley, R.J. (Ed.), Introduction to Fluvial Processes. Methuen, Suffolk, pp. 189–201.

Stoddart, D.R., 1969c. Climatic geomorphology: review and reassessment. In: Board, C. et al. (Eds), Progress in Geography. Arnold, London, pp. 160–222.

Stokes, S., 1997. Dating of desert sequences. In: Thomas, D.S.G. (Ed.), Arid Zone Geomorphology. Wiley, Chichester.

Strachan, A.D., Dearman, W.R., 1983. Engineering geological mapping in glaciated terrains. In: Eyles, N. (Ed.), Glacial Geology: An Introduction for Engineerings and Earth Scientists. Pergamon Press, Oxford, pp. 229–246.

Strahler, A.N., 1952. Dynamic basis of geomorphology. Bull. Geol. Soc. Am. 66, 923–938.

Strahler, A.N., 1965. Introduction to Physical Geography. Wiley, New York, p. 643.

Strakhov, N.M., 1967. Principles of Lithogenesis, Vol. 1. Oliver and Boyd, Edinburgh, p. 245.

Street, G.J., Anderson, A., 1993. Airborne electromagnetic surveys of the regolith. Explor. Geophys. 24, 795–800.

Stret-Perrott, F.A., Roberto, N., Metcalfe, S.L., 1985. Geomorphic implications of late Quaternary hydrological and climatic changes in the Northern Hemisphere tropics. In: Douglas, I., Spencer, T. (Eds), Environmental Change and Tropical Geomorphology. Allen and Unwin, London, pp. 165–183.

Stroeven, A., Wal, R.van de, Oerlemans, J., 1989. Historic front variations of the Rhone Glacier: simulation with an ice flow model. In: Oerlemans, J. (Ed.), Glacier Fluctuations and Climatic Change. Kluwer, Dordrecht, pp. 391–405.

Sueoka, T., 1988. Identification and classification of granitic residual soils using chemical weathering index, Geomechanics in Tropical Soils, Vol. 1. Balkema, Rotterdam, pp. 55–61.

Sugden, D.E., 1974. Landscapes of glacial erosion in Greenland and their relationship to ice, topographic and bedrock conditions. In: Brown, E.H., Waters, R.S. (Eds), Progress in Geomorphology, Vol. 7, pp. 177–195, Institute British Geographers.

Sugden, D.E., 1977. Reconstruction of the morphology, dynamics and thermal characteristics of the Laurentide ice sheet at its maximum. Arctic Alpine Res. 9, 27–47.

Sugden, D.E., 1978. Glacial erosion by the Laurentide ice sheet. J. Glaciol. 20, 367–391.

Sugden, D.E., 1982. Arctic and Antarctic. A Modern Geographical Systems. Blackwell, Oxford, p. 472.

Sugden, D.E., John, B.S., 1976. Glaciers and Landscape. Edward Arnold, London, p. 376.

Sugden, D.E., Glasser, N.F., Clapperton, C.M., 1992. Evolution of large roches moutonnées. Geogr. Ann. 74A, 253–264.

Summerfield, M.A., 1983. Silcrete. In: Goudie, A.S., Pye, K. (Eds), Chemical Sediments and Geomorphology: Precipitates and Residua in the Near Surface Environment. Academic Press, New York, pp. 59–91.

Sundquist, E.T., 1993. The global carbon dioxide budget. Science 259, 934–941.

Sung-Chiao, C., Xing, J., 1982. Origin and development of the Shamo gand deserts and the Goby stony deserts of China. In: Smiley, T. L. (Ed.). The geological history of the world's deserts. University of Uppsala Press, Uppsala, 79–91.

Sutherland, D.G., 1984. Geomorphology and mineral exploration: some samples for exploration for diamondiferous placer deposits. Z. Geomorphol. Suppl.bd. 51, 95–108.

Sutherland, D.G., 1985. Geomorphological controls on the distribution of placer deposits. J. Geol. Soc. Lond. 142, 727–737.

Swan, S.B.StC., 1972. Land surface evolution and related problems with reference to a humid tropical region: Johor, Malaya. Z. Geomorphol. 16, 160–181.

Sweet, M.L., Kocurek, G., 1990. An empirical model of aeolian dune lee-face airflow. Sedimentology 37, 1023–1038.

Sweeting, M.M., 1958. The karstlands of Jamaica. Geogr. J. 124, 184–199.

Sweeting, M.M., 1966. The weathering of limestones with particular reference to the Carboniferous Limestones of northern England. In: Dury, G.H. (Ed.), Essays in Geomorphology. Elsevier, Amsterdam, pp. 177–210.

Sweeting, M.M., 1972. Karst Landforms. Macmillan, London, p. 362.

Sweeting, M.M., 1995. Karst in China. Its Geomorphology and Environment. Springer, Berlin, p. 265.

Swinzow, G.K., 1969. Certain aspects of engineering geology in permafrost. Eng. Geol. 3, 177–215.

Swoboda-Colberg, N.G., Drever, J., 1993. Mineral dissolution rates in plot-scale field and laboratory experiments. Chem. Geol. 105, 51–69.

Syvitski, J.P.M., Burrell, D.C., Skei, J.M., 1987. Fjords: Processes and Products. Springer, New York, p. 379.

Taber, S., 1929. Frost heaving. J. Geol. 37, 428–461.

Takahashi, T., 1991. Debris Flow. A.A. Balkema, Rotterdam, p. 165.

Takei, A., Kobashi, S., Fukushima, Y., 1981. Erosion and sediment transport measurement in a weathered granite mountain areas. Symposium on Erosion and Sediment Transport Measurement, Florence 133, 493–502, International Association Hydrological Sciences.

Talbot, M.R., 1980. Environmental responses to climatic change in the West African Sahel over the past 20,000 years. In: Williams, M.A.J., Faure, H. (Eds), The Sahara and the Nile. Balkema, Rotterdam, pp. 37–62.

Talbot, M.R., Williams, M.A.J., 1978. Erosion of fixed dunes in the Sahel, Central Niger. Earth Surf. Processess 3, 107–113.

Tanner, W.F., 1961. An alternate approach to morphogenetic climates. Southeastern Geol. 2, 251–257.

Tardy, Y., Roquin, C., 1992. Geochemistry and evolution of lateritic landscapes. In: Martini, I.P., Chesworth, W. (Eds), Weathering, Soils and Paleosols. Elsevier, Amsterdam, pp. 407–443.

Tator, B.A., 1952. Pediment characteristics and terminology (Part I). Ann. Assoc. Am. Geogr. 42, 295–317.

Tator, B.A., 1953. Pediment characteristics and terminology (Part II). Ann. Assoc. Am. Geogr. 43, 47–53.

Taylor, G.R., Eggleton, R.A., 2001. Regolith Geology and Geomorphology. Wiley, Chichester, p. 375.

Taylor, G.R., Eggleton, R.A., Holzhauer, C.C., Maconachie, L.A., Gordon, M., Brown, M.C., McQueen, K.G., 1992. Cool climate lateritic and bauxitic weathering. J. Geol. 100, 669–677.

Taylor, K.C., Lamorey, G.W., Doyle, G.A., Alley, R.B., Grootes, P.M., Mayewskill, P.A., White, J.W.C., Barlow, L.K., 1993. The "flickering switch" of the Late Pleistocene climatic change. Nature, 361, 432–436.

Tchakerian, V.P., 1994. Palaeoclimatic interpretations from desert dunes and sediments. In: Abrahams, A.D., Parsons, A.J. (Eds), Geomorphology of Desert Environments. Chapman & Hall, London, pp. 631–643.

Tchakerian, V.P., 1999a. Playa. In: Mares, M.A. (Ed.), Encyclopaedia of Deserts. University of Oklahoma Press, Norman, pp. 443–444.

Tchakerian, V.P., 1999b. Sebkha. In: Mares, M.A. (Ed.), Encyclopaedia of Deserts. University of Oklahoma Press, Norman, p. 485.

Temple, P.H., Rapp, A., 1972. Landslides in the Mgeta area, western Uluguru Mountains, Tanzania. Geogr. Ann. 54A, 157–193.

Thomas, D.S., 1988. The biogeomorphology of arid and semi-arid environments. In: Viles, H. (Ed.), Biogeomorphology. Blackwell, Oxford, pp. 193–221.

Thomas, D.S.G., 1989. The nature of arid environments. In: Thomas, D.S.G. (Ed.), Arid Zone Geomorphology. Belhaven Press, London, pp. 1–8.

Thomas, D.S.G., (Ed.), 1989. Arid Zone Geomorphology. Belhaven Press, London, p. 372.

Thomas, D.S.G., 1992. Desert dune activity: concepts and significance. J. Arid Environ. 22, 31–38.

Thomas, D.S.G., (Ed.), 1997a. Arid Zone Geomorphology. Process, Form and Change in Drylands. Wiley, Chichester, p. 713.

Thomas, D.S.G., 1997b. Arid environments: their nature and extent. In: Thomas, D.S.G. (Ed.), Arid Zone Geomorphology. Process, Form and Change in Drylands. Wiley, Chichester, pp. 3–12.

Thomas, D.S.G., 1997c. Reconstructing ancient arid environments. In: Thomas, D.S.G. (Ed.), Arid Zone Geomorphology. Wiley, Chichester, pp. 577–605.

Thomas, D.S.G., 1997d. Sand seas and aeolian bedforms. In: Thomas, D.S.G. (Ed.), Arid Zone Geomorphology: Process, Form and Change in Drylands. Wiley, Chichester, pp. 373–412.

Thomas, D.S.G., Goudie, A.S., 1984. Ancient ergs of the Southern Hemisphere. In: Vogel, J.C. (Ed.), Late Cenozoic Palaeoclimates of the Southern Hemisphere. Balkema, Rotterdam, pp. 407–418.

Thomas, D.S.G., Middleton, N.J., 1994. Desertification: Exploding the Myth. Wiley, Chichester, p. 194.

Thomas, D.S.G., Shaw, P.A., 1991. The Kalahari Environment. Cambridge University Press, Cambridge, MA, p. 284.

Thomas, M.F., 1965. Some aspects of the geomorphology of domes and tors in Nigeria. Z. Geomorphol. 9, 63–81.

Thomas, M.F., 1966. Some geomorphological implications of deep weathering patterns in crystalline rocks in Nigeria. Trans. Inst. Br. Geogr. 40, 173–193.

Thomas, M.F., 1974. Tropical Geomorphology. A Study of Weathering and Landform Development in Warm Climates. Macmillan, London, p. 329.

Thomas, M.F., 1976. Criteria for the recognition of climatically induced variations in granite landforms. In: Derbyshire, E. (Ed.), Geomorphology and Climate. Wiley, London, pp. 411–445.

Thomas, M.F., 1978. The study of inselbergs. Z. Geomorphol. Suppl.bd. 31, 1–41.

Thomas, M.F., 1983. Contemporary denudation systems and the effects of climatic change in the humid tropics – some problems from Sierra Leone. In: Briggs, D.J., Waters, R.S. (Eds), Studies in Quaternary Geomorphology. Geo Books, Norwich, pp. 195–214.

Thomas, M.F., 1986. Savanna. In: Fookes, P.G., Vaughan, P.R. (Eds), A Handbook of Engineering Geology. Survey University Press, Glasgow, pp. 125–136.

Thomas, M.F., 1989. The role of etch processes in landform development: I. Etching concepts and the formation of relief. Z. Geomorphol. 33, 129–142.

Thomas, M.F., 1994. Geomorphology in the Tropics. A Study of Weathering and Denudation in Low Latitudes. Wiley, Chichester, p. 460.

Thomas, M.F., Goudie, A.S., (Eds), 1985. Dambos: small channelless valleys in the tropics. Characteristics, formation, utilisation. Z. Geomorphol. Suppl.bd., 52, p. 228.

Thomas, M.F., Thorp, M.B., 1985. Environmental change and episodic explanation in the humid tropics of Sierra Leona: the Koidn etchplain. In: Douglas, I., Spencer, T. (Eds), Environmental Change and Tropical Geomorphology. Allen and Unwin, London, pp. 239–267.

Thomas, M.F., Thorp, M.B., 1992. Landscape dynamics and surface deposits arising from late Quaternary fluctuations in the forest–savanna boundary. In: Furley, P.A., Proctor, J., Ratter, J.A. (Eds), Nature and Dynamics of Forest–Savanna Boundaries. Chapman & Hall, London, pp. 215–253.

Thomas, M.F., Thorp, M.B., 1993. The geomorphology of some Quaternary placer deposits. Z. Geomorphol. Suppl.bd. 87, 183–194.

Thomas, M.F., Thorp, M.B., 1995. Geomorphic response to rapid climatic and hydrologic change during the Late Pleistocene and Early Holecene in the humid and sub-humid tropics. Quaternary Sci. Rev. 14, 193–207.

Thorarinsson, S., 1953. Some new aspects of the Grimsvötn problem. J. Glaciol. 2, 267–274.

Thorbecke, F., 1927. Morphologie der klimazonen. Düsseldorfer Geogr. Vorträge Erörterungen, 1–100, Part 3.

Thorn, C.E., 1988a. Introduction to Theoretical Geomorphology. Unwin Hyman, Boston, p. 247.

Thorn, C.E., 1988b. Nivation: a geomorphic chimera. In: Clark, M.J. (Ed.), Advances in Periglacial Geomorphology. Wiley, Chichester, pp. 3–31.

Thorn, C.E., 1992. Periglacial geomorphology: what, where, when? In: Dixon, J.C., Abrahams, A.D. (Eds), Periglacial Geomorphology. Wiley, Chichester, pp. 1–30.

Thorn, C.E., Hall, K., 1980. Nivation: an arctic–alpine comparison and reappraisal. J. Glaciol. 25, 109–124.

Thorn, C.E., Hall, K., 2002. Nivation and cryoplanation: the case for scrutiny and integration. Prog. Phys. Geogr. 26, 533–550.

Thornbury, W.D., 1954. Principles of Geomorphology. Wiley, New York, p. 618.

Thornes, J., 1976. Semiarid Erosional Systems. In: London School of Economic, Geographical Papers 7, London, p. 79.

Thornes, J.B., 1994. Catchment and channel hydrology. In: Abrahams, A.P., Parsons, A.J. (Eds), Geomorphology of Desert Environments. Chapman & Hall, London, pp. 257–287.

Thornthwaite, C.W., 1948. An approach towards a rational classification of climate. Geogr. Rev. 38, 55–94.

Torri, D., Sfalanga, M., Del Sette, M., 1987. Splash detachment: runoff depth and soil cohesion. Catena 14, 149–155.

Toy, T.J., 1977. Hillslope form and climate. Geol. Soc. Am. Bull. 88, 16–22.

Trendall, A.F., 1962. The formation of "apparent peneplains" by a process of combined lateritisation and surface wash. Z. Geomorphol. 6, 183–197.

Trescasses, J. J., 1973. L'èvolution géochimique supergène des roches ultiabasiques en zone tropicale et la formation des egisements nichélifere, de Novella Calédonia. Thèse. Strasbourg. p. 347.

Tricart, J., 1952. Cours de Géomorphologie. Géomorphologie Climatique. CDU, Paris, p. 193.

Tricart, J., 1956. Geomorphologic dynamique du Delta du Senegal. Revue de Géomorphologie Dynamique 7, 65–86.

Tricart, J., 1965. Principes et Méthodes de la Géomorphologie. Masson, Paris, p. 496.

Tricart, J., 1967. Le Modelé des Régions Périglaciaires. SEDES, Paris, p. 512.

Tricart, J., 1969. Le Modelé des Régions Sèches. SEDES, Paris, p. 472.

Tricart, J., 1974a. Existence de périodes sèches au Quaternaire en Amazonie et dans les régions voisines. Revue de Géomorphologic Dynamique 23, 145–158.

Tricart, J., 1974b. Le Modelé des Régions Chaudes: Forêts et Savanes. SEDES, Paris, p. 345.

Tricart, J., 1975. Influence des oscillations climatiques récents sur le modelé en Amazonie Orientale (région de Santarém) d'aprés des images de radar latéral. Z. Geomorphol. 19, 140–163.

Tricart, J., 1979. El concepto de "pluvial". Actas de la III Reunión Nacional del Grupo Español de Trabajo del Cuaternario, Madrid, pp. 7–20.

Tricart, J., 1982. El Pantanal: Un ejemplo del impacto geomorfológico sobre el ambiente, Informaciones Geográficas, Vol. 29. Universidad de Chile, Santiago, pp. 81–97.

Tricart, J., 1985. Evidence of Upper Pleistocene dry climates in northern South America. In: Douglas, I., Spencer, T. (Eds), Environmental Change and Tropical Geomorphology. Allen and Unwin, London, pp. 197–217.

Tricart, J., Alfonsi, P., 1981. Actions éoliennes récents aux abords du delta de l'Orénoque. Bull. AGG 476, 75–82.

Tricart, J., Cailleux, A., 1962. Le Modelé Glaciaire et Nival. SEDES, Paris, p. 508.

Tricart, J., Cailleux, A., 1965. Introduction à la Géomorphologie Climatique. SEDES, Paris, p. 306.

Tricart, J., Raynal, R., Besancon, J., 1972. Cônes rocheux, pediments, glacis. Ann. Géogr. 43, 1–24.

Troeh, F.R., 1965. Landform equations fitted to topographic maps. Am. J. Sci. 263, 616–627.

Troll, C., 1944. Strukturböden, solifluktion und frostklimate der Erde. Geologische Rundschau 34, 545–694.

Trombe, F., 1952. Traité de Spéléologie. Payot, Paris, p. 376.

Trudgill, S.T., 1976. Rock weathering in climate: quantitative and experimental aspects. In: Derbyshire, E. (Ed.), Geomorphology and Climate. Wiley, London, pp. 59–99.

Trudgill, S., 1985. Limestone Geomorphology. Longman, London, p. 196.

Tseo, G., 1993. Two types of longitudinal dune fields and possible mechanisms for their development. Earth Surf. Processes Landforms 18, 627–643.

Tsoar, H., 1983a. Dynamic processes acting on a longitudinal (seif) dune. Sedimentology 30, 567–578.

Tsoar, H., 1983b. Wind tunnel modelling of echo and climbing dunes. In: Brookfield, M.E., Ahlbrandt, T.S. (Eds), Eolian Sediments and Processes. Elsevier, Amsterdam, pp. 247–259.

Tsoar, H., Moller, J.T., 1986. The role of vegetation in the formation of linear dunes. In: Nickling, W.G. (Ed.), Aeolian Geomorphology. Allen and Unwin, Boston, pp. 75–95.

Tsoar, H., Pye, K., 1987. Dust transport and the question of desert loess formation. Sedimentology 34, 139–154.

Tsukamoto, Y., Kusakabe, D., 1984. Vegetative influences on debris slides occurrences on steep slopes in Japan. Proc. Symp. On Effects Forest Land Use on Erosion and Slope Stability. Environment and Policy Institute. Honolulu, Hawaii.

Tucker, M.E., 1978. Gypsum crust (gypcrete) and patterned ground from northern Iraq. Z. Geomorphol. 22, 89–100.

Tufnell, L., 1972. Ploughing blocks with special reference to north-west England. Biuletyn Peryglacjalny 21, 237–270.

Tufnell, L., 1984. Glacier Hazards. Longman, London, p. 97.

Twidale, C.R., 1962. Steepened margins of inselbergs from north-western Eyre Peninsula, South Australia. Z. Geomorphol. NF 6, 51–69.

Twidale, C.R., 1964. A contribution to the general theory of domed inselbergs. Trans. Inst. Br. Geogr. 34, 91–113.

Twidale, C.R., 1967. Hillslopes and pediments in the Flinders Ranges, South Australia. In: Jennings, J.N., Mabbutt, J.A. (Eds), Landform Studies from Australia and New Guinea. Cambridge University Press, Cambridge, MA, pp. 95–117.

Twidale, C.R., 1968a. Geomorphology with Special Reference to Australia. Nelson, Melbourne.

Twidale, C.R., 1968b. Inselberg. In: Fairbridge, R.W. (Ed.), Encyclopedia of Geomorphology. Reinhold, New York, pp. 556–559.

Twidale, C.R., 1973. On the origin of the sheet jointing. Rock Mech. 5, 163–187.

Twidale, C.R., 1976a. Analysis of landforms. Wiley, Sydney, p. 572.

Twidale, C.R., 1976b. On the survival of palaeoforms. Am. J. Sci. 276, 77–94.

Twidale, C.R., 1978. On the origin of Ayers Rock, Central Australia. Z. Geomorphol. Suppl.bd. 31, 177–206.

Twidale, C.R., 1981. Granite Inselbergs. Geogr. J. 147, 54–71.

Twidale, C.R., 1982a. The evolution of bornhardts. Am. Sci. 70, 268–276.

Twidale, C.R., 1982b. Granite Landforms. Elsevier, Amsterdam, p. 372.

Twidale, C.R., 1983. Pediments, peneplains and ultiplains. Revue de Géomorphologie Dynamique 32, 1–35.

Twidale, C.R., 1990. The origin and implications of some erosional landforms. J. Geol. 98, 343–364.

Twidale, C.R., 1992. King of the plains: Lester King's contribution to Geomorphology. Geomorphology 5, 491–509.

Twidale, C.R., Bourne, J.A., 1975. Episodic exposure of inselbergs. Bull. Geol. Soc. Am. 86, 1473–1481.

Twidale, C.R., Bourne, J.A., 1978. Bornhardts. Z. Geomorphol. Suppl.bd. 31, 111–137.

Twidale, C.R., Corbin, E.M., 1963. Gnammas. Revue de Géomorphologie Dynamique 14, 1–20.

Twidale, C.R., Lageat, Y., 1994. Climatic geomorphology: a critique. Prog. Phys. Geogr. 18, 319–334.

UNEP, 1992. World Atlas of Desertification. Arnold, London, p. 69.

UNESCO, 1981. Avalanche Atlas. UNESCO, Paris, p. 265.

UNESCO–FAO, 1977. Desertification; An Overview, Desertification. Its Causes and Consequences. Pergamon Press, Oxford, pp. 1–61.

USGS, 1983. Permafrost. In: Tank, R.W. (Ed.), Environmental Geology. Oxford University Press, Oxford, pp. 163–170.

Valeton, I., 1972. Bauxites. Elsevier, Amsterdam, p. 226.

Valeton, I., 1994. Element concentration and formation of ore deposits by weathering. Catena 21, 99–129.

Vandenberghe, J., Thorn, C.E., 2002. Progress in periglacial research. Prog. Phys. Geogr. 26, 475–477.

Vandenberghe, J., Woo, M., 2002. Modern and ancient periglacial river types. Prog. Phys. Geogr. 26, 479–506.

Van der Wateren, D.F.M., 1987. Structural geology and sedimentology of the Dammer Berge push moraine, FRG. In: Van der Meer, J.J.M. (Ed.), Tills and Glaciotectonics. Balkema, Rotterdam, pp. 157–182.

Van der Wateren, D.F.M., 1992. Structural geology and sedimentology of push moraines. Processes of soft sediment deformation in a glacial environment and the distribution of glaciotectonic styles. Dissertation, Amsterdam, p. 230.

Van der Wateren, F.M., 1995. Processes of glaciotectonism. In: Menzies, J. (Ed.), Modern Glacial Environments. Processes, Dynamics and Sediments. Butterworth-Heinemann, Oxford, pp. 307–335.

Vaniman, D.T., Chipera, S.J., Bish, D.L., 1994. Pedogenesis of siliceous calcretes at Yucca Mountains, Nevada. Geoderma 63, 1–17.

Varnes, D.J., 1958. Landslide types and processes. In: Eckel, E.B. (Ed.), Landslides and Engineering Practice, Highway Research Board. National Academy of Sciences, Washington, DC, pp. 20–47, Special Report, 29.

Varnes, D.J., 1978. Slope movement types and processes. In: Schuster, R.L., Krizek, R.J. (Eds), Landslide Analysis and Control. National Academy of Sciences, Washington, DC, pp. 11–33, Special Report No. 176.

Veder, C., 1981. Landslides and their Stabilization. Springer, New York, p. 247.

Verger, F., 1964. Mottereaux et gilgais. Ann. Géogr. 73, 413–430.

Verstappen, H.T., 1964. Karst morphology of the Star Mountains (Central New Guinea) and its relation to lithology and climate. Z. Geomorphol. 8, 40–49.

Verstappen, H.Th., 1968. On the origin of longitudinal (seif) dunes. Z. Geomorphol. NF 12, 200–220.

Verstappen, H.Th., 1970. Aeolian geomorphology of the Thar Desert and palaeo-climates. Z. Geomorphol. Suppl.bd. 10, 104–120.

Veste, M., 1995. Structures of geomorphological and ecological units and ecosystems processes in the linear dune system near Nizzana/Negev. Bielefelder Ökologische Beitrage 8, 85–96.

Vidal Romani, J.R., Twidale, C.R., 1998. Formas y Paisajes Graníticos, Universidade da Coruña. Servicio de Publications, A Coruña, p. 411.

Vidal Romaní, J.R., Twidale, C.R., Campbell, E.M., Centeno, J.D., 1995. Pruebas morfológicas y estructurales sobre el origen de las fracturas por descamación. Cadernos Laboratorio Xeolóxico de Laxe 20, 307–346.

Vilaplana, J.M., 1987. Guia dels Paisatges Granítics dels Països Catalans. Kapel SA, Barcelona, p. 182.

Viles, H.A., 1984. Biokarst: review and prospect. Prog. Phys. Geogr. 8, 523–542.

Viles, H.A., 1995. Ecological perspectives on rock surface weathering: towards a conceptual model. Geomorphology 13, 21–35.

Vischer, S.S., 1945. Climatic maps of geological interest. Bull. Geol. Soc. Am. 56, 713–736.

Vivian, R., 1979. Les Glaciers sont Vivants. Ed. Denoël, Paris, p. 240.

Vogt, J., 1959. Aspects de l'évolution morphologique récente de l'Ouest Africain. Ann. Géogr. 68, 193–206.

Vogt, J., 1966. Le complexe de la stone-line. Mise au point. Bulletin du Bureau de Recherches Géologiques et Minières 4, 3–51.

Vogt, T., 1984. Croûtes Calcaires: types et Genèse. Université Louis Pasteur. Institut de Géographie, Strasbourg, p. 239.

Voight, B., 1990. The 1985 Nevado del Ruiz Volcano catastrophe: anatomy and retrospection. J. Volcanol. Geotherm. Res. 42, 151–188.

Voight, B., Pariseau, G., 1978. Rockslides and avalanches: an introduction. In: Voight, B. (Ed.), Rockslides and Avalanches, 1. Natural Phenomena. Elsevier, Amsterdam, pp. 1–67.

Völkel, J., Grunert, J., 1990. To the problem of dune formation and dune weathering during the Late Pleistocene and Holocene in the southern Sahara and Sahel. Z. Geomorphol. 34, 117.

Wahrhaftig, C., Cox, A., 1959. Rock glaciers in the Alaska Range. Bull. Geol. Soc. Am. 70, 383–436.

Walcott, R.I., 1970. Isostatic response to loading of the crust in Canada. Can. J. Earth Sci. 7, 716–726.

Walder, J.S., 1982. Stability of sheet flow of water beneath temperate glaciers and implications for glacier surging. J. Glaciol. 28, 273–293.

Walker, G., 2004. Frozen time. Nature 429, 596–597.

Walker, H., 1973. The morphology of the North Slope. In: Britton, M. (Ed.), Alaska Arctic Tundra. AINA, Washington, DC, pp. 49–92.

Walker, H.J., 1986. Periglacial environments. In: Fookes, P.G., Vaughan, P.R. (Eds), A Handbook of Engineering Geomorphology. Surrey University Press, London, pp. 82–96.

Walker, J.G.C., Sloan, L.C., 1992. Something is wrong with climate theory. Geotimes, 16–18, June.

Walling, D.E., 1984. The sediment yields of African rivers. In: Walling, D.E., Foster, S.S.D., Wurzel, P. (Eds), Challenges in African Hydrology and Water Resources, International Association of Hydrological Sciences, No. 144, pp. 265–283.

Walling, D.E., Webb, B.W., 1986. Solutes in river systems. In: Trudgill, S.T. (Ed.), Solute Processes. Wiley, Chichester, pp. 251–327.

Waltham, A.C., 1989. Ground Subsidence. Blackie, Glasgow, p. 202.

Walther, J., 1915. Laterite in Westaustralien. Zeitschrift der Deustschen Geologischen Gessellschaft 62 (1–7), 46–53.

Ward, A.W., Greeley, R., 1984. Evolution of yardangs at Rogers Lake, California. Bull. Geol. Soc. Am. 95, 829–837.

Ward, R., 1978. Floods: A Geographical Perspective. Macmillan, London.

Warren, A., Knott, P., 1983. Desert dunes: a short review of needs in desert dune research and a recent study of micrometeorological dune-initiation mechanisms. In: Brookfield, M.E., Ahlbrandt, T.S. (Eds), Eolian Sediments and Processes. Elsevier, Amsterdam, pp. 343–352.

Washburn, A.L., 1956. Classification of patterned ground and review of suggested origins. Bull. Geol. Soc. Am. 67, 823–865.

Washburn, A.L., 1967. Instrumental observations on mass wasting in the Mesters Vig District, Northeast Greenland. Meddelelser om Gronland 166, p. 318.

Washburn, A.L., 1969. Weathering, frost action and patterned ground in the Mesters Vig District, Northeast Greenland. Meddelelser om Gronland 176, 303.

Washburn, A.L., 1979. Geocryology. A Survey of Periglacial Processes and Environments. Arnold, London, p. 406.

Wasson, R.J., 1977. Last-glacial alluvial fan sedimentation in the lower Derwent Valley, Tasmania. Sedimentology 24, 781–799.

Wasson, R.J., 1983. Dune sediment types, sand colour, sediment provenance and hydrology in the Strzelecki-Simpson dunefield, Australia. In: Brookfield, M.E., Ahlbrandt, T.S. (Eds), Eolian Sediments and Processes. Elsevier, Amsterdam, pp. 165–195.

Wasson, R.J., 1984. Late Quaternary palaeo-environments in the desert dunnefields of Australia. In: Vogel, J.C. (Ed.), Late Cenozoic Palaeoenvironmental of the Southern Hemisphere. Balkema, Rotterdam, pp. 419–432.

Wasson, R.J., Rajaguru, S.N., Misra, V.N., Agarwal, D.P., Dhir, R.P., Singhvi, A.K., Kameswara Rao, K., 1983. Geomorphology, Late Quaternary stratigraphy and palaeoclimatology of the Thar dune field. Z. Geomorphol. Suppl.bd. 45, 117–151.

Watson, A., 1979. Gypsum crusts in deserts. J. Arid Environ. 2, 3–20.

Watson, A., 1983. Gypsum crusts. In: Goudie, A.S., Pye, K. (Eds), Chemical Sediments and Geomorphology: Precipitates and Residua in the Near Surface Environment. Academic Press, New York, pp. 133–161.

Watson, A., 1985. Structure, chemistry and origins of gypsum crusts in southern Tunisia and the central Namib Desert. Sedimentology 32, 855–875.

Watson, A., 1989a. Desert crusts and varnishes. In: Thomas, D.S.G. (Ed.), Arid Zone Geomorphology. Hallstead Press, New York, pp. 25–55.

Watson, A., 1989b. Windflow characteristics and aeolian entrainment. In: Thomas, D.S.G. (Ed.), Arid Zone Geomorphology. Belhaven Press, London, pp. 209–231.

Watson, A., 1990. The control of blowing sand and mobile desert dunes. In: Goudie, A. (Ed.), Techniques for Desert Reclamation. Wiley, New York, pp. 35–85.

Watson, A., Price-Williams, D., Goudie, A.S., 1983. Palaeoenvironmental interpretation of colluvial sediments and palaeosols of the Late Pleistocene hypothermal in southern Africa. Palaeogeogr. Palaeoclim. Palaeoecol. 45, 225–250.

Watts, S.H., 1980. Quaternary pedogenic calcretes from the Kalahari (southern Africa): mineralogy, genesis and diagenesis. Sedimentology 27, 661–686.

Wayland, E.J., 1933. Peneplains and some other erosional platforms, Annual Report and Bulletin, Protectorate of Uganda Geological Survey, Department of Mines, Note 1, pp. 77–79.

Weaver, A.J., 2003. The science of climatic change. Geosci. Can. 30, 91–109.

Webb, R.H., Wilshire, H.G., (Eds), 1983. Environmental Effects of Off-Road Vehicles. Springer, New York.

Weertman, J., 1957. On the sliding of glaciers. J. Glaciol. 3, 33–38.

Weertman, J., 1983. Creep deformation of ice. Annu. Rev. Earth Planet. Sci. 11, 215–240.

Wells, S.G., Harvey, A.M., 1987. Sedimentologic and geomorphic variations in storm generated alluvial fans, Howgill Fells, Norhwest England. Bull. Geol. Soc. Am. 98, 182–198.

Wells, S.G., Metadden, L.D., Dohrenwend, J.C., 1987. Influence of late Quaternary climatic changes on geomorphic and pedogenic processes on a desert piedmont, eastern Mojave Desert, California. Quaternary Res. 27, 130–146.

Wesson, R.L., 1983. The eruption of Mount St Helens – entering the era of real-time geology. In: Tank, R.W. (Ed.), Environmental Geology. Oxford University Press, New York, pp. 50–62.

Whitaker, C.R., 1973. A Bibliography of Pediments. GeoAbstracts, Norwich, p. 95.

Whitaker, C.R., 1979. The use of the term 'pediment' and related terminology. Z. Geomorphol. 23, 427–439.

White, W.A., 1972. Deep erosion by continental ice sheets. Bull. Geol. Soc. Am. 83, 1037–1056.

Whiteman, C.A., 1995. Processes of terrestrial deposition. In: Menzies, J. (Ed.), Modern Glacial Environments: Processes, Dynamics and Sediments. Butterworth-Heineman, Oxford, pp. 293–308.

Whitney, M.I., 1979. Electron micrography of mineral surfaces subject to wind-blast erosion. Bull. Geol. Soc. Am. 90, 917–934.

Whitney, M.I., 1985. Yardangs. J. Geol. Educ. 33, 93–96.

Whitney, M.I., Dietrich, R.V., 1973. Ventifact sculpture by windblown dust. Bull. Geol. Soc. Am. 84, 2561–2581.

Wickham, S.S., Johnson, W.H., 1981. The Tiskilwa till, a regional view of this origin and depositional processes. Ann. Glaciol. 2, 176–182.

Wilford, G.E., Wall, J.A.D., 1965. Karst topography in Sarawak. J. Trop. Geogr. 21, 44–70.

Wilhelmy, H., 1958. Klimamorphologie der Massengesteine. Georg Westerman Verlag, Braunschweig, p. 238.

Wilson, A.F., 1984. Origin of quartz-free gold nuggets and supergene gold found in laterites and soils – a review and some new observations. Aust. J. Earth Sci. 31, 303–316.

Wilson, I.G., 1971. Desert sandflow basins and a model for the development of ergs. Geogr. J. 137, 180–199.

Wilson, I.G., 1972. Aeolian bedforms – their development and origins. Sedimentology 19, 173–210.

Wilson, I.G., 1973. Ergs. Sediment. Geol. 10, 77–106.

Wilson, L., 1968. Morphogenetic classification. In: Fairbridge, R.W. (Ed.), The Encyclopedia of Geomorphology. Dowden, Hutchinson and Ross, Stroudsburg, PA, pp. 717–729.

Wilson, L., 1969. Les relations entre les processus géomorphologiques et le climat moderne comme méthode de paleoclimatologie. Revue de Géographie Physique et de Géologie Dynamique 11, 301–314.

Wilson, L., 1973. Variations in mean annual sediment yield as a function of mean annual precipitation. Am. J. Sci. 273, 335–349.

Willets, B., 1983. Transport by wind of granular materials of different grain shapes and densities. Sedimentology 30, 669–679.

Williams, G.E., 1973. Late Quaternary piedmont sedimentation, soil formation and paleoclimates in arid south Australia. Z. Geomorphol. 17, 102–125.

Williams, J.J., Butterfield, G.R., Clark, D.G., 1994. Aerodynamic entrainment thresholds: effect of boundary layer flow conditions. Sedimentology 41, 309–328.

Williams, M.A.J., 1968. Termites and soil development near Brocks Creek, Northern Territory. Aust. J. Sci. 31, 135–154.

Williams, M.A.J., 1978. Termites, soils and landscape equilibrium in the Northern Territory of Australia. In: Davis, J.L., Williams, M.A.J. (Eds), Landform Evolution in Australasia. Australian National University Press, Canberra, pp. 128–141.

Williams, M.A.J., 1994. Cenozoic climatic changes in deserts: a synthesis. In: Abrahams, A.D., Parsons, A.J. (Eds), Geomorphology of Desert Environments. Chapman & Hall, London, pp. 644–670.

Williams, P.J., Smith, M.W., 1989. The Frozen Earth. Fundamentals of Geocryology. Cambridge University Press, Cambridge, p. 306.

Williams, P.W., 1969. The geomorphic effects of ground water. In: Chorley, R.J. (Ed.), Water, Earth and Man. Methuen, London, pp. 269–284.

Williams, P.W., 1972. Morphometric analysis of polygonal karst in New Guinea. Bull. Geol. Soc. Am. 83, 761–796.

Williams, R.B.G., Robinson, D.A., 1981. Weathering of sandstone by the combined action of frost and salt. Earth Surf. Processes Landforms 6, 1–9.

Willis, B., 1936. East African Plateaus and Rift Valleys – Studies in Comparative Seismology. Carnegie Institute, Publication, Washington, DC, p. 470.

Winkler, E.M., Singer, P.C., 1972. Crystallisation pressure of salts in stone and concrete. Bull. Geol. Soc. Am. 83, 3509–3514.

Wintle, A.G., 1993. Luminescence dating of aeolian sands – an overview. In: Pye, K. (Ed.), Dynamics and Environmental Context of Aeolian Sedimentary Systems. Geological Society of London, London, pp. 49–58, Special Paper 72.

Wirthmann, A., 1964. Die Landformen der Edge-Inseö in Südost-Spitzbergen. Ergebnisse der Stanforland-Expedition 2, 53.

Wirthmann, A., 2000. Geomorphology of the Tropics. Springer, Berlin, p. 314.

Wolman, M.G., Gerson, R., 1978. Relative scales of time and effectiveness of climate in watershed geomorphology. Earth Surf. Processes 3, 189–208.

Wood, A., 1942. The development of hillside slopes. Proc. Geol. Assoc. 53, 128–140.

Woodward, J.C., 1995. Pattern of erosion and suspended yield in Mediterranean river basins. In: Foster, I.D.L., Gurnell, A.W., Webb, B.W. (Eds), Sediment and Water Quality in River Catchments. Wiley, Chichester, pp. 365–389.

Wopfner, H., 1978. Silcretes of northern South Australia and adjacent regions. In: Langford Smith, T. (Ed.), Department of Geography, University of New England, pp. 93–141.

World Glacier Monitoring Service, 1989. World Glacier Inventory, IAHS, (ICSI). UNEP–UNESCO.

Worsley, P., 1974. Recent "annual" moraine ridges at Austre Okstindebreen, North Norway. J. Glaciol. 13, 265–277.

Wright, V.P., Tucker, M.E., (Eds), 1991. Calcretes, Reprint Series of the International Association of Sedimentologits, Vol. 2. Blackwell, Oxford, p. 352.

Yair, A., Gerson, R., 1974. Mode and rate of escarpment retreat in an extremely arid environment (Sharm el Sheikh, southern Sinai Peninsula). Z. Geomorphol. 21, 106–121.

Yair, A., Lavee, H., Bryan, R.B., Adar, E., 1980. Runoff and erosion processes and rates in the Zian Valley badlands, northern Negev, Israel. Earth Surf. Processes 5, 205–225.

Yazawa, D., Toya, H., Kaizuka, S., 1971. Alluvial Fans. Kokon Shoin, Tokyo, p. 318.

Young, A., 1964. Slope profile analysis. Z. Geomorphol. Suppl.bd. 5, 17–27.

Young, A., 1970. Slope form in part of the Mato Grosso, Brasil. Geogr. J. 136, 383–392.

Young, A., 1972. Slopes. Oliver and Boyd, Edingurgh, p. 278.

Young, A., 1976. Tropical Soils and Soil Survey. Cambridge University Press, London, p. 468.

Young, R.A., 1985. Geomorphic evolution of the Colorado Plateau margin in West-Central Arizona: a tectonic model to distinguish between the causes of rapid, symmetrical scarp retreat and scarp dissection. In: Morisawa, M., Hack, J.T. (Eds), Tectonic Geomorphology. Allen and Unwin, Boston, pp. 261–278.

Young, R.M., 1987. Salt as an agent in the development of cavernous weathering. Geology 15, 962–966.

Yuan, D., 1987. New observations on tower karst. In: Gardiner, V. (Ed.), International Geomorphology 1986, Proceedings of First International Conference on Geomorphology, Part II. Wiley, Chichester, pp. 1109–1123.

Zachar, D., 1982. Soil Erosion. Elsevier, Amsterdam, p. 547.

Zakaria, A.S., 1977. Controls upon the mineral outputs from three small catchments in New England. Unpublished Ph.D. thesis, University of England, Australia.

Zaruba, Q., Mencl, V., 1969. Landslides and their Control. Elsevier, Amsterdam, p. 213.

Zazo, C., Dabrio, C.J., Borja, F., Goy, J.L., Lezine, A.M., Lario, J., Polo, M.D., Hoyos, M., Boersma, J.R., 1999. Pleistocene and Holocene aeolian facies along the Huelva Coast (southern Spain): climatic and neotectonic implications, Geologie en Mijabouw (en prensa).

Zekri, S., Albisu, L.M., Aragües, R., Herrero, J., 1990. Impacto económico de la salinidad de los suelos en la agricultura de Bárdenas I. Comunicaciones Instituto Nacional de Investigaciones Agrarias, Madrid, Serie Economía 36, p. 129.

Zhende, Z., 1984. In: El-Baz, F. (Ed.), Aeolian landforms in the Taklimakan Desert. Martinus Nijhoff, The Hague, pp. 133–144.

Zoltai, S.C., 1971. Southern limit of permafrost features in peat landforms, Manitoba and Saskatchewan. Geol. Assoc. Can. Spec. Paper 9, 305–310.

Zuidam, R.A.van, 1976a. Geomorphological Development of the Zaragoza Region, Spain. International Institute for Aerial Survey and Earth Sciences (ITC), Enschede, p. 211.

Zuidam, R.A.van, 1976b. Periglacial-like features in the Zaragoza region, Spain. Z. Geomorphol. 20, 227–234.

Subject Index